Graduate Texts in Mathematics 222

Graduate Texts in Mathematics

Graduate Texts in Mathematics bridge the gap between passive study and creative understanding, offering graduate-level introductions to advanced topics in mathematics. The volumes are carefully written as teaching aids and highlight characteristic features of the theory. Although these books are frequently used as textbooks in graduate courses, they are also suitable for individual study.

More information about this series at http://www.springer.com/series/136

Brian C. Hall

Lie Groups, Lie Algebras, and Representations

An Elementary Introduction

Second Edition

 Springer

Brian C. Hall
Department of Mathematics
University of Notre Dame
Notre Dame, IN, USA

ISSN 0072-5285   ISSN 2197-5612   (electronic)
Graduate Texts in Mathematics
ISBN 978-3-319-13466-6   ISBN 978-3-319-13467-3   (eBook)
DOI 10.1007/978-3-319-13467-3

Library of Congress Control Number: 2015935277

Springer Cham Heidelberg New York Dordrecht London

Printed on acid-free paper

Springer International Publishing AG Switzerland is part of Springer Science+Business Media (www.
springer.com)

For Carla

Contents

Preface

This text treats Lie groups, Lie algebras, and their representations. My pedagogical goals are twofold. First, I strive to develop the theory of Lie groups in an elementary fashion, with minimal prerequisites. In particular, in Part I, I develop the theory of (matrix) Lie groups and their Lie algebras using only linear algebra, without requiring any knowledge of manifold theory. Second, I strive to provide more motivation and intuition for the proofs, often using a figure, than in some of the classic texts on the subject. At the same time, I aim to be fully rigorous; an explanation or figure is a supplement to and not a replacement for a traditional proof.

Although Lie theory is widely used in both mathematics and physics, there is often a wide gulf between the presentations of the subject in the two disciplines: Physics books get down to business quickly but are often imprecise in definitions and statements of theorems, whereas math books are more rigorous but often have a high barrier to entry. It is my hope that this book will be useful to both mathematicians and physicists. In particular, the matrix approach in Part I allows for definitions that are precise but comprehensible. Although I do not delve into the details of how Lie algebras are used in particle theory, I do include an extended discussion of the representations of $SU(3)$, which has obvious applications to that field. (My recent book, Quantum Theory for Mathematicians [Hall], also aims to bridge a gap between the mathematics and physics literatures, and it contains some discussion of Lie-theoretic issues in quantum mechanics. The emphasis there, however, is on nonrelativistic quantum mechanics and not on quantum field theory.)

Content of the Book Part I of the text covers the general theory of matrix Lie groups (i.e., closed subgroups of $GL(n;\mathbb{C})$) and their Lie algebras. Chapter 1 introduces numerous examples of matrix Lie groups and examines their topological properties. After discussing the matrix exponential in Chapter 2, I turn to Lie algebras in Chapter 3, examining both abstract Lie algebras and Lie algebras associated with matrix Lie groups. Chapter 3 shows, among other things, that every matrix Lie group is an embedded submanifold of $GL(n;\mathbb{C})$ and, thus, a Lie group. In Chapter 4, I consider elementary representation theory. Finally, Chapter 5 covers the Baker–Campbell–Hausdorff formula and its consequences. I use this formula (in place of the more traditional Frobenius theorem) to establish some of the deeper results about the relationship between Lie groups and Lie algebras.

Part II of the text covers semisimple Lie algebras and their representations. I begin with an entire chapter on the representation theory of $\mathsf{sl}(3; \mathbb{C})$, that is, the complexification of the Lie algebra of the group $\mathsf{SU}(3)$. On the one hand, this example can be treated in an elementary way, simply by writing down a basis and calculating. On the other hand, this example allows the reader to see the machinery of roots, weights, and the Weyl group in action in a simple example, thus motivating the general version of these structures. For the general case, I use an unconventional definition of "semisimple," namely that a complex Lie algebra is semisimple if it has trivial center and is the complexification of the Lie algebra of a compact group. I show that every such Lie algebra decomposes as a direct sum of simple algebras, and is thus semisimple in the conventional sense. Actually, every complex Lie algebra that is semisimple in the conventional sense has a "compact real form," so that my definition of semisimple is equivalent to the standard one—but I do not prove this claim. As with the choice to consider matrix Lie groups in Part I, this (apparent) reduction in scope allows for a rapid development of the structure of semisimple Lie algebras. After developing the necessary properties of root systems in Chapter 8, I give the classification of representations in Chapter 9, as expressed in the theorem of the highest weight. Finally, Chapter 10 gives several additional properties of the representations, including complete reducibility, the Weyl character formula, and the Kostant multiplicity formula.

Finally, Part III of the book presents the compact-group approach to representation theory. Chapter 11 gives a proof of the torus theorem and establishes the equivalence between the Lie-group and Lie-algebra definitions of the Weyl group. This chapter does, however, make use of some of the manifold theory that I avoided previously. The reader who is unfamiliar with manifold theory but willing to take a few things on faith should be able to proceed on to Chapter 12, where I develop the Weyl character formula and the theorem of the highest weight from the compact group point of view. In particular, Chapter 12 gives a self-contained construction of the representations, independent of the Lie-algebraic argument in Chapter 9. Lastly, in Chapter 13, I examine the fundamental group of a compact group from two different perspectives, one that treats the classical groups by induction on the dimension and one that is based on the torus theorem and uses the structure of the root system. This chapter shows, among other things, that for a simply connected compact group, the integral elements from the group point of view coincide with the integral elements from the Lie algebra point of view. This result shows that for simply connected compact groups, the theorem of the highest weight for the group is equivalent to the theorem of the highest weight for the Lie algebra.

The first four chapters of the book cover elementary Lie theory and could be used for an undergraduate course. At the graduate level, one could pass quickly through Part I and then cover either Part II or Part III, depending on the interests of the instructor. Although I have tried to explain and motivate the results in Parts II and III of the book, using figures whenever possible, the material there is unquestionably more challenging than in Part I. Nevertheless, I hope that the explicit working out of the case of the Lie algebra $\mathsf{sl}(3; \mathbb{C})$ (or, equivalently, the group $\mathsf{SU}(3)$) in Chapter 6 will give the reader a good sense of the flavor of the results in the subsequent chapters.

In recent years, there have been several other books on Lie theory that use the matrix-group approach. Of these, the book of Rossmann [Ross] is most similar in style to my own. The first three chapters of [Ross] cover much of the same material as the first four chapters of this book. Although the organization of my book is, I believe, substantially different from that of other books on the subject, I make no claim to originality in any of the proofs. I myself learned most of the material here from the books of Bröcker and tom Dieck [BtD], Humphreys [Hum], and Miller [Mill].

New Features of Second Edition This second edition of the book is substantially expanded from the first edition. Part I has been reorganized but covers mostly the same material as in the first edition. In Part II, however, at least half of the material is new. Chapter 8 now provides a complete derivation of all relevant properties of root systems. In Chapter 9, the construction of the finite-dimensional representations of a semisimple Lie algebra has been fleshed out, with the definition of the universal enveloping algebra, a proof of the Poincaré–Birkhoff–Witt theorem, and a proof of the existence of Verma modules. Chapter 10 is mostly new and includes complete proofs of the Weyl character formula, the Weyl dimension formula, and the Kostant multiplicity formula. Part III, on the structure and representation theory of compact groups, is new in this edition.

I have also included many more figures in the second edition. The black-and-white images were created in Mathematica, while the color images in Sect. 8.9 were modeled in the Zometool system (www.zometool.com) and rendered in Scott Vorthmann's vZome program (vzome.com). I thank Paul Hildebrandt for assisting me with construction of the Zometool models and Scott Vorthmann for going above and beyond in assisting me with use of vZome.

Acknowledgments I am grateful for the input of many people on various versions of this text, which has improved it immensely. Contributors to the first printing of the first edition include Ed Bueler, Wesley Calvert, Tom Goebeler, Ruth Gornet, Keith Hubbard, Wicharn Lewkeeratiyutkul, Jeffrey Mitchell, Ambar Sengupta, and Erdinch Tatar. For the second printing of the first edition, contributors include Moshe Adrian, Kamthorn Chailuek, Paul Gibson, Keith Hubbard, Dennis Muhonen, Jason Quinn, Rebecca Weber, and Reed Wickner. Additional corrections to the first edition since the second printing appeared are due to Kate Brenneman, Edward Burkard, Moritz Firsching, Nathan Gray, Ishan Mata, Jean-Renaud Pycke, and Jason Quinn. Contributors to the second edition include Matt Cecil, Alexander Diaz-Lopez, Todd Kemp, Ben Lewis, George McNinch, and Ambar Sengupta. Thanks to Jonathan Conder, Christopher Gilbreth, Ian Iscoe, Benjamin Lewis, Brian Stoyell-Mulholland, and Reed Wickner for additional input on the second printing of the second edition. Please write to me with questions or corrections at bhall@nd.edu. For further information, click on the "Book" tab of my web site: www.nd.edu/~bhall/.

Notre Dame, IN   Brian C. Hall
January 2015

Part I
General Theory

Chapter 1
Matrix Lie Groups

1.1 Definitions

A Lie group is, roughly speaking, a *continuous group*, that is, a group described by several real parameters. In this book, we consider matrix Lie groups, which are Lie groups realized as groups of matrices. As an example, consider the set of all 2×2 real matrices with determinant 1, customarily denoted $\mathsf{SL}(2; \mathbb{R})$. Since the determinant of a product is the product of the determinants, this set forms a group under the operation of matrix multiplication. If we think of the set of *all* 2×2 matrices, with entries a, b, c, d, as $\mathbb{R}^4$, then $\mathsf{SL}(2; \mathbb{R})$ is the set of points in $\mathbb{R}^4$ for which the smooth function $ad - bc$ has the value 1.

Suppose f is a smooth function on $\mathbb{R}^k$ and we consider the set E where $f(\mathbf{x})$ equals some constant value c. If, at each point $\mathbf{x}_0$ in E, at least one of the partial derivatives of f is nonzero, then the implicit function theorem tells us that we can solve the equation $f(\mathbf{x}) = c$ near $\mathbf{x}_0$ for one of the variables as a function of the other $k - 1$ variables. Thus, E is a smooth "surface" (or *embedded submanifold*) in $\mathbb{R}^k$ of dimension $k - 1$. In the case of $\mathsf{SL}(2; \mathbb{R})$ inside $\mathbb{R}^4$, we note that the partial derivatives of $ad - bc$ with respect to a, b, c, and d are $d, -c, -b$, and a, respectively. Thus, at each point where $ad - bc = 1$, at least one of these partial derivatives is nonzero, and we conclude that $\mathsf{SL}(2; \mathbb{R})$ is a smooth surface of dimension 3. Thus, $\mathsf{SL}(2; \mathbb{R})$ is a Lie group of dimension 3.

For other groups of matrices (such as the ones we will encounter later in this section), one could use a similar approach. The analysis is, however, more complicated because most of the groups are defined by setting several different

---

A previous version of this book was inadvertently published without the middle initial of the author's name as "Brian Hall". For this reason an erratum has been published, correcting the mistake in the previous version and showing the correct name as Brian C. Hall (see DOI http://dx.doi.org/10.1007/978-3-319-13467-3_14). The version readers currently see is the corrected version. The Publisher would like to apologize for the earlier mistake.

© Springer International Publishing Switzerland 2015
B.C. Hall, *Lie Groups, Lie Algebras, and Representations*, Graduate
Texts in Mathematics 222, DOI 10.1007/978-3-319-13467-3_1

smooth functions equal to constants. One therefore has to check that these functions are "independent" in the sense of the implicit function theorem, which means that their gradient vectors have to be linearly independent at each point in the group.

We will use an alternative approach that makes all such analysis unnecessary. We consider groups G of matrices that are *closed* in the sense of Definition 1.4. To each such G, we will associate in Chapter 3 a "Lie algebra" $\mathfrak{g}$, which is a real vector space. A general result (Corollary 3.45) will then show that G is a smooth manifold whose dimension is equal the dimension of $\mathfrak{g}$ as a vector space.

This chapter makes use of various standard results from linear algebra, which are summarized in Appendix A.

Definition 1.1. The **general linear group** over the real numbers, denoted $\mathsf{GL}(n;\mathbb{R})$, is the group of all $n \times n$ invertible matrices with real entries. The general linear group over the complex numbers, denoted $\mathsf{GL}(n;\mathbb{C})$, is the group of all $n \times n$ invertible matrices with complex entries.

Definition 1.2. Let $M_n(\mathbb{C})$ denote the space of all $n \times n$ matrices with complex entries.

We may identify $M_n(\mathbb{C})$ with $\mathbb{C}^{n^2}$ and use the standard notion of convergence in $\mathbb{C}^{n^2}$. Explicitly, this means the following.

Definition 1.3. Let A_m be a sequence of complex matrices in $M_n(\mathbb{C})$. We say that A_m **converges** to a matrix A if each entry of A_m converges (as $m \to \infty$) to the corresponding entry of A (i.e., if $(A_m)_{jk}$ converges to A_{jk} for all $1 \leq j, k \leq n$).

We now consider *subgroups* of $\mathsf{GL}(n;\mathbb{C})$, that is, subsets G of $\mathsf{GL}(n;\mathbb{C})$ such that the identity matrix is in G and such that for all A and B in G, the matrices AB and A^{-1} are also in G.

Definition 1.4. A **matrix Lie group** is a subgroup G of $\mathsf{GL}(n;\mathbb{C})$ with the following property: If A_m is any sequence of matrices in G, and A_m converges to some matrix A, then either A is in G or A is not invertible.

The condition on G amounts to saying that G is a closed subset of $\mathsf{GL}(n;\mathbb{C})$. (This does not necessarily mean that G is closed in $M_n(\mathbb{C})$.) Thus, Definition 1.4 is equivalent to saying that a matrix Lie group is a **closed subgroup** of $\mathsf{GL}(n;\mathbb{C})$. Throughout the book, all topological properties of a matrix Lie group G will be considered with respect to the topology G inherits as a subset of $M_n(\mathbb{C}) \cong \mathbb{C}^{n^2}$.

The condition that G be a *closed* subgroup, as opposed to merely a subgroup, should be regarded as a technicality, in that most of the *interesting* subgroups of $\mathsf{GL}(n;\mathbb{C})$ have this property. Most of the matrix Lie groups G we will consider have the stronger property that if A_m is any sequence of matrices in G, and A_m converges to some matrix A, then $A \in G$ (i.e., that G is closed in $M_n(\mathbb{C})$).

An example of a subgroup of $\mathsf{GL}(n;\mathbb{C})$ which is not closed (and hence is not a matrix Lie group) is the set of all $n \times n$ invertible matrices with rational entries. This set is, in fact, a subgroup of $\mathsf{GL}(n;\mathbb{C})$, but not a closed subgroup. That is, one can

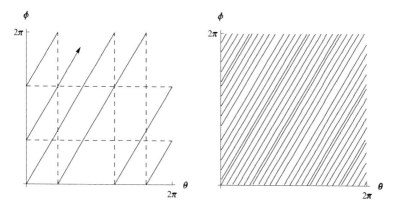

Fig. 1.1 A small portion of the group G inside $\bar{G}$ (*left*) and a larger portion (*right*)

(easily) have a sequence of invertible matrices with rational entries converging to an invertible matrix with some irrational entries. (In fact, *every* real invertible matrix is the limit of some sequence of invertible matrices with rational entries.)

Another example of a group of matrices which is not a matrix Lie group is the following subgroup of $\mathsf{GL}(2;\mathbb{C})$. Let a be an irrational real number and let

$$G = \left\{ \left(\begin{array}{cc} e^{it} & 0 \\ 0 & e^{ita} \end{array} \right) \middle| t \in \mathbb{R} \right\}. \tag{1.1}$$

Clearly, G is a subgroup of $\mathsf{GL}(2;\mathbb{C})$. According to Exercise 10, the closure of G is the group

$$\bar{G} = \left\{ \left(\begin{array}{cc} e^{i\theta} & 0 \\ 0 & e^{i\phi} \end{array} \right) \middle| \theta, \phi \in \mathbb{R} \right\}.$$

The group G inside $\bar{G}$ is known as an "irrational line in a torus"; see Figure 1.1.

1.2   Examples

Mastering the subject of Lie groups involves not only learning the general theory but also familiarizing oneself with examples. In this section, we introduce some of the most important examples of (matrix) Lie groups. Among these are the *classical groups*, consisting of the general and special linear groups, the unitary and orthogonal groups, and the symplectic groups. The classical groups, and their associated Lie algebras, will be key examples in Parts II and III of the book.

1.2.1   General and Special Linear Groups

The general linear groups (over $\mathbb{R}$ or $\mathbb{C}$) are themselves matrix Lie groups. Of course, $GL(n; \mathbb{C})$ is a subgroup of itself. Furthermore, if A_m is a sequence of matrices in $GL(n; \mathbb{C})$ and A_m converges to A, then by the definition of $GL(n; \mathbb{C})$, either A is in $GL(n; \mathbb{C})$, or A is not invertible.

Moreover, $GL(n; \mathbb{R})$ is a subgroup of $GL(n; \mathbb{C})$, and if $A_m \in GL(n; \mathbb{R})$ and A_m converges to A, then the entries of A are real. Thus, either A is not invertible or $A \in GL(n; \mathbb{R})$.

The **special linear group** (over $\mathbb{R}$ or $\mathbb{C}$) is the group of $n \times n$ invertible matrices (with real or complex entries) having determinant one. Both of these are subgroups of $GL(n; \mathbb{C})$. Furthermore, if A_n is a sequence of matrices with determinant one and A_n converges to A, then A also has determinant one, because the determinant is a continuous function. Thus, $SL(n; \mathbb{R})$ and $SL(n; \mathbb{C})$ are matrix Lie groups.

1.2.2   Unitary and Orthogonal Groups

An $n \times n$ complex matrix A is said to be **unitary** if the column vectors of A are orthonormal, that is, if

$$\sum_{l=1}^{n} \overline{A_{lj}} A_{lk} = \delta_{jk}. \tag{1.2}$$

We may rewrite (1.2) as

$$\sum_{l=1}^{n} (A^*)_{jl} A_{lk} = \delta_{jk}, \tag{1.3}$$

where δ_{jk} is the Kronecker delta, equal to 1 if $j = k$ and equal to zero if $j \neq k$. Here A^* is the **adjoint** of A, defined by

$$(A^*)_{jk} = \overline{A_{kj}}.$$

Equation (1.3) says that $A^* A = I$; thus, we see that A is unitary if and only if $A^* = A^{-1}$. In particular, every unitary matrix is invertible.

The adjoint operation on matrices satisfies $(AB)^* = B^* A^*$. From this, we can see that if A and B are unitary, then

$$(AB)^*(AB) = B^* A^* AB = B^{-1} A^{-1} AB = I,$$

showing that AB is also unitary. Furthermore, since $(AA^{-1})^* = I^* = I$, we see that $(A^{-1})^* A^* = I$, which shows that $(A^{-1})^* = (A^*)^{-1}$. Thus, if A is unitary, we have

$$(A^{-1})^* A^{-1} = (A^*)^{-1} A^{-1} = (AA^*)^{-1} = I,$$

showing that A^{-1} is again unitary.

Thus, the collection of unitary matrices is a subgroup of $\mathsf{GL}(n; \mathbb{C})$. We call this group the **unitary group** and we denote it by $\mathsf{U}(n)$. We may also define the **special unitary group** $\mathsf{SU}(n)$, the subgroup of $\mathsf{U}(n)$ consisting of unitary matrices with determinant 1. It is easy to check that both $\mathsf{U}(n)$ and $\mathsf{SU}(n)$ are closed subgroups of $\mathsf{GL}(n; \mathbb{C})$ and thus matrix Lie groups.

Meanwhile, let $\langle \cdot, \cdot \rangle$ denote the standard inner product on $\mathbb{C}^n$, given by

$$\langle x, y \rangle = \sum_j \overline{x_j} y_j.$$

(Note that we put the conjugate on the *first* factor in the inner product.) By Proposition A.8, we have

$$\langle x, Ay \rangle = \langle A^* x, y \rangle$$

for all $x, y \in \mathbb{C}^n$. Thus,

$$\langle Ax, Ay \rangle = \langle A^* Ax, y \rangle,$$

from which we can see that if A is unitary, then A preserves the inner product on $\mathbb{C}^n$, that is,

$$\langle Ax, Ay \rangle = \langle x, y \rangle$$

for all x and y. Conversely, if A preserves the inner product, we must have $\langle A^* Ax, y \rangle = \langle x, y \rangle$ for all x, y. It is not hard to see that this condition holds only if $A^* A = I$. Thus, an equivalent characterization of unitarity is that A is unitary if and only if A preserves the standard inner product on $\mathbb{C}^n$.

Finally, for any matrix A, we have that $\det A^* = \overline{\det A}$. Thus, if A is unitary, we have

$$\det(A^* A) = |\det A|^2 = \det I = 1.$$

Hence, for all unitary matrices A, we have $|\det A| = 1$.

In a similar fashion, an $n \times n$ real matrix A is said to be **orthogonal** if the column vectors of A are orthonormal. As in the unitary case, we may give equivalent versions of this condition. The only difference is that if A is real, A^* is the same as the **transpose** A^{tr} of A, given by

$$(A^{tr})_{jk} = A_{kj}.$$

Thus, A is orthogonal if and only if $A^{tr} = A^{-1}$, and this holds if and only if A preserves the inner product on $\mathbb{R}^n$. Since $\det(A^{tr}) = \det A$, if A is orthogonal, we have

$$\det(A^{tr}A) = \det(A)^2 = \det(I) = 1,$$

so that $\det(A) = \pm 1$. The collection of all orthogonal matrices forms a closed subgroup of $\mathsf{GL}(n; \mathbb{C})$, which we call the **orthogonal group** and denote by $\mathsf{O}(n)$. The set of $n \times n$ orthogonal matrices with determinant one is the **special orthogonal group**, denoted $\mathsf{SO}(n)$. Geometrically, elements of $\mathsf{SO}(n)$ are rotations, while the elements of $\mathsf{O}(n)$ are either rotations or combinations of rotations and reflections.

Consider now the bilinear form $(\cdot, \cdot)$ on $\mathbb{C}^n$ defined by

$$(x, y) = \sum_j x_j y_j. \tag{1.4}$$

This form is *not* an inner product (Sect. A.6) because, for example, it is symmetric rather than conjugate-symmetric. The set of all $n \times n$ complex matrices A which preserve this form (i.e., such that $(Ax, Ay) = (x, y)$ for all $x, y \in \mathbb{C}^n$) is the **complex orthogonal group** $\mathsf{O}(n; \mathbb{C})$, and it is a subgroup of $\mathsf{GL}(n; \mathbb{C})$. Since there are no conjugates in the definition of the form $(\cdot, \cdot)$, we have

$$(x, Ay) = (A^{tr}x, y),$$

for all $x, y \in \mathbb{C}^n$, where on the right-hand side of the above relation, we have A^{tr} rather than A^*. Repeating the arguments for the case of $\mathsf{O}(n)$, but now allowing complex entries in our matrices, we find that an $n \times n$ complex matrix A is in $\mathsf{O}(n; \mathbb{C})$ if and only if $A^{tr}A = I$, that $\mathsf{O}(n; \mathbb{C})$ is a matrix Lie group, and that $\det A = \pm 1$ for all A in $\mathsf{O}(n; \mathbb{C})$. Note that $\mathsf{O}(n; \mathbb{C})$ is *not* the same as the unitary group $\mathsf{U}(n)$. The group $\mathsf{SO}(n; \mathbb{C})$ is defined to be the set of all A in $\mathsf{O}(n; \mathbb{C})$ with $\det A = 1$ and it is also a matrix Lie group.

1.2.3 Generalized Orthogonal and Lorentz Groups

Let n and k be positive integers, and consider $\mathbb{R}^{n+k}$. Define a symmetric bilinear form $[\cdot, \cdot]_{n,k}$ on $\mathbb{R}^{n+k}$ by the formula

$$[x, y]_{n,k} = x_1 y_1 + \cdots + x_n y_n - x_{n+1} y_{n+1} - \cdots - x_{n+k} y_{n+k} \tag{1.5}$$

The set of $(n + k) \times (n + k)$ real matrices A which preserve this form (i.e., such that $[Ax, Ay]_{n,k} = [x, y]_{n,k}$ for all $x, y \in \mathbb{R}^{n+k}$) is the **generalized orthogonal group** $\mathsf{O}(n; k)$. It is a subgroup of $\mathsf{GL}(n + k; \mathbb{R})$ and a matrix Lie group (Exercise 1). Of particular interest in physics is the **Lorentz group** $\mathsf{O}(3; 1)$. We also define $\mathsf{SO}(n; k)$ to be the subgroup of $\mathsf{O}(n; k)$ consisting of elements with determinant 1.

If A is an $(n + k) \times (n + k)$ real matrix, let $A^{(j)}$ denote the jth column vector of A, that is,

$$A^{(j)} = \begin{pmatrix} A_{1,j} \\ \vdots \\ A_{n+k,j} \end{pmatrix}.$$

Note that $A^{(j)}$ is equal to Ae_j, that is, the result of applying A to the jth standard basis element e_j. Then A will belong to $O(n; k)$ if and only if $[Ae_j, Ae_l] = [e_j, e_l]$ for all $1 \leq j, l \leq n + k$. Explicitly, this means that $A \in O(n; k)$ if and only if the following conditions are satisfied:

$$\begin{aligned} \left[A^{(j)}, A^{(l)}\right]_{n,k} &= 0 & j \neq l, \\ \left[A^{(j)}, A^{(j)}\right]_{n,k} &= 1 & 1 \leq j \leq n, \\ \left[A^{(j)}, A^{(j)}\right]_{n,k} &= -1 & n + 1 \leq j \leq n + k. \end{aligned} \tag{1.6}$$

Let g denote the $(n+k)\times(n+k)$ diagonal matrix with ones in the first n diagonal entries and minus ones in the last k diagonal entries:

$$g = \begin{pmatrix} 1 & & & & & \\ & \ddots & & & & \\ & & 1 & & & \\ & & & -1 & & \\ & & & & \ddots & \\ & & & & & -1 \end{pmatrix}.$$

Then A is in $O(n; k)$ if and only if $A^{tr} g A = g$ (Exercise 1). Taking the determinant of this equation gives $(\det A)^2 \det g = \det g$, or $(\det A)^2 = 1$. Thus, for any A in $O(n; k)$, $\det A = \pm 1$.

1.2.4   Symplectic Groups

Consider the skew-symmetric bilinear form B on $\mathbb{R}^{2n}$ defined as follows:

$$\omega(x, y) = \sum_{j=1}^{n}(x_j y_{n+j} - x_{n+j} y_j). \tag{1.7}$$

The set of all $2n \times 2n$ real matrices A which preserve ω (i.e., such that $\omega(Ax, Ay) = \omega(x, y)$ for all $x, y \in \mathbb{R}^{2n}$) is the **real symplectic group** $\mathsf{Sp}(n; \mathbb{R})$, and it is a closed subgroup of $\mathsf{GL}(2n; \mathbb{R})$. (Some authors refer to the group we have just defined as $\mathsf{Sp}(2n; \mathbb{R})$ rather than $\mathsf{Sp}(n; \mathbb{R})$.) If Ω is the $2n \times 2n$ matrix

$$\Omega = \begin{pmatrix} 0 & I \\ -I & 0 \end{pmatrix}, \tag{1.8}$$

then

$$\omega(x, y) = \langle x, \Omega y \rangle .$$

From this, it is not hard to show that a $2n \times 2n$ real matrix A belongs to $\mathsf{Sp}(n; \mathbb{R})$ if and only if

$$- \Omega A^{tr} \Omega = A^{-1}. \tag{1.9}$$

(See Exercise 2.) Taking the determinant of this identity gives $\det A = (\det A)^{-1}$, i.e., $(\det A)^2 = 1$. This shows that $\det A = \pm 1$, for all $A \in \mathsf{Sp}(n; \mathbb{R})$. In fact, $\det A = 1$ for all $A \in \mathsf{Sp}(n; \mathbb{R})$, although this is not obvious.

One can define a bilinear form ω on $\mathbb{C}^{2n}$ by the same formula as in (1.7) (with no conjugates). Over $\mathbb{C}$, we have the relation

$$\omega(z, w) = (z, \Omega w),$$

where $(\cdot, \cdot)$ is the complex bilinear form in (1.4). The set of $2n \times 2n$ complex matrices which preserve this form is the **complex symplectic group** $\mathsf{Sp}(n; \mathbb{C})$. A $2n \times 2n$ complex matrix A is in $\mathsf{Sp}(n; \mathbb{C})$ if and only if (1.9) holds. (Note: This condition involves A^{tr}, *not* A^*.) Again, we can easily show that each $A \in \mathsf{Sp}(n; \mathbb{C})$ satisfies $\det A = \pm 1$ and, again, it is actually the case that $\det A = 1$. Finally, we have the **compact symplectic group** $\mathsf{Sp}(n)$ defined as

$$\mathsf{Sp}(n) = \mathsf{Sp}(n; \mathbb{C}) \cap \mathsf{U}(2n).$$

That is to say, $\mathsf{Sp}(n)$ is the group of $2n \times 2n$ matrices that preserve *both* the inner product and the bilinear form ω. For more information about $\mathsf{Sp}(n)$, see Sect. 1.2.8.

1.2.5 The Euclidean and Poincaré Groups

The **Euclidean group** $\mathsf{E}(n)$ is the group of all transformations of $\mathbb{R}^n$ that can be expressed as a composition of a translation and an orthogonal linear transformation. We write elements of $\mathsf{E}(n)$ as pairs $\{x, R\}$ with $x \in \mathbb{R}^n$ and $R \in \mathsf{O}(n)$, and we let $\{x, R\}$ act on $\mathbb{R}^n$ by the formula

$$\{x, R\} \, y = Ry + x.$$

Since

$$\{x_1, R_1\}\{x_2, R_2\} y = R_1(R_2 y + x_2) + x_1 = R_1 R_2 y + (x_1 + R_1 x_2),$$

the product operation for $\mathsf{E}(n)$ is the following:

$$\{x_1, R_1\}\{x_2, R_2\} = \{x_1 + R_1 x_2, R_1 R_2\}. \tag{1.10}$$

The inverse of an element of $\mathsf{E}(n)$ is given by

$$\{x, R\}^{-1} = \{-R^{-1}x, R^{-1}\}.$$

The group $\mathsf{E}(n)$ is not a subgroup of $\mathsf{GL}(n; \mathbb{R})$, since translations are not linear maps. However, $\mathsf{E}(n)$ is isomorphic to the (closed) subgroup of $\mathsf{GL}(n + 1; \mathbb{R})$ consisting of matrices of the form

$$\begin{pmatrix} & & x_1 \\ R & & \vdots \\ & & x_n \\ 0 \cdots 0 & 1 \end{pmatrix}, \tag{1.11}$$

with $R \in \mathsf{O}(n)$. (The reader may easily verify that matrices of the form (1.11) multiply according to the formula in (1.10).)

We similarly define the Poincaré group $\mathsf{P}(n; 1)$ (also known as the inhomogeneous Lorentz group) to be the group of all transformations of $\mathbb{R}^{n+1}$ of the form

$$T = T_x A$$

with $x \in \mathbb{R}^{n+1}$ and $A \in \mathsf{O}(n; 1)$. This group is isomorphic to the group of $(n + 2) \times (n + 2)$ matrices of the form

$$\begin{pmatrix} & & x_1 \\ A & & \vdots \\ & & x_{n+1} \\ 0 \cdots 0 & 1 \end{pmatrix} \tag{1.12}$$

with $A \in \mathsf{O}(n; 1)$.

1.2.6 The Heisenberg Group

The set of all 3×3 real matrices A of the form

$$A = \begin{pmatrix} 1 & a & b \\ 0 & 1 & c \\ 0 & 0 & 1 \end{pmatrix}, \tag{1.13}$$

where a, b, and c are arbitrary real numbers, is the **Heisenberg group**. It is easy to check that the product of two matrices of the form (1.13) is again of that form, and, clearly, the identity matrix is of the form (1.13). Furthermore, direct computation shows that if A is as in (1.13), then

$$A^{-1} = \begin{pmatrix} 1 & -a & ac-b \\ 0 & 1 & -c \\ 0 & 0 & 1 \end{pmatrix}.$$

Thus, H is a subgroup of $\mathsf{GL}(3;\mathbb{R})$. The Heisenberg group is a model for the Heisenberg–Weyl commutation relations in physics and also serves as a illuminating example for the Baker–Campbell–Hausdorff formula (Sect. 5.2). See also Exercise 8.

1.2.7 The Groups $\mathbb{R}^*$, $\mathbb{C}^*$, S^1, $\mathbb{R}$, and $\mathbb{R}^n$

Several important groups which are not defined as groups of matrices can be thought of as such. The group $\mathbb{R}^*$ of non-zero real numbers under multiplication is isomorphic to $\mathsf{GL}(1;\mathbb{R})$. Similarly, the group $\mathbb{C}^*$ of nonzero complex numbers under multiplication is isomorphic to $\mathsf{GL}(1;\mathbb{C})$ and the group S^1 of complex numbers with absolute value one is isomorphic to $\mathsf{U}(1)$.

The group $\mathbb{R}$ under addition is isomorphic to $\mathsf{GL}(1;\mathbb{R})^+$ (1×1 real matrices with positive determinant) via the map $x \to [e^x]$. The group $\mathbb{R}^n$ (with vector addition) is isomorphic to the group of diagonal real matrices with positive diagonal entries, via the map

$$(x_1, \ldots, x_n) \to \begin{pmatrix} e^{x_1} & & 0 \\ & \ddots & \\ 0 & & e^{x_n} \end{pmatrix}.$$

1.2.8 The Compact Symplectic Group

Of the groups introduced in the preceding subsections, the compact symplectic group $\mathsf{Sp}(n) := \mathsf{Sp}(n;\mathbb{C}) \cap \mathsf{U}(2n)$ is the most mysterious. In this section, we attempt to understand the structure of $\mathsf{Sp}(n)$ and to show that it can be understood as being the "unitary group over the quaternions."

Since the definition of $\mathsf{Sp}(n)$ involves unitarity, it is convenient to express the bilinear form ω on $\mathbb{C}^{2n}$ in terms of the inner product $\langle \cdot, \cdot \rangle$, rather than in terms of the bilinear form $(\cdot, \cdot)$, as we did in Sect. 1.2.4. To this end, define a *conjugate-linear* map $J : \mathbb{C}^{2n} \to \mathbb{C}^{2n}$ by

$$J(\alpha, \beta) = (-\bar{\beta}, \bar{\alpha}),$$

where α and β are in $\mathbb{C}^n$ and (α, β) is in $\mathbb{C}^{2n}$. We can easily check that for all $z, w \in \mathbb{C}^{2n}$, we have

$$\omega(z, w) = \langle Jz, w \rangle .$$

Recall that we take our inner product to be conjugate linear in the first factor; since J is also conjugate linear, $\langle Jz, w \rangle$ is actually linear in z. We may easily check that

$$\langle Jz, w \rangle = -\overline{\langle z, Jw \rangle} = -\langle Jw, z \rangle$$

for all $z, w \in \mathbb{C}^{2n}$ and that

$$J^2 = -I.$$

Proposition 1.5. *If U belongs to $\mathsf{U}(2n)$ then U belongs to $\mathsf{Sp}(n)$ if and only if U commutes with J.*

Proof. Fix some U in $\mathsf{U}(2n)$. Then for z and w in $\mathbb{C}^{2n}$, we have, on the one hand,

$$\omega(Uz, Uw) = \langle JUz, Uw \rangle = \langle U^* JUz, w \rangle = \left(U^{-1} JUz, w \right),$$

and, on the other hand,

$$\omega(z, w) = \langle Jz, w \rangle .$$

From this it is easy to check that U preserves ω if and only if

$$U^{-1} JU = J,$$

which is equivalent to $JU = UJ$. $\square$

The preceding result can be used to give a different perspective on the definition of $\mathsf{Sp}(n)$, as follows. The **quaternion algebra** $\mathbb{H}$ is the four-dimensional associative algebra over $\mathbb{R}$ spanned by elements 1 (the identity), $\mathbf{i}$, $\mathbf{j}$, and $\mathbf{k}$ satisfying

$$\mathbf{i}^2 = \mathbf{j}^2 = \mathbf{k}^2 = -1$$

and

$$\mathbf{ij} = \mathbf{k}; \quad \mathbf{ji} = -\mathbf{k};$$
$$\mathbf{jk} = \mathbf{i}; \quad \mathbf{kj} = -\mathbf{i};$$
$$\mathbf{ki} = \mathbf{j}; \quad \mathbf{ik} = -\mathbf{j}.$$

We may realize the quaternion algebra inside $M_2(\mathbb{C})$ by identifying 1 with the identity matrix and setting

$$\mathbf{i} = \begin{pmatrix} i & 0 \\ 0 & -i \end{pmatrix}; \quad \mathbf{j} = \begin{pmatrix} 0 & 1 \\ -1 & 0 \end{pmatrix}; \quad \mathbf{k} = \begin{pmatrix} 0 & i \\ i & 0 \end{pmatrix}.$$

The algebra $\mathbb{H}$ is then the space of real linear combinations of $I, \mathbf{i}, \mathbf{j}$, and $\mathbf{k}$.

Now, since J is conjugate linear, we have

$$J(iz) = -iJ(z)$$

for all $z \in \mathbb{C}^{2n}$; that is, $iJ = -Ji$. Thus, if we define K to be iJ, we have

$$K^2 = iJiJ = -J(i)^2 J = J^2 = -I,$$

and one can easily check that iI, J, and K satisfy the same commutation relations as $\mathbf{i}, \mathbf{j}$, and $\mathbf{k}$. We can therefore make $\mathbb{C}^{2n}$ into a "vector space" over the noncommutative algebra $\mathbb{H}$ by setting

$$\mathbf{i} \cdot z = iz$$
$$\mathbf{j} \cdot z = Jz$$
$$\mathbf{k} \cdot z = iJz.$$

Now, if U belongs to $\mathsf{Sp}(n)$, then U commutes with multiplication by i and with J (Proposition 1.5) and thus, also, with $K := iJ$. Thus, U is actually "quaternion linear." A $2n \times 2n$ matrix U therefore belongs to $\mathsf{Sp}(n)$ if and only if U is quaternion linear and preserves the norm. Thus, we may think of $\mathsf{Sp}(n)$ as the "unitary group over the quaternions." The compact symplectic group then fits naturally with the orthogonal groups (norm-preserving maps over $\mathbb{R}$) and the unitary groups (norm-preserving maps over $\mathbb{C}$).

Every $U \in \mathsf{U}(2n)$ has an orthonormal basis of eigenvectors, with eigenvalues having absolute value 1. We now determine the additional properties the eigenvectors and eigenvalues must satisfy in order for U to be in $\mathsf{Sp}(n) = \mathsf{U}(2n) \cap \mathsf{Sp}(n; \mathbb{C})$.

Theorem 1.6. *If $U \in \mathsf{Sp}(n)$, then there exists an orthonormal basis $u_1, \ldots, u_n$, $v_1, \ldots v_n$ for $\mathbb{C}^{2n}$ such that the following properties hold: First, $J u_j = v_j$; second, for some real numbers $\theta_1, \ldots, \theta_n$, we have*

$$U u_j = e^{i\theta_j} u_j$$
$$U v_j = e^{-i\theta_j} v_j;$$

and third,

$$\omega(u_j, u_k) = \omega(v_j, v_k) = 0$$
$$\omega(u_j, v_k) = \delta_{jk}.$$

Conversely, if there exists an orthonormal basis with these properties, U belongs to $\mathsf{Sp}(n)$.

Lemma 1.7. *Suppose V is a complex subspace of $\mathbb{C}^{2n}$ that is invariant under the conjugate-linear map J. Then the orthogonal complement $V^\perp$ of V (with respect to the inner product $\langle \cdot, \cdot \rangle$) is also invariant under J. Furthermore, V and $V^\perp$ are orthogonal with respect to ω; that is,*

$$\omega(z, w) = 0$$

for all $z \in V$ and $w \in V^\perp$.

Proof. If $w \in V^\perp$ then for all $z \in V$, we have

$$\langle Jw, z \rangle = -\langle Jz, w \rangle = 0,$$

because Jz is again in V. Thus, $V^\perp$ is invariant under J. Then if $z \in V$ and $w \in V^\perp$, we have

$$\omega(z, w) = \langle Jz, w \rangle = 0,$$

because Jz is again in V. $\qquad\square$

Proof of Theorem 1.6. Consider U in $\mathsf{Sp}(n; \mathbb{C}) \cap \mathsf{U}(2n)$, choose an eigenvector for U, normalized to be a unit vector, and call it u_1. Since U preserves the norms of vectors, the eigenvalue λ_1 for u_1 must be of the form $e^{i\theta_1}$, for some $\theta_1 \in \mathbb{R}$. If we set $v_1 = Ju_1$, then since J is conjugate linear and commutes with U (Proposition 1.5), we have

$$Uv_1 = J(Uu_1) = J(e^{i\theta_1} u_1) = e^{-i\theta_1} v_1.$$

That is to say, v_1 is an eigenvector for U with eigenvalue $e^{-i\theta_1}$. Furthermore,

$$\langle v_1, u_1 \rangle = \langle Ju_1, u_1 \rangle = \omega(u_1, u_1) = 0,$$

since ω is a skew-symmetric form. On the other hand,

$$\omega(u_1, v_1) = \langle Ju_1, v_1 \rangle = \langle Ju_1, Ju_1 \rangle = 1,$$

since J preserves the magnitude of vectors.

Now, since $J^2 = -I$, we can easily check that the span V of u_1 and $v_1 = Ju_1$ is invariant under J. Thus, by Lemma 1.7, $V^\perp$ is also invariant under J and is

ω-orthogonal to V. Meanwhile, V is invariant under both U and $U^* = U^{-1}$. Thus, by Proposition A.10, $V^\perp$ is invariant under both $U^{**} = U$ and U^*. Since U preserves $V^\perp$, the restriction of U to $V^\perp$ will have an eigenvector, which we can normalize to be a unit vector and call u_2. If we let $v_2 = Ju_2$, then we have all the same properties for u_2 and v_2 as for u_1 and v_1. Furthermore, u_2 and v_2 are orthogonal—with respect to both $\langle \cdot, \cdot \rangle$ and $\omega(\cdot, \cdot)$—to u_1 and v_1. We can then proceed on in a similar fashion to obtain the full set of vectors $u_1, \ldots, u_n$ and $v_1, \ldots, v_n$. (If $u_1, \ldots, u_k$ and $v_1, \ldots, v_k$ have been chosen, we take u_{k+1} and $v_{k+1} := Ju_{k+1}$ in the orthogonal complement of the span of $u_1, \ldots, u_k$ and $v_1, \ldots, v_k$.)

The other direction of the theorem is left to the reader (Exercise 6). □

1.3 Topological Properties

In this section, we investigate three important topological properties of matrix Lie groups, each of which is satisfied by some groups but not others.

1.3.1 Compactness

The first property we consider is compactness.

Definition 1.8. A matrix Lie group $G \subset \mathsf{GL}(n; \mathbb{C})$ is said to be **compact** if it is compact in the usual topological sense as a subset of $M_n(\mathbb{C}) \cong \mathbb{R}^{2n^2}$.

In light of the Heine–Borel theorem (Theorem 2.41 in [Rud1]), a matrix Lie group G is compact if and only if it is closed (as a subset of $M_n(\mathbb{C})$, not just as a subset of $\mathsf{GL}(n; \mathbb{C})$) and bounded. Explicitly, this means that G is compact if and only if (1) whenever $A_m \in G$ and $A_m \to A$, then A is in G, and (2) there exists a constant C such that for all $A \in G$, we have $|A_{jk}| \le C$ for all $1 \le j, k \le n$.

The following groups are compact: $\mathsf{O}(n)$ and $\mathsf{SO}(n)$, $\mathsf{U}(n)$ and $\mathsf{SU}(n)$, and $\mathsf{Sp}(n)$. Each of these groups is easily seen to be closed in $M_n(\mathbb{C})$ and each satisfies the bound $|A_{jk}| \le 1$, since in each case, the columns of $A \in G$ are required to be unit vectors. Most of the other groups we have considered are noncompact. The special linear group $\mathsf{SL}(n; \mathbb{R})$, for example, is unbounded (except in the trivial case $n = 1$), because for all $m \ne 0$, the matrix

$$A_m = \begin{pmatrix} m & & & & \\ & \frac{1}{m} & & & \\ & & 1 & & \\ & & & \ddots & \\ & & & & 1 \end{pmatrix}$$

has determinant one.

1.3.2 Connectedness

The second property we consider is connectedness.

Definition 1.9. A matrix Lie group G is said to be **connected** if for all A and B in G, there exists a continuous path $A(t)$, $a \leq t \leq b$, lying in G with $A(a) = A$ and $A(b) = B$. For any matrix Lie group G, the **identity component** of G, denoted G_0, is the set of $A \in G$ for which there exists a continuous path $A(t)$, $a \leq t \leq b$, lying in G with $A(a) = I$ and $A(b) = A$.

The property we have called "connected" in Definition 1.9 is what is called **path connected** in topology, which is not (in general) the same as connected. However, we will eventually prove that a matrix Lie group is connected (in the usual topological sense) if and only if it is path-connected. Thus, in a slight abuse of terminology, we shall continue to refer to the above property as connectedness. (See the remarks following Corollary 3.45.)

To show that a matrix Lie group G is connected, it suffices to show that each $A \in G$ can be connected to the identity by a continuous path lying in G.

Proposition 1.10. *If G is a matrix Lie group, the identity component G_0 of G is a normal subgroup of G.*

We will see in Sect. 3.7 that G_0 is closed and hence a matrix Lie group.

Proof. If A and B are any two elements of G_0, then there are continuous paths $A(t)$ and $B(t)$ connecting I to A and to B in G. Then the path $A(t)B(t)$ is a continuous path connecting I to AB in G, and $(A(t))^{-1}$ is a continuous path connecting I to A^{-1} in G. Thus, both AB and A^{-1} belong to G_0, showing that G_0 is a subgroup of G. Now suppose A is in G_0 and B is any element of G. Then there is a continuous path $A(t)$ connecting I to A in G, and the path $BA(t)B^{-1}$ connects I to BAB^{-1} in G. Thus, $BAB^{-1} \in G_0$, showing that G_0 is normal. $\qquad\square$

Note that because matrix multiplication and matrix inversion are continuous on $\mathsf{GL}(n;\mathbb{C})$, it follows that if $A(t)$ and $B(t)$ are continuous, then so are $A(t)B(t)$ and $A(t)^{-1}$. The continuity of the matrix product is obvious. The continuity of the inverse follows from the formula for the inverse in terms of cofactors; this formula is continuous as long as we remain in the set of invertible matrices where the determinant in the denominator is nonzero.

Proposition 1.11. *The group $\mathsf{GL}(n;\mathbb{C})$ is connected for all $n \geq 1$.*

Proof. We make use of the result that every matrix is similar to an upper triangular matrix (Theorem A.4). That is to say, we can express any $A \in M_n(\mathbb{C})$ in the form $A = CBC^{-1}$, where

$$B = \begin{pmatrix} \lambda_1 & & * \\ & \ddots & \\ 0 & & \lambda_n \end{pmatrix}.$$

If A is invertible, each λ_j must be nonzero. Let $B(t)$ be obtained by multiplying the part of B above the diagonal by $(1-t)$, for $0 \leq t \leq 1$, and let $A(t) = CB(t)C^{-1}$. Then $A(t)$ is a continuous path lying in $\mathsf{GL}(n;\mathbb{C})$ which starts at A and ends at CDC^{-1}, where D is the diagonal matrix with diagonal entries $\lambda_1, \ldots, \lambda_n$. We can now define paths $\lambda_j(t)$ connecting λ_j to 1 in $\mathbb{C}^*$ as t goes from 1 to 2, and we can define $A(t)$ on the interval $1 \leq t \leq 2$ by

$$A(t) = C \begin{pmatrix} \lambda_1(t) & & 0 \\ & \ddots & \\ 0 & & \lambda_n(t) \end{pmatrix} C^{-1}.$$

Then $A(t)$, $0 \leq t \leq 2$, is a continuous path in $\mathsf{GL}(n;\mathbb{C})$ connecting A to I.   □

An alternative proof of this result is given in Exercise 12.

Proposition 1.12. *The group* $\mathsf{SL}(n;\mathbb{C})$ *is connected for all* $n \geq 1$.

Proof. The proof is almost the same as for $\mathsf{GL}(n;\mathbb{C})$, except that we must make sure our path connecting $A \in \mathsf{SL}(n;\mathbb{C})$ to I lies entirely in $\mathsf{SL}(n;\mathbb{C})$. We can ensure this by choosing $\lambda_n(t)$, in the second part of the preceding proof, to be equal to $(\lambda_1(t) \cdots \lambda_{n-1}(t))^{-1}$.   □

Proposition 1.13. *The groups* $\mathsf{U}(n)$ *and* $\mathsf{SU}(n)$ *are connected, for all* $n \geq 1$.

Proof. By Theorem A.3, every unitary matrix has an orthonormal basis of eigenvectors, with eigenvalues having absolute value 1. Thus, each $U \in \mathsf{U}(n)$ can be written as $U_1 D U_1^{-1}$, where $U_1 \in \mathsf{U}(n)$ and D is diagonal with diagonal entries $e^{i\theta_1}, \ldots, e^{i\theta_n}$. We may then define

$$U(t) = U_1 \begin{pmatrix} e^{i(1-t)\theta_1} & & 0 \\ & \ddots & \\ 0 & & e^{i(1-t)\theta_n} \end{pmatrix} U_1^{-1}, \quad 0 \leq t \leq 1.$$

It is easy to see that $U(t)$ is in $\mathsf{U}(n)$ for all t, and $U(t)$ connects U to I. A slight modification of this argument, as in the proof of Proposition 1.12, shows that $\mathsf{SU}(n)$ is connected.   □

The group $\mathsf{SO}(n)$ is also connected; see Exercise 13.

1.3.3 Simple Connectedness

The last topological property we consider is simple connectedness.

Definition 1.14. A matrix Lie group G is said to be **simply connected** if it is connected and, in addition, every loop in G can be shrunk continuously to a point in G.

More precisely, assume that G is connected. Then G is simply connected if for every continuous path $A(t)$, $0 \leq t \leq 1$, lying in G and with $A(0) = A(1)$, there exists a continuous function $A(s, t)$, $0 \leq s, t \leq 1$, taking values in G and having the following properties: (1) $A(s, 0) = A(s, 1)$ for all s, (2) $A(0, t) = A(t)$, and (3) $A(1, t) = A(1, 0)$ for all t.

One should think of $A(t)$ as a loop and $A(s, t)$ as a family of loops, parameterized by the variable s which shrinks $A(t)$ to a point. Condition 1 says that for each value of the parameter s, we have a loop; Condition 2 says that when $s = 0$ the loop is the specified loop $A(t)$; and Condition 3 says that when $s = 1$ our loop is a point. The condition of simple connectedness is important because for simply connected groups, there is a particularly close relationship between the group and the Lie algebra. (See Sect. 5.7.)

Proposition 1.15. *The group* $\mathsf{SU}(2)$ *is simply connected.*

Proof. Exercise 5 shows that $\mathsf{SU}(2)$ may be thought of (topologically) as the three-dimensional sphere S^3 sitting inside $\mathbb{R}^4$. It is well known that S^3 is simply connected; see, for example, Proposition 1.14 in [Hat].   □

If a matrix Lie group G is *not* simply connected, the degree to which it fails to be simply connected is encoded in the fundamental group of G. (See Sect. 13.1.) Sections 13.2 and 13.3 analyze several additional examples. It is shown there, for example, that $\mathsf{SU}(n)$ is simply connected for all n.

1.3.4 The Topology of $\mathsf{SO}(3)$

We conclude this section with an analysis of the topological structure of the group $\mathsf{SO}(3)$. We begin by describing *real projective spaces*.

Definition 1.16. The **real projective space** of dimension n, denoted $\mathbb{R}P^n$, is the set of lines through the origin in $\mathbb{R}^{n+1}$. Since each line through the origin intersects the unit sphere exactly twice, we may think of $\mathbb{R}P^n$ as the unit sphere S^n with "antipodal" points u and $-u$ identified.

Using the second description, we think of points in $\mathbb{R}P^n$ as pairs $\{u, -u\}$, with $u \in S^n$. There is a natural map $\pi : S^n \to \mathbb{R}P^n$, given by

$$\pi(u) = \{u, -u\}.$$

We may define a distance function on $\mathbb{R}P^n$ by defining

$$d(\{u, -u\}, \{v, -v\}) = \min(d(u, v), d(u, -v), d(-u, v), d(-u, -v))$$
$$= \min(d(u, v), d(u, -v)).$$

(The second equality holds because $d(x, y) = d(-x, -y)$.) With this metric, $\mathbb{R}P^n$ is locally isometric to S^n, since if u and v are nearby points in S^n, we have $d(\{u, -u\}, \{v, -v\}) = d(u, v)$.

It is known that $\mathbb{R}P^n$ is not simply connected. (See, for example, Example 1.43 in [Hat].) Indeed, suppose u is any unit vector in $\mathbb{R}^{n+1}$ and $B(t)$ is any path in S^n connecting u to $-u$. Then

$$A(t) := \pi(B(t))$$

is a loop in $\mathbb{R}P^n$, and this loop cannot be shrunk continuously to a point in $\mathbb{R}P^n$. To prove this claim, suppose that a map $A(s, t)$ as in Definition 1.14 exists. Then $A(s, t)$ can be "lifted" to a continuous map $B(s, t)$ into S^n such that $B(0, t) = B(t)$ and such that $A(s, t) = \pi(B(s, t))$. (See Proposition 1.30 in [Hat].) Since $A(s, 0) = A(s, 1)$ for all s, we must have $B(s, 0) = \pm B(s, 1)$. But by construction, $B(0, 0) = -B(0, 1)$. If order for $B(s, t)$ to be continuous in s, we must then have $B(s, 0) = -B(s, 1)$ for all s. It follows that $B(1, t)$ is a nonconstant path in S^n. It is then easily verified that $A(1, t) = \pi(B(1, t))$ cannot be constant, contradicting our assumption about $A(s, t)$.

Let D^n denote the closed upper hemisphere in S^n, that is, the set of points $u \in S^n$ with $u_{n+1} \geq 0$. Then π maps D^n onto $\mathbb{R}P^n$, since at least one of u and $-u$ is in D^n. The restriction of π to D^n is injective except on the equator, that is, the set of $u \in S^n$ with $u_{n+1} = 0$. If u is in the equator, then $-u$ is also in the equator, and $\pi(-u) = \pi(u)$. Thus, we may also think of $\mathbb{R}P^n$ as the upper hemisphere D^n, with antipodal points on the equator identified (Figure 1.2).

We may now make one last identification using the projection P of $\mathbb{R}^{n+1}$ onto $\mathbb{R}^n$. (That is to say, P is the map sending $(x_1, \ldots, x_n, x_{n+1})$ to $(x_1, \ldots, x_n)$.) The restriction of P to D^n is a continuous bijection between D^n and the closed unit ball B^n in $\mathbb{R}^n$, with the equator in D^n mapping to the boundary of the ball. Thus, our

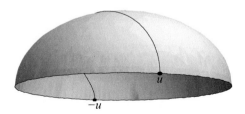

Fig. 1.2 The space $\mathbb{R}P^n$ is the upper hemisphere with antipodal points on the equator identified. The indicated path from u to $-u$ corresponds to a loop in $\mathbb{R}P^n$ that cannot be shrunk to a point

last model of $\mathbb{R}P^n$ is the closed unit ball $B^n \subset \mathbb{R}^n$, with antipodal points on the boundary of B^n identified.

We now turn to a topological analysis of $\mathsf{SO}(3)$.

Proposition 1.17. *There is a continuous bijection between* $\mathsf{SO}(3)$ *and* $\mathbb{R}P^3$.

Since $\mathbb{R}P^3$ is not simply connected, it follows that $\mathsf{SO}(3)$ is not simply connected, either.

Proof. If v is a unit vector in $\mathbb{R}^3$, let $R_{v,\theta}$ be the element of $\mathsf{SO}(3)$ consisting of a "right-handed" rotation by angle θ in the plane orthogonal to v. That is to say, let $v^\perp$ denote the plane orthogonal to v and choose an orthonormal basis (u_1, u_2) for $v^\perp$ in such a way that the linear map taking the orthonormal basis (u_1, u_2, v) to the standard basis (e_1, e_2, e_3) has positive determinant. We use the basis (u_1, u_2) to identify $v^\perp$ with $\mathbb{R}^2$, and the rotation is then in the counterclockwise direction in $\mathbb{R}^2$. It is easily seen that $R_{-v,\theta}$ is the same as $R_{v,-\theta}$. It is also not hard to show (Exercise 14) that every element of $\mathsf{SO}(3)$ can be expressed as $R_{v,\theta}$, for some v and θ with $-\pi \leq \theta \leq \pi$. Furthermore, we can arrange that $0 \leq \theta \leq \pi$ by replacing v with $-v$ if necessary.

If $R = I$, then $R = R_{v,0}$ for any unit vector v. If R is a rotation by angle π about some axis v, then R can be expressed both as $R_{v,\pi}$ and as $R_{-v,\pi}$. It is not hard to see that if $R \neq I$ and R is not a rotation by angle π, then R has a *unique* representation as $R_{v,\theta}$ with $0 < \theta < \pi$.

Now let B^3 denote the closed ball of radius π in $\mathbb{R}^3$ and consider the map $\Phi : B^3 \to \mathsf{SO}(3)$ given by

$$\Phi(u) = R_{\hat{u},\|u\|}, \quad u \neq 0,$$

$$\Phi(0) = I.$$

Here, $\hat{u} = u/\|u\|$ is the unit vector in the u-direction. The map Φ is continuous, even at I, since $R_{v,\theta}$ approaches the identity as θ approaches zero, regardless of how v is behaving. The discussion in the preceding paragraph shows that Φ maps B^3 onto $\mathsf{SO}(3)$. The map Φ is injective except that "antipodal" points on the boundary of B^3 have the same image: $R_{v,\pi} = R_{-v,\pi}$. Thus, Φ descends to a continuous, injective map of $\mathbb{R}P^3$ onto $\mathsf{SO}(3)$. Since both $\mathbb{R}P^3$ and $\mathsf{SO}(3)$ are compact, Theorem 4.17 in [Rud1] tells us that the inverse map is also continuous, meaning that $\mathsf{SO}(3)$ is homeomorphic to $\mathbb{R}P^3$. $\qquad\square$

For a different approach to proving Proposition 1.17, see the discussion following Proposition 1.19.

1.4   Homomorphisms

We now look at the notion of homomorphisms for matrix Lie groups.

Definition 1.18. Let G and H be matrix Lie groups. A map Φ from G to H is called a **Lie group homomorphism** if (1) Φ is a group homomorphism and (2) Φ is continuous. If, in addition, Φ is one-to-one and onto and the inverse map Φ^{-1} is continuous, then Φ is called a **Lie group isomorphism**.

The condition that Φ be continuous should be regarded as a technicality, in that it is very difficult to give an example of a group homomorphism between two matrix Lie groups which is not continuous. In fact, if $G = \mathbb{R}$ and $H = \mathbb{C}^*$, then any group homomorphism from G to H which is even measurable (a very weak condition) must be continuous. (See Exercise 17 in Chapter 9 of [Rud2].)

Note that the inverse of a Lie group isomorphism is continuous (by definition) and a group homomorphism (by elementary group theory), and thus a Lie group isomorphism. If G and H are matrix Lie groups and there exists a Lie group isomorphism from G to H, then G and H are said to be **isomorphic**, and we write $G \cong H$.

The simplest interesting example of a Lie group homomorphism is the determinant, which is a homomorphism of $\mathsf{GL}(n; \mathbb{C})$ into $\mathbb{C}^*$. Another simple example is the map $\Phi : \mathbb{R} \to \mathsf{SO}(2)$ given by

$$\Phi(\theta) = \begin{pmatrix} \cos\theta & -\sin\theta \\ \sin\theta & \cos\theta \end{pmatrix}.$$

This map is clearly continuous, and calculation (using standard trigonometric identities) shows that it is a homomorphism.

An important topic for us will be the relationship between the groups $\mathsf{SU}(2)$ and $\mathsf{SO}(3)$, which are almost, but not quite, isomorphic. Specifically, we now show that there exists a Lie group homomorphism $\Phi : \mathsf{SU}(2) \to \mathsf{SO}(3)$ that is *two*-to-one and onto. Consider the space V of all 2×2 complex matrices X which are self-adjoint (i.e., $X^* = X$) and have trace zero. Elements of V are precisely the matrices of the form

$$X = \begin{pmatrix} x_1 & x_2 + ix_3 \\ x_2 - ix_3 & -x_1 \end{pmatrix}, \tag{1.14}$$

with $x_1, x_2, x_3 \in \mathbb{R}$. If we identify V with $\mathbb{R}^3$ by means of the coordinates $x_1, x_2,$ and x_3 in (1.14), then the standard inner product on $\mathbb{R}^3$ can be computed as

$$\langle X_1, X_2 \rangle = \frac{1}{2}\text{trace}(X_1 X_2).$$

That is to say,

$$\frac{1}{2}\text{trace}\left(\begin{pmatrix} x_1 & x_2 + ix_3 \\ x_2 - ix_3 & -x_1 \end{pmatrix} \begin{pmatrix} x_1' & x_2' + ix_3' \\ x_2' - ix_3' & -x_1' \end{pmatrix} \right)$$
$$= x_1 x_1' + x_2 x_2' + x_3 x_3',$$

as one may easily check by direct calculation.

For each $U \in \mathsf{SU}(2)$, define a linear map $\Phi_U : V \to V$ by

$$\Phi_U(X) = UXU^{-1}.$$

Since U is unitary,

$$(UXU^{-1})^* = (U^{-1})^*XU^* = UXU^{-1},$$

showing that UXU^{-1} is again in V.

It is easy to see that $\Phi_{U_1 U_2} = \Phi_{U_1}\Phi_{U_2}$. Furthermore,

$$\frac{1}{2}\mathrm{trace}((UX_1U^{-1})(UX_2U^{-1})) = \frac{1}{2}\mathrm{trace}(UX_1X_2U^{-1})$$

$$= \frac{1}{2}\mathrm{trace}(X_1X_2),$$

since the trace is invariant under conjugation. Thus, each Φ_U preserves the inner product $\mathrm{trace}(X_1X_2)/2$ on V. It follows that the map $U \mapsto \Phi_U$ is a homomorphism of $\mathsf{SU}(2)$ into the group of orthogonal linear transformations of $V \cong \mathbb{R}^3$, that is, into $\mathsf{O}(3)$. Since $\mathsf{SU}(2)$ is connected (Proposition 1.13), Φ_U must actually lie in $\mathsf{SO}(3)$ for all $U \in \mathsf{SU}(2)$. Thus, Φ (i.e., the map $U \mapsto \Phi_U$) is a homomorphism of $\mathsf{SU}(2)$ into $\mathsf{SO}(3)$, which is easily seen to be continuous. Since $(-I)X(-I)^{-1} = X$, we see that Φ_{-I} is the identity element of $\mathsf{SO}(3)$.

Suppose, for example, that U is the matrix

$$U = \begin{pmatrix} e^{i\theta/2} & 0 \\ 0 & e^{-i\theta/2} \end{pmatrix}.$$

Then by direct calculation, we obtain

$$U \begin{pmatrix} x_1 & x_2 + ix_3 \\ x_2 - ix_3 & -x_1 \end{pmatrix} U^{-1} = \begin{pmatrix} x_1' & x_2' + ix_3' \\ x_2' - ix_3' & -x_1' \end{pmatrix}, \tag{1.15}$$

where $x_1' = x_1$ and

$$x_2' + ix_3' = e^{i\theta}(x_2 + ix_3)$$

$$= (x_2 \cos\theta - x_3 \sin\theta) + i(x_2 \sin\theta + x_3 \cos\theta). \tag{1.16}$$

In this case, then, Φ_U is a rotation by angle θ in the (x_2, x_3)-plane. Note that even though the diagonal entries of U are $e^{\pm i\theta/2}$, the map Φ_U is a rotation by angle θ, not $\theta/2$.

Proposition 1.19. *The map* $U \mapsto \Phi_U$ *is a 2-1 and onto map of* $\mathsf{SU}(2)$ *to* $\mathsf{SO}(3)$, *with kernel equal to* $\{I, -I\}$.

Since $\mathsf{SU}(2)$ is homeomorphic to S^3, the proposition gives another way of seeing that $\mathsf{SO}(3)$ is homeomorphic to $\mathbb{R}P^3$, that is, S^3 with antipodal points identified. This result was obtained in a different way in Proposition 1.17.

It is not hard to show that Φ is a "covering map" in the topological sense (Section 1.3 of [Hat]). Since $\mathsf{SU}(2)$ is simply connected (Proposition 1.15) and the map is 2-1, it follows by the theory of covering maps (e.g., Theorem 1.38 in [Hat]) that $\mathsf{SO}(3)$ cannot be simply connected and, indeed, it must have fundamental group $\mathbb{Z}/2$. See Chapter 13 for general information about fundamental groups and for a computation of the fundamental group of $\mathsf{SO}(n)$, $n > 2$.

Proof. Exercise 16 shows that the kernel of Φ is precisely the set $\{I, -I\}$. To see that Φ maps *onto* $\mathsf{SO}(3)$, let R be a rotation of $V \cong \mathbb{R}^3$. By Exercise 14, there exists an "axis" $X \in V$ such that R is a rotation by some angle θ in the plane orthogonal to X. If we express X in the form

$$X = U_0 \begin{pmatrix} x_1 & 0 \\ 0 & -x_1 \end{pmatrix} U_0^{-1}$$

with $U_0 \in \mathsf{U}(2)$, then the plane orthogonal to X in V is the space of matrices of the form

$$X' = U_0 \begin{pmatrix} 0 & x_2 + ix_3 \\ x_2 - ix_3 & 0 \end{pmatrix} U_0^{-1}. \tag{1.17}$$

If we now take

$$U = U_0 \begin{pmatrix} e^{i\theta/2} & 0 \\ 0 & e^{-i\theta/2} \end{pmatrix} U_0^{-1},$$

we can easily see that $UXU^{-1} = X$. On the other hand, the calculations in (1.15) and (1.16) show that $UX'U^{-1}$ is of the same form as in (1.17), but with (x_2, x_3) rotated by angle θ. Thus, Φ_U is a rotation by angle θ in the plane perpendicular to X, showing that Φ_U coincides with R.   □

It is possible, if not terribly useful, to calculate Φ explicitly. If you write an element of $\mathsf{SU}(2)$ as in Exercise 5, you (or your computer) may calculate

$$\begin{pmatrix} \alpha & -\bar{\beta} \\ \beta & \bar{\alpha} \end{pmatrix} \begin{pmatrix} x_1 & x_2 + ix_3 \\ x_2 - ix_3 & -x_1 \end{pmatrix} \begin{pmatrix} \bar{\alpha} & \bar{\beta} \\ -\beta & \alpha \end{pmatrix}$$
$$= \begin{pmatrix} x_1' & x_2' + ix_3' \\ x_2' - ix_3' & -x_3' \end{pmatrix}$$

explicitly. Then (x'_1, x'_2, x'_3) will depend linearly on (x_1, x_2, x_3) and you can express (x'_1, x'_2, x'_3) as a matrix applied to (x_1, x_2, x_3), with the result that

$$\Phi_U = \begin{pmatrix} |\alpha|^2 - |\beta|^2 & -2\operatorname{Re}(\alpha\beta) & 2\operatorname{Im}(\alpha\beta) \\ 2\operatorname{Re}(\alpha\bar\beta) & \operatorname{Re}(\alpha^2 - \beta^2) & \operatorname{Im}(\beta^2 - \alpha^2) \\ 2\operatorname{Im}(\alpha\bar\beta) & \operatorname{Im}(\alpha^2 + \beta^2) & \operatorname{Re}(\alpha^2 + \beta^2) \end{pmatrix}.$$

If we take $\alpha = e^{i\theta/2}$ and $\beta = 0$, we may see directly that Φ_U is a rotation by angle θ in the (x_2, x_3)-plane, as we saw already in (1.15) and (1.16).

1.5   Lie Groups

A Lie group is a smooth manifold equipped with a group structure such that the operations of group multiplication and inversion are smooth. As the terminology suggests, every matrix Lie group is a Lie group. (See Corollary 3.45 in Chapter 3.) The reverse is not true: Not every Lie group is isomorphic to a matrix Lie group. Nevertheless, we have restricted our attention in this book to matrix Lie groups, in order to minimize prerequisites and keep the discussion as concrete as possible. Most of the interesting examples of Lie groups are, in any case, matrix Lie groups.

A **manifold** is an object M that looks locally like a piece of $\mathbb{R}^n$. More precisely, an n-dimensional manifold is a second-countable, Hausdorff topological space with the property that each $m \in M$ has a neighborhood that is homeomorphic to an open subset of $\mathbb{R}^n$. A two-dimensional torus, for example, looks locally but not globally like $\mathbb{R}^2$ and is, thus, a two-dimensional manifold. A **smooth manifold** is a manifold M together with a collection of local coordinates covering M such that the change-of-coordinates map between two overlapping coordinate systems is smooth.

Definition 1.20. A **Lie group** is a smooth manifold G which is also a group and such that the group product

$$G \times G \to G$$

and the inverse map $G \to G$ are smooth.

Example 1.21. Let

$$G = \mathbb{R} \times \mathbb{R} \times S^1 = \{(x, y, u) | x \in \mathbb{R}, y \in \mathbb{R}, u \in S^1 \subset \mathbb{C}\},$$

equipped with the group product given by

$$(x_1, y_1, u_1) \cdot (x_2, y_2, u_2) = (x_1 + x_2, y_1 + y_2, e^{ix_1 y_2} u_1 u_2).$$

Then G is a Lie group.

Proof. It is easily checked that this operation is associative; the product of three elements with either grouping is

$$(x_1 + x_2 + x_3, y_1 + y_2 + y_3, e^{i(x_1 y_2 + x_1 y_3 + x_2 y_3)} u_1 u_2 u_3).$$

There is an identity element in G, namely $e = (0, 0, 1)$ and each element (x, y, u) has an inverse given by $(-x, -y, e^{ixy} u^{-1})$. Thus, G is, in fact, a group. Furthermore, both the group product and the map that sends each element to its inverse are clearly smooth, showing that G is a Lie group.   □

Although there is nothing about matrices in the definition of the group G in Example 1.21, we may ask whether G is *isomorphic to* some matrix Lie group. This turns out to be false. As shown in Sect. 4.8, there is no continuous, injective homomorphism of G into any $\mathsf{GL}(n; \mathbb{C})$. We conclude, then, that not every Lie group is isomorphic to a matrix Lie group. Nevertheless, most of the interesting examples of Lie groups *are* matrix Lie groups.

Let us now think briefly about how we might show that every matrix Lie group is a Lie group. We will prove in Sect. 3.7 that every matrix Lie group is an "embedded submanifold" of $M_n(\mathbb{C}) \cong \mathbb{R}^{2n^2}$. The operations of matrix multiplication and inversion are smooth on $M_n(\mathbb{C})$ (after restricting to the open subset of invertible matrices in the case of inversion). Thus, the restriction of these operations to a matrix Lie group $G \subset M_n(\mathbb{C})$ is also smooth, making G into a Lie group.

It is customary to call a map Φ between two Lie groups a Lie group homomorphism if Φ is a group homomorphism and Φ is *smooth*, whereas we have (in Definition 1.18) required only that Φ be continuous. We will show, however, that every continuous homomorphism between matrix Lie groups is automatically smooth, so that there is no conflict of terminology. See Corollary 3.50 to Theorem 3.42. Finally, we note that since every matrix Lie group G is a manifold, G must be locally path connected. It then follows by a standard topological argument that G is connected if and only if it is path connected.

1.6 Exercises

1. Let $[\cdot, \cdot]_{n,k}$ be the symmetric bilinear form on $\mathbb{R}^{n+k}$ defined in (1.5). Let g be the $(n + k) \times (n + k)$ diagonal matrix with first n diagonal entries equal to one and last k diagonal entries equal to minus one:

$$g = \begin{pmatrix} I_n & 0 \\ 0 & -I_k \end{pmatrix}.$$

Show that for all $x, y \in \mathbb{R}^{n+k}$,

$$[x, y]_{n,k} = \langle x, gy \rangle.$$

Show that a $(n + k) \times (n + k)$ real matrix A belongs to $\mathsf{O}(n; k)$ if and only if $gA^{tr}g = A^{-1}$.

2. Let ω be the skew-symmetric bilinear form on $\mathbb{R}^{2n}$ given by (1.7). Let Ω be the $2n \times 2n$ matrix

$$\Omega = \begin{pmatrix} 0 & I \\ -I & 0 \end{pmatrix}.$$

Show that for all $x, y \in \mathbb{R}^{2n}$, we have

$$\omega(x, y) = \langle x, \Omega y \rangle.$$

Show that a $2n \times 2n$ matrix A belongs to $\mathsf{Sp}(n; \mathbb{R})$ if and only if $-\Omega A^{tr} \Omega = A^{-1}$.

Note: A similar analysis applies to $\mathsf{Sp}(n; \mathbb{C})$.

3. Show that the symplectic group $\mathsf{Sp}(1; \mathbb{R}) \subset \mathsf{GL}(2; \mathbb{R})$ is equal to $\mathsf{SL}(2; \mathbb{R})$. Show that $\mathsf{Sp}(1; \mathbb{C}) = \mathsf{SL}(2; \mathbb{C})$ and that $\mathsf{Sp}(1) = \mathsf{SU}(2)$.

4. Show that a matrix R belongs to $\mathsf{SO}(2)$ if and only if it can be expressed in the form

$$\begin{pmatrix} \cos\theta & -\sin\theta \\ \sin\theta & \cos\theta \end{pmatrix}$$

for some $\theta \in \mathbb{R}$. Show that a matrix R belongs to $\mathsf{O}(2)$ if and only if it is of one of the two forms:

$$A = \begin{pmatrix} \cos\theta & -\sin\theta \\ \sin\theta & \cos\theta \end{pmatrix} \quad \text{or} \quad A = \begin{pmatrix} \cos\theta & \sin\theta \\ \sin\theta & -\cos\theta \end{pmatrix}.$$

Hint: Recall that for A to be in $\mathsf{O}(2)$, the columns of A must be orthonormal.

5. Show that if α and β are arbitrary complex numbers satisfying $|\alpha|^2 + |\beta|^2 = 1$, then the matrix

$$A = \begin{pmatrix} \alpha & -\overline{\beta} \\ \beta & \overline{\alpha} \end{pmatrix}$$

is in $\mathsf{SU}(2)$. Show that every $A \in \mathsf{SU}(2)$ can be expressed in this form for a unique pair (α, β) satisfying $|\alpha|^2 + |\beta|^2 = 1$.

6. Suppose U belongs to $M_{2n}(\mathbb{C})$ and U has an orthonormal basis of eigenvectors satisfying the conditions in Theorem 1.6. Show that U belongs to $\mathsf{Sp}(n)$.

Hint: Start by showing that U is unitary. Then show that $\omega(Uz, Uw) = \omega(z, w)$ if z and w belong to the basis $u_1, \ldots, u_n, v_1, \ldots, v_n$.

7. Using Theorem 1.6, show that $\mathsf{Sp}(n)$ is connected and that every element of $\mathsf{Sp}(n)$ has determinant 1.

8. Determine the center $Z(H)$ of the Heisenberg group H. Show that the quotient group $H/Z(H)$ is commutative.

9. Suppose a is an irrational real number. Show that the set E_a of numbers of the form $e^{2\pi i n a}$, $n \in \mathbb{Z}$, is dense in the unit circle S^1.

   Hint: Show that if we divide S^1 into N equally sized "bins" of length $2\pi/N$, there is at least one bin that contains infinitely many elements of E_a. Then use the fact that E_a is a subgroup of S^1.

10. Let a be an irrational real number and let G be the following subgroup of $\mathsf{GL}(2; \mathbb{C})$:

$$G = \left\{ \begin{pmatrix} e^{it} & 0 \\ 0 & e^{ita} \end{pmatrix} \middle| t \in \mathbb{R} \right\}.$$

    Show that

$$\overline{G} = \left\{ \begin{pmatrix} e^{i\theta} & 0 \\ 0 & e^{i\phi} \end{pmatrix} \middle| \theta, \phi \in \mathbb{R} \right\},$$

    where $\overline{G}$ denotes the closure of the set G inside the space of 2×2 matrices.
    Hint: Use Exercise 9.

11. A subset E of a matrix Lie group G is called **discrete** if for each A in E there is a neighborhood U of A in G such that U contains no point in E except for A. Suppose that G is a connected matrix Lie group and N is a discrete normal subgroup of G. Show that N is contained in the center of G.

12. This problem gives an alternative proof of Proposition 1.11, namely that $\mathsf{GL}(n; \mathbb{C})$ is connected. Suppose A and B are invertible $n \times n$ matrices. Show that there are only finitely many complex numbers λ for which $\det(\lambda A + (1 - \lambda) B) = 0$. Show that there exists a continuous path $A(t)$ of the form $A(t) = \lambda(t) A + (1 - \lambda(t)) B$ connecting A to B and such that $A(t)$ lies in $\mathsf{GL}(n; \mathbb{C})$. Here, $\lambda(t)$ is a continuous path in the plane with $\lambda(0) = 0$ and $\lambda(1) = 1$.

13. Show that $\mathsf{SO}(n)$ is connected, using the following outline.
    For the case $n = 1$, there is nothing to show, since a 1×1 matrix with determinant one must be [1]. Assume, then, that $n \geq 2$. Let e_1 denote the unit vector with entries $1, 0, \ldots, 0$ in $\mathbb{R}^n$. For every unit vector $v \in \mathbb{R}^n$, show that there exists a continuous path $R(t)$ in $\mathsf{SO}(n)$ with $R(0) = I$ and $R(1)v = e_1$. (Thus, any unit vector can be "continuously rotated" to e_1.)
    Now, show that any element R of $\mathsf{SO}(n)$ can be connected to a block-diagonal matrix of the form

$$\begin{pmatrix} 1 & \\ & R_1 \end{pmatrix}$$

    with $R_1 \in \mathsf{SO}(n - 1)$ and proceed by induction.

14. If R is an element of $\mathsf{SO}(3)$, show that R must have an eigenvector v with eigenvalue 1. Show that R maps the plane orthogonal to v into itself. Conclude that R is a rotation by some angle θ around the "axis" v.

    Hint: Since $\mathsf{SO}(3) \subset \mathsf{SU}(3)$, every (real or complex) eigenvalue of R must have absolute value 1. Since, also, R is real, any nonreal eigenvalues of R come in conjugate pairs.

15. Let R be an element of $\mathsf{SO}(n)$.

    (a) Suppose $v \in \mathbb{C}^n$ is an eigenvector for R with eigenvalue $\lambda \in \mathbb{C}$, and suppose λ is not real. Let $V \subset \mathbb{R}^n$ be the two-dimensional span of $(v + \bar{v})/2$ and $(v - \bar{v})/(2i)$. Show that V is invariant under R and that the restriction of R to V has determinant 1.

    (b) Suppose that a subspace $V \subset \mathbb{R}^n$ is invariant under both R and R^{-1}. Show that the orthogonal complement $V^\perp$ of V is also invariant under both R and R^{-1}.

    (c) Show that if $n = 2k$, there exists $S \in \mathsf{SO}(n)$ such that

$$R = S \begin{pmatrix} \cos\theta_1 & -\sin\theta_1 & & \\ \sin\theta_1 & \cos\theta_1 & & \\ & & \ddots & \\ & & & \cos\theta_k & -\sin\theta_k \\ & & & \sin\theta_k & \cos\theta_k \end{pmatrix} S^{-1}$$

and that if $n = 2k + 1$, there exists $S \in \mathsf{SO}(n)$ such that

$$R = S \begin{pmatrix} \cos\theta_1 & -\sin\theta_1 & & & \\ \sin\theta_1 & \cos\theta_1 & & & \\ & & \ddots & & \\ & & & \cos\theta_k & -\sin\theta_k & \\ & & & \sin\theta_k & \cos\theta_k & \\ & & & & & 1 \end{pmatrix} S^{-1}.$$

    That is, in a suitable orthonormal basis, R is block diagonal with 2×2 blocks of the indicated form, with a single 1×1 block if n is odd.

    Hint: Show that the number of eigenvalues of R equal to -1 is even.

16. (a) Show that if a matrix A commutes with every matrix X of the form (1.14), then A commutes with every element of $M_2(\mathbb{C})$. Conclude that A must be a multiple of the identity.

    (b) Show that the kernel of the map $U \mapsto \Phi_U$ in Proposition 1.19 is precisely the set $\{I, -I\}$.

17. Suppose $G \subset \mathsf{GL}(n_1; \mathbb{C})$ and $H \subset \mathsf{GL}(n_2; \mathbb{C})$ are matrix Lie groups and that $\Phi : G \to H$ is a Lie group homomorphism. Then the image of G under Φ is a subgroup of H and thus of $\mathsf{GL}(n_2; \mathbb{C})$. Is the image of G under Φ necessarily a matrix Lie group? Prove or give a counter-example.

18. Show that every continuous homomorphism Φ from $\mathbb{R}$ to S^1 is of the form $\Phi(x) = e^{iax}$ for some $a \in \mathbb{R}$.

    Hint: Since Φ is continuous, there is some $\varepsilon > 0$ such that if $|x| < \varepsilon$, then $\Phi(x)$ belongs to the right half of the unit circle.

Chapter 2
The Matrix Exponential

2.1 The Exponential of a Matrix

The exponential of a matrix plays a crucial role in the theory of Lie groups. The exponential enters into the definition of the Lie algebra of a matrix Lie group (Sect. 3.3) and is the mechanism for passing information from the Lie algebra to the Lie group.

If X is an $n \times n$ matrix, we define the **exponential** of X, denoted e^X or $\exp X$, by the usual power series

$$e^X = \sum_{m=0}^{\infty} \frac{X^m}{m!}, \tag{2.1}$$

where X^0 is defined to be the identity matrix I and where X^m is the repeated matrix product of X with itself.

Proposition 2.1. *The series (2.1) converges for all* $X \in M_n(\mathbb{C})$ *and* e^X *is a continuous function of* X.

Our proof will use the notion of the norm of a matrix $X \in M_n(\mathbb{C})$, which we define by thinking of $M_n(\mathbb{C})$ as $\mathbb{C}^{n^2}$.

A previous version of this book was inadvertently published without the middle initial of the author's name as "Brian Hall". For this reason an erratum has been published, correcting the mistake in the previous version and showing the correct name as Brian C. Hall (see DOI http://dx.doi.org/10.1007/978-3-319-13467-3_14). The version readers currently see is the corrected version. The Publisher would like to apologize for the earlier mistake.

© Springer International Publishing Switzerland 2015
B.C. Hall, *Lie Groups, Lie Algebras, and Representations*, Graduate
Texts in Mathematics 222, DOI 10.1007/978-3-319-13467-3_2

Definition 2.2. For any $X \in M_n(\mathbb{C})$, we define

$$\|X\| = \left(\sum_{j,k=1}^{n} |X_{jk}|^2 \right)^{1/2}.$$

The quantity $\|X\|$ is called the **Hilbert–Schmidt** norm of X.

The Hilbert–Schmidt norm may be computed in a basis-independent way as

$$\|X\| = (\text{trace}(X^* X))^{1/2}. \tag{2.2}$$

(See Sect. A.6.) This norm satisfies the inequalities

$$\|X + Y\| \leq \|X\| + \|Y\|, \tag{2.3}$$

$$\|XY\| \leq \|X\| \, \|Y\| \tag{2.4}$$

for all $X, Y \in M_n(\mathbb{C})$. The first of these inequalities is the triangle inequality for $\mathbb{C}^{n^2}$ and the second follows from the Cauchy–Schwarz inequality (Exercise 1). If X_m is a sequence of matrices, then it is easy to see that X_m converges to a matrix X in the sense of Definition 1.3 if and only if $\|X_m - X\| \to 0$ as $m \to \infty$.

Proof of Proposition 2.1. In light of (2.4), we see that

$$\|X^m\| \leq \|X\|^m$$

for all $m \geq 1$, and, hence,

$$\sum_{m=0}^{\infty} \left\| \frac{X^m}{m!} \right\| \leq \|I\| + \sum_{m=1}^{\infty} \frac{\|X\|^m}{m!} < \infty.$$

Thus, the series (2.1) converges absolutely.

To show continuity, note that since X^m is a continuous function of X, the partial sums of (2.1) are continuous. By the Weierstrass M-test, the series (2.1) converges uniformly on each set of the form $\{\|X\| \leq R\}$. Thus, e^X is continuous on each such set, and, thus, continuous on all of $M_n(\mathbb{C})$. $\qquad\square$

We now list some elementary properties of the matrix exponential.

Proposition 2.3. *Let X and Y be arbitrary $n \times n$ matrices. Then we have the following:*

1. $e^0 = I$.
2. $\left(e^X \right)^* = e^{X^*}$.
3. e^X *is invertible and* $\left(e^X \right)^{-1} = e^{-X}$.
4. $e^{(\alpha + \beta)X} = e^{\alpha X} e^{\beta X}$ *for all α and β in $\mathbb{C}$.*
5. *If $XY = YX$, then $e^{X+Y} = e^X e^Y = e^Y e^X$.*
6. *If C is in $\mathsf{GL}(N;\mathbb{C})$, then $e^{CXC^{-1}} = Ce^X C^{-1}$.*

Although $e^{X+Y} = e^X e^Y$ when X and Y commute, this identity fails in general. This is an important point, which we will return to in the Lie product formula in Sect. 2.4 and the Baker–Campbell–Hausdorff formula in Chapter 5.

Proof. Point 1 is obvious and Point 2 follows from taking term-by-term adjoints of the series for e^X. Points 3 and 4 are special cases of Point 5. To verify Point 5, we simply multiply the two power series term by term, which is permitted because both series converge absolutely. Multiplying out $e^X e^Y$ and collecting terms where the power of X plus the power of Y equals m, we obtain

$$e^X e^Y = \sum_{m=0}^{\infty} \sum_{k=0}^{m} \frac{X^k}{k!} \frac{Y^{m-k}}{(m-k)!} = \sum_{m=0}^{\infty} \frac{1}{m!} \sum_{k=0}^{m} \frac{m!}{k!(m-k)!} X^k Y^{m-k}. \tag{2.5}$$

Now, because (and *only* because) X and Y commute,

$$(X+Y)^m = \sum_{k=0}^{m} \frac{m!}{k!(m-k)!} X^k Y^{m-k},$$

and, thus, (2.5) becomes

$$e^X e^Y = \sum_{m=0}^{\infty} \frac{1}{m!} (X+Y)^m = e^{X+Y}.$$

To prove Point 6, simply note that

$$\left(CXC^{-1} \right)^m = CX^m C^{-1}$$

and, thus, the two sides of Point 6 are equal term by term.   □

Proposition 2.4. *Let X be a $n \times n$ complex matrix. Then e^{tX} is a smooth curve in $M_n(\mathbb{C})$ and*

$$\frac{d}{dt} e^{tX} = X e^{tX} = e^{tX} X.$$

In particular,

$$\frac{d}{dt} e^{tX} \bigg|_{t=0} = X.$$

Results that hold for the exponential of numbers may or may not hold for the matrix exponential. Although Proposition 2.4 is what one would expect from the scalar case, it should be noted that, in general, the derivative of e^{X+tY} is not equal to $e^{X+tY} Y$. See Sect. 5.4.

Proof. Differentiate the power series for e^{tX} term by term. This is permitted because, for each j and k, $(e^{tX})_{jk}$ is given by a convergent power series in t, and one can differentiate a power series term by term inside its radius of convergence (e.g., Theorem 12 in Chapter 4 of [Pugh]).   □

2.2 Computing the Exponential

We consider here methods for exponentiating general matrices. A special method for exponentiating 2×2 matrices is described in Exercises 6 and 7. Suppose that $X \in M_n(\mathbb{C})$ has n linearly independent eigenvectors $v_1, \ldots, v_n$ with eigenvalues $\lambda_1, \ldots, \lambda_n$. Let C be the $n \times n$ matrix whose columns are $v_1, \ldots, v_n$ and let D be the diagonal matrix with diagonal entries $\lambda_1, \ldots, \lambda_n$. Then $X = CDC^{-1}$. It is easily verified that e^D is the diagonal matrix with diagonal entries $e^{\lambda_1}, \ldots, e^{\lambda_n}$, and thus, by Proposition 2.3, we have

$$e^X = C \begin{pmatrix} e^{\lambda_1} & & 0 \\ & \ddots & \\ 0 & & e^{\lambda_n} \end{pmatrix} C^{-1}.$$

Meanwhile, if X is nilpotent (i.e., $X^k = 0$ for some k), then the series that defines e^X terminates. Finally, according to Theorem A.6, every matrix X can be written (uniquely) in the form $X = S + N$, with S diagonalizable, N nilpotent, and $SN = NS$. Then, since N and S commute,

$$e^X = e^{S+N} = e^S e^N.$$

Example 2.5. Consider the matrices

$$X_1 = \begin{pmatrix} 0 & -a \\ a & 0 \end{pmatrix}; \quad X_2 = \begin{pmatrix} 0 & a & b \\ 0 & 0 & c \\ 0 & 0 & 0 \end{pmatrix}; \quad X_3 = \begin{pmatrix} a & b \\ 0 & a \end{pmatrix}.$$

Then

$$e^{X_1} = \begin{pmatrix} \cos a & -\sin a \\ \sin a & \cos a \end{pmatrix}$$

and

$$e^{X_2} = \begin{pmatrix} 1 & a & b + ac/2 \\ 0 & 1 & c \\ 0 & 0 & 1 \end{pmatrix}.$$

and

$$e^{X_3} = \begin{pmatrix} e^a & e^a b \\ 0 & e^a \end{pmatrix}.$$

Proof. The eigenvectors of X_1 are $(1, i)$ and $(i, 1)$, with eigenvalues $-ia$ and ia, respectively. Thus,

$$e^{X_1} = \begin{pmatrix} 1 & i \\ i & 1 \end{pmatrix} \begin{pmatrix} e^{-ia} & 0 \\ 0 & e^{ia} \end{pmatrix} \begin{pmatrix} 1/2 & -i/2 \\ -i/2 & 1/2 \end{pmatrix},$$

which simplifies to the claimed result. Meanwhile, X_2^2 has the value ac in the upper right-hand corner and all other entries equal to zero, whereas $X_2^3 = 0$. Thus, $e^{X_2} = 1 + X_2 + X_2^2/2$, which reduces to the claimed result. Finally,

$$X_3 = \begin{pmatrix} a & 0 \\ 0 & a \end{pmatrix} + \begin{pmatrix} 0 & b \\ 0 & 0 \end{pmatrix},$$

where the two terms clearly commute and the second term is nilpotent. Thus, we obtain

$$e^{X_3} = \begin{pmatrix} e^a & 0 \\ 0 & e^a \end{pmatrix} \begin{pmatrix} 1 & b \\ 0 & 1 \end{pmatrix},$$

which reduces to the claimed result.   □

The matrix exponential is used in the elementary theory of differential equations, to solve systems of linear equations. Consider a first-order differential equation of the form

$$\frac{d\mathbf{v}}{dt} = X\mathbf{v},$$

$$\mathbf{v}(0) = \mathbf{v}_0,$$

where $\mathbf{v}(t) \in \mathbb{R}^n$ and X is a fixed $n \times n$ matrix. The (unique) solution of this equation is given by

$$\mathbf{v}(t) = e^{tX}\mathbf{v}_0,$$

as may be easily verified using Proposition 2.4. Curves of the form $t \mapsto e^{tX}\mathbf{v}_0$, with $\mathbf{v}_0$ fixed, trace out the flow along the vector field $\mathbf{v} \mapsto X\mathbf{v}$.

Let us consider the two matrices

$$X_4 = \begin{pmatrix} 1 & 2 \\ -2 & 1 \end{pmatrix}; \quad X_5 = \begin{pmatrix} 1 & 2 \\ 2 & 1 \end{pmatrix}. \tag{2.6}$$

Figure 2.1 plots several curves of this form for each matrix (see Exercise 8).

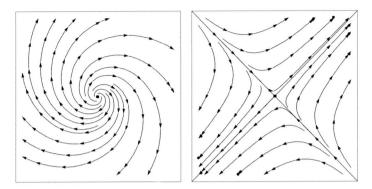

Fig. 2.1 Curves of the form $t \mapsto e^{tX_4} \mathbf{v}_0$ (*left*) and $t \mapsto e^{tX_5} \mathbf{v}_0$ (*right*)

2.3   The Matrix Logarithm

We now wish to define a matrix logarithm, which should be an inverse function
(to the extent possible) to the matrix exponential. Let us recall the situation for the
logarithm of complex numbers, in order to see what is reasonable to expect in the
matrix case. Since e^z is never zero, only nonzero numbers can have a logarithm.
Every nonzero complex number can be written as e^z for some z, but the z is not
unique and cannot be defined continuously on $\mathbb{C}^*$. In the matrix case, e^X is invertible
for all $X \in M_n(\mathbb{C})$. We will see (Theorem 2.10) that every invertible matrix can be
written as e^X, for some $X \in M_n(\mathbb{C})$, but the X is not unique.

   The simplest way to define the matrix logarithm is by a power series. We recall
how this works in the complex case.

Lemma 2.6. *The function*

$$\log z = \sum_{m=1}^{\infty} (-1)^{m+1} \frac{(z-1)^m}{m} \tag{2.7}$$

is defined and holomorphic in a circle of radius 1 about $z = 1$.
   For all z with $|z - 1| < 1$,

$$e^{\log z} = z.$$

For all u with $|u| < \log 2$, we have $|e^u - 1| < 1$ and

$$\log e^u = u.$$

Proof. The usual logarithm for real, positive numbers satisfies

$$\frac{d}{dx}\log(1-x) = \frac{-1}{1-x} = -\left(1 + x + x^2 + \cdots\right)$$

for $|x| < 1$. Integrating term by term and noting that $\log 1 = 0$ gives

$$\log(1-x) = -\left(x + \frac{x^2}{2} + \frac{x^3}{3} + \cdots\right).$$

Taking $z = 1 - x$ (so that $x = 1 - z$), we have

$$\log z = -\left((1-z) + \frac{(1-z)^2}{2} + \frac{(1-z)^3}{3} + \cdots\right)$$

$$= \sum_{m=1}^{\infty} (-1)^{m+1} \frac{(z-1)^m}{m}. \tag{2.8}$$

The series (2.8) has radius of convergence 1 and defines a holomorphic function on the set $\{|z-1| < 1\}$, which coincides with the usual logarithm for real z in the interval $(0, 2)$. Now, $\exp(\log z) = z$ for $z \in (0, 2)$ and since both sides of this identity are holomorphic in z, the identity continues to hold on the whole set $\{|z - 1| < 1\}$.

On the other hand, if $|u| < \log 2$, then

$$|e^u - 1| = \left|u + \frac{u^2}{2!} + \cdots\right| \leq |u| + \frac{|u|^2}{2!} + \cdots = e^{|u|} - 1 < 1.$$

Thus, $\log(\exp u)$ makes sense for all such u. Since $\log(\exp u) = u$ for real u with $|u| < \log 2$, it follows by holomorphicity that $\log(\exp u) = u$ for all complex numbers with $|u| < \log 2$. ☐

Definition 2.7. For an $n \times n$ matrix A, define $\log A$ by

$$\log A = \sum_{m=1}^{\infty} (-1)^{m+1} \frac{(A - I)^m}{m} \tag{2.9}$$

whenever the series converges.

Since the complex-valued series (2.7) has radius of convergence 1 and since $\|(A - I)^m\| \leq \|A - I\|^m$ for $m \geq 1$, the matrix-valued series (2.9) will converge if $\|A - I\| < 1$. Even if $\|A - I\| > 1$, the series might converge, for example, if $A - I$ is nilpotent (see Exercise 9).

Theorem 2.8. *The function*

$$\log A = \sum_{m=1}^{\infty} (-1)^{m+1} \frac{(A-I)^m}{m}$$

is defined and continuous on the set of all $n \times n$ complex matrices A with $\|A - I\| < 1$.
  For all $A \in M_n(\mathbb{C})$ with $\|A - I\| < 1$,

$$e^{\log A} = A.$$

For all $X \in M_n(\mathbb{C})$ with $\|X\| < \log 2$, $\|e^X - I\| < 1$ and

$$\log e^X = X.$$

  Although it might seem plausible that $\log(e^X)$ should be equal to X whenever the series for the logarithm is convergent, this claim is false (even over $\mathbb{C}$). If, for example, $X = 2\pi i I$, then $e^X = e^{2\pi i} I = I$. Then $e^X - I = 0$, so that $\log(e^X)$ is defined and equal to 0. In this case, $\log(e^X)$ is defined but not equal to X. Thus, the assumption that $\|X\| < \log 2$ cannot be replaced by, say, the assumption that $\|e^X - I\| < 1$.
  The proof of the theorem will establish a variant of the result, as follows. If A is diagonalizable and all the eigenvalues λ of A satisfy $|\lambda - 1| < 1$, then $\log A$ is defined and $e^{\log A} = A$. Similarly, if X is diagonalizable and all the eigenvalues λ of X satisfy $|\lambda| < \log 2$, then $\log e^X$ is defined and $\log e^X = X$.

Proof. Since $\|(A - I)^m\| \leq \|(A - I)\|^m$ and since the series (2.7) has radius of convergence 1, the series (2.9) converges absolutely for all A with $\|A - I\| < 1$. The proof of continuity is essentially the same as for the exponential.
  Suppose now that A satisfies $\|A - I\| < 1$. If A is diagonalizable with eigenvalues $z_1, \ldots, z_n$, then we can express A in the form CDC^{-1} with D diagonal, in which case

$$(A-I)^m = C \begin{pmatrix} (z_1 - 1)^m & & 0 \\ & \ddots & \\ 0 & & (z_n - 1)^m \end{pmatrix} C^{-1}.$$

Since $\|A - I\| < 1$, each eigenvalue z_j of A must satisfy $|z_j - 1| < 1$ (Exercise 2). Thus,

$$\sum_{m=1}^{\infty} (-1)^{m+1} \frac{(A-I)^m}{m} = C \begin{pmatrix} \log z_1 & & 0 \\ & \ddots & \\ 0 & & \log z_n \end{pmatrix} C^{-1},$$

and by Lemma 2.6,

$$
e^{\log A} = C \begin{pmatrix} e^{\log z_1} & & 0 \\ & \ddots & \\ 0 & & e^{\log z_n} \end{pmatrix} C^{-1} = A.
$$

If A is not diagonalizable, we approximate A by a sequence A_m of diagonalizable matrices (Exercise 4) and appeal to the continuity of the logarithm and exponential functions. Thus, $\exp(\log A) = A$ for all A with $\|A - I\| < 1$.

Now, the same argument as in the complex case shows that if $\|X\| < \log 2$, then $\|e^X - I\| < 1$. The proof that $\log(e^X) = X$ is then very similar to the proof that $\exp(\log A) = A$. $\qquad \square$

Proposition 2.9. *There exists a constant c such that for all $n \times n$ matrices B with $\|B\| < 1/2$, we have*

$$
\|\log(I + B) - B\| \le c \, \|B\|^2.
$$

Proof. Note that

$$
\log(I + B) - B = \sum_{m=2}^{\infty} (-1)^{m+1} \frac{B^m}{m} = B^2 \sum_{m=2}^{\infty} (-1)^{m+1} \frac{B^{m-2}}{m}
$$

so that if $\|B\| < 1/2$, we have

$$
\|\log(I + B) - B\| \le \|B\|^2 \sum_{m=2}^{\infty} \frac{\left(\frac{1}{2}\right)^{m-2}}{m},
$$

which is an estimate of the desired form. $\qquad \square$

We may restate the proposition in a more concise way by saying that

$$
\log(I + B) = B + O(\|B\|^2),
$$

where $O(\|B\|^2)$ denotes a quantity of order $\|B\|^2$ (i.e., a quantity that is bounded by a constant times $\|B\|^2$ for all sufficiently small values of $\|B\|$).

We conclude this section with a result that, although we will not use it elsewhere, is worth recording. The proof is sketched in Exercises 9 and 10.

Theorem 2.10. *Every invertible $n \times n$ matrix can be expressed as e^X for some $X \in M_n(\mathbb{C})$.*

2.4 Further Properties of the Exponential

In this section, we give several additional results involving the exponential of a matrix that will be important in our study of Lie algebras.

Theorem 2.11 (Lie Product Formula). *For all* $X, Y \in M_n(\mathbb{C})$, *we have*

$$e^{X+Y} = \lim_{m \to \infty} \left(e^{\frac{X}{m}} e^{\frac{Y}{m}} \right)^m.$$

There is a version of this result, known as the Trotter product formula, which holds for suitable unbounded operators on an infinite-dimensional Hilbert space. See, for example, Theorem 20.1 in [Hall].

Proof. If we multiply the power series for $e^{\frac{X}{m}}$ and $e^{\frac{Y}{m}}$, all but three of the terms will involve $1/m^2$ or higher powers of $1/m$. Thus,

$$e^{\frac{X}{m}} e^{\frac{Y}{m}} = I + \frac{X}{m} + \frac{Y}{m} + O\left(\frac{1}{m^2}\right).$$

Now, since $e^{\frac{X}{m}} e^{\frac{Y}{m}} \to I$ as $m \to \infty$, $e^{\frac{X}{m}} e^{\frac{Y}{m}}$ is in the domain of the logarithm for all sufficiently large m. By Proposition 2.9,

$$
\begin{aligned}
\log\left(e^{\frac{X}{m}} e^{\frac{Y}{m}}\right) &= \log\left(I + \frac{X}{m} + \frac{Y}{m} + O\left(\frac{1}{m^2}\right)\right) \\
&= \frac{X}{m} + \frac{Y}{m} + O\left(\left\|\frac{X}{m} + \frac{Y}{m} + O\left(\frac{1}{m^2}\right)\right\|^2\right) \\
&= \frac{X}{m} + \frac{Y}{m} + O\left(\frac{1}{m^2}\right).
\end{aligned}
$$

Exponentiating the logarithm then gives

$$e^{\frac{X}{m}} e^{\frac{Y}{m}} = \exp\left(\frac{X}{m} + \frac{Y}{m} + O\left(\frac{1}{m^2}\right)\right)$$

and, therefore,

$$\left(e^{\frac{X}{m}} e^{\frac{Y}{m}}\right)^m = \exp\left(X + Y + O\left(\frac{1}{m}\right)\right).$$

Thus, by the continuity of the exponential, we conclude that

$$\lim_{m \to \infty} \left(e^{\frac{X}{m}} e^{\frac{Y}{m}}\right)^m = \exp(X + Y),$$

which is the Lie product formula.   □

Recall (Sect. A.5) that the trace of matrix is defined as the sum of its diagonal entries and that similar matrices have the same trace.

Theorem 2.12. *For any $X \in M_n(\mathbb{C})$, we have*

$$\det\left(e^X\right) = e^{\text{trace}(X)}.$$

Proof. If X is diagonalizable with eigenvalues $\lambda_1, \ldots, \lambda_n$, then e^X is diagonalizable with eigenvalues $e^{\lambda_1}, \ldots, e^{\lambda_n}$. Thus, $\text{trace}(X) = \sum_j \lambda_j$ and

$$\det(e^X) = e^{\lambda_1} \cdots e^{\lambda_n} = e^{\lambda_1 + \cdots + \lambda_n} = e^{\text{trace}(X)}.$$

If X is not diagonalizable, we can approximate it by matrices that are diagonalizable (Exercise 4). □

Definition 2.13. A function $A : \mathbb{R} \to \mathsf{GL}(n; \mathbb{C})$ is called a **one-parameter subgroup** of $\mathsf{GL}(n; \mathbb{C})$ if

1. A is continuous,
2. $A(0) = I$,
3. $A(t + s) = A(t)A(s)$ for all $t, s \in \mathbb{R}$.

Theorem 2.14 (One-Parameter Subgroups). *If $A(\cdot)$ is a one-parameter subgroup of $\mathsf{GL}(n; \mathbb{C})$, there exists a unique $n \times n$ complex matrix X such that*

$$A(t) = e^{tX}.$$

By taking $n = 1$, and noting that $\mathsf{GL}(1; \mathbb{C}) \cong \mathbb{C}^*$, this theorem provides a method of solving Exercise 18 in Chapter 1.

Lemma 2.15. *Fix some ε with $\varepsilon < \log 2$. Let $B_{\varepsilon/2}$ be the ball of radius $\varepsilon/2$ around the origin in $M_n(\mathbb{C})$, and let $U = \exp(B_{\varepsilon/2})$. Then every $B \in U$ has a unique square root C in U, given by $C = \exp(\frac{1}{2} \log B)$.*

Proof. It is evident that C is a square root of B and that C is in U. To establish uniqueness, suppose $C' \in U$ satisfies $(C')^2 = B$. Let $Y = \log C'$; then $\exp(Y) = C'$ and

$$\exp(2Y) = (C')^2 = B = \exp(\log B).$$

We have that $Y \in B_{\varepsilon/2}$ and, thus, $2Y \in B_\varepsilon$, and also that $\log B \in B_{\varepsilon/2} \subset B_\varepsilon$. Since, by Theorem 2.8, exp is injective on B_ε and $\exp(2Y) = \exp(\log B)$, we must have $2Y = \log B$. Thus, $C' = \exp(\frac{1}{2} \log B) = C$. □

Proof of Theorem 2.14. The uniqueness is immediate, since if there is such an X, then $X = \frac{d}{dt} A(t)\big|_{t=0}$. To prove existence, let U be as in Lemma 2.15, which is an open set in $\mathsf{GL}(n; \mathbb{C})$. The continuity of A guarantees that there exists $t_0 > 0$ such that $A(t) \in U$ for all t with $|t| \leq t_0$. Define

$$X = \frac{1}{t_0} \log(A(t_0)),$$

so that $t_0 X = \log(A(t_0))$. Then $t_0 X \in B_{\varepsilon/2}$ and

$$e^{t_0 X} = A(t_0).$$

Now, $A(t_0/2)$ is again in U and $A(t_0/2)^2 = A(t_0)$. But by Lemma 2.15, $A(t_0)$ has a *unique* square root in U, and that unique square root is $\exp(t_0 X/2)$. Thus,

$$A(t_0/2) = \exp(t_0 X/2).$$

Applying this argument repeatedly, we conclude that

$$A(t_0/2^k) = \exp(t_0 X/2^k)$$

for all positive integers k. Then for any integer m, we have

$$A(m t_0/2^k) = A(t_0/2^k)^m = \exp(m t_0 X/2^k).$$

It follows that $A(t) = \exp(tX)$ for all real numbers t of the form $t = m t_0/2^k$, and the set of such t's is dense in $\mathbb{R}$. Since both $\exp(tX)$ and $A(t)$ are continuous, it follows that $A(t) = \exp(tX)$ for all real numbers t. □

Proposition 2.16. *The exponential map is an infinitely differentiable map of $M_n(\mathbb{C})$ into $M_n(\mathbb{C})$.*

We will compute the derivative of the matrix exponential in Chapter 5.

Proof. Note that for each j and k, the quantity $(X^m)_{jk}$ is a homogeneous polynomial of degree m in the entries of X. Thus, the series for the function $(X^m)_{jk}$ has the form of a multivariable power series on $M_n(\mathbb{C}) \cong \mathbb{R}^{2n^2}$. Since the series converges on all of $\mathbb{R}^{2n^2}$, it is permissible to differentiate the series term by term as many times as we like. (Apply Theorem 12 in Chapter 4 of [Pugh] in each of the n^2 variables with the other variables fixed.) □

2.5 The Polar Decomposition

The polar decomposition for a nonzero complex number z states that z can be written uniquely as $z = up$, where $|u| = 1$ and p is real and positive. (If $z = 0$, the decomposition still exists, with $p = 0$, but u is not unique.) Since p is real and positive, it can be written as $p = e^x$ for a unique real number x. This gives an unconventional form of the polar decomposition for z, namely

$$z = ue^x, \tag{2.10}$$

with $x \in \mathbb{R}$ and $|u| = 1$. Although it is customary to leave p as a positive real number and to write u as $u = e^{i\theta}$, the decomposition in (2.10) is more convenient for us because x, unlike θ, is unique.

We wish to establish a similar polar decomposition first for $\mathsf{GL}(n; \mathbb{C})$ and then for various subgroups thereof. If P is a self-adjoint $n \times n$ matrix (i.e., $P^* = P$), we say that P is **positive** if $\langle v, Pv \rangle > 0$ for all nonzero $v \in \mathbb{C}^n$. It is easy to check that a self-adjoint matrix P is positive if and only if all the eigenvalues of P are positive. Suppose now that A is an invertible $n \times n$ matrix. We wish to write A as $A = UP$ where U is unitary and P is self-adjoint and positive. We will then write the self-adjoint, positive matrix P as $P = e^X$ where X is self-adjoint but not necessarily positive.

Theorem 2.17.

1. *Every $A \in \mathsf{GL}(n; \mathbb{C})$ can be written uniquely in the form*

$$A = UP$$

   where U is unitary and P is self-adjoint and positive.
2. *Every self-adjoint positive matrix P can be written uniquely in the form*

$$P = e^X$$

   with X self-adjoint. Conversely, if X is self-adjoint, then e^X is self-adjoint and positive.
3. *If we decompose each $A \in \mathsf{GL}(n; \mathbb{C})$ (uniquely) as*

$$A = Ue^X$$

   with U unitary and X self-adjoint, then U and X depend continuously on A.

Lemma 2.18. *If Q is a self-adjoint, positive matrix, then Q has a unique positive, self-adjoint square root.*

Proof. Since Q has an orthonormal basis of eigenvectors, Q can be written as

$$Q = U \begin{pmatrix} \lambda_1 & & \\ & \ddots & \\ & & \lambda_n \end{pmatrix} U^{-1}$$

with U unitary. Since Q is self-adjoint and positive, each λ_j is positive. Thus, we can construct a square root of Q as

$$Q^{1/2} = U \begin{pmatrix} \lambda_1^{1/2} & & \\ & \ddots & \\ & & \lambda_n^{1/2} \end{pmatrix} U^{-1}, \qquad (2.11)$$

and $Q^{1/2}$ will still be self-adjoint and positive, establishing the existence of the square root.

If P is a self-adjoint, positive matrix, the eigenspaces of P^2 are precisely the same as the eigenspaces of P, with the eigenvalues of P^2 being, of course, the squares of the eigenvalues of P. The point here is that because the function $x \mapsto x^2$ is injective on positive real numbers, eigenspaces with distinct eigenvalues remain with distinct eigenvalues after squaring. Looking at this claim the other way around, if a positive, self-adjoint matrix Q is to have a positive self-adjoint square root P, the eigenspaces of P must be the same as the eigenspaces of Q, and the eigenvalues of P must be the positive square roots of the eigenvalues of Q. Thus, P is uniquely determined by Q. □

Proof of Theorem 2.17. For the existence of the decomposition in Point 1, note that if $A = UP$, then $A^*A = PU^*UP = P^2$. Now, for any matrix A, the matrix A^*A is self-adjoint. If, in addition, A is invertible, then for all nonzero $v \in \mathbb{C}^n$, we have

$$\langle v, A^*Av \rangle = \langle Av, Av \rangle > 0,$$

showing that A^*A is positive. For all invertible A, then, let us *define* P by

$$P = (A^*A)^{1/2},$$

where $(\cdot)^{1/2}$ is the unique self-adjoint positive square root of Lemma 2.18. We then define

$$U = AP^{-1} = A[(A^*A)^{1/2}]^{-1}.$$

Since P is, by construction, self-adjoint and positive, and since $A = UP$ by the definition of U, it remains only to check that U is unitary. To that end, we check that

$$U^*U = [(A^*A)^{1/2}]^{-1} A^* A [(A^*A)^{1/2}]^{-1},$$

since the inverse of a positive self-adjoint matrix is self-adjoint. Since A^*A is the square of $(A^*A)^{1/2}$, we see that $U^*U = I$, showing that U is unitary.

For the uniqueness of the decomposition, we have already noted that if $A = UP$, then $P^2 = A^*A$, where A^*A is self-adjoint and positive. Thus, the uniqueness of P follows from the uniqueness in Lemma 2.18. The uniqueness of U then follows, since if $A = UP$, then $U = AP^{-1}$.

The existence and uniqueness of the decomposition in Point 2 are proved in precisely the same way as in Lemma 2.18, with the logarithm function (which is a bijection between $(0, \infty)$ and $\mathbb{R}$) replacing the square root function. The same sort of reasoning shows that for any self-adjoint X, the matrix e^X is self-adjoint and positive. We refer to the matrix X as the "logarithm" of P, even though X is not, in general, computable by the series in Definition 2.7.

Finally, we address the continuity claim in Point 3. We first show that the logarithm X of a self-adjoint, positive matrix P depends continuously on P. To see this, note that if the eigenvalues of P are between 0 and 2, then by the remarks just before the proof of Theorem 2.8, the power series for log P will converge to X, in which case, continuity follows by the same argument as in the proof of Proposition 2.1. In general, fix some positive, self-adjoint matrix P_0. Fix a small neighborhood V of P_0 and choose a large positive number a. For $P \in V$, write $P = e^a(e^{-a}P)$. Then the unique self-adjoint logarithm X of P may be computed as

$$X = aI + \log(e^{-a}P).$$

If a is large enough, then for all $P \in V$, the (positive) eigenvalues of $e^{-a}P$ will all be less than 2, and the series for $\log(e^{-a}P)$ will converge and depend continuously on P, showing that X depends continuously on P.

We then note that the square root operation on self-adjoint, positive matrices P is also continuous, since $P^{1/2}$ may be computed as $\exp(\log(P)/2)$. It is then evident from the formulas for U and P that these quantities depend continuously on A. $\square$

We now establish polar decompositions for $\mathsf{GL}(n;\mathbb{R})$, $\mathsf{SL}(n;\mathbb{C})$, and $\mathsf{SL}(n;\mathbb{R})$.

Proposition 2.19.

1. Every $A \in \mathsf{GL}(n;\mathbb{R})$ can be written uniquely as

$$A = Re^X,$$

where R is in $\mathsf{O}(n)$ and X is real and symmetric.
2. Every $A \in \mathsf{SL}(n;\mathbb{C})$ can be written uniquely as

$$A = Ue^X,$$

where U is in $\mathsf{SU}(n)$ and X is self-adjoint with trace zero.
3. Every $A \in \mathsf{SL}(n;\mathbb{R})$ can be written uniquely as

$$A = Re^X,$$

where R is in $\mathsf{SO}(n)$ and X is real and symmetric and has trace zero.

Proof. If A is real, then A^*A is real and symmetric. Now, a real, symmetric matrix can be diagonalized over $\mathbb{R}$. Thus, P, which is the unique self-adjoint positive square root of A^*A (constructed as in (2.11)), is real. Then $U = AP^{-1}$ is real and unitary, hence in $\mathsf{O}(n)$.

Meanwhile, if $A \in \mathsf{SL}(n;\mathbb{C})$ and we write $A = Ue^X$ with $U \in \mathsf{U}(n)$ and X self-adjoint, then $\det(A) = \det(U)e^{\operatorname{trace}(X)}$. Now, $|\det(U)| = 1$, and $e^{\operatorname{trace}(X)}$ is real and positive. Thus, by the uniqueness of the polar decomposition for nonzero complex numbers, we must have $\det(U) = 1$ and $\operatorname{trace}(X) = 0$. The case of $A \in \mathsf{SL}(n;\mathbb{R})$ follows by combining the results of the two previous cases. $\square$

2.6   Exercises

1. The Cauchy–Schwarz inequality from elementary analysis tells us that for all
   $u = (u_1, \ldots, u_n)$ and $v = (v_1, \ldots, v_n)$ in $\mathbb{C}^n$, we have

$$|u_1 v_1 + \cdots + u_n v_n|^2 \leq \left(\sum_{j=1}^n |u_j|^2 \right) \left(\sum_{k=1}^n |v_k|^2 \right).$$

   Use this to verify that $\|XY\| \leq \|X\| \|Y\|$ for all $X, Y \in M_n(\mathbb{C})$, where $\|\cdot\|$ is
   the Hilbert–Schmidt norm in Definition 2.2.
2. Show that for $X \in M_n(\mathbb{C})$ and any orthonormal basis $\{u_1, \ldots, u_n\}$ of $\mathbb{C}^n$,
   $\|X\|^2 = \sum_{j,k=1}^n |\langle u_j, X u_k \rangle|^2$, where $\|X\|$ is as in Definition 2.2. Now show
   that if v is an eigenvector for X with eigenvalue λ, then $|\lambda| \leq \|X\|$.
3. *The product rule.* Recall that a matrix-valued function $A(t)$ is said to be smooth
   if each $A_{jk}(t)$ is smooth. The derivative of such a function is defined as

$$\left(\frac{dA}{dt} \right)_{jk} = \frac{dA_{jk}}{dt}$$

   or, equivalently,

$$\frac{d}{dt} A(t) = \lim_{h \to 0} \frac{A(t+h) - A(t)}{h}.$$

   Let $A(t)$ and $B(t)$ be two such functions. Prove that $A(t)B(t)$ is again smooth
   and that

$$\frac{d}{dt} [A(t)B(t)] = \frac{dA}{dt} B(t) + A(t) \frac{dB}{dt}.$$

4. Using Theorem A.4, show that every $n \times n$ complex matrix A is the limit of a
   sequence of diagonalizable matrices.
   Hint: If an $n \times n$ matrix has n distinct eigenvalues, it is necessarily
   diagonalizable.
5. For any a and d in $\mathbb{C}$, define the expression $(e^a - e^d)/(a - d)$ in the obvious
   way for $a \neq d$ and by means of the limit

$$\lim_{a \to d} \frac{e^a - e^d}{a - d} = e^a$$

   when $a = d$. Show that for any $a, b, d \in \mathbb{C}$, we have

$$\exp \begin{pmatrix} a & b \\ 0 & d \end{pmatrix} = \begin{pmatrix} e^a & b \frac{e^a - e^d}{a - d} \\ 0 & e^d \end{pmatrix}.$$

Hint: Show that if $a \neq d$, then

$$\begin{pmatrix} a & b \\ 0 & d \end{pmatrix}^m = \begin{pmatrix} a^m & b\frac{a^m - d^m}{a - d} \\ 0 & b^m \end{pmatrix}$$

for every positive integer m.

6. Show that every 2×2 matrix X with $\mathrm{trace}(X) = 0$ satisfies

$$X^2 = -\det(X)I.$$

If X is 2×2 with trace zero, show by direct calculation using the power series for the exponential that

$$e^X = \cos\left(\sqrt{\det X}\right) I + \frac{\sin\sqrt{\det X}}{\sqrt{\det X}} X, \qquad (2.12)$$

where $\sqrt{\det X}$ is either of the two (possibly complex) square roots of $\det X$. Use this to give an alternative computation of the exponential e^{X_1} in Example 2.5.
Note: The value of the coefficient of X in (2.12) is to be interpreted as 1 when $\det X = 0$, in accordance with the limit $\lim_{\theta \to 0} \sin \theta / \theta = 1$.

7. Use the result of Exercise 6 to compute the exponential of the matrix

$$X = \begin{pmatrix} 4 & 3 \\ -1 & 2 \end{pmatrix}.$$

Hint: Reduce the calculation to the trace-zero case.

8. Consider the two matrices X_4 and X_5 in (2.6). Compute e^{tX_4} and e^{tX_5} either by diagonalization or by the method in Exercises 6 and 7. Show that curves of the form $t \mapsto e^{tX_4} \mathbf{v}_0$, with $\mathbf{v}_0 \neq 0$, spiral out to infinity. Show that for $\mathbf{v}_0$ outside of a certain one-dimensional subspace of $\mathbb{R}^2$, curves of the form $t \mapsto e^{tX_5} \mathbf{v}_0$ tend to infinity in the direction of $(1, 1)$ or the direction of $(-1, -1)$.

9. A matrix A is said to be **unipotent** if $A - I$ is nilpotent (i.e., if A is of the form $A = I + N$, with N nilpotent). Note that $\log A$ is defined whenever A is unipotent, because the series in Definition 2.7 terminates.

   (a) Show that if A is unipotent, then $\log A$ is nilpotent.
   (b) Show that if X is nilpotent, then e^X is unipotent.
   (c) Show that if A is unipotent, then $\exp(\log A) = A$ and that if X is nilpotent, then $\log(\exp X) = X$.

   Hint: Let $A(t) = I + t(A - I)$. Show that $\exp(\log(A(t)))$ depends polynomially on t and that $\exp(\log(A(t))) = A(t)$ for all sufficiently small t.

10. Show that every invertible $n \times n$ matrix A can be written as $A = e^X$ for some
    $X \in M_n(\mathbb{C})$.

    Hint: Theorem A.5 implies that A is similar to a block-diagonal matrix in which
    each block is of the form $\lambda I + N_\lambda$, with N_λ being nilpotent. Use this result and
    Exercise 9.

11. Show that for all $X \in M_n(\mathbb{C})$, we have

$$\lim_{m \to \infty} \left[I + \frac{X}{m} \right]^m = e^X.$$

    Hint: Use the matrix logarithm.

Chapter 3
Lie Algebras

3.1 Definitions and First Examples

We now introduce the "abstract" notion of a Lie algebra. In Sect. 3.3, we will associate to each matrix Lie group a Lie algebra. It is customary to use lowercase Gothic (Fraktur) characters such as $\mathfrak{g}$ and $\mathfrak{h}$ to refer to Lie algebras.

Definition 3.1. A **finite-dimensional real or complex Lie algebra** is a finite-dimensional real or complex vector space $\mathfrak{g}$, together with a map $[\cdot, \cdot]$ from $\mathfrak{g} \times \mathfrak{g}$ into $\mathfrak{g}$, with the following properties:

1. $[\cdot, \cdot]$ is bilinear.
2. $[\cdot, \cdot]$ is skew symmetric: $[X, Y] = -[Y, X]$ for all $X, Y \in \mathfrak{g}$.
3. The **Jacobi identity** holds:

$$[X, [Y, Z]] + [Y, [Z, X]] + [Z, [X, Y]] = 0$$

for all $X, Y, Z \in \mathfrak{g}$.

Two elements X and Y of a Lie algebra $\mathfrak{g}$ **commute** if $[X, Y] = 0$. A Lie algebra $\mathfrak{g}$ is **commutative** if $[X, Y] = 0$ for all $X, Y \in \mathfrak{g}$.

The map $[\cdot, \cdot]$ is referred to as the **bracket** operation on $\mathfrak{g}$. Note also that Condition 2 implies that $[X, X] = 0$ for all $X \in \mathfrak{g}$. The bracket operation on a Lie algebra is not, in general associative; nevertheless, the Jacobi identity can be viewed as a substitute for associativity.

A previous version of this book was inadvertently published without the middle initial of the author's name as "Brian Hall". For this reason an erratum has been published, correcting the mistake in the previous version and showing the correct name as Brian C. Hall (see DOI http://dx.doi.org/10.1007/978-3-319-13467-3_14). The version readers currently see is the corrected version. The Publisher would like to apologize for the earlier mistake.

© Springer International Publishing Switzerland 2015   49
B.C. Hall, *Lie Groups, Lie Algebras, and Representations*, Graduate
Texts in Mathematics 222, DOI 10.1007/978-3-319-13467-3_3

Example 3.2. Let $\mathfrak{g} = \mathbb{R}^3$ and let $[\cdot, \cdot] : \mathbb{R}^3 \times \mathbb{R}^3 \to \mathbb{R}^3$ be given by

$$[x, y] = x \times y,$$

where $x \times y$ is the cross product (or vector product). Then $\mathfrak{g}$ is a Lie algebra.

Proof. Bilinearity and skew symmetry are standard properties of the cross product. To verify the Jacobi identity, it suffices (by bilinearity) to verify it when $x = e_j$, $y = e_k$, and $z = e_l$, where e_1, e_2, and e_3 are the standard basis elements for $\mathbb{R}^3$. If j, k, and l are all equal, each term in the Jacobi identity is zero. If j, k, and l are all different, the cross product of any two of e_j, e_k, and e_l is equal to a multiple of the third, so again, each term in the Jacobi identity is zero. It remains to consider the case in which two of j, k, l are equal and the third is different. By re-ordering the terms in the Jacobi identity as necessary, it suffices to verify the identity

$$[e_j, [e_j, e_k]] + [e_j, [e_k, e_j]] + [e_k, [e_j, e_j]] = 0. \tag{3.1}$$

The first two terms in (3.1) are negatives of each other and the third is zero.   □

Example 3.3. Let $\mathcal{A}$ be an associative algebra and let $\mathfrak{g}$ be a subspace of $\mathcal{A}$ such that $XY - YX \in \mathfrak{g}$ for all $X, Y \in \mathfrak{g}$. Then $\mathfrak{g}$ is a Lie algebra with bracket operation given by

$$[X, Y] = XY - YX.$$

Proof. The bilinearity and skew symmetry of the bracket are evident. To verify the Jacobi identity, note that each double bracket generates four terms, for a total of 12 terms. It is left to the reader to verify that the product of X, Y, and Z in each of the six possible orderings occurs twice, once with a plus sign and once with a minus sign.   □

If we look carefully at the proof of the Jacobi identity, we see that the two occurrences of, say, XYZ occur with different groupings, once as $X(YZ)$ and once as $(XY)Z$. Thus, associativity of the algebra $\mathcal{A}$ is essential. For any Lie algebra, the Jacobi identity means that the bracket operation *behaves as if* it were $XY - YX$ in some associative algebra, even if it is not actually defined this way. Indeed, we will prove in Chapter 9 that every Lie algebra $\mathfrak{g}$ can be embedded into an associative algebra $\mathcal{A}$ in such a way that the bracket becomes $XY - YX$. (This claim follows from Theorem 9.9, the Poincaré–Birkhoff–Witt theorem.)

Of particular interest to us is the case in which $\mathcal{A}$ is the space $M_n(\mathbb{C})$ of all $n \times n$ complex matrices.

Example 3.4. Let $\mathsf{sl}(n; \mathbb{C})$ denote the space of all $X \in M_n(\mathbb{C})$ for which $\operatorname{trace}(X) = 0$. Then $\mathsf{sl}(n; \mathbb{C})$ is a Lie algebra with bracket $[X, Y] = XY - YX$.

Proof. For any X and Y in $M_n(\mathbb{C})$, we have

$$\operatorname{trace}(XY - YX) = \operatorname{trace}(XY) - \operatorname{trace}(YX) = 0.$$

This holds, in particular, if X and Y have trace zero. Thus, Example 3.3 applies.   □

Definition 3.5. A **subalgebra** of a real or complex Lie algebra $\mathfrak{g}$ is a subspace $\mathfrak{h}$ of $\mathfrak{g}$ such that $[H_1, H_2] \in \mathfrak{h}$ for all H_1 and $H_2 \in \mathfrak{h}$. If $\mathfrak{g}$ is a complex Lie algebra and $\mathfrak{h}$ is a real subspace of $\mathfrak{g}$ which is closed under brackets, then $\mathfrak{h}$ is said to be a **real subalgebra** of $\mathfrak{g}$.

A subalgebra $\mathfrak{h}$ of a Lie algebra $\mathfrak{g}$ is said to be an **ideal** in $\mathfrak{g}$ if $[X, H] \in \mathfrak{h}$ for all X in $\mathfrak{g}$ and H in $\mathfrak{h}$.

The **center** of a Lie algebra $\mathfrak{g}$ is the set of all $X \in \mathfrak{g}$ for which $[X, Y] = 0$ for all $Y \in \mathfrak{g}$.

Definition 3.6. If $\mathfrak{g}$ and $\mathfrak{h}$ are Lie algebras, then a linear map $\phi : \mathfrak{g} \to \mathfrak{h}$ is called a **Lie algebra homomorphism** if $\phi([X, Y]) = [\phi(X), \phi(Y)]$ for all $X, Y \in \mathfrak{g}$. If, in addition, ϕ is one-to-one and onto, then ϕ is called a **Lie algebra isomorphism**. A Lie algebra isomorphism of a Lie algebra with itself is called a **Lie algebra automorphism**.

Definition 3.7. If $\mathfrak{g}$ is a Lie algebra and X is an element of $\mathfrak{g}$, define a linear map $\text{ad}_X : \mathfrak{g} \to \mathfrak{g}$ by

$$\text{ad}_X(Y) = [X, Y].$$

The map $X \mapsto \text{ad}_X$ is the **adjoint map** or **adjoint representation**.

Although $\text{ad}_X(Y)$ is just $[X, Y]$, the alternative "ad" notation can be useful. For example, instead of writing

$$[X, [X, [X, [X, Y]]]],$$

we can now write

$$(\text{ad}_X)^4 (Y).$$

This sort of notation will be essential in Chapter 5. We can view ad (that is, the map $X \mapsto \text{ad}_X$) as a linear map of $\mathfrak{g}$ into $\text{End}(\mathfrak{g})$, the space of linear operators on $\mathfrak{g}$. The Jacobi identity is then equivalent to the assertion that ad_X is a *derivation* of the bracket:

$$\text{ad}_X([Y, Z]) = [\text{ad}_X(Y), Z] + [Y, \text{ad}_X(Z)]. \tag{3.2}$$

Proposition 3.8. *If $\mathfrak{g}$ is a Lie algebra, then*

$$\text{ad}_{[X,Y]} = \text{ad}_X \text{ad}_Y - \text{ad}_Y \text{ad}_X = [\text{ad}_X, \text{ad}_Y];$$

that is, ad: $\mathfrak{g} \to \text{End}(\mathfrak{g})$ is a Lie algebra homomorphism.

Proof. Observe that

$$\text{ad}_{[X,Y]}(Z) = [[X, Y], Z],$$

whereas

$$[\mathrm{ad}_X, \mathrm{ad}_Y](Z) = [X, [Y, Z]] - [Y, [X, Z]].$$

Thus, we want to show that

$$[[X, Y], Z] = [X, [Y, Z]] - [Y, [X, Z]],$$

which is equivalent to the Jacobi identity.   □

Definition 3.9. If $\mathfrak{g}_1$ and $\mathfrak{g}_2$ are Lie algebras, the **direct sum** of $\mathfrak{g}_1$ and $\mathfrak{g}_2$ is the vector space direct sum of $\mathfrak{g}_1$ and $\mathfrak{g}_2$, with bracket given by

$$[(X_1, X_2), (Y_1, Y_2)] = ([X_1, Y_1], [X_2, Y_2]). \tag{3.3}$$

If $\mathfrak{g}$ is a Lie algebra and $\mathfrak{g}_1$ and $\mathfrak{g}_2$ are subalgebras, we say that $\mathfrak{g}$ **decomposes as the Lie algebra direct sum** of $\mathfrak{g}_1$ and $\mathfrak{g}_2$ if $\mathfrak{g}$ is the direct sum of $\mathfrak{g}_1$ and $\mathfrak{g}_2$ as vector spaces and $[X_1, X_2] = 0$ for all $X_1 \in \mathfrak{g}_1$ and $X_2 \in \mathfrak{g}_2$.

It is straightforward to verify that the bracket in (3.3) makes $\mathfrak{g}_1 \oplus \mathfrak{g}_2$ into a Lie algebra. If $\mathfrak{g}$ decomposes as a Lie algebra direct sum of subalgebras $\mathfrak{g}_1$ and $\mathfrak{g}_2$, it is easy to check that $\mathfrak{g}$ is isomorphic as a Lie algebra to the "abstract" direct sum of $\mathfrak{g}_1$ and $\mathfrak{g}_2$. (This would not be the case without the assumption that every element of $\mathfrak{g}_1$ commutes with every element of $\mathfrak{g}_2$.)

Definition 3.10. Let $\mathfrak{g}$ be a finite-dimensional real or complex Lie algebra, and let $X_1, \ldots, X_N$ be a basis for $\mathfrak{g}$ (as a vector space). Then the unique constants c_{jkl} such that

$$[X_j, X_k] = \sum_{l=1}^{N} c_{jkl} X_l \tag{3.4}$$

are called the **structure constants** of $\mathfrak{g}$ (with respect to the chosen basis).

Although we will not have much occasion to use them, structure constants do appear frequently in the physics literature. The structure constants satisfy the following two conditions:

$$c_{jkl} + c_{kjl} = 0,$$

$$\sum_n (c_{jkn} c_{nlm} + c_{kln} c_{njm} + c_{ljn} c_{nkm}) = 0$$

for all j, k, l, m. The first of these conditions comes from the skew symmetry of the bracket, and the second comes from the Jacobi identity.

3.2 Simple, Solvable, and Nilpotent Lie Algebras

In this section, we consider various special types of Lie algebras. Recall from Definition 3.5 the notion of an ideal in a Lie algebra.

Definition 3.11. A Lie algebra $\mathfrak{g}$ is called **irreducible** if the only ideals in $\mathfrak{g}$ are $\mathfrak{g}$ and $\{0\}$. A Lie algebra $\mathfrak{g}$ is called **simple** if it is irreducible and $\dim \mathfrak{g} \geq 2$.

A one-dimensional Lie algebra is certainly irreducible, since it is has no nontrivial subspaces and therefore no nontrivial subalgebras and no nontrivial ideals. Nevertheless, such a Lie algebra is, by definition, not considered simple.

Note that a one-dimensional Lie algebra $\mathfrak{g}$ is necessarily commutative, since $[aX, bX] = 0$ for any $X \in \mathfrak{g}$ and any scalars a and b. On the other hand, if $\mathfrak{g}$ is commutative, then any subspace of $\mathfrak{g}$ is an ideal. Thus, the only way a commutative Lie algebra can be irreducible is if it is one dimensional. Thus, an equivalent definition of "simple" is that a Lie algebra is simple if it is *irreducible and noncommutative*.

There is an analogy between groups and Lie algebras, in which the role of subgroups is played by subalgebras and the role of normal subgroups is played by ideals. (For example, the kernel of a Lie algebra homomorphism is always an ideal, just as the kernel of a Lie group homomorphism is always a normal subgroup.) There is, however, an inconsistency in the terminology in the two fields. On the group side, *any* group with no nontrivial normal subgroups is called simple, including the most obvious example, a cyclic group of prime order. On the Lie algebra side, by contrast, the most obvious example of an algebra with no nontrivial ideals—namely, a one-dimensional algebra—is not called simple.

We will eventually see many examples of simple Lie algebras, but for now we content ourselves with a single example. Recall the Lie algebra $\mathsf{sl}(n; \mathbb{C})$ in Example 3.4.

Proposition 3.12. *The Lie algebra* $\mathsf{sl}(2; \mathbb{C})$ *is simple.*

Proof. We use the following basis for $\mathsf{sl}(2; \mathbb{C})$:

$$X = \begin{pmatrix} 0 & 1 \\ 0 & 0 \end{pmatrix}; \quad Y = \begin{pmatrix} 0 & 0 \\ 1 & 0 \end{pmatrix}; \quad H = \begin{pmatrix} 1 & 0 \\ 0 & -1 \end{pmatrix}.$$

Direct calculation shows that these basis elements have the following commutation relations: $[X, Y] = H$, $[H, X] = 2X$, and $[H, Y] = -2Y$. Suppose $\mathfrak{h}$ is an ideal in $\mathsf{sl}(2; \mathbb{C})$ and that $\mathfrak{h}$ contains an element $Z = aX + bH + cY$, where a, b, and c are not all zero. We will show, then, that $\mathfrak{h} = \mathsf{sl}(2; \mathbb{C})$. Suppose first that $c \neq 0$. Then the element

$$[X, [X, Z]] = [X, [-2bX + cH]] = -2cX$$

is a nonzero multiple of X. Since $\mathfrak{h}$ is an ideal, we conclude that $X \in \mathfrak{h}$. But $[Y, X]$ is a nonzero multiple of H and $[Y, [Y, X]]$ is a nonzero multiple of Y, showing that Y and H also belong to $\mathfrak{h}$, from which we conclude that $\mathfrak{h} = \mathsf{sl}(2; \mathbb{C})$.

Suppose next that $c = 0$ but $b \neq 0$. Then $[X, Z]$ is a nonzero multiple of X and we may then apply the same argument in the previous paragraph to show that $\mathfrak{h} = \mathsf{sl}(2; \mathbb{C})$. Finally, if $c = 0$ and $b = 0$ but $a \neq 0$, then Z itself is a nonzero multiple of X and we again conclude that $\mathfrak{h} = \mathsf{sl}(2; \mathbb{C})$.   □

Definition 3.13. If $\mathfrak{g}$ is a Lie algebra, then the **commutator ideal** in $\mathfrak{g}$, denoted $[\mathfrak{g}, \mathfrak{g}]$, is the space of linear combinations of commutators, that is, the space of elements Z in $\mathfrak{g}$ that can be expressed as

$$Z = c_1 [X_1, Y_1] + \cdots + c_m [X_m, Y_m]$$

for some constants c_j and vectors $X_j, Y_j \in \mathfrak{g}$.

For any X and Y in $\mathfrak{g}$, the commutator $[X, Y]$ is in $[\mathfrak{g}, \mathfrak{g}]$. This holds, in particular, if X is in $[\mathfrak{g}, \mathfrak{g}]$, showing that $[\mathfrak{g}, \mathfrak{g}]$ is an ideal in $\mathfrak{g}$.

Definition 3.14. For any Lie algebra $\mathfrak{g}$, we define a sequence of subalgebras $\mathfrak{g}_0, \mathfrak{g}_1, \mathfrak{g}_2, \ldots$ of $\mathfrak{g}$ inductively as follows: $\mathfrak{g}_0 = \mathfrak{g}$, $\mathfrak{g}_1 = [\mathfrak{g}_0, \mathfrak{g}_0]$, $\mathfrak{g}_2 = [\mathfrak{g}_1, \mathfrak{g}_1]$, etc. These subalgebras are called the **derived series** of $\mathfrak{g}$. A Lie algebra $\mathfrak{g}$ is called **solvable** if $\mathfrak{g}_j = \{0\}$ for some j.

It is not hard to show, using the Jacobi identity and induction on j, that each $\mathfrak{g}_j$ is an ideal in $\mathfrak{g}$.

Definition 3.15. For any Lie algebra $\mathfrak{g}$, we define a sequence of ideals $\mathfrak{g}^j$ in $\mathfrak{g}$ inductively as follows. We set $\mathfrak{g}^0 = \mathfrak{g}$ and then define $\mathfrak{g}^{j+1}$ to be the space of linear combinations of commutators of the form $[X, Y]$ with $X \in \mathfrak{g}$ and $Y \in \mathfrak{g}^j$. These algebras are called the **upper central series** of $\mathfrak{g}$. A Lie algebra $\mathfrak{g}$ is said to be **nilpotent** if $\mathfrak{g}^j = \{0\}$ for some j.

Equivalently, $\mathfrak{g}^j$ is the space spanned by all jth-order commutators,

$$[X_1, [X_2, [X_3, \ldots [X_j, X_{j+1}] \ldots]]].$$

Note that every jth-order commutator is also a $(j - 1)$th-order commutator, by setting $\tilde{X}_j = [X_j, X_{j+1}]$. Thus, $\mathfrak{g}^{j-1} \supset \mathfrak{g}^j$. For every $X \in \mathfrak{g}$ and $Y \in \mathfrak{g}^j$, we have $[X, Y] \in \mathfrak{g}^{j+1} \subset \mathfrak{g}^j$, showing that $\mathfrak{g}^j$ is an ideal in $\mathfrak{g}$. Furthermore, it is clear that $\mathfrak{g}_j \subset \mathfrak{g}^j$ for all j; thus, if $\mathfrak{g}$ is nilpotent, $\mathfrak{g}$ is also solvable.

Proposition 3.16. *If $\mathfrak{g} \subset M_3(\mathbb{R})$ denotes the space of 3×3 upper triangular matrices with zeros on the diagonal, then $\mathfrak{g}$ satisfies the assumptions of Example 3.3. The Lie algebra $\mathfrak{g}$ is a nilpotent Lie algebra.*

Proof. We will use the following basis for $\mathfrak{g}$,

$$X = \begin{pmatrix} 0 & 1 & 0 \\ 0 & 0 & 0 \\ 0 & 0 & 0 \end{pmatrix}; \quad Y = \begin{pmatrix} 0 & 0 & 0 \\ 0 & 0 & 1 \\ 0 & 0 & 0 \end{pmatrix}; \quad Z = \begin{pmatrix} 0 & 0 & 1 \\ 0 & 0 & 0 \\ 0 & 0 & 0 \end{pmatrix}. \tag{3.5}$$

Direct calculation then establishes the following commutation relations: $[X, Y] = Z$ and $[X, Z] = [Y, Z] = 0$. In particular, the bracket of two elements of $\mathfrak{g}$ is again in $\mathfrak{g}$, so that $\mathfrak{g}$ is a Lie algebra. Then $[\mathfrak{g}, \mathfrak{g}]$ is the span of Z and $[\mathfrak{g}, [\mathfrak{g}, \mathfrak{g}]] = 0$, showing that $\mathfrak{g}$ is nilpotent. □

Proposition 3.17. *If* $\mathfrak{g} \subset M_2(\mathbb{C})$ *denotes the space of* 2×2 *matrices of the form*

$$\begin{pmatrix} a & b \\ 0 & c \end{pmatrix}$$

with a, b, and c in $\mathbb{C}$, *then* $\mathfrak{g}$ *satisfies the assumptions of Example 3.3. The Lie algebra* $\mathfrak{g}$ *is solvable but not nilpotent.*

Proof. Direct calculation shows that

$$\left[\begin{pmatrix} a & b \\ 0 & c \end{pmatrix}, \begin{pmatrix} d & e \\ 0 & f \end{pmatrix} \right] = \begin{pmatrix} 0 & h \\ 0 & 0 \end{pmatrix}, \tag{3.6}$$

where $h = ae + bf - bd - ce$, showing that $\mathfrak{g}$ is a Lie subalgebra of $M_2(\mathbb{C})$. Furthermore, the commutator ideal $[\mathfrak{g}, \mathfrak{g}]$ is one dimensional and hence commutative. Thus, $\mathfrak{g}_2 = \{0\}$, showing that $\mathfrak{g}$ is solvable. On the other hand, consider the following elements of $\mathfrak{g}$:

$$H = \begin{pmatrix} 1 & 0 \\ 0 & -1 \end{pmatrix}; \quad X = \begin{pmatrix} 0 & 1 \\ 0 & 0 \end{pmatrix}.$$

Using (3.6), we can see that $[H, X] = 2X$, and thus that

$$[H, [H, [H, \cdots [H, X] \cdots]]]$$

is a nonzero multiple of X, showing that $\mathfrak{g}^j \neq \{0\}$ for all j. □

3.3   The Lie Algebra of a Matrix Lie Group

In this section, we associate to each matrix Lie group G a Lie algebra $\mathfrak{g}$. Many questions involving a group can be studied by transferring them to the Lie algebra, where we can use tools of linear algebra. We begin by defining $\mathfrak{g}$ as a set, and then proceed to give $\mathfrak{g}$ the structure of a Lie algebra.

Definition 3.18. Let G be a matrix Lie group. The **Lie algebra of** G, denoted $\mathfrak{g}$, is the set of all matrices X such that e^{tX} is in G for all real numbers t.

Equivalently, X is in $\mathfrak{g}$ if and only if the entire one-parameter subgroup (Definition 2.13) generated by X lies in G. Note that merely having e^X in G does not guarantee that X is in $\mathfrak{g}$. Even though G is a subgroup of $\mathsf{GL}(n;\mathbb{C})$ (and not necessarily of $\mathsf{GL}(n;\mathbb{R})$), we do *not* require that e^{tX} be in G for all complex numbers t, but only for all *real* numbers t. We will show in Sect. 3.7 that every matrix Lie group is an embedded submanifold of $\mathsf{GL}(n;\mathbb{C})$. We will then show (Corollary 3.46) that $\mathfrak{g}$ is the *tangent space to G at the identity*.

We will now establish various basic properties of the Lie algebra $\mathfrak{g}$ of a matrix Lie group G. In particular, we will see that there is a bracket operation on $\mathfrak{g}$ that makes $\mathfrak{g}$ into a Lie algebra in the sense of Definition 3.1.

Proposition 3.19. *Let G be a matrix Lie group, and X an element of its Lie algebra. Then e^X is an element of the identity component G_0 of G.*

Proof. By definition of the Lie algebra, e^{tX} lies in G for all real t. However, as t varies from 0 to 1, e^{tX} is a continuous path connecting the identity to e^X. □

Theorem 3.20. *Let G be a matrix Lie group with Lie algebra $\mathfrak{g}$. If X and Y are elements of $\mathfrak{g}$, the following results hold.*

1. $AXA^{-1} \in \mathfrak{g}$ *for all $A \in G$.*
2. $sX \in \mathfrak{g}$ *for all real numbers s.*
3. $X + Y \in \mathfrak{g}$.
4. $XY - YX \in \mathfrak{g}$.

It follows from this result and Example 3.3 that the Lie algebra of a matrix Lie group is a real Lie algebra, with bracket given by $[X, Y] = XY - YX$. For X and Y in $\mathfrak{g}$, we refer to $[X, Y] = XY - YX \in \mathfrak{g}$ as the **bracket** or **commutator** of X and Y.

Proof. For Point 1, we observe that, by Proposition 2.3,

$$e^{t(AXA^{-1})} = Ae^{tX}A^{-1} \in G$$

for all t, showing that AXA^{-1} is in $\mathfrak{g}$. For Point 2, we observe that $e^{t(sX)} = e^{(ts)X}$, which must be in G for all $t \in \mathbb{R}$ if X is in $\mathfrak{g}$. For Point 3 we use the Lie product formula, which says that

$$e^{t(X+Y)} = \lim_{m \to \infty} \left(e^{tX/m} e^{tY/m} \right)^m .$$

Now, $\left(e^{tX/m} e^{tY/m} \right)^m$ is in G for all m. Since G is closed, the limit (which is invertible) must be again in G. This shows that $X + Y$ is again in $\mathfrak{g}$.

Finally, for Point 4, we use the product rule (Exercise 3 in Chapter 2) and Proposition 2.4 to compute

$$\frac{d}{dt}\left(e^{tX}Ye^{-tX}\right)\bigg|_{t=0} = (XY)e^{0} + (e^{0}Y)(-X)$$

$$= XY - YX.$$

Now, by Point 1, $e^{tX}Ye^{-tX}$ is in $\mathfrak{g}$ for all t. Furthermore, by Points 2 and 3, $\mathfrak{g}$ is a real subspace of $M_n(\mathbb{C})$, from which it follows that $\mathfrak{g}$ is a (topologically) closed subset of $M_n(\mathbb{C})$. Thus,

$$XY - YX = \lim_{h \to 0} \frac{e^{hX}Ye^{-hX} - Y}{h}$$

belongs to $\mathfrak{g}$. $\square$

Note that even if the elements of G have complex entries, the Lie algebra $\mathfrak{g}$ of G is not necessarily a complex vector space, since Point 2 holds, in general, only for $s \in \mathbb{R}$. Nevertheless, it may happen in certain cases that $\mathfrak{g}$ is a complex vector space.

Definition 3.21. A matrix Lie group G is said to be **complex** if its Lie algebra $\mathfrak{g}$ is a complex subspace of $M_n(\mathbb{C})$, that is, if $iX \in \mathfrak{g}$ for all $X \in \mathfrak{g}$.

Examples of complex groups are $\mathsf{GL}(n;\mathbb{C})$, $\mathsf{SL}(n;\mathbb{C})$, $\mathsf{SO}(n;\mathbb{C})$, and $\mathsf{Sp}(n;\mathbb{C})$, as the calculations in Sect. 3.4 will show.

Proposition 3.22. *If G is commutative then $\mathfrak{g}$ is commutative.*

We will see in Sect. 3.7 that if G is connected and $\mathfrak{g}$ is commutative, G must be commutative.

Proof. For any two matrices $X, Y \in M_n(\mathbb{C})$, the commutator of X and Y may be computed as

$$[X, Y] = \frac{d}{dt}\left(\frac{d}{ds}e^{tX}e^{sY}e^{-tX}\bigg|_{s=0}\right)\bigg|_{t=0}. \tag{3.7}$$

If G is commutative and X and Y belong to $\mathfrak{g}$, then e^{tX} commutes with e^{sY} and the expression in parentheses on the right hand side of (3.7) is independent of t, so that $[X, Y] = 0$. $\square$

3.4 Examples

Physicists are accustomed to using the map $t \mapsto e^{itX}$ rather than $t \mapsto e^{tX}$. Thus, the physicists' expressions for the Lie algebras of matrix Lie groups will differ by a factor of i from the expressions we now derive.

Proposition 3.23. *The Lie algebra of* $\mathsf{GL}(n;\mathbb{C})$ *is the space* $M_n(\mathbb{C})$ *of all* $n \times n$ *matrices with complex entries. Similarly, the Lie algebra of* $\mathsf{GL}(n;\mathbb{R})$ *is equal to* $M_n(\mathbb{R})$. *The Lie algebra of* $\mathsf{SL}(n;\mathbb{C})$ *consists of all* $n \times n$ *complex matrices with trace zero, and the Lie algebra of* $\mathsf{SL}(n;\mathbb{R})$ *consists of all* $n \times n$ *real matrices with trace zero.*

We denote the Lie algebras of these groups as $\mathsf{gl}(n;\mathbb{C})$, $\mathsf{gl}(n;\mathbb{R})$, $\mathsf{sl}(n;\mathbb{C})$, and $\mathsf{sl}(n;\mathbb{R})$, respectively.

Proof. If $X \in M_n(\mathbb{C})$, then e^{tX} is invertible, so that X belongs to the Lie algebra of $\mathsf{GL}(n;\mathbb{C})$. If $X \in M_n(\mathbb{R})$, then e^{tX} is invertible and real, so that X is in the Lie algebra of $\mathsf{GL}(n;\mathbb{R})$. Conversely, if e^{tX} is real for all real t, then $X = d(e^{tX})/dt\big|_{t=0}$ must also real. If $X \in M_n(\mathbb{C})$ has trace zero, then by Theorem 2.12, $\det(e^{tX}) = 1$, showing that X is in the Lie algebra of $\mathsf{SL}(n;\mathbb{C})$. Conversely, if $\det(e^{tX}) = e^{t\,\mathrm{trace}(X)} = 1$ for all real t, then

$$\mathrm{trace}(X) = \frac{d}{dt} e^{t\,\mathrm{trace}(X)}\bigg|_{t=0} = 0.$$

Finally, if X is real and has trace zero, then e^{tX} is real and has determinant 1 for all real t, showing that X is in the Lie algebra of $\mathsf{SL}(n;\mathbb{R})$. Conversely, if e^{tX} is real and has determinant 1 for all real t, the preceding arguments show that X must be real and have trace zero.   □

Proposition 3.24. *The Lie algebra of* $\mathsf{U}(n)$ *consists of all complex matrices satisfying* $X^* = -X$ *and the Lie algebra of* $\mathsf{SU}(n)$ *consists of all complex matrices satisfying* $X^* = -X$ *and* $\mathrm{trace}(X) = 0$. *The Lie algebra of the orthogonal group* $\mathsf{O}(n)$ *consists of all real matrices* X *satisfying* $X^{tr} = -X$ *and the Lie algebra of* $\mathsf{SO}(n)$ *is the same as that of* $\mathsf{O}(n)$.

The Lie algebras of $\mathsf{U}(n)$ and $\mathsf{SU}(n)$ are denoted $\mathsf{u}(n)$ and $\mathsf{su}(n)$, respectively. The Lie algebra of $\mathsf{SO}(n)$ (which is the same as that of $\mathsf{O}(n)$) is denoted $\mathsf{so}(n)$.

Proof. A matrix U is unitary if and only if $U^* = U^{-1}$. Thus, e^{tX} is unitary if and only if

$$\left(e^{tX}\right)^* = \left(e^{tX}\right)^{-1} = e^{-tX}. \tag{3.8}$$

By Point 2 of Proposition 2.3, $\left(e^{tX}\right)^* = e^{tX^*}$, and so (3.8) becomes

$$e^{tX^*} = e^{-tX}. \tag{3.9}$$

The condition (3.9) holds for all real t if and only if $X^* = -X$. Thus, the Lie algebra of $\mathsf{U}(n)$ consists precisely of matrices X such that $X^* = -X$. As in the proof of Proposition 3.23, adding the "determinant 1" condition at the group level adds the "trace 0" condition at the Lie algebra level.

An exactly similar argument over $\mathbb{R}$ shows that a real matrix X belongs to the Lie algebra of $\mathsf{O}(n)$ if and only if $X^{tr} = -X$. Since any such matrix has trace$(X) = 0$ (since the diagonal entries of X are all zero), we see that every element of the Lie algebra of $\mathsf{O}(n)$ is also in the Lie algebra of $\mathsf{SO}(n)$. □

Proposition 3.25. *If g is the matrix in Exercise 1 of Chapter 1, then the Lie algebra of $\mathsf{O}(n;k)$ consists precisely of those real matrices X such that*

$$gX^{tr}g = -X,$$

and the Lie algebra of $\mathsf{SO}(n;k)$ is the same as that of $\mathsf{O}(n;k)$. If Ω is the matrix (1.8), then the Lie algebra of $\mathsf{Sp}(n;\mathbb{R})$ consists precisely of those real matrices X such that

$$\Omega X^{tr}\Omega = X,$$

and the Lie algebra of $\mathsf{Sp}(n;\mathbb{C})$ consists precisely of those complex matrices X satisfying the same condition. The Lie algebra of $\mathsf{Sp}(n)$ consists precisely of those complex matrices X such that $\Omega X^{tr}\Omega = X$ and $X^* = -X$.

The verification of Proposition 3.25 is similar to our previous computations and is omitted. The Lie algebra of $\mathsf{SO}(n;k)$ (which is the same as that of $\mathsf{O}(n;k)$) is denoted $\mathsf{so}(n;k)$, whereas the Lie algebras of the symplectic groups are denoted $\mathsf{sp}(n;\mathbb{R})$, $\mathsf{sp}(n;\mathbb{C})$, and $\mathsf{sp}(n)$.

Proposition 3.26. *The Lie algebra of the Heisenberg group H in Sect. 1.2.6 is the space of all matrices of the form*

$$X = \begin{pmatrix} 0 & a & b \\ 0 & 0 & c \\ 0 & 0 & 0 \end{pmatrix}, \tag{3.10}$$

with $a, b, c \in \mathbb{R}$.

Proof. If X is strictly upper triangular, it is easy to verify that X^m will be strictly upper triangular for all positive integers m. Thus, for X as in (3.10), we will have $e^{tX} = I + B$ with B strictly upper triangular, showing that $e^{tX} \in H$. Conversely, if e^{tX} belongs to H for all real t, then all of the entries of e^{tX} on or below the diagonal are independent of t. Thus, $X = d(e^{tX})/dt|_{t=0}$ will be of the form in (3.10). □

We leave it as an exercise to determine the Lie algebras of the Euclidean and Poincaré groups.

Example 3.27. The following elements form a basis for the Lie algebra $\mathsf{su}(2)$:

$$E_1 = \frac{1}{2}\begin{pmatrix} i & 0 \\ 0 & -i \end{pmatrix}; \quad E_2 = \frac{1}{2}\begin{pmatrix} 0 & i \\ i & 0 \end{pmatrix}; \quad E_3 = \frac{1}{2}\begin{pmatrix} 0 & -1 \\ 1 & 0 \end{pmatrix}.$$

These elements satisfy the commutation relations $[E_1, E_2] = E_3$, $[E_2, E_3] = E_1$, and $[E_3, E_1] = E_2$. The following elements form a basis for the Lie algebra $so(3)$:

$$F_1 = \begin{pmatrix} 0 & 0 & 0 \\ 0 & 0 & -1 \\ 0 & 1 & 0 \end{pmatrix}, \quad F_2 = \begin{pmatrix} 0 & 0 & 1 \\ 0 & 0 & 0 \\ -1 & 0 & 0 \end{pmatrix}, \quad F_3 = \begin{pmatrix} 0 & -1 & 0 \\ 1 & 0 & 0 \\ 0 & 0 & 0 \end{pmatrix},$$

These elements satisfy the commutation relations $[F_1, F_2] = F_3$, $[F_2, F_3] = F_1$, and $[F_3, F_1] = F_2$.

Note that the listed relations completely determine all commutation relations among, say, E_1, E_2, and E_3, since by the skew symmetry of the bracket, we must have $[E_1, E_1] = 0$, $[E_2, E_1] = -E_3$, and so on. Since E_1, E_2, and E_3 satisfy the same commutation relations as F_1, F_2, and F_3, the two Lie algebras are isomorphic.

Proof. Direct calculation from Proposition 3.24.   □

3.5   Lie Group and Lie Algebra Homomorphisms

The following theorem tells us that a Lie group homomorphism between two Lie groups gives rise in a natural way to a map between the corresponding Lie algebras. It will follow (Exercise 8) that isomorphic Lie groups have isomorphic Lie algebras.

Theorem 3.28. *Let G and H be matrix Lie groups, with Lie algebras $\mathfrak{g}$ and $\mathfrak{h}$, respectively. Suppose that $\Phi : G \to H$ is a Lie group homomorphism. Then there exists a unique real-linear map $\phi : \mathfrak{g} \to \mathfrak{h}$ such that*

$$\Phi(e^X) = e^{\phi(X)} \tag{3.11}$$

for all $X \in \mathfrak{g}$. The map ϕ has following additional properties:

1. *$\phi\left(AXA^{-1}\right) = \Phi(A)\phi(X)\Phi(A)^{-1}$, for all $X \in \mathfrak{g}$, $A \in G$.*
2. *$\phi([X, Y]) = [\phi(X), \phi(Y)]$, for all $X, Y \in \mathfrak{g}$.*
3. *$\phi(X) = \frac{d}{dt}\Phi(e^{tX})\big|_{t=0}$, for all $X \in \mathfrak{g}$.*

In practice, given a Lie group homomorphism Φ, the way one goes about computing ϕ is by using Property 3. In the language of manifolds, Property 3 says that ϕ is the derivative (or differential) of Φ at the identity. By Point 2, $\phi : \mathfrak{g} \to \mathfrak{h}$ is a Lie algebra homomorphism. Thus, every Lie group homomorphism gives rise to a Lie algebra homomorphism. In Chapter 5, we will investigate the reverse question: If ϕ is a homomorphism between the Lie algebras of two Lie groups, is there an associated Lie group homomorphism Φ?

Proof. The proof is similar to the proof of Theorem 3.20. Since Φ is a continuous group homomorphism, $\Phi(e^{tX})$ will be a one-parameter subgroup of H, for each $X \in \mathfrak{g}$. Thus, by Theorem 2.14, there is a unique matrix Z such that

$$\Phi(e^{tX}) = e^{tZ} \tag{3.12}$$

for all $t \in \mathbb{R}$. We define $\phi(X) = Z$ and check that ϕ has the required properties. First, by putting $t = 1$ in (3.12), we see that $\Phi(e^X) = e^{\phi(X)}$ for all $X \in \mathfrak{g}$. Next, if $\Phi(e^{tX}) = e^{tZ}$ for all t, then $\Phi(e^{tsX}) = e^{tsZ}$, showing that $\phi(sX) = s\phi(X)$. Using the Lie product formula and the continuity of Φ, we then compute that

$$e^{t\phi(X+Y)} = \Phi\left(\lim_{m\to\infty} \left(e^{tX/m} e^{tY/m} \right)^m \right)$$

$$= \lim_{m\to\infty} \left(\Phi(e^{tX/m}) \Phi(e^{tY/m}) \right)^m .$$

Thus,

$$e^{t\phi(X+Y)} = \lim_{m\to\infty} \left(e^{t\phi(X)/m} e^{t\phi(Y)/m} \right)^m = e^{t(\phi(X)+\phi(Y))}.$$

Differentiating this result at $t = 0$ shows that $\phi(X + Y) = \phi(X) + \phi(Y)$.

We have thus obtained a real-linear map ϕ satisfying (3.11). If there were another real-linear map ϕ' with this property, we would have

$$e^{t\phi(X)} = e^{t\phi'(X)} = \Phi(e^{tX})$$

for all $t \in \mathbb{R}$. Differentiating this result at $t = 0$ shows that $\phi(X) = \phi'(X)$.

We now verify the remaining claimed properties of ϕ. For any $A \in G$, we have

$$e^{t\phi(AXA^{-1})} = e^{\phi(tAXA^{-1})} = \Phi(e^{tAXA^{-1}}).$$

Thus,

$$e^{t\phi(AXA^{-1})} = \Phi(A)\Phi(e^{tX})\Phi(A)^{-1}$$

$$= \Phi(A)e^{t\phi(X)}\Phi(A)^{-1}.$$

Differentiating this identity at $t = 0$ gives Point 1.

Meanwhile, for any X and Y in $\mathfrak{g}$, we have, as in the proof of Theorem 3.20,

$$\phi([X,Y]) = \phi\left(\frac{d}{dt} e^{tX} Y e^{-tX} \Big|_{t=0} \right)$$

$$= \frac{d}{dt} \phi(e^{tX} Y e^{-tX}) \Big|_{t=0},$$

where we have used the fact that a derivative commutes with a linear transformation. Thus,

$$\phi\left([X,Y]\right) = \left.\frac{d}{dt}\Phi(e^{tX})\phi(Y)\Phi(e^{-tX})\right|_{t=0}$$

$$= \left.\frac{d}{dt}e^{t\phi(X)}\phi(Y)e^{-t\phi(X)}\right|_{t=0}$$

$$= [\phi(X),\phi(Y)],$$

establishing Point 2. Finally, since $\Phi(e^{tX}) = e^{\phi(tX)} = e^{t\phi(X)}$, we can compute $\phi(X)$ as in Point 3.   □

Example 3.29. Let $\Phi : \mathsf{SU}(2) \to \mathsf{SO}(3)$ be the homomorphism in Proposition 1.19. Then the associated Lie algebra homomorphism $\phi : \mathsf{su}(2) \to \mathsf{so}(3)$ satisfies

$$\phi(E_j) = F_j, \quad j = 1,2,3,$$

where $\{E_1, E_2, E_3\}$ and $\{F_1, F_2, F_3\}$ are the bases for $\mathsf{su}(2)$ and $\mathsf{so}(3)$, respectively, given in Example 3.27.

Since ϕ maps a basis for $\mathsf{su}(2)$ to a basis for $\mathsf{so}(3)$, we see that ϕ is a Lie algebra *isomorphism*, even though Φ is not a Lie group isomorphism (since $\ker(\Phi) = \{I, -I\}$).

Proof. If X is in $\mathsf{su}(2)$ and Y is in the space V in (1.14), then

$$\left.\frac{d}{dt}\Phi(e^{tX})Y\right|_{t=0} = \left.\frac{d}{dt}e^{tX}Ye^{-tX}\right|_{t=0} = [X,Y].$$

Thus, $\phi(X)$ is the linear map of $V \cong \mathbb{R}^3$ to itself given by $Y \mapsto [X, Y]$. If, say, $X = E_1$, then direct computation shows that

$$\left[E_1, \begin{pmatrix} x_1 & x_2 + ix_3 \\ x_2 - ix_3 & -x_1 \end{pmatrix}\right] = \begin{pmatrix} x_1' & x_2' + ix_3' \\ x_2' - ix_3' & -x_1' \end{pmatrix},$$

where $(x_1', x_2', x_3') = (0, -x_3, x_2)$. Since

$$\begin{pmatrix} 0 \\ -x_3 \\ x_2 \end{pmatrix} = \begin{pmatrix} 0 & 0 & 0 \\ 0 & 0 & -1 \\ 0 & 1 & 0 \end{pmatrix} \begin{pmatrix} x_1 \\ x_2 \\ x_3 \end{pmatrix}, \tag{3.13}$$

we conclude that $\phi(E_1)$ is the 3×3 matrix appearing on the right-hand side of (3.13), which is precisely F_1. The computation of $\phi(E_2)$ and $\phi(E_3)$ is similar and is left to the reader.   □

Proposition 3.30. *Suppose that G, H, and K are matrix Lie groups and $\Phi : H \to K$ and $\Psi : G \to H$ are Lie group homomorphisms. Let $\Lambda : G \to K$ be the composition of Φ and Ψ and let ϕ, ψ, and λ be the Lie algebra maps associated to Φ, Ψ, and Λ, respectively. Then we have*

$$\lambda = \phi \circ \psi.$$

Proof. For any $X \in \mathfrak{g}$,

$$\Lambda(e^{tX}) = \Phi(\Psi(e^{tX})) = \Phi(e^{t\psi(X)}) = e^{t\phi(\psi(X))}.$$

Thus, $\lambda(X) = \phi(\psi(X))$.   □

Proposition 3.31. *If $\Phi : G \to H$ is a Lie group homomorphism and $\phi : \mathfrak{g} \to \mathfrak{h}$ is the associated Lie algebra homomorphism, then the kernel of Φ is a closed, normal subgroup of G and the Lie algebra of the kernel is given by*

$$\mathrm{Lie}(\ker(\Phi)) = \ker(\phi).$$

Proof. The usual algebraic argument shows that $\ker(\Phi)$ is normal subgroup of G. Since, also, Φ is continuous, $\ker(\Phi)$ is closed. If $X \in \ker(\phi)$, then

$$\Phi(e^{tX}) = e^{t\phi(X)} = I,$$

for all $t \in \mathbb{R}$, showing that X is in the Lie algebra of $\ker(\Phi)$. In the other direction, if e^{tX} lies in $\ker(\Phi)$ for all $t \in \mathbb{R}$, then

$$e^{t\phi(X)} = \Phi(e^{tX}) = I$$

for all t. Differentiating this relation with respect to t at $t = 0$ gives that $\phi(X) = 0$, showing that $X \in \ker(\phi)$.   □

Definition 3.32 (The Adjoint Map).   Let G be a matrix Lie group, with Lie algebra $\mathfrak{g}$. Then for each $A \in G$, define a linear map $\mathrm{Ad}_A : \mathfrak{g} \to \mathfrak{g}$ by the formula

$$\mathrm{Ad}_A(X) = AXA^{-1}.$$

Proposition 3.33. *Let G be a matrix Lie group, with Lie algebra $\mathfrak{g}$. Let $\mathsf{GL}(\mathfrak{g})$ denote the group of all invertible linear transformations of $\mathfrak{g}$. Then the map $A \to \mathrm{Ad}_A$ is a homomorphism of G into $\mathsf{GL}(\mathfrak{g})$. Furthermore, for each $A \in G$, Ad_A satisfies $\mathrm{Ad}_A([X, Y]) = [\mathrm{Ad}_A(X), \mathrm{Ad}_A(Y)]$ for all $X, Y \in \mathfrak{g}$.*

Proof. Easy. Note that Point 1 of Theorem 3.20 guarantees that $\mathrm{Ad}_A(X)$ is actually in $\mathfrak{g}$ for all $X \in \mathfrak{g}$.   □

Since $\mathfrak{g}$ is a real vector space with some dimension k, $\mathsf{GL}(\mathfrak{g})$ is essentially the same as $\mathsf{GL}(k; \mathbb{R})$. Thus, we will regard $\mathsf{GL}(\mathfrak{g})$ as a matrix Lie group. It is easy to show that $\mathrm{Ad} : G \to \mathsf{GL}(\mathfrak{g})$ is continuous, and so is a Lie group homomorphism. By Theorem 3.28, there is an associated real linear map $X \to \mathrm{ad}_X$ from the Lie algebra of G to the Lie algebra of $\mathsf{GL}(\mathfrak{g})$ (i.e., from $\mathfrak{g}$ to $\mathfrak{gl}(\mathfrak{g})$), with the property that

$$e^{\mathrm{ad}_X} = \mathrm{Ad}_{e^X}.$$

Here, $\mathfrak{gl}(\mathfrak{g})$ is the Lie algebra of $\mathsf{GL}(\mathfrak{g})$, namely the space of all linear maps of $\mathfrak{g}$ to itself.

Proposition 3.34. *Let G be a matrix Lie group, let $\mathfrak{g}$ be its Lie algebra, and let $\mathrm{Ad} : G \to \mathsf{GL}(\mathfrak{g})$ be as in Proposition 3.33. Let $\mathrm{ad} : \mathfrak{g} \to \mathfrak{gl}(\mathfrak{g})$ be the associated Lie algebra map. Then for all $X, Y \in \mathfrak{g}$*

$$\mathrm{ad}_X(Y) = [X, Y]. \tag{3.14}$$

The proposition shows that our usage of the notation ad_X in this section is consistent with that in Definition 3.7.

Proof. By Point 3 of Theorem 3.28, ad can be computed as follows:

$$\mathrm{ad}_X = \frac{d}{dt}\mathrm{Ad}_{e^{tX}}\bigg|_{t=0}.$$

Thus,

$$\mathrm{ad}_X(Y) = \frac{d}{dt}e^{tX}Ye^{-tX}\bigg|_{t=0} = [X, Y],$$

as claimed.   □

We have proved, as a consequence of Theorem 3.28 and Proposition 3.34, the following result, which we will make use of later.

Proposition 3.35. *For any X in $M_n(\mathbb{C})$, let $\mathrm{ad}_X : M_n(\mathbb{C}) \to M_n(\mathbb{C})$ be given by $\mathrm{ad}_X Y = [X, Y]$. Then for any Y in $M_n(\mathbb{C})$, we have*

$$e^X Y e^{-X} = \mathrm{Ad}_{e^X}(Y) = e^{\mathrm{ad}_X}(Y),$$

where

$$e^{\mathrm{ad}_X}(Y) = Y + [X, Y] + \frac{1}{2}[X, [X, Y]] + \cdots.$$

This result can also be proved by direct calculation—see Exercise 14.

3.6   The Complexification of a Real Lie Algebra

In studying the representations of a matrix Lie group G (as we will do in later chapters), it is often useful to pass to the Lie algebra $\mathfrak{g}$ of G, which is, in general, only a real Lie algebra. It is then often useful to pass to an associated complex Lie algebra, called the *complexification* of $\mathfrak{g}$.

Definition 3.36. If V is a finite-dimensional real vector space, then the **complexification** of V, denoted $V_{\mathbb{C}}$, is the space of formal linear combinations

$$v_1 + iv_2,$$

with $v_1, v_2 \in V$. This becomes a real vector space in the obvious way and becomes a complex vector space if we define

$$i(v_1 + iv_2) = -v_2 + iv_1.$$

We could more pedantically define $V_{\mathbb{C}}$ to be the space of ordered pairs (v_1, v_2) with $v_1, v_2 \in V$, but this is notationally cumbersome. It is straightforward to verify that the above definition really makes $V_{\mathbb{C}}$ into a complex vector space. We will regard V as a real subspace of $V_{\mathbb{C}}$ in the obvious way.

Proposition 3.37. *Let* $\mathfrak{g}$ *be a finite-dimensional real Lie algebra and* $\mathfrak{g}_{\mathbb{C}}$ *its complexification. Then the bracket operation on* $\mathfrak{g}$ *has a unique extension to* $\mathfrak{g}_{\mathbb{C}}$ *that makes* $\mathfrak{g}_{\mathbb{C}}$ *into a complex Lie algebra. The complex Lie algebra* $\mathfrak{g}_{\mathbb{C}}$ *is called the* **complexification** *of the real Lie algebra* $\mathfrak{g}$.

Proof. The uniqueness of the extension is obvious, since if the bracket operation on $\mathfrak{g}_{\mathbb{C}}$ is to be bilinear, then it must be given by

$$[X_1 + iX_2, Y_1 + iY_2] = ([X_1, Y_1] - [X_2, Y_2]) + i \left([X_1, Y_2] + [X_2, Y_1]\right). \quad (3.15)$$

To show existence, we must now check that (3.15) is really bilinear and skew symmetric and that it satisfies the Jacobi identity. It is clear that (3.15) is *real* bilinear, and skew-symmetric. The skew symmetry means that if (3.15) is complex linear in the first factor, it is also complex linear in the second factor. Thus, we need only show that

$$[i(X_1 + iX_2), Y_1 + iY_2] = i[X_1 + iX_2, Y_1 + iY_2]. \quad (3.16)$$

The left-hand side of (3.16) is

$$[-X_2 + iX_1, Y_1 + iY_2] = (-[X_2, Y_1] - [X_1, Y_2]) + i \left([X_1, Y_1] - [X_2, Y_2]\right),$$

whereas the right-hand side of (3.16) is

$$i\,\{([X_1, Y_1] - [X_2, Y_2]) + i\,([X_2, Y_1] + [X_1, Y_2])\}$$
$$= (-[X_2, Y_1] - [X_1, Y_2]) + i\,([X_1, Y_1] - [X_2, Y_2]),$$

and, indeed, these expressions are equal.

It remains to check the Jacobi identity. Of course, the Jacobi identity holds if X, Y, and Z are in $\mathfrak{g}$. Furthermore, for all $X, Y, Z \in \mathfrak{g}_{\mathbb{C}}$, the expression

$$[X, [Y, Z]] + [Y, [Z, X]] + [Z, [X, Y]]$$

is complex-linear in X with Y and Z fixed. Thus, the Jacobi identity continues to hold if X is in $\mathfrak{g}_{\mathbb{C}}$ and Y and Z are in $\mathfrak{g}$. The same argument then shows that the Jacobi identity holds when X and Y are in $\mathfrak{g}_{\mathbb{C}}$ and Z is in $\mathfrak{g}$. Applying this argument one more time establishes the Jacobi identity for $\mathfrak{g}_{\mathbb{C}}$ in general. □

Proposition 3.38. *Suppose that $\mathfrak{g} \subset M_n(\mathbb{C})$ is a real Lie algebra and that for all nonzero X in $\mathfrak{g}$, the element iX is not in $\mathfrak{g}$. Then the "abstract" complexification $\mathfrak{g}_{\mathbb{C}}$ of $\mathfrak{g}$ in Definition 3.36 is isomorphic to the set of matrices in $M_n(\mathbb{C})$ that can be expressed in the form $X + iY$ with X and Y in $\mathfrak{g}$.*

Proof. Consider the map from $\mathfrak{g}_{\mathbb{C}}$ into $M_n(\mathbb{C})$ sending the formal linear combination $X + iY$ to the linear combination $X + iY$ of matrices. This map is easily seen to be a complex Lie algebra homomorphism. If $\mathfrak{g}$ satisfies the assumption in the statement of the proposition, this map is also injective and thus an isomorphism of $\mathfrak{g}_{\mathbb{C}}$ with $\mathfrak{g} + i\mathfrak{g} \subset M_n(\mathbb{C})$. □

Using the proposition, we easily obtain the following list of isomorphisms:

$$\begin{aligned}
\mathsf{gl}(n;\mathbb{R})_{\mathbb{C}} &\cong \mathsf{gl}(n;\mathbb{C}),\\
\mathsf{u}(n)_{\mathbb{C}} &\cong \mathsf{gl}(n;\mathbb{C}),\\
\mathsf{su}(n)_{\mathbb{C}} &\cong \mathsf{sl}(n;\mathbb{C}),\\
\mathsf{sl}(n;\mathbb{R})_{\mathbb{C}} &\cong \mathsf{sl}(n;\mathbb{C}),\\
\mathsf{so}(n)_{\mathbb{C}} &\cong \mathsf{so}(n;\mathbb{C}),\\
\mathsf{sp}(n;\mathbb{R})_{\mathbb{C}} &\cong \mathsf{sp}(n;\mathbb{C}),\\
\mathsf{sp}(n)_{\mathbb{C}} &\cong \mathsf{sp}(n;\mathbb{C}).
\end{aligned} \qquad (3.17)$$

Let us verify just one example, that of $\mathsf{u}(n)$. If $X^* = -X$, then $(iX)^* = iX$. Thus, X and iX cannot both be in $\mathsf{u}(n)$ unless X is zero. Furthermore, every X in $M_n(\mathbb{C})$ can be expressed as $X = X_1 + iX_2$, where $X_1 = (X - X^*)/2$ and $X_2 = (X + X^*)/(2i)$ are both in $\mathsf{u}(n)$. This shows that $\mathsf{u}(n)_{\mathbb{C}} \cong \mathsf{gl}(n;\mathbb{C})$.

Although both $\mathsf{su}(2)_{\mathbb{C}}$ and $\mathsf{sl}(2;\mathbb{R})_{\mathbb{C}}$ are isomorphic to $\mathsf{sl}(2;\mathbb{C})$, the Lie algebra $\mathsf{su}(2)$ is *not* isomorphic to $\mathsf{sl}(2;\mathbb{R})$. See Exercise 11.

Proposition 3.39. *Let $\mathfrak{g}$ be a real Lie algebra, $\mathfrak{g}_{\mathbb{C}}$ its complexification, and $\mathfrak{h}$ an arbitrary complex Lie algebra. Then every real Lie algebra homomorphism of $\mathfrak{g}$ into $\mathfrak{h}$ extends uniquely to a complex Lie algebra homomorphism of $\mathfrak{g}_{\mathbb{C}}$ into $\mathfrak{h}$.*

This result is the **universal property** of the complexification of a real Lie algebra.

Proof. The unique extension is given by $\pi(X + iY) = \pi(X) + i\pi(Y)$ for all $X, Y \in \mathfrak{g}$. It is easy to check that this map is, indeed, a homomorphism of complex Lie algebras. □

3.7 The Exponential Map

Definition 3.40. If G is a matrix Lie group with Lie algebra $\mathfrak{g}$, then the **exponential map** for G is the map

$$\exp : \mathfrak{g} \to G.$$

That is to say, the exponential map for G is the matrix exponential restricted to the Lie algebra $\mathfrak{g}$ of G. We have shown (Theorem 2.10) that every matrix in $\mathsf{GL}(n; \mathbb{C})$ is the exponential of some $n \times n$ matrix. Nevertheless, if $G \subset \mathsf{GL}(n; \mathbb{C})$ is a closed subgroup, there may exist A in G such that there is no X *in the Lie algebra $\mathfrak{g}$ of G* with $\exp X = A$.

Example 3.41. There does not exist a matrix $X \in \mathsf{sl}(2; \mathbb{C})$ with

$$e^X = \begin{pmatrix} -1 & 1 \\ 0 & -1 \end{pmatrix}, \tag{3.18}$$

even though the matrix on the right-hand side of (3.18) is in $\mathsf{SL}(2; \mathbb{C})$.

Proof. If $X \in \mathsf{sl}(2; \mathbb{C})$ has distinct eigenvalues, then X is diagonalizable and e^X will also be diagonalizable, unlike the matrix on the right-hand side of (3.18). If $X \in \mathsf{sl}(2; \mathbb{C})$ has a repeated eigenvalue, this eigenvalue must be 0 or the trace of X would not be zero. Thus, there is a nonzero vector v with $Xv = 0$, from which it follows that $e^X v = e^0 v = v$. We conclude that e^X has 1 as an eigenvalue, unlike the matrix on the right-hand side of (3.18). □

We see, then, that the exponential map for a matrix Lie group G does not necessarily map $\mathfrak{g}$ onto G. Furthermore, the exponential map may not be one-to-one on $\mathfrak{g}$, as may be seen, for example, from the case $\mathfrak{g} = \mathsf{su}(2)$. Nevertheless, it provides a crucial mechanism for passing information between the group and the Lie algebra. Indeed, we will see (Corollary 3.44) that the exponential map is *locally* one-to-one and onto, a result that will be essential later.

Theorem 3.42. *For $0 < \varepsilon < \log 2$, let $U_\varepsilon = \{X \in M_n(\mathbb{C}) \,|\, \|X\| < \varepsilon\}$ and let $V_\varepsilon = \exp(U_\varepsilon)$. Suppose $G \subset \mathsf{GL}(n;\mathbb{C})$ is a matrix Lie group with Lie algebra $\mathfrak{g}$. Then there exists $\varepsilon \in (0, \log 2)$ such that for all $A \in V_\varepsilon$, A is in G if and only if $\log A$ is in $\mathfrak{g}$.*

The condition $\varepsilon < \log 2$ guarantees (Theorem 2.8) that for all $X \in U_\varepsilon$, $\log(e^X)$ is defined and equal to X. Note that if $X = \log A$ is in $\mathfrak{g}$, then $A = e^X$ is in G. Thus, the content of the theorem is that for some ε, having A in $V_\varepsilon \cap G$ implies that $\log A$ must be in $\mathfrak{g}$. See Figure 3.1.

We begin with a lemma.

Lemma 3.43. *Suppose B_m are elements of G and that $B_m \to I$. Let $Y_m = \log B_m$, which is defined for all sufficiently large m. Suppose that Y_m is nonzero for all m and that $Y_m / \|Y_m\| \to Y \in M_n(\mathbb{C})$. Then Y is in $\mathfrak{g}$.*

Proof. For any $t \in \mathbb{R}$, we have $(t / \|Y_m\|) Y_m \to tY$. Note that since $B_m \to I$, we have $\|Y_m\| \to 0$. Thus, we can find integers k_m such that $k_m \|Y_m\| \to t$. We have, then,

$$e^{k_m Y_m} = \exp\left[(k_m \|Y_m\|) \frac{Y_m}{\|Y_m\|} \right] \to e^{tY}.$$

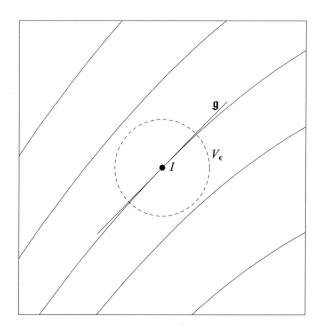

Fig. 3.1 If $A \in V_\varepsilon$ belongs to G, then $\log A$ belongs to $\mathfrak{g}$

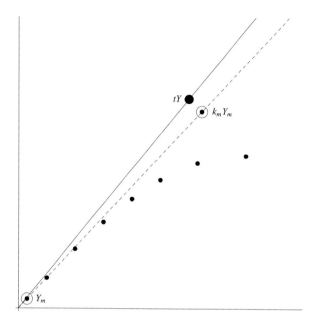

Fig. 3.2 The points $k_m Y_m$ are converging to tY

However,

$$e^{k_m Y_m} = (e^{Y_m})^{k_m} = (B_m)^{k_m} \in G$$

and G is closed, and we conclude that $e^{tY} \in G$. This shows that $Y \in \mathfrak{g}$. (See Figure 3.2.)   □

Proof of Theorem 3.42. Let us think of $M_n(\mathbb{C})$ as $\mathbb{C}^{n^2} \cong \mathbb{R}^{2n^2}$ and let D denote the orthogonal complement of $\mathfrak{g}$ with respect to the usual inner product on $\mathbb{R}^{2n^2}$. Consider the map $\Phi : M_n(\mathbb{C}) \to M_n(\mathbb{C})$ given by

$$\Phi(Z) = e^X e^Y,$$

where $Z = X + Y$ with $X \in \mathfrak{g}$ and $Y \in D$. Since (Proposition 2.16) the exponential is continuously differentiable, the map Φ is also continuously differentiable, and we may compute that

$$\frac{d}{dt} \Phi(tX, 0) \bigg|_{t=0} = X,$$

$$\frac{d}{dt} \Phi(0, tY) \bigg|_{t=0} = Y.$$

This calculation shows that the derivative of Φ at the point $0 \in \mathbb{R}^{2n^2}$ is the identity. (Recall that the derivative at a point of a function from $\mathbb{R}^{2n^2}$ to itself is a linear map of $\mathbb{R}^{2n^2}$ to itself.) Since the derivative of Φ at the origin is invertible, the inverse function theorem says that Φ has a continuous local inverse, defined in a neighborhood of I.

We need to prove that for some ε, if $A \in V_\varepsilon \cap G$, then $\log A \in \mathfrak{g}$. If this were not the case, we could find a sequence A_m in G such that $A_m \to I$ as $m \to \infty$ and such that for all m, $\log A_m \notin \mathfrak{g}$. Using the local inverse of the map Φ, we can write A_m (for all sufficiently large m) as

$$ A_m = e^{X_m} e^{Y_m}, \quad X_m \in \mathfrak{g}, \ Y_m \in D, $$

with X_m and Y_m tending to zero as m tends to infinity. We must have $Y_m \neq 0$, since otherwise we would have $\log A_m = X_m \in \mathfrak{g}$. Since e^{X_m} and A_m are in G, we see that

$$ B_m := e^{-X_m} A_m = e^{Y_m} $$

is in G.

Since the unit sphere in D is compact, we can choose a subsequence of the Y_m's (still called Y_m) so that $Y_m / \|Y_m\|$ converges to some $Y \in D$, with $\|Y\| = 1$. Then, by the lemma, $Y \in \mathfrak{g}$. This is a contradiction, because D is the orthogonal complement of $\mathfrak{g}$. Thus, there must be some ε such that $\log A \in \mathfrak{g}$ for all A in $V_\varepsilon \cap G$. $\square$

3.8 Consequences of Theorem 3.42

In this section, we derive several consequences of the main result of the last section, Theorem 3.42.

Corollary 3.44. *If G is a matrix Lie group with Lie algebra $\mathfrak{g}$, there exists a neighborhood U of 0 in $\mathfrak{g}$ and a neighborhood V of I in G such that the exponential map takes U homeomorphically onto V.*

Proof. Let ε be such that Theorem 3.42 holds and set $U = U_\varepsilon \cap \mathfrak{g}$ and $V = V_\varepsilon \cap G$. The theorem implies that exp takes U onto V. Furthermore, exp is a homeomorphism of U onto V, since there is a continuous inverse map, namely, the restriction of the matrix logarithm to V. $\square$

Corollary 3.45. *Let G be a matrix Lie group with Lie algebra $\mathfrak{g}$ and let k be the dimension of $\mathfrak{g}$ as a real vector space. Then G is a smooth embedded submanifold of $M_n(\mathbb{C})$ of dimension k and hence a Lie group.*

It follows from the corollary that G is locally path connected: every point in G has a neighborhood U that is homeomorphic to a ball in $\mathbb{R}^k$ and hence path connected. It then follows that G is connected (in the usual topological sense) if and only if it is path connected. (See, for example, Proposition 3.4.25 of [Run].)

Proof. Let $\varepsilon \in (0, \log 2)$ be such that Theorem 3.42 holds. Then for any $A_0 \in G$, consider the neighborhood $A_0 V_\varepsilon$ of A_0 in $M_n(\mathbb{C})$. Note that $A \in A_0 V_\varepsilon$ if and only if $A_0^{-1} A \in V_\varepsilon$. Define a local coordinate system on $A_0 V_\varepsilon$ by writing each $A \in A_0 V_\varepsilon$ as $A = A_0 e^X$, for $X \in U_\varepsilon \subset M_n(\mathbb{C})$. It follows from Theorem 3.42 that (for $A \in A_0 V_\varepsilon$) $A \in G$ if and only if $X \in \mathfrak{g}$. Thus, in this local coordinate system defined near A_0, the group G looks like the subspace $\mathfrak{g}$ of $M_n(\mathbb{C})$. Since we can find such local coordinates near any point A_0 in G, we conclude that G is an embedded submanifold of $M_n(\mathbb{C})$.

Now, the operation of matrix multiplication is clearly smooth. Furthermore, by the formula for the inverse of a matrix in terms of cofactors, the map $A \mapsto A^{-1}$ is also smooth on $\mathsf{GL}(n; \mathbb{C})$. The restrictions of these maps to G are then also smooth, showing that G is a Lie group. $\square$

Corollary 3.46. *Suppose $G \subset \mathsf{GL}(n; \mathbb{C})$ is a matrix Lie group with Lie algebra $\mathfrak{g}$. Then a matrix X is in $\mathfrak{g}$ if and only if there exists a smooth curve γ in $M_n(\mathbb{C})$ with $\gamma(t) \in G$ for all t and such that $\gamma(0) = I$ and $d\gamma/dt|_{t=0} = X$. Thus, $\mathfrak{g}$ is the* **tangent space at the identity** *to G.*

This result is illustrated in Figure 3.1.

Proof. If X is in $\mathfrak{g}$, then we may take $\gamma(t) = e^{tX}$ and then $\gamma(0) = I$ and $d\gamma/dt|_{t=0} = X$. In the other direction, suppose that $\gamma(t)$ is a smooth curve in G with $\gamma(0) = I$. For all sufficiently small t, we can write $\gamma(t) = e^{\delta(t)}$, where δ is a smooth curve in $\mathfrak{g}$. Now, the derivative of $\delta(t)$ at $t = 0$ is the same as the derivative of $t \mapsto t\delta'(0)$ at $t = 0$. Thus, by the chain rule, we have

$$\gamma'(0) = \left. \frac{d}{dt} e^{\delta(t)} \right|_{t=0} = \left. \frac{d}{dt} e^{t\delta'(0)} \right|_{t=0} = \delta'(0).$$

Since $\delta(t)$ belongs to $\mathfrak{g}$ for all sufficiently small t, we conclude (as in the proof of Theorem 3.20) that $\delta'(0) = \gamma'(0)$ belongs to $\mathfrak{g}$. $\square$

Corollary 3.47. *If G is a connected matrix Lie group, every element A of G can be written in the form*

$$A = e^{X_1} e^{X_2} \cdots e^{X_m} \tag{3.19}$$

for some $X_1, X_2, \ldots, X_m$ in $\mathfrak{g}$.

Even if G is connected, it is in general *not* possible to write every $A \in G$ as single exponential, $A = \exp X$, with $X \in \mathfrak{g}$. (See Example 3.41.) We begin with a simple analytic lemma.

Lemma 3.48. *Suppose $A : [a, b] \to \mathsf{Gl}(n; \mathbb{C})$ is a continuous map. Then for all $\varepsilon > 0$ there exists $\delta > 0$ such that if $s, t \in [a, b]$ satisfy $|s - t| < \delta$, then*

$$\left\| A(s) A(t)^{-1} - I \right\| < \varepsilon.$$

Proof. We note that

$$\|A(s)A(t)^{-1} - I\| = \|(A(s) - A(t))A(t)^{-1}\|$$
$$\leq \|A(s) - A(t)\|\,\|A(t)^{-1}\|. \qquad (3.20)$$

Since $[a,b]$ is compact and the map $t \mapsto \|A(t)^{-1}\|$ is continuous, there is a constant C such that $\|A(t)^{-1}\| \leq C$ for all $t \in [a,b]$. Furthermore, since $[a,b]$ is compact, Theorem 4.19 in [Rud1] tells us that the map A is actually *uniformly* continuous on $[a,b]$. Thus, for any $\varepsilon > 0$, there exists $\delta > 0$ such that when $|s - t| < \delta$, we have $\|A(s) - A(t)\| < \varepsilon/C$. Thus, in light of (3.20), we have the desired δ. $\qquad \square$

Proof of Corollary 3.47. Let V_ε be as in Theorem 3.42. For any $A \in G$, choose a continuous path $A : [0,1] \to G$ with $A(0) = I$ and $A(1) = A$. By Lemma 3.48, we can find some $\delta > 0$ such that if $|s - t| < \delta$, then $A(s)A(t)^{-1} \in V_\varepsilon$. Now divide $[0,1]$ into m pieces, where $1/m < \delta$. Then for $j = 1,\ldots,m$, we see that $A((j-1)/m)^{-1}A(j/m)$ belongs to V_ε, so that

$$A((j-1)/m)^{-1}A(j/m) = e^{X_j}$$

for some elements $X_1,\ldots,X_m$ of $\mathfrak{g}$. Thus,

$$A = A(0)^{-1}A(1)$$
$$= A(0)^{-1}A(1/m)A(1/m)^{-1}A(2/m)\cdots A((m-1)/m)^{-1}A(1)$$
$$= e^{X_1}e^{X_2}\cdots e^{X_m},$$

as claimed. $\qquad \square$

Corollary 3.49. *Let G and H be matrix Lie groups with Lie algebras $\mathfrak{g}$ and $\mathfrak{h}$, respectively, and assume G is connected. Suppose Φ_1 and Φ_2 are Lie group homomorphisms of G into H and that ϕ_1 and ϕ_2 be the associated Lie algebra homomorphisms of $\mathfrak{g}$ into $\mathfrak{h}$. Then if $\phi_1 = \phi_2$, we have $\Phi_1 = \Phi_2$.*

Proof. Let g be any element of G. Since G is connected, Corollary 3.47 tells us that every element of G can be written as $e^{X_1}e^{X_2}\cdots e^{X_m}$, with $X_j \in \mathfrak{g}$. Thus,

$$\Phi_1(e^{X_1}\cdots e^{X_m}) = e^{\phi_1(X_1)}\cdots e^{\phi_1(X_m)}$$
$$= e^{\phi_2(X_1)}\cdots e^{\phi_2(X_m)}$$
$$= \Phi_2(e^{X_1}\cdots e^{X_m}),$$

as claimed. $\qquad \square$

Corollary 3.50. *Every continuous homomorphism between two matrix Lie groups is smooth.*

Proof. For all $g \in G$, we write nearby elements $h \in G$ as $h = ge^X$, with $X \in \mathfrak{g}$, so that

$$\Phi(h) = \Phi(g)\Phi(e^X) = \Phi(g)e^{\phi(X)}.$$

This relation says that in the exponential coordinates near g, the map Φ is a composition of a linear map, the exponential map, and multiplication on the left by $\Phi(g)$, all of which are smooth. This shows that Φ is smooth near g. □

Corollary 3.51. *If G is a* connected *matrix Lie group and the Lie algebra $\mathfrak{g}$ of G is commutative, then G is commutative.*

This result is a partial converse to Proposition 3.22.

Proof. Since $\mathfrak{g}$ is commutative, any two elements of G, when written as in Corollary 3.47, will commute. □

Corollary 3.52. *If $G \subset M_n(\mathbb{C})$ is a matrix Lie group, the identity component G_0 of G is a* closed *subgroup of $\mathsf{GL}(N;\mathbb{C})$ and thus a matrix Lie group. Furthermore, the Lie algebra of G_0 is the same as the Lie algebra of G.*

Proof. Suppose that $\langle A_m \rangle$ is a sequence in G_0 converging to some $A \in \mathsf{GL}(n;\mathbb{C})$. Then certainly $A \in G$, since G is closed. Furthermore, $A_m A^{-1}$ lies in G for all m and $A_m A^{-1} \to I$ as $m \to \infty$. If m is large enough, Theorem 3.42 tells us that $A_m A^{-1} = e^X$ for some $X \in \mathfrak{g}$, so that $A = e^{-X} A_m$. Since $A_m \in G_0$, there is a path joining I to A_m in G. But we can then join A_m to $e^{-X} A_m = A$ by the path $e^{-tX} A_m$, $0 \leq t \leq 1$. By combining these two paths, we can join I to A in G, showing that A belongs to G_0.

Now, since $G_0 \subset G$, the Lie algebra of G_0 is contained in the Lie algebra of G. In the other direction, if e^{tX} lies in G for all t, then it actually lies in G_0, since any point $e^{t_0 X}$ on the curve e^{tX} can be connected to I in G, using the curve e^{tX} itself, for $0 \leq t \leq t_0$. □

3.9 Exercises

1. (a) If $\mathfrak{g}$ is a Lie algebra, show that the center of $\mathfrak{g}$ is an ideal in $\mathfrak{g}$.
   (b) If $\mathfrak{g}$ and $\mathfrak{h}$ are Lie algebras and $\phi : \mathfrak{g} \to \mathfrak{h}$ is a Lie algebra homomorphism, show that the kernel of ϕ is an ideal in $\mathfrak{g}$.
2. Classify up to isomorphism all one-dimensional and two-dimensional real Lie algebras. There is one isomorphism class of one-dimensional algebras and two isomorphism classes of two-dimensional algebras.
3. Let $\mathfrak{g}$ denote the space of $n \times n$ upper triangular matrices with zeros on the diagonal. Show that $\mathfrak{g}$ is a nilpotent Lie algebra under the bracket given by $[X, Y] = XY - YX$.
4. Give an example of a matrix Lie group G and a matrix X such that $e^X \in G$, but $X \notin \mathfrak{g}$.

5. If $G_1 \subset GL(n_1; \mathbb{C})$ and $G_2 \subset GL(n_2; \mathbb{C})$ are matrix Lie groups and $G_1 \times G_2$ is their direct product (regarded as a subgroup of $GL(n_1 + n_2; \mathbb{C})$ in the obvious way), show that the Lie algebra of $G_1 \times G_2$ is isomorphic to $\mathfrak{g}_1 \oplus \mathfrak{g}_2$.

6. Let G and H be matrix Lie groups with $H \subset G$, so that the Lie algebra $\mathfrak{h}$ of H is a subalgebra of the Lie algebra $\mathfrak{g}$ of G.

   (a) Show that if H is a normal subgroup of G, then $\mathfrak{h}$ is an ideal in $\mathfrak{g}$.
   (b) Show that if G and H are connected and $\mathfrak{h}$ is an ideal in $\mathfrak{g}$, then H is a normal subgroup of G.

7. Suppose $G \subset GL(n; \mathbb{C})$ is a matrix Lie group with Lie algebra $\mathfrak{g}$. Suppose that A is in G and that $\| A - I \| < 1$, so that the power series for $\log A$ is convergent. Is it necessarily the case that $\log A$ is in $\mathfrak{g}$? Prove or give a counterexample.

8. Show that two isomorphic matrix Lie groups have isomorphic Lie algebras.

9. Write out explicitly the general form of a 4×4 real matrix in $\mathsf{so}(3; 1)$ (see Proposition 3.25).

10. Show that there is an invertible linear map $\phi : \mathsf{su}(2) \to \mathbb{R}^3$ such that $\phi([X, Y]) = \phi(X) \times \phi(Y)$ for all $X, Y \in \mathsf{su}(2)$, where $\times$ denotes the cross product (vector product) on $\mathbb{R}^3$.

11. Show that $\mathsf{su}(2)$ and $\mathsf{sl}(2; \mathbb{R})$ are not isomorphic Lie algebras, even though $\mathsf{su}(2)_{\mathbb{C}} \cong \mathsf{sl}(2; \mathbb{R})_{\mathbb{C}} \cong \mathsf{sl}(2; \mathbb{C})$.

12. Show the groups $SU(2)$ and $SO(3)$ are not isomorphic, even though the associated Lie algebras $\mathsf{su}(2)$ and $\mathsf{so}(3)$ are isomorphic.

13. Let G be a matrix Lie group and let $\mathfrak{g}$ be its Lie algebra. For each $A \in G$, show that Ad_A is a Lie algebra automorphism of $\mathfrak{g}$.

14. Let X and Y be $n \times n$ matrices. Show by induction that

$$(\mathrm{ad}_X)^m (Y) = \sum_{k=0}^{m} \binom{m}{k} X^k Y (-X)^{m-k},$$

where

$$(\mathrm{ad}_X)^m (Y) = \underbrace{[X, \cdots [X, [X, Y]] \cdots]}_{m}.$$

Now, show by direct computation that

$$e^{\mathrm{ad}_X} (Y) = \mathrm{Ad}_{e^X}(Y) = e^X Y e^{-X}.$$

Assume it is legal to multiply power series term by term. (This result was obtained indirectly in Proposition 3.35.)

Hint: Use Pascal's triangle.

15. If G is a matrix Lie group, a **component** of G is the collection of all points that can be connected to a fixed $A \in G$ by a continuous path in G. Show that if G is compact, G has only finitely many components.

Hint: Suppose G had infinitely many components and consider a sequence A_j with each element of the sequence in a different component. Extract a convergent subsequence and $B_k = A_{j_k}$ and consider $B_k^{-1} B_{k+1}$.

16. Suppose that G is a connected, *commutative* matrix Lie group with Lie algebra $\mathfrak{g}$. Show that the exponential map for G maps $\mathfrak{g}$ onto G.

17. Suppose G is a connected matrix Lie group with Lie algebra $\mathfrak{g}$ and that A is an element of G. Show that A belongs to the center of G if and only if $\mathrm{Ad}_A(X) = X$ for all $X \in \mathfrak{g}$.

18. Show that the exponential map from the Lie algebra of the Heisenberg group to the Heisenberg group is one-to-one and onto.

19. Show that the exponential map from $\mathsf{u}(n)$ to $\mathsf{U}(n)$ is onto, but not one-to-one. *Hint*: Every unitary matrix has an orthonormal basis of eigenvectors.

20. Suppose X is a 2×2 real matrix with trace zero, and assume X has a nonreal eigenvalue.

    (a) Show that the eigenvalues of X must be of the form $ia, -ia$ with a a nonzero real number.

    (b) Show that the corresponding eigenvectors of X can be chosen to be complex conjugates of each other, say, v and $\bar{v}$.

    (c) Show that there exists an invertible *real* matrix C such that

    $$X = C \begin{pmatrix} 0 & a \\ -a & 0 \end{pmatrix} C^{-1}.$$

    Hint: Use v and $\bar{v}$ to construct a real basis for $\mathbb{R}^2$, and determine the matrix X in this basis.

21. Suppose A is a 2×2 real matrix with determinant one, and assume A has a nonreal eigenvalue. Show that there exists a real number θ that is not an integer multiple of π and an invertible real matrix C such that

    $$A = C \begin{pmatrix} \cos\theta & \sin\theta \\ -\sin\theta & \cos\theta \end{pmatrix} C^{-1}.$$

22. Show that the image of the exponential map for $\mathsf{SL}(2; \mathbb{R})$ consists of precisely those matrices $A \in \mathsf{SL}(2; \mathbb{R})$ such that trace $(A) > -2$, together with the matrix $-I$ (which has trace -2). To do this, consider the possibilities for the eigenvalues of a matrix in the Lie algebra $\mathsf{sl}(2; \mathbb{R})$ and in the group $\mathsf{SL}(2; \mathbb{R})$. In the Lie algebra, show that the eigenvalues are of the form $(\lambda, -\lambda)$ or $(i\lambda, -i\lambda)$, with λ real. In the group, show that the eigenvalues are of the form $(a, 1/a)$ or $(-a, -1/a)$, with a real and positive, or of the form $(e^{i\theta}, e^{-i\theta})$, with θ real. The case of a repeated eigenvalue $((0,0)$ in the Lie algebra and $(1, 1)$ or $(-1, -1)$ in the group) will have to be treated separately using the Jordan canonical form (Sect. A.4).

    Hint: You may assume that if a real matrix X has real eigenvalues, then X is similar *over the reals* to its Jordan canonical form. Then use the two previous exercises.

Chapter 4
Basic Representation Theory

4.1 Representations

If V is a finite-dimensional real or complex vector space, let $\mathsf{GL}(V)$ denote the group of invertible linear transformations of V. If we choose a basis for V, we can identify $\mathsf{GL}(V)$ with $\mathsf{GL}(n; \mathbb{R})$ or $\mathsf{GL}(n; \mathbb{C})$. Any such identification gives rise to a topology on $\mathsf{GL}(V)$, which is easily seen to be independent of the choice of basis. With this discussion in mind, we think of $\mathsf{GL}(V)$ as a matrix Lie group. Similarly, we let $\mathsf{gl}(V) = \mathrm{End}(V)$ denote the space of all linear operators from V to itself, which forms a Lie algebra under the bracket $[X, Y] = XY - YX$.

Definition 4.1. Let G be a matrix Lie group. A **finite-dimensional complex representation** of G is a Lie group homomorphism

$$\Pi : G \to \mathsf{GL}(V),$$

where V is a finite-dimensional complex vector space (with $\dim(V) \geq 1$). A **finite-dimensional real representation** of G is a Lie group homomorphism Π of G into $\mathsf{GL}(V)$, where V is a finite-dimensional real vector space.

If $\mathfrak{g}$ is a real or complex Lie algebra, then a **finite-dimensional complex representation** of $\mathfrak{g}$ is a Lie algebra homomorphism π of $\mathfrak{g}$ into $\mathsf{gl}(V)$, where V is a finite-dimensional complex vector space. If $\mathfrak{g}$ is a *real* Lie algebra, then a **finite-dimensional real representation** of $\mathfrak{g}$ is a Lie algebra homomorphism π of $\mathfrak{g}$ into $\mathsf{gl}(V)$, where V is a finite-dimensional real vector space.

A previous version of this book was inadvertently published without the middle initial of the author's name as "Brian Hall". For this reason an erratum has been published, correcting the mistake in the previous version and showing the correct name as Brian C. Hall (see DOI http://dx.doi.org/10.1007/978-3-319-13467-3_14). The version readers currently see is the corrected version. The Publisher would like to apologize for the earlier mistake.

© Springer International Publishing Switzerland 2015
B.C. Hall, *Lie Groups, Lie Algebras, and Representations*, Graduate
Texts in Mathematics 222, DOI 10.1007/978-3-319-13467-3_4

If Π or π is a one-to-one homomorphism, the representation is called **faithful**.

One should think of a representation as a **linear action** of a group or Lie algebra on a vector space (since, say, to every $g \in G$, there is associated an operator $\Pi(g)$, which acts on the vector space V). If the homomorphism $\Pi : G \to \mathsf{GL}(V)$ is fixed, we will occasionally use the alternative notation

$$g \cdot v \qquad\qquad (4.1)$$

in place $\Pi(g)v$. We will often use terminology such as "Let Π be a representation of G acting on the space V."

If a representation Π is a faithful representation of a matrix Lie group G, then $\{\Pi(A) \,|\, A \in G\}$ is a group of matrices that is isomorphic to the original group G. Thus, Π allows us to *represent* G as a group of matrices. This is the motivation for the term "representation." (Of course, we still call Π a representation even if it is not faithful.) Despite the origin of the term, the goal of representation theory is *not* simply to represent a group as a group of matrices. After all, the groups we study in this book are already matrix groups! Rather, the goal is to determine (up to isomorphism) *all* the ways a fixed group can act as a group of matrices.

Linear actions of groups on vector spaces arise naturally in many branches of both mathematics and physics. A typical example would be a linear differential equation in three-dimensional space which has rotational symmetry, such as the equations that describe the energy states of a hydrogen atom in quantum mechanics. Since the equation is rotationally invariant, the space of solutions is invariant under rotations and thus constitutes a representation of the rotation group $\mathsf{SO}(3)$. The representation theory of $\mathsf{SO}(3)$ (or of its Lie algebra) is very helpful in narrowing down what the space of solutions can be. See, for example, Chapter 18 in [Hall].

Definition 4.2. Let Π be a finite-dimensional real or complex representation of a matrix Lie group G, acting on a space V. A subspace W of V is called **invariant** if $\Pi(A)w \in W$ for all $w \in W$ and all $A \in G$. An invariant subspace W is called **nontrivial** if $W \neq \{0\}$ and $W \neq V$. A representation with no nontrivial invariant subspaces is called **irreducible**.

The terms **invariant**, **nontrivial**, and **irreducible** are defined analogously for representations of Lie algebras.

Even if $\mathfrak{g}$ is a real Lie algebra, we will consider mainly *complex* representations of $\mathfrak{g}$. It should be emphasized that if we are speaking about complex representations of a real Lie algebra $\mathfrak{g}$ acting on a complex subspace V, an invariant subspace W is required to be a complex subspace of V.

Definition 4.3. Let G be a matrix Lie group, let Π be a representation of G acting on the space V, and let Σ be a representation of G acting on the space W. A linear map $\phi : V \to W$ is called an **intertwining map** of representations if

$$\phi(\Pi(A)v) = \Sigma(A)\phi(v)$$

for all $A \in G$ and all $v \in V$. The analogous property defines intertwining maps of representations of a Lie algebra.

If ϕ is an intertwining map of representations and, in addition, ϕ is invertible, then ϕ is said to be an **isomorphism** of representations. If there exists an isomorphism between V and W, then the representations are said to be **isomorphic**.

If we use the "action" notation of (4.1), the defining property of an intertwining map may be written as

$$\phi(A \cdot v) = A \cdot \phi(v)$$

for all $A \in G$ and $v \in V$. That is to say, ϕ should commute with the action of G. A typical problem in representation theory is to determine, up to isomorphism, all of the irreducible representations of a particular group or Lie algebra. In Sect. 4.6, we will determine all the finite-dimensional complex irreducible representations of the Lie algebra $\mathsf{sl}(2; \mathbb{C})$.

After identifying $\mathsf{GL}(V)$ with $\mathsf{GL}(n; \mathbb{R})$ or $\mathsf{GL}(n; \mathbb{C})$, Theorem 3.28 has the following consequence.

Proposition 4.4. *Let G be a matrix Lie group with Lie algebra $\mathfrak{g}$ and let Π be a (finite-dimensional real or complex) representation of G, acting on the space V. Then there is a unique representation π of $\mathfrak{g}$ acting on the same space such that*

$$\Pi(e^X) = e^{\pi(X)}$$

for all $X \in \mathfrak{g}$. The representation π can be computed as

$$\pi(X) = \frac{d}{dt} \Pi(e^{tX}) \Big|_{t=0}$$

and satisfies

$$\pi(AXA^{-1}) = \Pi(A)\pi(X)\Pi(A)^{-1}$$

for all $X \in \mathfrak{g}$ and all $A \in G$.

Given a matrix Lie group G with Lie algebra $\mathfrak{g}$, we may ask whether *every* representation π of $\mathfrak{g}$ comes from a representation Π of G. As it turns out, this is not true in general, but is true if G is simply connected. See Sect. 4.7 for examples of this phenomenon and Sect. 5.7 for the general result.

Proposition 4.5.

1. *Let G be a connected matrix Lie group with Lie algebra $\mathfrak{g}$. Let Π be a representation of G and π the associated representation of $\mathfrak{g}$. Then Π is irreducible if and only if π is irreducible.*
2. *Let G be a connected matrix Lie group, let Π_1 and Π_2 be representations of G, and let π_1 and π_2 be the associated Lie algebra representations. Then π_1 and π_2 are isomorphic if and only if Π_1 and Π_2 are isomorphic.*

Proof. For Point 1, suppose first that Π is irreducible. We then want to show that π is irreducible. So, let W be a subspace of V that is invariant under $\pi(X)$ for all $X \in \mathfrak{g}$. We want to show that W is either $\{0\}$ or V. Now, suppose A is an element of G. Since G is assumed connected, Corollary 3.47 tells us that A can be written as $A = e^{X_1} \cdots e^{X_m}$ for some $X_1, \ldots, X_m$ in $\mathfrak{g}$. Since W is invariant under $\pi(X_j)$ it will also be invariant under $\exp(\pi(X_j)) = I + \pi(X_j) + \pi(X_j)^2/2 + \cdots$ and, hence, under

$$\Pi(A) = \Pi(e^{X_1} \cdots e^{X_m}) = \Pi(e^{X_1}) \cdots \Pi(e^{X_m})$$

$$= e^{\pi(X_1)} \cdots e^{\pi(X_m)}.$$

Since Π is irreducible and W is invariant under each $\Pi(A)$, W must be either $\{0\}$ or V. This shows that π is irreducible.

In the other direction, assume that π is irreducible and that W is an invariant subspace for Π. Then W is invariant under $\Pi(\exp tX)$ for all $X \in \mathfrak{g}$ and, hence, under

$$\pi(X) = \left. \frac{d}{dt} \Pi(e^{tX}) \right|_{t=0}.$$

Thus, since π is irreducible, W is $\{0\}$ or V, and we conclude that Π is irreducible. This establishes Point 1 of the proposition.

Point 2 of the proposition is similar and is left as an exercise to the reader (Exercise 1).   □

Proposition 4.6. *Let $\mathfrak{g}$ be a real Lie algebra and $\mathfrak{g}_{\mathbb{C}}$ its complexification. Then every finite-dimensional complex representation π of $\mathfrak{g}$ has a unique extension to a complex-linear representation of $\mathfrak{g}_{\mathbb{C}}$, also denoted π. Furthermore, π is irreducible as a representation of $\mathfrak{g}_{\mathbb{C}}$ if and only if it is irreducible as a representation of $\mathfrak{g}$.*

Of course, the extension of π to $\mathfrak{g}_{\mathbb{C}}$ is given by $\pi(X + iY) = \pi(X) + i\pi(Y)$ for all $X, Y \in \mathfrak{g}$.

Proof. The existence and uniqueness of the extension follow from Proposition 3.39. The claim about irreducibility holds because a complex subspace W of V is invariant under $\pi(X + iY)$, with X and Y in $\mathfrak{g}$, if and only if it is invariant under the operators $\pi(X)$ and $\pi(Y)$. Thus, the representation of $\mathfrak{g}$ and its extension to $\mathfrak{g}_{\mathbb{C}}$ have precisely the same invariant subspaces.   □

Definition 4.7. If V is a finite-dimensional inner product space and G is a matrix Lie group, a representation $\Pi : G \to \mathsf{GL}(V)$ is **unitary** if $\Pi(A)$ is a unitary operator on V for every $A \in G$.

Proposition 4.8. *Suppose G is a matrix Lie group with Lie algebra $\mathfrak{g}$. Suppose V is a finite-dimensional inner product space, Π is a representation of G acting on V, and π is the associated representation of $\mathfrak{g}$. If Π is unitary, then $\pi(X)$ is skew self-adjoint for all $X \in \mathfrak{g}$. Conversely, if G is connected and $\pi(X)$ is skew self-adjoint for all $X \in \mathfrak{g}$, then Π is unitary.*

In a slight abuse of notation, we will say that a representation π of a real Lie algebra $\mathfrak{g}$ acting on a finite-dimensional inner product space is **unitary** if $\pi(X)$ is skew self-adjoint for all $X \in \mathfrak{g}$.

Proof. The proof is similar to the computation of the Lie algebra of the unitary group $\mathsf{U}(n)$. If Π is unitary, then for all $X \in \mathfrak{g}$ we have

$$(e^{t\pi(X)})^* = \Pi(e^{tX})^* = \Pi(e^{tX})^{-1} = e^{-t\pi(X)}, \quad t \in \mathbb{R},$$

so that $e^{t\pi(X)^*} = e^{-t\pi(X)}$. Differentiating this relation with respect to t at $t = 0$ gives $\pi(X)^* = -\pi(X)$. In the other direction, if $\pi(X)^* = -\pi(X)$, then the above calculation shows that $\Pi(e^{tX}) = e^{t\pi(X)}$ is unitary. If G is connected, then by Corollary 3.47, every $A \in G$ is a product of exponentials, showing that $\Pi(A)$ is unitary. $\qquad\square$

4.2 Examples of Representations

A matrix Lie group G is, by definition, a subset of some $\mathsf{GL}(n; \mathbb{C})$. The inclusion map of G into $\mathsf{GL}(n; \mathbb{C})$ (i.e., the map $\Pi(A) = A$) is a representation of G, called the **standard representation** of G. If G happens to be contained in $\mathsf{GL}(n; \mathbb{R}) \subset \mathsf{GL}(n; \mathbb{C})$, then we can also think of the standard representation as a real representation. Thus, for example, the standard representation of $\mathsf{SO}(3)$ is the one in which $\mathsf{SO}(3)$ acts in the usual way on $\mathbb{R}^3$ and the standard representation of $\mathsf{SU}(2)$ is the one in which $\mathsf{SU}(2)$ acts on $\mathbb{C}^2$ in the usual way. Similarly, if $\mathfrak{g} \subset M_n(\mathbb{C})$ is a Lie algebra of matrices, the map $\pi(X) = X$ is called the **standard representation** of $\mathfrak{g}$.

Consider the one-dimensional complex vector space $\mathbb{C}$. For any matrix Lie group G, we can define the **trivial representation**, $\Pi : G \to \mathsf{GL}(1; \mathbb{C})$, by the formula

$$\Pi(A) = I$$

for all $A \in G$. Of course, this is an irreducible representation, since $\mathbb{C}$ has no nontrivial subspaces, let alone nontrivial invariant subspaces. If $\mathfrak{g}$ is a Lie algebra, we can also define the **trivial representation** of $\mathfrak{g}$, $\pi : \mathfrak{g} \to \mathsf{gl}(1; \mathbb{C})$, by

$$\pi(X) = 0$$

for all $X \in \mathfrak{g}$. This is an irreducible representation.

Recall the adjoint map of a group or Lie algebra, described in Definitions 3.32 and 3.7.

Definition 4.9. If G is a matrix Lie group with Lie algebra $\mathfrak{g}$, the **adjoint representation** of G is the map $\mathrm{Ad} : G \to \mathsf{GL}(\mathfrak{g})$ given by $A \mapsto \mathrm{Ad}_A$. Similarly, the **adjoint**

representation of a finite-dimensional Lie algebra $\mathfrak{g}$ is the map $\mathrm{ad} : \mathfrak{g} \to \mathfrak{gl}(\mathfrak{g})$ given by $X \mapsto \mathrm{ad}_X$.

If G is a matrix Lie group with Lie algebra $\mathfrak{g}$, then by Proposition 3.34, the Lie algebra representation associated to the adjoint representation of G is the adjoint representation of $\mathfrak{g}$. Note that in the case of $\mathsf{SO}(3)$, the standard representation and the adjoint representation are both three-dimensional real representations. In fact, these two representations are isomorphic (Exercise 2).

Example 4.10. Let V_m denote the space of homogeneous polynomials of degree m in two complex variables. For each $U \in \mathsf{SU}(2)$, define a linear transformation $\Pi_m(U)$ on the space V_m by the formula

$$[\Pi_m(U)f](z) = f(U^{-1}z), \quad z \in \mathbb{C}^2. \tag{4.2}$$

Then Π_m is a representation of $\mathsf{SU}(2)$.

Elements of V_m have the form

$$f(z_1, z_2) = a_0 z_1^m + a_1 z_1^{m-1} z_2 + a_2 z_1^{m-2} z_2^2 + \cdots + a_m z_2^m \tag{4.3}$$

with $z_1, z_2 \in \mathbb{C}$ and the a_j's arbitrary complex constants, from which we see that $\dim(V_m) = m + 1$. Explicitly, if f is as in (4.3), then

$$[\Pi_m(U)f](z_1, z_2) = \sum_{k=0}^{m} a_k (U_{11}^{-1} z_1 + U_{12}^{-1} z_2)^{m-k} (U_{21}^{-1} z_1 + U_{22}^{-1} z_2)^k.$$

By expanding out the right-hand side of this formula, we see that $\Pi_m(U)f$ is again a homogeneous polynomial of degree m. Thus, $\Pi_m(U)$ actually maps V_m into V_m.

To see that Π_m is actually a representation, compute that

$$\Pi_m(U_1)[\Pi_m(U_2)f](z) = [\Pi_m(U_2)f](U_1^{-1}z) = f(U_2^{-1}U_1^{-1}z)$$
$$= \Pi_m(U_1 U_2)f(z).$$

The inverse on the right-hand side of (4.2) is necessary in order to make Π_m a representation. We will see in Proposition 4.11 that each Π_m is irreducible and we will see in Sect. 4.6 that every finite-dimensional irreducible representation of $\mathsf{SU}(2)$ is isomorphic to one (and only one) of the Π_m's. (Of course, no two of the Π_m's are isomorphic, since they do not even have the same dimension.)

The associated representation π_m of $\mathsf{su}(2)$ can be computed as

$$(\pi_m(X)f)(z) = \frac{d}{dt} f(e^{-tX}z)\Big|_{t=0}.$$

Now, let $z(t) = (z_1(t), z_2(t))$ be the curve in $\mathbb{C}^2$ defined as $z(t) = e^{-tX}z$. By the chain rule, we have

$$\pi_m(X)f = \frac{\partial f}{\partial z_1}\frac{dz_1}{dt}\bigg|_{t=0} + \frac{\partial f}{\partial z_2}\frac{dz_2}{dt}\bigg|_{t=0}.$$

Since $dz/dt|_{t=0} = -Xz$, so we obtain

$$\pi_m(X)f = -\frac{\partial f}{\partial z_1}(X_{11}z_1 + X_{12}z_2) - \frac{\partial f}{\partial z_2}(X_{21}z_1 + X_{22}z_2). \tag{4.4}$$

We may then take the unique complex-linear extension of π to $\mathsf{sl}(2;\mathbb{C}) \cong \mathsf{su}(2)_\mathbb{C}$, as in Proposition 3.39. This extension is given by the same formula, but with $X \in \mathsf{sl}(2;\mathbb{C})$.

If X, Y, and H are the following basis elements for $\mathsf{sl}(2;\mathbb{C})$:

$$H = \begin{pmatrix} 1 & 0 \\ 0 & -1 \end{pmatrix}; \quad X = \begin{pmatrix} 0 & 1 \\ 0 & 0 \end{pmatrix}; \quad Y = \begin{pmatrix} 0 & 0 \\ 1 & 0 \end{pmatrix},$$

then applying formula (4.4) gives

$$\pi_m(H) = -z_1\frac{\partial}{\partial z_1} + z_2\frac{\partial}{\partial z_2}$$

$$\pi_m(X) = -z_2\frac{\partial}{\partial z_1}$$

$$\pi_m(Y) = -z_1\frac{\partial}{\partial z_2}.$$

Applying these operators to a basis element $z_1^{m-k}z_2^k$ for V_m gives

$$\pi_m(H)(z_1^{m-k}z_2^k) = (-m + 2k)z_1^{m-k}z_2^k$$

$$\pi_m(X)(z_1^{m-k}z_2^k) = -(m - k)\,z_1^{m-k-1}z_2^{k+1},$$

$$\pi_m(Y)(z_1^{m-k}z_2^k) = -kz_1^{m-k+1}z_2^{k-1}. \tag{4.5}$$

Thus, $z_1^{m-k}z_2^k$ is an eigenvector for $\pi_m(H)$ with eigenvalue $-m + 2k$, while $\pi_m(X)$ and $\pi_m(Y)$ have the effect of shifting the exponent k of z_2 up or down by one. Note that since $\pi_m(X)$ increases the value of k, this operator increases the eigenvalue of $\pi_m(H)$ by 2, whereas $\pi_m(Y)$ decreases the eigenvalue of $\pi_m(H)$ by 2.

Fig. 4.1 The *black dots* indicate the nonzero terms for a vector w in the space V_6. Applying $\pi_6(X)^4$ to w gives a nonzero multiple of z_2^6

Proposition 4.11. *For each $m \geq 0$, the representation π_m is irreducible.*

Proof. It suffices to show that every nonzero invariant subspace of V_m is equal to V_m. So, let W be such a space and let w be a nonzero element of W. Then w can be written in the form

$$w = a_0 z_1^m + a_1 z_1^{m-1} z_2 + a_2 z_1^{m-2} z_2^2 + \cdots + a_m z_2^m$$

with at least one of the a_k's being nonzero. Let k_0 be the smallest value of k for which $a_k \neq 0$ and consider

$$\pi_m(X)^{m-k_0} w.$$

(See Figure 4.1.)

Since each application of $\pi_m(X)$ raises the power of z_2 by 1, $\pi_m(X)^{m-k_0}$ will kill all the terms in w except the $a_{k_0} z_1^{m-k_0} z_2^{k_0}$ term. On the other hand, since $\pi_m(X)(z_1^{m-k} z_2^k)$ is zero only if $k = m$, we see that $\pi_m(X)^{m-k_0} w$ is a *nonzero* multiple of z_2^m. Since W is assumed invariant, W must contain this multiple of z_2^m and so also z_2^m itself. Now, for $0 \leq k \leq m$, it follows from (4.5) that $\pi_m(Y)^k z_2^m$ is a *nonzero* multiple of $z_1^k z_2^{m-k}$. Therefore, W must also contain $z_1^k z_2^{m-k}$ for all $0 \leq k \leq m$. Since these elements form a basis for V_m, we see that $W = V_m$, as desired.   □

4.3   New Representations from Old

One way of generating representations is to take some representations one knows and combine them in some fashion. In this section, we will consider three standard methods of obtaining new representations from old, namely direct sums of representations, tensor products of representations, and dual representations.

4.3.1   Direct Sums

Definition 4.12. Let G be a matrix Lie group and let $\Pi_1, \Pi_2, \ldots, \Pi_m$ be representations of G acting on vector spaces $V_1, V_2, \ldots, V_m$. Then the **direct sum** of $\Pi_1, \Pi_2, \ldots, \Pi_m$ is a representation $\Pi_1 \oplus \cdots \oplus \Pi_m$ of G acting on the space $V_1 \oplus \cdots \oplus V_m$, defined by

$$[\Pi_1 \oplus \cdots \oplus \Pi_m(A)] (v_1, \ldots, v_m) = (\Pi_1(A)v_1, \ldots, \Pi_m(A)v_m)$$

for all $A \in G$.

Similarly, if $\mathfrak{g}$ is a Lie algebra, and $\pi_1, \pi_2, \ldots, \pi_m$ are representations of $\mathfrak{g}$ acting on $V_1, V_2, \ldots, V_m$, then we define the **direct sum** of $\pi_1, \pi_2, \ldots, \pi_m$, acting on $V_1 \oplus \cdots \oplus V_m$ by

$$[\pi_1 \oplus \cdots \oplus \pi_m(X)] (v_1, \ldots, v_m) = (\pi_1(X)v_1, \ldots, \pi_m(X)v_m)$$

for all $X \in \mathfrak{g}$.

It is straightforward to check that, say, $\Pi_1 \oplus \cdots \oplus \Pi_m$ is really a representation of G.

4.3.2 Tensor Products

Let U and V be finite-dimensional real or complex vector spaces. We wish to define the **tensor product** of U and V, which will be a new vector space $U \otimes V$ "built" out of U and V. We will discuss the idea of this first and then give the precise definition.

We wish to consider a formal "product" of an element u of U with an element v of V, denoted $u \otimes v$. The *space* $U \otimes V$ is then the space of linear combinations of such products, that is, the space of elements of the form

$$a_1 u_1 \otimes v_1 + a_2 u_2 \otimes v_2 + \cdots + a_n u_n \otimes v_n. \tag{4.6}$$

Of course, if "$\otimes$" is to be interpreted as a product, then it should be bilinear:

$$(u_1 + a u_2) \otimes v = u_1 \otimes v + a u_2 \otimes v,$$
$$u \otimes (v_1 + a v_2) = u \otimes v_1 + a u \otimes v_2.$$

We do not assume that the product is commutative. That is to say, even if $U = V$ so that $U \otimes V$ and $V \otimes U$ are the same space, $u \otimes v$ will not, in general, equal $v \otimes u$.

Now, if $e_1, e_2, \ldots, e_n$ is a basis for U and $f_1, f_2, \ldots, f_m$ is a basis for V, then, using bilinearity, it is easy to see that any element of the form (4.6) can be written as a linear combination of the elements $e_j \otimes f_k$. In fact, it seems reasonable to expect that $\{e_j \otimes f_k \mid 1 \le j \le n, 1 \le k \le m\}$ should be a basis for the space $U \otimes V$. This, in fact, turns out to be the case.

Definition 4.13. If U and V are finite-dimensional real or complex vector spaces, then a **tensor product** of U with V is a vector space W, together with a bilinear map $\phi : U \times V \to W$ with the following property: If ψ is any bilinear map of $U \times V$ into a vector space X, there exists a unique linear map $\tilde{\psi}$ of W into X such that the following diagram commutes:

$$U \times V \xrightarrow{\phi} W$$
$$\psi \searrow \qquad \swarrow \tilde{\psi} \;.$$
$$X$$

Note that the *bilinear* map ψ from $U \times V$ into X turns into the *linear* map $\tilde{\psi}$ of W into X. This is one of the points of tensor products: Bilinear maps on $U \times V$ turn into linear maps on W.

Theorem 4.14. *If U and V are any finite-dimensional real or complex vector spaces, then a tensor product (W, ϕ) exists. Furthermore, (W, ϕ) is unique up to canonical isomorphism. That is, if (W_1, ϕ_1) and (W_2, ϕ_2) are two tensor products, then there exists a unique vector space isomorphism $\Phi : W_1 \to W_2$ such that the following diagram commutes:*

$$U \times V \xrightarrow{\phi_1} W_1$$
$$\phi_2 \searrow \qquad \swarrow \Phi \;.$$
$$W_2$$

Suppose that (W, ϕ) is a tensor product and that $e_1, e_2, \ldots, e_n$ is a basis for U and $f_1, f_2, \ldots, f_m$ is a basis for V. Then

$$\{\phi(e_j, f_k) \,|\, 1 \le j \le n, 1 \le k \le m\}$$

is a basis for W.

In particular, $\dim (U \otimes V) = (\dim U)(\dim V)$.

Proof. Exercise 7.   □

Notation 4.15 *Since the tensor product of U and V is essentially unique, we will let $U \otimes V$ denote an arbitrary tensor product space and we will write $u \otimes v$ instead of $\phi(u, v)$. In this notation, Theorem 4.14 says that*

$$\{e_j \otimes f_k \,|\, 1 \le j \le n, 1 \le k \le m\}$$

is a basis for $U \otimes V$, as expected.

The defining property of $U \otimes V$ is called the **universal property** of tensor products. To understand the significance of this property, suppose we want to define a linear map T from $U \otimes V$ into some other space. We could try to define T using bases for U and V, but then we would have to worry about whether T depends on the choice of basis. Instead, we could try to define T on elements of the form $u \otimes v$, with $u \in U$ and $v \in V$. While it follows from Theorem 4.14 that elements of this form span $U \otimes V$, the decomposition of an element of $U \otimes V$ as a linear combination of elements of the form $u \otimes v$ is far from unique. Thus, if we wish to define T on such elements, we have to worry whether T is well defined.

This is where the universal property comes in. If $\psi(u, v)$ is any bilinear expression in (u, v), the universal property says precisely that there is a unique linear map T $(= \tilde{\psi})$ such that

$$T(u \otimes v) = \psi(u, v).$$

Thus, we can construct a well-defined linear map T on $U \otimes V$ simply by defining it on elements of the form $u \otimes v$, provided that our definition of $T(u \otimes v)$ is bilinear in u and v. The following result is an application of this line of reasoning.

Proposition 4.16. *Let U and V be finite-dimensional real or complex vector spaces. Let $A : U \to U$ and $B : V \to V$ be linear operators. Then there exists a unique linear operator from $U \otimes V$ to $U \otimes V$, denoted $A \otimes B$, such that*

$$(A \otimes B)(u \otimes v) = (Au) \otimes (Bv)$$

for all $u \in U$ and $v \in V$. If A_1 and A_2 are linear operators on U and B_1 and B_2 are linear operators on V, then

$$(A_1 \otimes B_1)(A_2 \otimes B_2) = (A_1 A_2) \otimes (B_1 B_2). \tag{4.7}$$

Proof. Define a map ψ from $U \times V$ into $U \otimes V$ by

$$\psi(u, v) = (Au) \otimes (Bv).$$

Since A and B are linear and since $\otimes$ is bilinear, ψ is a bilinear map of $U \times V$ into $U \otimes V$. By the universal property, there is an associated linear map $\tilde{\psi} : U \otimes V \to U \otimes V$ such that

$$\tilde{\psi}(u \otimes v) = \psi(u, v) = (Au) \otimes (Bv).$$

Thus, $\tilde{\psi}$ is the desired map $A \otimes B$. An elementary calculation shows that the identity (4.7) holds on elements of the form $u \otimes v$. Since, by Theorem 4.14, such elements span $U \otimes V$, the identity holds in general. $\square$

We are now ready to define tensor products of representations. There are two different approaches to this, both of which are important. The first approach starts with a representation of a group G acting on a space U and a representation of another group H acting on a space V and produces a representation of the product group $G \times H$ acting on the space $U \otimes V$. The second approach starts with two different representations of the same group G, acting on spaces U and V, and produces a representation of G acting on $U \otimes V$. Both of these approaches can be adapted to apply to Lie algebras.

Definition 4.17. Let G and H be matrix Lie groups. Let Π_1 be a representation of G acting on a space U and let Π_2 be a representation of H acting on a space V.

Then the **tensor product** of Π_1 and Π_2 is a representation $\Pi_1 \otimes \Pi_2$ of $G \times H$ acting on $U \otimes V$ defined by

$$(\Pi_1 \otimes \Pi_2)(A, B) = \Pi_1(A) \otimes \Pi_2(B)$$

for all $A \in G$ and $B \in H$.

Using Proposition 4.16, it is easy to check that $\Pi_1 \otimes \Pi_2$ is, in fact, a representation of $G \times H$.

Now, if G and H are matrix Lie groups (i.e., G is a closed subgroup of $\mathsf{GL}(n; \mathbb{C})$ and H is a closed subgroup of $\mathsf{GL}(m; \mathbb{C})$), then $G \times H$ can be regarded in an obvious way as a closed subgroup of $\mathsf{GL}(n + m; \mathbb{C})$. Thus, the direct product of matrix Lie groups can be regarded as a matrix Lie group. It is easy to check that the Lie algebra of $G \times H$ is isomorphic to the direct sum of the Lie algebra of G and the Lie algebra of H.

Proposition 4.18. *Let G and H be matrix Lie groups with Lie algebras $\mathfrak{g}$ and $\mathfrak{h}$, respectively. Let Π_1 and Π_2 be representations of G and H, respectively, and consider the representation $\Pi_1 \otimes \Pi_2$ of $G \times H$. If $\pi_1 \otimes \pi_2$ denotes the associated representation of $\mathfrak{g} \oplus \mathfrak{h}$, then*

$$(\pi_1 \otimes \pi_2)(X, Y) = \pi_1(X) \otimes I + I \otimes \pi_2(Y)$$

for all $X \in \mathfrak{g}$ and $Y \in \mathfrak{h}$.

Proof. Suppose that $u(t)$ is a smooth curve in U and $v(t)$ is a smooth curve in V. Then, by repeating the proof of the product rule for scalar-valued functions (or by calculating everything in a basis), we have

$$\frac{d}{dt}(u(t) \otimes v(t)) = \frac{du}{dt} \otimes v(t) + u(t) \otimes \frac{dv}{dt}.$$

This being the case, we compute as follows:

$$(\pi_1 \otimes \pi_2)(X, Y)(u \otimes v)$$

$$= \left. \frac{d}{dt} \Pi_1(e^{tX})u \otimes \Pi_2(e^{tY})v \right|_{t=0}$$

$$= \left(\left. \frac{d}{dt} \Pi_1(e^{tX})u \right|_{t=0} \right) \otimes v + u \otimes \left(\left. \frac{d}{dt} \Pi_2(e^{tY})v \right|_{t=0} \right).$$

This establishes the claimed form of $(\pi_1 \otimes \pi_2)(X, Y)$ on elements of the form $u \otimes v$, which span $U \otimes V$. $\square$

The proposition motivates the following definition.

Definition 4.19. Let $\mathfrak{g}$ and $\mathfrak{h}$ be Lie algebras and let π_1 and π_2 be representations of $\mathfrak{g}$ and $\mathfrak{h}$, acting on spaces U and V. Then the **tensor product** of π_1 and π_2, denoted $\pi_1 \otimes \pi_2$, is a representation of $\mathfrak{g} \oplus \mathfrak{h}$ acting on $U \otimes V$, given by

$$(\pi_1 \otimes \pi_2)(X, Y) = \pi_1(X) \otimes I + I \otimes \pi_2(Y)$$

for all $X \in \mathfrak{g}$ and $Y \in \mathfrak{h}$.

It is easy to check that this indeed defines a representation of $\mathfrak{g} \oplus \mathfrak{h}$. Note that if we defined $(\pi_1 \otimes \pi_2)(X, Y) = \pi_1(X) \otimes \pi_2(Y)$, this would *not* be a representation of $\mathfrak{g} \oplus \mathfrak{h}$, since this is expression is not linear in (X, Y).

We now define a variant of the above definitions in which we take the tensor product of two representations of the same group G and regard the result as a representation of G rather than of $G \times G$.

Definition 4.20. Let G be a matrix Lie group and let Π_1 and Π_2 be representations of G, acting on spaces V_1 and V_2. Then the **tensor product** representation of G, acting on $V_1 \otimes V_2$, is defined by

$$(\Pi_1 \otimes \Pi_2)(A) = \Pi_1(A) \otimes \Pi_2(A)$$

for all $A \in G$. Similarly, if π_1 and π_2 are representations of a Lie algebra $\mathfrak{g}$, we define a tensor product representation of $\mathfrak{g}$ on $V_1 \otimes V_2$ by

$$(\pi_1 \otimes \pi_2)(X) = \pi_1(X) \otimes I + I \otimes \pi_2(X).$$

It is easy to check that $\Pi_1 \otimes \Pi_2$ and $\pi_1 \otimes \pi_2$ are actually representations of G and $\mathfrak{g}$, respectively. The notation is, unfortunately, ambiguous, since if Π_1 and Π_2 are representations of the same group G, we can regard $\Pi_1 \otimes \Pi_2$ either as a representation of G or as a representation of $G \times G$. We must, therefore, be careful to specify which way we are thinking about $\Pi_1 \otimes \Pi_2$.

If Π_1 and Π_2 are *irreducible* representations of a group G, then $\Pi_1 \otimes \Pi_2$ will typically not be irreducible when viewed as a representation of G. One can, then, attempt to decompose $\Pi_1 \otimes \Pi_2$ as a direct sum of irreducible representations. This process is called the **Clebsch–Gordan** theory or, in the physics literature, "addition of angular momentum." See Exercise 12 and Appendix C for more information about this topic.

4.3.3   Dual Representations

Suppose that π is a representation of a Lie algebra $\mathfrak{g}$ acting on a finite-dimensional vector space V. Let V^* denote the dual space of V, that is, the space of linear functionals on V (Sect. A.7). If A is a linear operator on V, let A^{tr} denote the dual or transpose operator on V^*, given by

$$(A^{tr}\phi)(v) = \phi(Av)$$

for $\phi \in V^*$, $v \in V$. If $v_1, \ldots, v_n$ is a basis for V, then there is a naturally associated "dual basis" $\phi_1, \ldots, \phi_n$ with the property that $\phi_j(v_k) = \delta_{jk}$. The matrix for A^{tr} in the dual basis is then simply the transpose (*not* the conjugate transpose!) of the matrix of A in the original basis. If A and B are linear operators on V, it is easily verified that

$$(AB)^{tr} = B^{tr} A^{tr}. \tag{4.8}$$

Definition 4.21. Suppose G is a matrix Lie group and Π is a representation of G acting on a finite-dimensional vector space V. Then the **dual representation** Π^* to Π is the representation of G acting on V^* and given by

$$\Pi^*(g) = [\Pi(g^{-1})]^{tr}. \tag{4.9}$$

If π is a representation of a Lie algebra $\mathfrak{g}$ acting on a finite-dimensional vector space V, then π^* is the representation of $\mathfrak{g}$ acting on V^* and given by

$$\pi^*(X) = -\pi(X)^{tr}. \tag{4.10}$$

Using (4.8), it is easy to check that both Π^* and π^* are actually representations. (Here the inverse on the right-hand side of (4.9) and the minus sign on the right-hand side of (4.10) are essential.) The dual representation is also called **contragredient representation**.

Proposition 4.22. *If Π is a representation of a matrix Lie group G, then (1) Π^* is irreducible if and only if Π is irreducible and (2) $(\Pi^*)^*$ is isomorphic to Π. Similar statements apply to Lie algebra representations.*

Proof. See Exercise 6.   □

4.4   Complete Reducibility

Much of representation theory is concerned with studying *irreducible* representations of a group or Lie algebra. In favorable cases, knowing the irreducible representations leads to a description of all representations.

Definition 4.23. A finite-dimensional representation of a group or Lie algebra is said to be **completely reducible** if it is isomorphic to a direct sum of a finite number of irreducible representations.

Definition 4.24. A group or Lie algebra is said to have the **complete reducibility property** if every finite-dimensional representation of it is completely reducible.

As it turns out, most groups and Lie algebras do not have the complete reducibility property. Nevertheless, many interesting example groups and Lie algebras do have this property, as we will see in this section and Sect. 10.3.

Example 4.25. Let $\Pi : \mathbb{R} \to \mathsf{GL}(2; \mathbb{C})$ be given by

$$\Pi(x) = \begin{pmatrix} 1 & x \\ 0 & 1 \end{pmatrix}.$$

Then Π is not completely reducible.

Proof. Direct calculation shows that Π is, in fact, a representation of $\mathbb{R}$. If $\{e_1, e_2\}$ is the standard basis for $\mathbb{C}^2$, then clearly the span of e_1 is an invariant subspace. We now claim that $\langle e_1 \rangle$ is the *only* nontrivial invariant subspace for Π. To see this, suppose V is a nonzero invariant subspace and suppose V contains a vector not in the span of e_1, say, $v = ae_1 + be_2$ with $b \neq 0$. Then

$$\Pi(1)v - v = be_1$$

also belongs to V. Thus, e_1 and $e_2 = (v - ae_1)/b$ belong to V, showing that $V = \mathbb{C}^2$. We conclude, then, that $\mathbb{C}^2$ does not decompose as a direct sum of irreducible invariant subspaces. □

Proposition 4.26. *If V is a completely reducible representation of a group or Lie algebra, then the following properties hold.*

1. *For every invariant subspace U of V, there is another invariant subspace W such that V is the direct sum of U and W.*
2. *Every invariant subspace of V is completely reducible.*

Proof. For Point 1, suppose that V decomposes as

$$V = U_1 \oplus U_2 \oplus \cdots \oplus U_k,$$

where the U_j's are irreducible invariant subspaces, and that U is any invariant subspace of V. If U is all of V, then we can take $W = \{0\}$ and we are done. If $U \neq V$, there must be some j_1 such that U_{j_1} is not contained in U. Since U_{j_1} is irreducible, it follows that the invariant subspace $U_{j_1} \cap U$ must be $\{0\}$. Suppose now that $U + U_{j_1} = V$. If so, the sum is direct (since $U_{j_1} \cap U = \{0\}$) and we are done.

If $U + U_{j_1} \neq V$, there is some j_2 such that $U + U_{j_1}$ does not contain U_{j_2}, in which case, $(U + U_{j_1}) \cap U_{j_2} = \{0\}$. Proceeding on in the same way, we must eventually obtain some family $j_1, j_2, \ldots, j_l$ such that $U + U_{j_1} + \cdots + U_{j_l} = V$ and the sum is direct. Then $W := U_{j_1} + \cdots + U_{j_l}$ is the desired complement to U.

For Point 2, suppose U is an invariant subspace of V. We first establish that U has the "invariant complement property" in Point 1. Suppose, then, that X is another invariant subspace of V with $X \subset U$. By Point 1, we can find invariant subspace Y

such that $V = X \oplus Y$. Let $Z = Y \cap U$, which is then an invariant subspace. We want to show that $U = X \oplus Z$. For all $u \in U$, we can write $u = x + y$ with $x \in X$ and $y \in Y$. But since $X \subset U$, we have $x \in U$ and therefore $y = u - x \in U$. Thus, $y \in Z = Y \cap U$. We have shown, then, that every $u \in U$ can be written as the sum of an element of X and an element of Z. Furthermore, $X \cap Z \subset X \cap Y = \{0\}$, so actually U is the *direct* sum of X and Z.

We may now easily show that U is completely reducible. If U is irreducible, we are done. If not, U has a nontrivial invariant subspace X and thus U decomposes as $U = X \oplus Z$ for some invariant subspace Z. If X and Z are irreducible, we are done, and if not, we proceed on in the same way. Since U is finite dimensional, this process must eventually terminate with U being decomposed as a direct sum of irreducibles.   □

Proposition 4.27. *If G is a matrix Lie group and Π is a finite-dimensional unitary representation of G, then Π is completely reducible. Similarly, if $\mathfrak{g}$ is a real Lie algebra and π is a finite-dimensional "unitary" representation of $\mathfrak{g}$ (meaning that $\pi(X)^* = -\pi(X)$ for all $X \in \mathfrak{g}$), then π is completely reducible.*

Proof. Let V denote the Hilbert space on which Π acts and let $\langle \cdot, \cdot \rangle$ denote the inner product on V. If $W \subset V$ is an invariant subspace, let $W^\perp$ be the orthogonal complement of W, so that V is the direct sum of W and $W^\perp$. We claim that $W^\perp$ is also an invariant subspace for Π or π.

To see this, note that since Π is unitary, $\Pi(A)^* = \Pi(A)^{-1} = \Pi(A^{-1})$ for all $A \in G$. Then, for any $w \in W$ and any $v \in W^\perp$, we have

$$\langle \Pi(A)v, w \rangle = \langle v, \Pi(A)^* w \rangle = \langle v, \Pi(A^{-1})w \rangle$$
$$= \langle v, w' \rangle = 0.$$

In the last step, we have used that $w' = \Pi(A^{-1})w$ is in W, since W is invariant. This shows that $\Pi(A)v$ is orthogonal to every element of W, as claimed. A similar argument, with $\Pi(A^{-1})$ replaced by $-\pi(X)$, shows that the orthogonal complement of an invariant subspace for π is also invariant.

We have established, then, that for a unitary representation, the orthogonal complement of an invariant subspace is again invariant. Suppose now that V is not irreducible. Then we can find an invariant subspace W that is neither $\{0\}$ nor V, and we decompose V as $W \oplus W^\perp$. Then W and $W^\perp$ are both invariant subspaces and thus unitary representations of G in their own right. Then W is either irreducible or it splits as an orthogonal direct sum of invariant subspaces, and similarly for $W^\perp$. We continue this process, and since V is finite dimensional, it cannot go on forever, and we eventually arrive at a decomposition of V as a direct sum of irreducible invariant subspaces.   □

Theorem 4.28. *If G is a compact matrix Lie group, every finite-dimensional representation of G is completely reducible.*

See also Sect. 10.3 for a similar result for semisimple Lie algebras. The argument below is sometimes called "Weyl's unitarian trick" for the role of unitarity in the proof. We require a notion of integration over matrix Lie groups that is invariant under the right action of the group. One way to construct such a right-invariant integral is to construct a right-invariant measure on G, known as a Haar measure. It is, however, simpler to introduce the integral by means of a right-invariant differential form on G. (See Appendix B for a quick introduction to the notion of differential forms.)

If $G \subset M_n(\mathbb{C})$ is a matrix Lie group, then the tangent space to G at the identity is the Lie algebra $\mathfrak{g}$ of G (Corollary 3.46). It is then easy to see that the tangent space $T_A G$ at any point $A \in G$ is the space of vectors of the form XA with $X \in \mathfrak{g}$. If the dimension of $\mathfrak{g}$ as a real vector space is k, choose a nonzero k-linear, alternating form $\alpha_I : \mathfrak{g}^k \to \mathbb{R}$. (Such a form exists and is unique up to multiplication by a constant.) Then we may define a k-linear, alternating form $\alpha_A : (T_A G)^k \to \mathbb{R}$ by setting

$$\alpha_A(Y_1, \ldots, Y_k) = \alpha_I(Y_1 A^{-1}, \ldots, Y_k A^{-1})$$

for all $Y_1, \ldots, Y_k \in T_A G$. That is to say, we define α in an arbitrary nonzero fashion at the identity, and we then use the right action of G to "transport" α to every other point in G. The resulting family of functionals is a k-form on G.

Once such an α has been constructed, we can use it to construct an orientation on G, by decreeing that an ordered basis $Y_1, \ldots, Y_k$ for $T_A G$ is positively oriented if $\alpha_A(Y_1, \ldots, Y_k) > 0$. If $f : G \to \mathbb{R}$ is a smooth function, we can integrate the k-form $f\alpha$ over nice domains in G. If G is compact, we may $f\alpha$ integrate over all of G, leading to a notion of integration, which we denote as

$$\int_G f(A)\alpha(A).$$

Since the orientation on G was defined in terms of the k-form α itself, it is not hard to see that if $f(A) > 0$ for all A, then $\int_G f\alpha > 0$. Furthermore, since the form α was constructed using the right action of G, it is easily seen to be invariant under that action. As a result, the notion of integration of a function over a compact group G is invariant under the right action of G: For all $B \in G$, we have

$$\int_G f(AB)\alpha(A) = \int_G f(A)\alpha(A).$$

Proof of Theorem 4.28. Choose an arbitrary inner product $\langle \cdot, \cdot \rangle$ on V, and then define a map $\langle \cdot, \cdot \rangle_G : V \times V \to \mathbb{C}$ by the formula

$$\langle v, w \rangle_G = \int_G \langle \Pi(A)v, \Pi(A)w \rangle \, \alpha(A).$$

It is easy to check that $\langle \cdot, \cdot \rangle_G$ is an inner product; in particular, the positivity of $\langle \cdot, \cdot \rangle_G$ holds because $\langle \Pi(A)v, \Pi(A)v \rangle > 0$ for all A if $v \neq 0$. We now compute that for each $B \in G$, we have

$$\langle \Pi(B)v, \Pi(B)w \rangle_G = \int_G \langle \Pi(A)\Pi(B)v, \Pi(A)\Pi(B)w \rangle \, \alpha(A)$$

$$= \int_G \langle \Pi(AB)v, \Pi(AB)w \rangle \, \alpha(A)$$

$$= \int_G \langle \Pi(A)v, \Pi(A)w \rangle \, \alpha(A)$$

$$= \langle v, w \rangle_G \, ,$$

where we have used the right invariance of the integral in the third equality. This computation shows that for each $B \in G$, the operator $\Pi(B)$ is unitary with respect to $\langle \cdot, \cdot \rangle_G$. Thus, by Proposition 4.27, Π is completely reducible.   □

Note that compactness of the group G is needed to ensure that the integral defining $\langle \cdot, \cdot \rangle_G$ is convergent.

4.5   Schur's Lemma

It is desirable to be able to state Schur's lemma simultaneously for groups and Lie algebras. In order to do so, we need to indulge in a common abuse of notation. If, say, Π is a representation of G acting on a space V, we will refer to V as the representation, without explicit reference to Π.

Theorem 4.29 (Schur's Lemma).

1. *Let V and W be irreducible real or complex representations of a group or Lie algebra and let $\phi : V \to W$ be an intertwining map. Then either $\phi = 0$ or ϕ is an isomorphism.*
2. *Let V be an irreducible complex representation of a group or Lie algebra and let $\phi : V \to V$ be an intertwining map of V with itself. Then $\phi = \lambda I$, for some $\lambda \in \mathbb{C}$.*
3. *Let V and W be irreducible complex representations of a group or Lie algebra and let $\phi_1, \phi_2 : V \to W$ be nonzero intertwining maps. Then $\phi_1 = \lambda \phi_2$, for some $\lambda \in \mathbb{C}$.*

It is important to note that the last two points in the theorem hold only over $\mathbb{C}$ (or some other algebraically closed field) and not over $\mathbb{R}$. See Exercise 8.

Before proving Schur's lemma, we obtain two corollaries of it.

Corollary 4.30. *Let Π be an irreducible complex representation of a matrix Lie group G. If A is in the center of G, then $\Pi(A) = \lambda I$, for some $\lambda \in \mathbb{C}$. Similarly, if π is an irreducible complex representation of a Lie algebra $\mathfrak{g}$ and if X is in the center of $\mathfrak{g}$, then $\pi(X) = \lambda I$.*

Proof. We prove the group case; the proof of the Lie algebra case is similar. If A is in the center of G, then for all $B \in G$,

$$\Pi(A)\Pi(B) = \Pi(AB) = \Pi(BA) = \Pi(B)\Pi(A).$$

However, this says exactly that $\Pi(A)$ is an intertwining map of the space with itself. Thus, by Point 2 of Schur's lemma, $\Pi(A)$ is a multiple of the identity. □

Corollary 4.31. *An irreducible complex representation of a commutative group or Lie algebra is one dimensional.*

Proof. Again, we prove only the group case. If G is commutative, the center of G is all of G, so by the previous corollary $\Pi(A)$ is a multiple of the identity for each $A \in G$. However, this means that *every* subspace of V is invariant! Thus, the only way that V can fail to have a nontrivial invariant subspace is if it is one dimensional. □

We now provide the proof of Schur's lemma.

Proof of Theorem 4.29. As usual, we will prove just the group case; the proof of the Lie algebra case requires only the obvious notational changes. For Point 1, if $v \in \ker(\phi)$, then

$$\phi(\Pi(A)v) = \Sigma(A)\phi(v) = 0.$$

This shows that $\ker \phi$ is an invariant subspace of V. Since V is irreducible, we must have $\ker \phi = 0$ or $\ker \phi = V$. Thus, ϕ is either one-to-one or zero.

Suppose ϕ is one-to-one. Then the image of ϕ is a nonzero subspace of W. On the other hand, the image of ϕ is invariant, for if $w \in W$ is of the form $\phi(v)$ for some $v \in V$, then

$$\Sigma(A)w = \Sigma(A)\phi(v) = \phi(\Pi(A)v).$$

Since W is irreducible and image(V) is nonzero and invariant, we must have image(V) = W. Thus, ϕ is either zero or one-to-one and onto.

For Point 2, suppose V is an irreducible *complex* representation and that $\phi :$ $V \to V$ is an intertwining map of V to itself, that is that $\phi\Pi(A) = \Pi(A)\phi$ for all $A \in G$. Since we are working over an algebraically closed field, ϕ must have at least one eigenvalue $\lambda \in \mathbb{C}$. If U denotes the corresponding eigenspace for ϕ, then Proposition A.2 tells us that each $\Pi(A)$ maps U to itself, meaning that U is an invariant subspace. Since λ is an eigenvalue, $U \neq 0$, and so we must have $U = V$, which means that $\phi = \lambda I$ on all of V.

For Point 3, if $\phi_2 \neq 0$, then by Point 1, ϕ_2 is an isomorphism. Then $\phi_1 \circ \phi_2^{-1}$ is an intertwining map of W with itself. Thus, by Point 2, $\phi_1 \circ \phi_2^{-1} = \lambda I$, whence $\phi_1 = \lambda \phi_2$. □

4.6   Representations of sl(2; ℂ)

In this section, we will compute (up to isomorphism) all of the finite-dimensional irreducible complex representations of the Lie algebra sl(2; ℂ). This computation is important for several reasons. First, sl(2; ℂ) is the complexification of su(2), which in turn is isomorphic to so(3), and the representations of so(3) are of physical significance. Indeed, the computation we will perform in the proof of Theorem 4.32 is found in every standard textbook on quantum mechanics, under the heading "angular momentum." Second, the representation theory of su(2) is an illuminating example of how one uses commutation relations to determine the representations of a Lie algebra. Third, in determining the representations of a semisimple Lie algebra $\mathfrak{g}$ (Chapters 6 and 7), we will make frequent use of the representation theory of sl(2; ℂ), applying it to various subalgebras of $\mathfrak{g}$ that are isomorphic to sl(2; ℂ).

We use the following basis for sl(2; ℂ):

$$X = \begin{pmatrix} 0 & 1 \\ 0 & 0 \end{pmatrix}; \quad Y = \begin{pmatrix} 0 & 0 \\ 1 & 0 \end{pmatrix}; \quad H = \begin{pmatrix} 1 & 0 \\ 0 & -1 \end{pmatrix},$$

which have the commutation relations

$$\begin{aligned} [H, X] &= 2X, \\ [H, Y] &= -2Y, \\ [X, Y] &= H. \end{aligned} \tag{4.11}$$

If V is a finite-dimensional complex vector space and A, B, and C are operators on V satisfying $[A, B] = 2B$, $[A, C] = -2C$, and $[B, C] = A$, then because of the skew symmetry and bilinearity of brackets, the unique linear map $\pi : \text{sl}(2; ℂ) \rightarrow \text{gl}(V)$ satisfying

$$\pi(H) = A, \quad \pi(X) = B, \quad \pi(Y) = C$$

will be a representation of sl(2; ℂ).

Theorem 4.32. *For each integer $m \geq 0$, there is an irreducible complex representation of* sl(2; ℂ) *with dimension $m+1$. Any two irreducible complex representations of* sl(2; ℂ) *with the same dimension are isomorphic. If π is an irreducible complex representation of* sl(2; ℂ) *with dimension $m + 1$, then π is isomorphic to the representation π_m described in Sect. 4.2.*

Our goal is to show that any finite-dimensional irreducible representation of sl(2; ℂ) "looks like" one of the representations π_m coming from Example 4.10. In that example, the space V_m is spanned by eigenvectors for $\pi_m(H)$ and the operators $\pi_m(X)$ and $\pi_m(Y)$ act by shifting the eigenvalues up or down in increments of 2. We now introduce a simple but crucial lemma that allows us to develop a similar structure in an arbitrary irreducible representation of sl(2; ℂ).

Lemma 4.33. *Let u be an eigenvector of $\pi(H)$ with eigenvalue $\alpha \in \mathbb{C}$. Then we have*

$$\pi(H)\pi(X)u = (\alpha + 2)\pi(X)u.$$

Thus, either $\pi(X)u = 0$ or $\pi(X)u$ is an eigenvector for $\pi(H)$ with eigenvalue $\alpha + 2$. Similarly,

$$\pi(H)\pi(Y)u = (\alpha - 2)\pi(Y)u,$$

so that either $\pi(Y)u = 0$ or $\pi(Y)u$ is an eigenvector for $\pi(H)$ with eigenvalue $\alpha - 2$.

Proof. We know that $[\pi(H), \pi(X)] = \pi([H, X]) = 2\pi(X)$. Thus,

$$\begin{aligned}
\pi(H)\pi(X)u &= \pi(X)\pi(H)u + 2\pi(X)u \\
&= \pi(X)(\alpha u) + 2\pi(X)u \\
&= (\alpha + 2)\pi(X)u.
\end{aligned}$$

The argument with $\pi(X)$ replaced by $\pi(Y)$ is similar.   □

Proof of Theorem 4.32. Let π be an irreducible representation of $sl(2; \mathbb{C})$ acting on a finite-dimensional complex vector space V. Our strategy is to diagonalize the operator $\pi(H)$. Since we are working over $\mathbb{C}$, the operator $\pi(H)$ must have at least one eigenvector. Let u be an eigenvector for $\pi(H)$ with eigenvalue α. Applying Lemma 4.33 repeatedly, we see that

$$\pi(H)\pi(X)^k u = (\alpha + 2k)\pi(X)^k u.$$

Since operator on a finite-dimensional space can have only finitely many eigenvalues, the $\pi(X)^k u$'s cannot all be nonzero. Thus, there is some $N \geq 0$ such that

$$\pi(X)^N u \neq 0$$

but

$$\pi(X)^{N+1} u = 0.$$

If we set $u_0 = \pi(X)^N u$ and $\lambda = \alpha + 2N$, then,

$$\pi(H)u_0 = \lambda u_0, \tag{4.12}$$

$$\pi(X)u_0 = 0. \tag{4.13}$$

Let us then define

$$u_k = \pi(Y)^k u_0$$

for $k \geq 0$. By Lemma 4.33, we have

$$\pi(H)u_k = (\lambda - 2k)\, u_k. \tag{4.14}$$

Now, it is easily verified by induction (Exercise 3) that

$$\pi(X)u_k = k[\lambda - (k-1)]u_{k-1} \quad (k \geq 1). \tag{4.15}$$

Furthermore, since $\pi(H)$ can have only finitely many eigenvalues, the u_k's cannot all be nonzero. There must, therefore, be a non-negative integer m such that

$$u_k = \pi(Y)^k u_0 \neq 0$$

for all $k \leq m$, but

$$u_{m+1} = \pi(Y)^{m+1} u_0 = 0.$$

If $u_{m+1} = 0$, then $\pi(X)u_{m+1} = 0$ and so, by (4.15),

$$0 = \pi(X)u_{m+1} = (m+1)(\lambda - m)u_m.$$

Since u_m and $m+1$ are nonzero, we must have $\lambda - m = 0$. Thus, λ must coincide with the non-negative integer m.

Thus, for every irreducible representation (π, V), there exists an integer $m \geq 0$ and nonzero vectors $u_0, \ldots, u_m$ such that

$$\pi(H)u_k = (m - 2k)u_k$$

$$\pi(Y)u_k = \begin{cases} u_{k+1} & \text{if } k < m \\ 0 & \text{if } k = m \end{cases}$$

$$\pi(X)u_k = \begin{cases} k(m - (k-1))u_{k-1} & \text{if } k > 0 \\ 0 & \text{if } k = 0 \end{cases}. \tag{4.16}$$

The vectors $u_0, \ldots, u_m$ must be linearly independent, since they are eigenvectors of $\pi(H)$ with distinct eigenvalues (Proposition A.1). Moreover, the $(m+1)$-dimensional span of $u_0, \ldots, u_m$ is explicitly invariant under $\pi(H)$, $\pi(X)$, and $\pi(Y)$ and, hence, under $\pi(Z)$ for all $Z \in \mathsf{sl}(2; \mathbb{C})$. Since π is irreducible, this space must be all of V.

We have shown that every irreducible representation of $\mathsf{sl}(2; \mathbb{C})$ is of the form (4.16). Conversely, if we *define* $\pi(H)$, $\pi(X)$, and $\pi(Y)$ by (4.16) (where the u_k's are basis elements for some $(m+1)$-dimensional vector space), it is not hard to check that operators defined as in (4.16) really do satisfy the $\mathsf{sl}(2; \mathbb{C})$ commutation relations (Exercise 4). Furthermore, we may prove irreducibility of this representation in the same way as in the proof of Proposition 4.11.

The preceding analysis shows that every irreducible representation of dimension $m + 1$ must have the form in (4.16), which shows that any two such representations are isomorphic. In particular, the $(m + 1)$-dimensional representation π_m described in Sect. 4.2 must be isomorphic to (4.16).

This completes the proof of Theorem 4.32. □

As mentioned earlier in this section, the representation theory of $\mathsf{sl}(2; \mathbb{C})$ plays a key role in the representation theory of other Lie algebras, such as $\mathsf{sl}(3; \mathbb{C})$, because these Lie algebras contain subalgebras isomorphic to $\mathsf{sl}(2; \mathbb{C})$. For such applications, we need a few results about finite-dimensional representations of $\mathsf{sl}(2; \mathbb{C})$ that are not necessarily irreducible.

Theorem 4.34. *If (π, V) is a finite-dimensional representation of $\mathsf{sl}(2; \mathbb{C})$, not necessarily irreducible, the following results hold.*

1. *Every eigenvalue of $\pi(H)$ is an integer. Furthermore, if v is an eigenvector for $\pi(H)$ with eigenvalue λ and $\pi(X)v = 0$, then λ is a non-negative integer.*
2. *The operators $\pi(X)$ and $\pi(Y)$ are nilpotent.*
3. *If we define $S : V \to V$ by*

$$S = e^{\pi(X)} e^{-\pi(Y)} e^{\pi(X)},$$

then S satisfies

$$S\pi(H)S^{-1} = -\pi(H).$$

4. *If an integer k is an eigenvalue for $\pi(H)$, so is each of the numbers*

$$-|k|, -|k| + 2, \ldots, |k| - 2, |k|.$$

Since $\mathsf{SU}(2)$ is simply connected, Theorem 5.6 will tell us that the representations of $\mathsf{sl}(2; \mathbb{C}) \cong \mathsf{su}(2)_{\mathbb{C}}$ are in one-to-one correspondence with the representations of $\mathsf{SU}(2)$. Since $\mathsf{SU}(2)$ is compact, Theorem 4.28 then tells us that every representation of $\mathsf{sl}(2; \mathbb{C})$ is completely reducible. If we decompose V as a direct sum of irreducibles, it is easy to prove the theorem for each summand separately. It is, however, preferable to give a proof of the theorem that does not rely on Theorem 5.6, which in turn relies on the Baker–Campbell–Hausdorff formula.

See also Exercise 13 for a different approach to the first part of Point 1, and Exercise 14 for a different approach to Point 3.

Proof. For Point 1, suppose v is an eigenvector of $\pi(H)$ with eigenvalue λ. Then there is some $N \geq 0$ such that $\pi(X)^N v \neq 0$ but $\pi(X)^{N+1} v = 0$, where $\pi(X)^N v$ is an eigenvector of $\pi(H)$ with eigenvalue $\lambda + 2N$. The proof of Theorem 4.32 shows that $m := \lambda + 2N$ must be a non-negative integer, so that λ is an integer. If $\pi(X)v = 0$ then we take $N = 0$ and $\lambda = m$ is non-negative.

For Point 2, it follows from the SN decomposition (Sect. A.3) that $\pi(H)$ has a basis of generalized eigenvectors, that is, vectors v for which $(\pi(H) - \lambda I)^k v = 0$ for some positive integer k. But, using the commutation relation $[H, X] = 2X$ and induction on k, we can see that

$$[\pi(H) - (\lambda + 2)I]^k \pi(X) = \pi(X)[\pi(H) - \lambda I]^k.$$

Thus, if v is a generalized eigenvector for $\pi(H)$ with eigenvalue λ, then $\pi(X)v$ is either zero or a generalized eigenvector with eigenvalue $\lambda + 2$. Applying $\pi(X)$ repeatedly to a generalized eigenvector for $\pi(H)$ must eventually give zero, since $\pi(H)$ can have only finitely many generalized eigenvalues. Thus, $\pi(X)$ is nilpotent. A similar argument applies to $\pi(Y)$.

For Point 3, we note that

$$S\pi(H)S^{-1} = e^{\pi(X)}e^{-\pi(Y)}e^{\pi(X)}\pi(H)e^{-\pi(X)}e^{\pi(Y)}e^{-\pi(X)}. \tag{4.17}$$

By Proposition 3.35, we have

$$e^{\pi(X)}\pi(H)e^{-\pi(X)} = \mathrm{Ad}_{e^{\pi(X)}}(\pi(H)) = e^{\mathrm{ad}_{\pi(X)}}(\pi(H))$$

and similarly for the remaining products in (4.17).

Now, $\mathrm{ad}_X(X) = 0$, $\mathrm{ad}_X(H) = -2X$ and $\mathrm{ad}_X(Y) = H$, and similarly with π applied to each Lie algebra element. Thus,

$$e^{\mathrm{ad}_{\pi(X)}}(\pi(H)) = \pi(H) - 2\pi(X).$$

Meanwhile, $\mathrm{ad}_Y(X) = -H$, $\mathrm{ad}_Y(H) = 2Y$, and $\mathrm{ad}_Y(Y) = 0$. Thus,

$$e^{-\mathrm{ad}_{\pi(Y)}}(\pi(H) - 2\pi(X))$$

$$= \pi(H) - 2\pi(X) - 2\pi(Y) - 2\pi(H) + \frac{1}{2}4\pi(Y)$$

$$= -\pi(H) - 2\pi(X).$$

Finally,

$$e^{\mathrm{ad}_{\pi(X)}}(-\pi(H) - 2\pi(X)) = -\pi(H) - 2\pi(X) + 2\pi(X)$$

$$= -\pi(H),$$

which establishes Point 3.

For Point 4, assume first that k is non-negative and let v be an eigenvector for $\pi(H)$ with eigenvalue k. Then as in Point 1, there is then another eigenvector v_0 for $\pi(H)$ with eigenvalue $m := k + 2N \geq k$ and such that $\pi(X)v_0 = 0$. Then by the proof of Theorem 4.32, we obtain a chain of eigenvectors $v_0, v_1, \ldots, v_m$ for

$\pi(H)$ with eigenvalues ranging from m to $-m$ in increments of 2. These eigenvalues include all of the numbers $k, k-2, \ldots, -k$. If k is negative and v is an eigenvector for $\pi(H)$ with eigenvalue k, then Sv is an eigenvector for $\pi(H)$ with eigenvalue $|k|$. Hence, by the preceding argument, each of the numbers from $|k|$ to $-|k|$ in increments of 2 is an eigenvalue.   □

4.7   Group Versus Lie Algebra Representations

We know from Chapter 3 (Theorem 3.28) that every Lie group homomorphism gives rise to a Lie algebra homomorphism. In particular, every representation of a matrix Lie group gives rise to a representation of the associated Lie algebra. In Chapter 5, we will prove a partial converse to this result: If G is a *simply connected* matrix Lie group with Lie algebra $\mathfrak{g}$, every representation of $\mathfrak{g}$ comes from a representation of G. (See Theorem 5.6). Thus, for a simply connected matrix Lie group G, there is a natural one-to-one correspondence between the representations of G and the representations of $\mathfrak{g}$.

It is instructive to see how this general theory works out in the case of $\mathsf{SU}(2)$ (which is simply connected) and $\mathsf{SO}(3)$ (which is not). For every irreducible representation π of $\mathsf{su}(2)$, the complex-linear extension of π to $\mathsf{sl}(2; \mathbb{C})$ must be isomorphic (Theorem 4.32) to one of the representations π_m described in Sect. 4.2. Since those representations are constructed from representations of the group $\mathsf{SU}(2)$, we can see directly (without appealing to Theorem 5.6) that every irreducible representation of $\mathsf{su}(2)$ comes from a representation of $\mathsf{SU}(2)$.

Now, by Example 3.27, there is a Lie algebra isomorphism $\phi : \mathsf{su}(2) \to \mathsf{so}(3)$ such that $\phi(E_j) = F_j$, $j = 1, 2, 3$, where $\{E_1, E_2, E_3\}$ and $\{F_1, F_2, F_3\}$ are the bases listed in the example. Thus, the irreducible representations of $\mathsf{so}(3)$ are precisely of the form $\sigma_m = \pi_m \circ \phi^{-1}$. We wish to determine, for a particular m, whether or not there is a representation Σ_m of the group $\mathsf{SO}(3)$ such that $\Sigma_m(\exp X) = \exp(\sigma_m(X))$ for all X in $\mathsf{so}(3)$.

Proposition 4.35. *Let $\sigma_m = \pi_m \circ \phi^{-1}$ be an irreducible complex representation of the Lie algebra $\mathsf{so}(3)$ ($m \geq 0$). If m is even, there is a representation Σ_m of the group $\mathsf{SO}(3)$ such that $\Sigma_m(e^X) = e^{\sigma_m(X)}$ for all X in $\mathsf{so}(3)$. If m is odd, there is no such representation of $\mathsf{SO}(3)$.*

Note that the condition that m be even is equivalent to the condition that $\dim V_m = m + 1$ be odd. Thus, it is the odd-dimensional representations of the Lie algebra $\mathsf{so}(3)$ which come from group representations. In the physics literature, the representations of $\mathsf{su}(2) \cong \mathsf{so}(3)$ are labeled by the parameter $l = m/2$. In terms of this notation, a representation of $\mathsf{so}(3)$ comes from a representation of $\mathsf{SO}(3)$ if and only if l is an integer. The representations with l an integer are called "integer spin"; the others are called "half-integer spin."

For any m, one could attempt to construct Σ_m by the construction in the proof of Theorem 5.6. The construction is based on defining $\Sigma_m(A)$ along a path joining

I to *A* and then proving that the value of $\Sigma_m(A)$ is independent of the choice of path. The construction of Σ_m along a path goes through without change. Since $\mathsf{SO}(3)$ is not simply connected, however, two paths in $\mathsf{SO}(3)$ with the same endpoint are not necessarily homotopic with endpoints fixed and the proof of independence of the path breaks down. One can join the identity to itself, for example, either by the constant path or by the path consisting of rotations by angle $2\pi t$ in the (y, z)-plane, $0 \le t \le 1$. If one defines Σ_m along the constant path, one gets the value $\Sigma_m(I) = I$. If *m* is odd, however, and one defines Σ_m along the path of rotations in the (y, z)-plane, then one gets the value $\Sigma_m(I) = -I$, as the calculations in the proof of Proposition 4.35 will show. This strongly suggests (and Proposition 4.35 confirms) that when *m* is odd, there is no way to define Σ_m as a "single-valued" representation of $\mathsf{SO}(3)$.

An electron, for example, is a "spin one-half" particle, which means that it is described in quantum mechanics in a way that involves the two-dimensional representation σ_1 of $\mathsf{so}(3)$. In the physics literature, one finds statements to the effect that applying a 360° rotation to the wave function of the electron gives back the negative of the original wave function. This statement reflects that if one attempts to construct the nonexistent representation Σ_1 of $\mathsf{SO}(3)$, then when defining Σ_1 along a path of rotations in the (x, y)-plane, one gets that $\Sigma_1(I) = -I$.

Proof. Suppose, first, that *m* is odd and suppose that such a Σ_m existed. Computing as in Sect. 2.2, we see that

$$e^{2\pi F_1} = \begin{pmatrix} 1 & 0 & 0 \\ 0 & \cos 2\pi & -\sin 2\pi \\ 0 & \sin 2\pi & \cos 2\pi \end{pmatrix} = I.$$

Meanwhile, $\sigma_m(F_1) = \pi_m(\phi^{-1}(F_1)) = \pi_m(E_1)$, with $E_1 = iH/2$, where, as usual, *H* is the diagonal matrix with diagonal entries $(1, -1)$. We know that there is a basis $u_0, u_1, \ldots, u_m$ for V_m such that u_j is an eigenvector for $\pi_m(H)$ with eigenvalue $m - 2j$. This means that u_j is also an eigenvector for $\sigma_m(F_1) = i\pi_m(H)/2$, with eigenvalue $i(m - 2j)/2$. Thus, in the basis $\{u_j\}$, we have

$$\sigma_m(F_1) = \begin{pmatrix} \frac{i}{2}m & & & \\ & \frac{i}{2}(m-2) & & \\ & & \ddots & \\ & & & \frac{i}{2}(-m) \end{pmatrix}.$$

Since we are assuming *m* is odd, $m - 2j$ is an odd integer. Thus, $e^{2\pi\sigma_m(F_1)}$ has eigenvalues $e^{2\pi i(m-2j)/2} = -1$ in the basis $\{u_j\}$, showing that $e^{2\pi\sigma_m(F_1)} = -I$. Thus, on the one hand,

$$\Sigma_m(e^{2\pi F_1}) = \Sigma_m(I) = I,$$

whereas, on the other hand,

$$\Sigma_m(e^{2\pi F_1}) = e^{2\pi\sigma_m(F_1)} = -I.$$

Thus, there can be no such group representation Σ_m.

Suppose now that m is even. Recall that the Lie algebra isomorphism ϕ comes from the surjective group homomorphism Φ in Proposition 1.19, where $\ker(\Phi) = \{I, -I\}$. Let Π_m be the representation of $SU(2)$ in Example 4.10. Now, $e^{2\pi E_1} = -I$, and, thus,

$$\Pi_m(-I) = \Pi_m(e^{2\pi E_1}) = e^{\pi_m(2\pi E_1)}.$$

If, however m is even, then $e^{\pi_m(2\pi E_1)}$ is diagonal in the basis $\{u_j\}$ with eigenvalues $e^{2\pi i (m-2j)/2} = 1$, showing that $\Pi_m(-I) = e^{\pi_m(2\pi E_1)} = I$.

Now, for each $R \in SO(3)$, there is a unique pair of elements $\{U, -U\}$ such that $\Phi(U) = \Phi(-U) = R$. Since $\Pi_m(-I) = I$, we see that $\Pi_m(U) = \Pi_m(-U)$. It thus makes sense to define

$$\Sigma_m(R) = \Pi_m(U).$$

It is easy to see that Σ_m is a homomorphism. To see that Σ_m is continuous, the reader may first verify that if a homomorphism between matrix Lie groups is continuous at the identity, it is continuous everywhere. Thus, it suffices to observe that for R near the identity, U may be chosen to depend continuously on R, by setting $U = \exp(\phi^{-1}(\log R))$. Now, by construction, we have $\Pi_m = \Sigma_m \circ \Phi$. Thus, the Lie algebra representation σ_m associated to Σ_m satisfies $\pi_m = \sigma_m \circ \phi$ or $\sigma_m = \pi_m \circ \phi^{-1}$, showing that Σ_m is the desired representation of $SO(3)$. $\square$

4.8   A Nonmatrix Lie Group

In this section, we will show that the Lie group introduced in Sect. 1.5 is not isomorphic to a matrix Lie group. (The universal cover of $SL(2;\mathbb{R})$ is another example of a Lie group that is not a matrix Lie group; see Sect. 5.8.) The group in question is $G = \mathbb{R} \times \mathbb{R} \times S^1$, with the group product defined by

$$(x_1, y_1, u_1) \cdot (x_2, y_2, u_2) = (x_1 + x_2, y_1 + y_2, e^{ix_1 y_2} u_1 u_2).$$

Meanwhile, let H be the Heisenberg group and consider the map $\Phi : H \to G$ given by

$$\Phi\begin{pmatrix} 1 & a & b \\ 0 & 1 & c \\ 0 & 0 & 1 \end{pmatrix} = (a, c, e^{2\pi ib}).$$

Direct computation shows that Φ is a homomorphism. The kernel of Φ is the discrete normal subgroup N of H given by

$$N = \left\{ \begin{pmatrix} 1 & 0 & n \\ 0 & 1 & 0 \\ 0 & 0 & 1 \end{pmatrix} \middle| n \in \mathbb{Z} \right\}.$$

Now, suppose that Σ is any finite-dimensional representation of G. Then we can define an associated representation Π of H by $\Pi = \Sigma \circ \Phi$. Clearly, the kernel of any such representation of H must include the kernel N of Φ. Now, let $Z(H)$ denote the center of H, which is easily shown to be

$$Z(H) = \left\{ \begin{pmatrix} 1 & 0 & b \\ 0 & 1 & 0 \\ 0 & 0 & 1 \end{pmatrix} \middle| b \in \mathbb{R} \right\}.$$

Theorem 4.36. *Let Π be any finite-dimensional representation of H. If $\ker \Pi \supset N$, then $\ker \Pi \supset Z(H)$.*

Once this is established, we will be able to conclude that there are no faithful finite-dimensional representations of G. After all, if Σ is any finite-dimensional representation of G, then the kernel of $\Pi = \Sigma \circ \Phi$ will contain N and, thus, $Z(H)$, by the theorem. Thus, for all $b \in \mathbb{R}$,

$$\Pi \begin{pmatrix} 1 & 0 & b \\ 0 & 1 & 0 \\ 0 & 0 & 1 \end{pmatrix} = \Sigma(0, 0, e^{2\pi i b}) = I.$$

This means that the kernel of Σ contains all elements of the form $(0, 0, u)$ and Σ is not faithful. Thus, we obtain the following result.

Corollary 4.37. *The Lie group G has no faithful finite-dimensional representations. In particular, G is not isomorphic to any matrix Lie group.*

We now begin the proof of Theorem 4.36.

Lemma 4.38. *If X is a nilpotent matrix and $e^{tX} = I$ for some nonzero t, then $X = 0$.*

Proof. Since X is nilpotent, the power series for e^{tX} terminates after a finite number of terms. Thus, each entry of e^{tX} depends polynomially on t; that is, there exist polynomials $p_{jk}(t)$ such that $(e^{tX})_{jk} = p_{jk}(t)$. If $e^{tX} = I$ for some nonzero t, then $e^{ntX} = I$ for all $n \in \mathbb{Z}$, showing that $p_{jk}(nt) = \delta_{jk}$ for all n. However, a polynomial that takes on a certain value infinitely many times must be constant. Thus, actually, $e^{tX} = I$ for all t, from which it follows that $X = 0$. $\square$

Proof of Theorem 4.36. Let π be the associated representation of the Lie algebra $\mathfrak{h}$ of H. Let $\{X, Y, Z\}$ be the following basis for $\mathfrak{h}$:

$$X = \begin{pmatrix} 0 & 1 & 0 \\ 0 & 0 & 0 \\ 0 & 0 & 0 \end{pmatrix}, \quad Y = \begin{pmatrix} 0 & 0 & 0 \\ 0 & 0 & 1 \\ 0 & 0 & 0 \end{pmatrix}, \quad Z = \begin{pmatrix} 0 & 0 & 1 \\ 0 & 0 & 0 \\ 0 & 0 & 0 \end{pmatrix}. \tag{4.18}$$

These satisfy the commutation relations $[X, Y] = Z$ and $[X, Z] = [Y, Z] = 0$.

We now claim that $\pi(Z)$ must be nilpotent, or, equivalently, that all of the eigenvalues of $\pi(Z)$ are zero. Let λ be an eigenvalue for $\pi(Z)$ and let V_λ be the associated eigenspace. Then V_λ is certainly invariant under $\pi(Z)$. Furthermore, since $\pi(X)$ and $\pi(Y)$ commute with $\pi(Z)$, Proposition A.2 tells us that V_λ is invariant under $\pi(X)$ and $\pi(Y)$. Thus, the restriction of $\pi(Z)$ to V_λ—namely, λI—is the commutator of the restrictions to V_λ of $\pi(X)$ and $\pi(Y)$. Since the trace of a commutator is zero, we have $0 = \lambda \dim(V_\lambda)$ and λ must be zero.

Now, direct calculation shows that e^{nZ} belongs to N for all integers n. Thus, if Π is a representation of H and $\ker \Pi \supset N$, we have $\Pi(e^{nZ}) = I$ for all n. Since $\pi(Z)$ is nilpotent, Lemma 4.38 tells us that $\pi(Z)$ is zero and thus that $\Pi(e^{tZ}) = e^{t\pi(Z)} = I$ for all $t \in \mathbb{R}$. Since every element of $Z(H)$ is of the form e^{tZ} for some t, we have the desired conclusion.   □

4.9 Exercises

1. Prove Point 2 of Proposition 4.5.
2. Show that the adjoint representation and the standard representation are isomorphic representations of the Lie algebra $\mathsf{so}(3)$. Show that the adjoint and standard representations of the group $\mathsf{SO}(3)$ are isomorphic.
3. Using the commutation relation $[\pi(X), \pi(Y)] = \pi(H)$ and induction on k, verify the relation (4.15).
4. Define a vector space with basis $u_0, u_1, \ldots, u_m$. Now, define operators $\pi(H)$, $\pi(X)$, and $\pi(Y)$ by formula (4.16). Verify by direct computation that the operators defined by (4.16) satisfy the commutation relations (4.11) for $\mathsf{sl}(2; \mathbb{C})$.
   Hint: When dealing with $\pi(Y)$, treat the case of u_k, $k < m$, separately from the case of u_m, and similarly for $\pi(X)$.
5. Consider the standard representation of the Heisenberg group, acting on $\mathbb{C}^3$. Determine all subspaces of $\mathbb{C}^3$ which are invariant under the action of the Heisenberg group. Is this representation completely reducible?
6. Prove Proposition 4.22.
   Hint: There is a one-to-one correspondence between subspaces of V and subspaces of V^* as follows: For any subspace W of V, the *annihilator* of W is the subspace of all ϕ in V^* such that ϕ is zero on W. See Sect. A.7.
7. Prove Theorem 4.14.

Hints: For existence, choose bases $\{e_j\}$ and $\{f_k\}$ for U and V. Then define a space W which has as a basis $\{w_{jk} \mid 1 \leq j \leq n, 1 \leq k \leq m\}$. Define $\phi(e_j, f_k) = w_{jk}$ and extend by bilinearity. For uniqueness, use the universal property.

8. Let $SO(2)$ act on $\mathbb{R}^2$ in the obvious way. Show that $\mathbb{R}^2$ is an irreducible real representation under this action, but that Point 2 of Schur's lemma (Theorem 4.29) fails.

9. Suppose V is a finite-dimensional representation of a group or Lie algebra and that W is a nonzero invariant subspace of V. Show that there exists a nonzero *irreducible* invariant subspace for V that is contained in W.

10. Suppose that V_1 and V_2 are *nonisomorphic* irreducible representations of a group or Lie algebra, and consider the associated representation $V_1 \oplus V_2$. Regard V_1 and V_2 as subspaces of $V_1 \oplus V_2$ in the obvious way. Following the outline below, show that V_1 and V_2 are the only nontrivial invariant subspaces of $V_1 \oplus V_2$.

    (a) First assume that U is a nontrivial *irreducible* invariant subspace. Let $P_1 : V_1 \oplus V_2 \to V_1$ be the projection onto the first factor and let P_2 be the projection onto the second factor. Show that P_1 and P_2 are intertwining maps. Show that $U = V_1$ or $U = V_2$.

    (b) Using Exercise 9, show that V_1 and V_2 are the only nontrivial invariant subspaces of $V_1 \oplus V_2$.

11. Suppose that V is an irreducible finite-dimensional representation of a group or Lie algebra over $\mathbb{C}$, and consider the associated representation $V \oplus V$. Show that every nontrivial invariant subspace U of $V \oplus V$ is isomorphic to V and is of the form

$$U = \{(\lambda_1 v, \lambda_2 v) \mid v \in V\}$$

for some constants λ_1 and λ_2, not both zero.

12. Recall the spaces V_m introduced in Sect. 4.2, viewed as representations of the Lie algebra $\mathsf{sl}(2; \mathbb{C})$. In particular, consider the space V_1 (which has dimension 2).

    (a) Regard $V_1 \otimes V_1$ as a representation of $\mathsf{sl}(2; \mathbb{C})$, as in Definition 4.20. Show that this representation is not irreducible.

    (b) Now, view $V_1 \otimes V_1$ as a representation of $\mathsf{sl}(2; \mathbb{C}) \oplus \mathsf{sl}(2; \mathbb{C})$, as in Definition 4.19. Show that this representation is irreducible.

13. Let (Π, V) be a finite-dimensional representation of $SU(2)$ with associated representation π of $\mathsf{su}(2)$, which extends by complex linearity to $\mathsf{sl}(2; \mathbb{C})$. (Since $SU(2)$ is simply connected, Theorem 5.6 will show that every representation π of $\mathsf{sl}(2; \mathbb{C})$ arise in this way.) If H is the diagonal matrix with diagonal entries $(1, -1)$, show that $e^{2\pi i H} = I$ and use this to prove (independently of Theorem 4.34) that every eigenvalue of $\pi(H)$ is an integer.

14. Let (Π, V) be a finite-dimensional representation of $\mathsf{SU}(2)$ with associated representation π of $\mathsf{su}(2)$, which extends by complex linearity to $\mathsf{sl}(2; \mathbb{C})$. If X, Y, and H are the usual basis element for $\mathsf{sl}(2; \mathbb{C})$, compute $e^X e^{-Y} e^X$ and show that

$$e^X e^{-Y} e^X H (e^X e^{-Y} e^X)^{-1} = -H.$$

Use this result to give a different proof of Point 3 of Theorem 4.34.

Chapter 5
The Baker–Campbell–Hausdorff Formula and Its Consequences

5.1   The "Hard" Questions

Consider three elementary results from the preceding chapters of this book: (1) Every matrix Lie group G has a Lie algebra $\mathfrak{g}$. (2) A continuous homomorphism Φ between matrix Lie groups G and H gives rise to a Lie algebra homomorphism $\phi : \mathfrak{g} \to \mathfrak{h}$. (3) If G and H are matrix Lie groups and H is a subgroup of G, then the Lie algebra $\mathfrak{h}$ of H is a subalgebra of the Lie algebra $\mathfrak{g}$ of G. Each of these results goes in the "easy" direction, from a group notion to an associated Lie algebra notion. In this chapter, we attempt to go in the "hard" direction, from the Lie algebra to the Lie group. We will investigate three questions relating to the preceding three theorems.

- **Question 1**: Is every finite-dimensional, real Lie algebra the Lie algebra of some matrix Lie group?
- **Question 2**: Let G and H be matrix Lie groups with Lie algebras $\mathfrak{g}$ and $\mathfrak{h}$, respectively, and let $\phi : \mathfrak{g} \to \mathfrak{h}$ be a Lie algebra homomorphism. Does there exists a Lie group homomorphism $\Phi : G \to H$ such that $\Phi(e^X) = e^{\phi(X)}$ for all $X \in \mathfrak{g}$?
- **Question 3**: If G is a matrix Lie group with Lie algebra $\mathfrak{g}$ and $\mathfrak{h}$ is a subalgebra of $\mathfrak{g}$, is there a matrix Lie group $H \subset G$ whose Lie algebra is $\mathfrak{h}$?

The answer to Question 1 is yes; see Sect. 5.10. The answer to Question 2 is, in general, no, but is yes if G is simply connected; see Sect. 5.7. The answer to

---

A previous version of this book was inadvertently published without the middle initial of the author's name as "Brian Hall". For this reason an erratum has been published, correcting the mistake in the previous version and showing the correct name as Brian C. Hall (see DOI http://dx.doi.org/10.1007/978-3-319-13467-3_14). The version readers currently see is the corrected version. The Publisher would like to apologize for the earlier mistake.

© Springer International Publishing Switzerland 2015
B.C. Hall, *Lie Groups, Lie Algebras, and Representations*, Graduate
Texts in Mathematics 222, DOI 10.1007/978-3-319-13467-3_5

Question 3 is no, in general, but is yes if we allow H to be a "connected Lie subgroup" that is not necessarily closed; see Sect. 5.9.

Our tool for investigating these questions is the Baker–Campbell–Hausdorff formula, which expresses $\log(e^X e^Y)$, where X and Y are sufficiently small $n \times n$ matrices, in Lie-algebraic terms, that is, in terms of iterated commutators involving X and Y. The formula implies that all information about the product operation on a matrix Lie group, at least near the identity, is encoded in the Lie algebra. In the case of Questions 2 and 3 in the preceding paragraph, we will give a complete proof of the theorem that answers the question. In the case of Question 1, we will need to assume Ado's theorem, which asserts that every finite-dimensional real Lie algebra is isomorphic to an algebra of matrices.

5.2   An Illustrative Example

In this section, we prove one of the main theorems alluded to above (the answer to Question 2), in the case of the Heisenberg group. We introduce the problem in a general setting and then specialize to the Heisenberg case. Suppose G and H are matrix Lie groups with Lie algebras $\mathfrak{g}$ and $\mathfrak{h}$, respectively, and suppose $\phi : \mathfrak{g} \to \mathfrak{h}$ is a Lie algebra homomorphism. We would like to construct a Lie group homomorphism $\Phi : G \to H$ such that $\Phi(e^X) = e^{\phi(X)}$ for all X in $\mathfrak{g}$. In light of Theorem 3.42, we can define a map Φ from a neighborhood U of the identity in G into H as follows:

$$\Phi(A) = e^{\phi(\log A)}, \tag{5.1}$$

so that

$$\Phi(e^X) = e^{\phi(X)}, \tag{5.2}$$

at least for sufficiently small X.

A key issue is then to show that Φ, as defined near the identity by (5.1) or (5.2), is a "local homomorphism." Suppose $A = e^X$ and $B = e^Y$, where $X, Y \in \mathfrak{g}$ are small that e^X, e^Y, and $e^X e^Y$ are all in the domain of Φ. To compute $\Phi(AB) = \Phi(e^X e^Y)$, we need to express $e^X e^Y$ in the form e^Z, so that $\Phi(e^X e^Y)$ will equal $e^{\phi(Z)}$. The Baker–Campbell–Hausdorff formula states that for sufficiently small X and Y, we have

$$Z = \log(e^X e^Y)$$
$$= X + Y + \tfrac{1}{2}[X, Y] + \tfrac{1}{12}[X, [X, Y]] - \tfrac{1}{12}[Y, [X, Y]] + \cdots, \tag{5.3}$$

where the "$\cdots$" refers to additional terms involving iterated brackets of X and Y. (A precise statement of and a proof of the formula will be given in subsequent sections.) If ϕ is a Lie algebra homomorphism, then

$$\phi\left(\log(e^X e^Y)\right) = \phi(X) + \phi(Y) + \tfrac{1}{2}[\phi(X), \phi(Y)]$$
$$+ \tfrac{1}{12}[\phi(X), [\phi(X), \phi(Y)]] - \tfrac{1}{12}[\phi(Y), [\phi(X), \phi(Y)]] + \cdots$$
$$= \log(e^{\phi(X)} e^{\phi(Y)}). \tag{5.4}$$

It then follows that $\Phi(e^X e^Y) = e^{\phi(X)} e^{\phi(Y)} = \Phi(e^X)\Phi(e^Y)$.

In the general case, it requires considerable effort to prove the Baker–Campbell–Hausdorff formula and then to prove that, when G is simply connected, Φ can be extended to all of G. In the case of the Heisenberg group (which is simply connected), the argument can be greatly simplified.

Theorem 5.1. *Suppose X and Y are $n \times n$ complex matrices, and that X and Y commute with their commutator:*

$$[X, [X, Y]] = [Y, [X, Y]] = 0. \tag{5.5}$$

Then we have

$$e^X e^Y = e^{X+Y+\frac{1}{2}[X,Y]}.$$

This is the special case of (5.3) in which the series terminates after the $[X, Y]$ term. See Exercise 5 for another special case of the Baker–Campbell–Hausdorff formula.

Proof. Consider X and Y in $M_n(\mathbb{C})$ satisfying (5.5). We will prove that

$$e^{tX} e^{tY} = \exp\left(tX + tY + \frac{t^2}{2}[X, Y]\right),$$

which reduces to the desired result in the case $t = 1$. Since, by assumption, $[X, Y]$ commutes with X and Y, the above relation is equivalent to

$$e^{tX} e^{tY} e^{-\frac{t^2}{2}[X,Y]} = e^{t(X+Y)}. \tag{5.6}$$

Let us denote by $A(t)$ the left-hand side of (5.6) and by $B(t)$ the right-hand side. Our strategy will be to show that $A(t)$ and $B(t)$ satisfy the same differential equation, with the same initial conditions. We can see immediately that

$$\frac{dB}{dt} = B(t)(X + Y).$$

On the other hand, differentiating $A(t)$ by means of the product rule gives

$$\frac{dA}{dt} = e^{tX} X e^{tY} e^{-\frac{t^2}{2}[X,Y]} + e^{tX} e^{tY} Y e^{-\frac{t^2}{2}[X,Y]}$$
$$+ e^{tX} e^{tY} e^{-\frac{t^2}{2}[X,Y]}(-t [X, Y]). \tag{5.7}$$

(The correctness of the last term may be verified by differentiating $e^{-t^2[X,Y]/2}$ term by term.) Now, since Y commutes with $[X, Y]$, it also commute with $e^{-t^2[X,Y]/2}$. Thus, the second term on the right in (5.7) can be rewritten as

$$e^{tX} e^{tY} e^{-\frac{t^2}{2}[X,Y]} Y.$$

For the first term on the right-hand side of (5.7), we compute, using Proposition 3.35, that

$$Xe^{tY} = e^{tY} e^{-tY} Xe^{tY}$$

$$= e^{tY} \mathrm{Ad}_{e^{-tY}}(X)$$

$$= e^{tY} e^{-t\,\mathrm{ad}_Y}(X).$$

Since $[Y, [Y, X]] = -[Y, [X, Y]] = 0$, we have

$$e^{-t\,\mathrm{ad}_Y}(X) = X - t[Y, X] = X + t[X, Y],$$

with all higher terms being zero. We may then simplify (5.7) to

$$\frac{dA}{dt} = e^{tX} e^{tY} e^{-\frac{t^2}{2}[X,Y]}(X + Y) = A(t)(X + Y).$$

We see, then, that $A(t)$ and $B(t)$ satisfy the same differential equation, with the same initial condition $A(0) = B(0) = I$. Thus, by standard uniqueness results for (linear) ordinary differential equations, $A(t) = B(t)$ for all t.   □

Theorem 5.2. *Let H denote the Heisenberg group and $\mathfrak{h}$ its Lie algebra. Let G be a matrix Lie group with Lie algebra $\mathfrak{g}$ and let $\phi : \mathfrak{h} \to \mathfrak{g}$ be a Lie algebra homomorphism. Then there exists a unique Lie group homomorphism $\Phi : H \to G$ such that*

$$\Phi(e^X) = e^{\phi(X)}$$

for all $X \in \mathfrak{h}$.

Proof. Recall (Exercise 18 in Chapter 3) that the Heisenberg group has the special property that its exponential map is one-to-one and onto. Let "log" denote the inverse of this map and define $\Phi : H \to G$ by the formula

$$\Phi(A) = e^{\phi(\log A)}.$$

We will show that Φ is a group homomorphism.

If X and Y are in the Lie algebra of the Heisenberg group (3×3 strictly upper triangular matrices), direct computation shows that every entry of $[X, Y]$ is zero except possibly for the entry in the upper right corner. It is then easily seen that

$[X, Y]$ commutes with both X and Y. Since ϕ is a Lie algebra homomorphism, $\phi(X)$ and $\phi(Y)$ will also commute with their commutator. Thus, by Theorem 5.1, for any X and Y in the Lie algebra of the Heisenberg group, we have

$$\Phi\left(e^X e^Y\right) = \Phi\left(e^{X+Y+\frac{1}{2}[X,Y]}\right)$$

$$= e^{\phi(X)+\phi(Y)+\frac{1}{2}[\phi(X),\phi(Y)]}$$

$$= e^{\phi(X)} e^{\phi(Y)}$$

$$= \Phi(e^X)\Phi(e^Y).$$

Thus, Φ is a group homomorphism, which is continuous because each of exp, log, and ϕ is continuous. $\qquad\square$

5.3 The Baker–Campbell–Hausdorff Formula

The goal of the Baker–Campbell–Hausdorff formula (BCH formula) is to compute $\log(e^X e^Y)$. One may well ask, "Why do we not simply expand both exponentials and the logarithm in power series and multiply everything out?" While it is possible to do this, what is not clear is *why the answer is expressible in terms of commutators*. Consider, for example, the quadratic term in the expression for $\log(e^X e^Y)$, which will be a linear combination of X^2, Y^2, XY, and YX. For this term to be expressible in terms of commutators, it must be a multiple of $XY - YX$. Although direct computation verifies that this is, indeed, the case, it is far from obvious how to prove that a similar result occurs for all the higher terms.

We will actually state and prove an integral form of the BCH formula, rather than the series form (5.3). The integral version of the formula, along with the argument we present in Sect. 5.5, is actually due to Poincaré. (See [Poin1, Poin2] and Section 1.1.2.2 of [BF].)

Consider the function

$$g(z) = \frac{\log z}{1 - \frac{1}{z}},$$

which is defined and holomorphic in the disk $\{|z - 1| < 1\}$. Thus, $g(z)$ can be expressed as a series

$$g(z) = \sum_{m=0}^{\infty} a_m (z - 1)^m,$$

with radius of convergence one. If V is a finite-dimensional vector space, we may identify V with $\mathbb{C}^n$ by means of an arbitrary basis, so that the Hilbert–Schmidt norm

(Definition 2.2) of a linear operator on V can be defined. For any operator A on V with $\|A - I\| < 1$, we can define

$$g(A) = \sum_{m=0}^{\infty} a_m (A - I)^m.$$

We are now ready to state the integral form of the BCH formula.

Theorem 5.3 (Baker–Campbell–Hausdorff). *For all $n \times n$ complex matrices X and Y with $\|X\|$ and $\|Y\|$ sufficiently small, we have*

$$\log(e^X e^Y) = X + \int_0^1 g(e^{\operatorname{ad}_X} e^{t \operatorname{ad}_Y})(Y)\, dt. \tag{5.8}$$

The proof of this theorem is given in Sect. 5.5 of this chapter. Note that $e^{\operatorname{ad}_X} e^{t\operatorname{ad}_Y}$ and, hence, also $g(e^{\operatorname{ad}_X} e^{t\operatorname{ad}_Y})$ are linear operators on the space $M_n(\mathbb{C})$ of all $n \times n$ complex matrices. In (5.8), this operator is being applied to the matrix Y. The fact that X and Y are assumed small guarantees that $e^{\operatorname{ad}_X} e^{t\operatorname{ad}_Y}$ is close to the identity operator on $M_n(\mathbb{C})$ for $0 \leq t \leq 1$, so that $g(e^{\operatorname{ad}_X} e^{t\operatorname{ad}_Y})$ is well defined. Although the right-hand side of (5.8) is rather complicated to compute explicitly, we are not so much interested in the details of the formula but in the fact that it expresses $\log(e^X e^Y)$ (and hence $e^X e^Y$) in terms of the Lie-algebraic quantities ad_X and ad_Y.

5.4   The Derivative of the Exponential Map

In this section we prove a result that is useful in its own right and will play a key role in our proof of the BCH formula. Consider the directional derivative of exp at a point X in the direction of Y:

$$\left. \frac{d}{dt} e^{X+tY} \right|_{t=0}. \tag{5.9}$$

Unless X and Y commute, this derivative may not equal $e^X Y$. Nevertheless, since exp is continuously differentiable (Proposition 2.16), the directional derivatives in (5.9) will depend linearly on Y with X fixed. Now, the function $(1 - e^{-z})/z$ is an entire function of z (with a removable singularity at the origin) and is given by the power series

$$\frac{1 - e^{-z}}{z} = \sum_{k=0}^{\infty} (-1)^k \frac{z^k}{(k+1)!},$$

which has infinite radius of convergence. It thus makes sense to replace z by an arbitrary linear operator A on a finite-dimensional vector space.

Theorem 5.4 (Derivative of Exponential). *For all* $X, Y \in M_n(\mathbb{C})$, *we have*

$$\frac{d}{dt}e^{X+tY}\bigg|_{t=0} = e^X \left\{ \frac{I - e^{-\text{ad}_X}}{\text{ad}_X}(Y) \right\}$$

$$= e^X \left\{ Y - \frac{[X, Y]}{2!} + \frac{[X, [X, Y]]}{3!} - \cdots \right\}. \tag{5.10}$$

More generally, if $X(t)$ *is a smooth matrix-valued function, then*

$$\frac{d}{dt}e^{X(t)} = e^{X(t)} \left\{ \frac{I - e^{-\text{ad}_{X(t)}}}{\text{ad}_{X(t)}} \left(\frac{dX}{dt} \right) \right\}. \tag{5.11}$$

Our proof follows [Tuy].

Lemma 5.5. *If* Z *is a linear operator on a finite-dimensional vector space, then*

$$\lim_{m \to \infty} \frac{1}{m} \sum_{k=0}^{m-1} (e^{-Z/m})^k = \frac{1 - e^{-Z}}{Z}. \tag{5.12}$$

Proof. If we formally applied the formula for the sum of a finite geometric series to $e^{-Z/m}$, we would get

$$\frac{1}{m} \sum_{k=0}^{m-1} (e^{-Z/m})^k = \frac{1}{m} \frac{1 - e^{-Z}}{1 - e^{-Z/m}} \xrightarrow{m \to \infty} \frac{1 - e^{-Z}}{Z}.$$

To give a rigorous argument, we observe that

$$\frac{1 - e^{-x}}{x} = \int_0^1 e^{-tx} \, dt,$$

from which it follows that

$$\frac{1 - e^{-Z}}{Z} = \int_0^1 e^{-tZ} \, dt. \tag{5.13}$$

(The reader may check, using term-by-term integration of the series expansion of e^{-tZ}, that this formula for $(1 - e^{-Z})/Z$ agrees with our earlier definition.)

Since $(e^{-Z/m})^k = e^{-kZ/m}$, the left-hand side of (5.12) is a Riemann-sum approximation to the matrix-valued integral on the right-hand side of (5.13). These Riemann sums converge to the integral of e^{-tZ}—which is a continuous function of t—establishing (5.12).   □

Proof of Theorem 5.4. The formula (5.11) follows from (5.10) by applying the chain rule to the composition of exp and $X(t)$. Thus, it suffices to prove (5.10). For any $n \times n$ matrices X and Y, set

$$\Delta(X,Y) = \frac{d}{dt}e^{X+tY}\bigg|_{t=0}.$$

Since (Proposition 2.16) exp is a continuously differentiable map, $\Delta(X,Y)$ is jointly continuous in X and Y and is linear in Y for each fixed X.

Now, for every positive integer m, we have

$$e^{X+tY} = \left[\exp\left(\frac{X}{m} + t\frac{Y}{m}\right)\right]^m. \tag{5.14}$$

Applying the product rule, we will get m terms, where in each term, $m - 1$ of the factors in (5.14) are simply evaluated at $t = 0$ and the remaining factor is differentiated at $t = 0$. Thus,

$$\frac{d}{dt}e^{X+tY}\bigg|_{t=0} = \sum_{k=0}^{m-1}(e^{X/m})^{m-k-1}\left[\frac{d}{dt}\exp\left(\frac{X}{m} + t\frac{Y}{m}\right)\bigg|_{t=0}\right](e^{X/m})^k$$

$$= e^{(m-1)X/m}\sum_{k=0}^{m-1}(e^{X/m})^{-k}\Delta\left(\frac{X}{m},\frac{Y}{m}\right)(e^{X/m})^k$$

$$= e^{(m-1)X/m}\frac{1}{m}\sum_{k=0}^{m-1}\exp\left(-\frac{\mathrm{ad}_X}{m}\right)^k\left(\Delta\left(\frac{X}{m},Y\right)\right). \tag{5.15}$$

In the third equality, we have used the linearity of $\Delta(X,Y)$ in Y and the relationship between Ad and ad (Proposition 3.35).

We now wish to let m tend to infinity in (5.15). The factor in front tends to $\exp(X)$. Since $\Delta(X,Y)$ is jointly continuous in X and Y, the expression $\Delta(X/m,Y)$ tends to $\Delta(0,Y)$, where it is easily verified that $\Delta(0,Y) = Y$. Finally, applying Lemma 5.5 with $Z = \mathrm{ad}_X$, we see that

$$\lim_{m\to\infty}\frac{1}{m}\sum_{k=0}^{m-1}\exp\left(-\frac{\mathrm{ad}_X}{m}\right)^k = \frac{1 - e^{-\mathrm{ad}_X}}{\mathrm{ad}_X}.$$

Thus, by letting m tend to infinity in (5.15), we obtain the desired result. □

5.5   Proof of the BCH Formula

We now turn to the proof of Theorem 5.3. For sufficiently small X and Y in $M_n(\mathbb{C})$, let

$$Z(t) = \log(e^X e^{tY})$$

for $0 \leq t \leq 1$. Our goal is to compute $Z(1)$. Since $e^{Z(t)} = e^X e^{tY}$, we have

$$e^{-Z(t)} \frac{d}{dt} e^{Z(t)} = \left(e^X e^{tY} \right)^{-1} e^X e^{tY} Y = Y.$$

On the other hand, by Theorem 5.4,

$$e^{-Z(t)} \frac{d}{dt} e^{Z(t)} = \left\{ \frac{I - e^{-\mathrm{ad}_{Z(t)}}}{\mathrm{ad}_{Z(t)}} \right\} \left(\frac{dZ}{dt} \right).$$

Hence,

$$\left\{ \frac{I - e^{-\mathrm{ad}_{Z(t)}}}{\mathrm{ad}_{Z(t)}} \right\} \left(\frac{dZ}{dt} \right) = Y.$$

Now, if X and Y are small enough, $Z(t)$ will also be small, so that $[I - e^{-\mathrm{ad}_{Z(t)}}]/\mathrm{ad}_{Z(t)}$ will be close to the identity and thus invertible. In that case, we obtain

$$\frac{dZ}{dt} = \left\{ \frac{I - e^{-\mathrm{ad}_{Z(t)}}}{\mathrm{ad}_{Z(t)}} \right\}^{-1} (Y). \tag{5.16}$$

Meanwhile, if we apply the homomorphism "Ad" to the equation $e^{Z(t)} = e^X e^{tY}$, use the relationship between "Ad" and "ad," and take a logarithm, we obtain the following relations:

$$\mathrm{Ad}_{e^{Z(t)}} = \mathrm{Ad}_{e^X} \mathrm{Ad}_{e^{tY}}$$

$$e^{\mathrm{ad}_{Z(t)}} = e^{\mathrm{ad}_X} e^{t\,\mathrm{ad}_Y}$$

$$\mathrm{ad}_{Z(t)} = \log(e^{\mathrm{ad}_X} e^{t\,\mathrm{ad}_Y}).$$

Plugging the last two of these relations into (5.16) gives

$$\frac{dZ}{dt} = \left\{ \frac{I - (e^{\mathrm{ad}_X} e^{t\,\mathrm{ad}_Y})^{-1}}{\log(e^{\mathrm{ad}_X} e^{t\,\mathrm{ad}_Y})} \right\}^{-1} (Y). \tag{5.17}$$

Now, observe that

$$g(z) = \left\{ \frac{1 - z^{-1}}{\log z} \right\}^{-1}$$

so that (5.17) is the same as

$$\frac{dZ}{dt} = g(e^{\operatorname{ad}_X} e^{t \operatorname{ad}_Y})(Y). \qquad (5.18)$$

Noting that $Z(0) = X$ and integrating (5.18) gives

$$\log(e^X e^Y) = Z(1) = X + \int_0^1 g(e^{\operatorname{ad}_X} e^{t \operatorname{ad}_Y})(Y) \, dt,$$

which is the Baker–Campbell–Hausdorff formula.

5.6   The Series Form of the BCH Formula

Let us see how to get the first few terms of the series form of the BCH formula from
the integral form in Theorem 5.3. Using the Taylor series (2.7) for the logarithm, we
may easily compute that

$$g(z) = 1 + \frac{1}{2}(z - 1) - \frac{1}{6}(z - 1)^2 + \frac{1}{12}(z - 1)^3 - \cdots .$$

Meanwhile,

$$e^{\operatorname{ad}_X} e^{t \operatorname{ad}_Y} - I$$

$$= \left(I + \operatorname{ad}_X + \frac{(\operatorname{ad}_X)^2}{2} + \cdots \right) \left(I + t \operatorname{ad}_Y + \frac{t^2 (\operatorname{ad}_Y)^2}{2} + \cdots \right) - I$$

$$= \operatorname{ad}_X + t \operatorname{ad}_Y + t \operatorname{ad}_X \operatorname{ad}_Y + \frac{(\operatorname{ad}_X)^2}{2} + \frac{t^2 (\operatorname{ad}_Y)^2}{2} + \cdots .$$

Since $e^{\operatorname{ad}_X} e^{t \operatorname{ad}_Y} - I$ has no zeroth-order term, $(e^{\operatorname{ad}_X} e^{t \operatorname{ad}_Y} - I)^m$ will contribute
only terms of degree m or higher in ad_X and/or ad_Y. Computing up to degree 2 in
ad_X and ad_Y gives

$$g(e^{\operatorname{ad}_X} e^{t \operatorname{ad}_Y})$$

$$= I + \frac{1}{2} \left(\operatorname{ad}_X + t \operatorname{ad}_Y + t \operatorname{ad}_X \operatorname{ad}_Y + \frac{(\operatorname{ad}_X)^2}{2} + \frac{t^2 (\operatorname{ad}_Y)^2}{2} \right)$$

$$-\frac{1}{6}\left[(\mathrm{ad}_X)^2 + t^2(\mathrm{ad}_Y)^2 + t\,\mathrm{ad}_X\,\mathrm{ad}_Y + t\,\mathrm{ad}_Y\,\mathrm{ad}_X\right]$$

$$+ \ \text{higher-order terms.}$$

We now apply $g\left(e^{\mathrm{ad}_X}e^{t\,\mathrm{ad}_Y}\right)$ to Y and integrate. Computing to second order and noting that any term with ad_Y acting first is zero, we obtain:

$$\log(e^X e^Y)$$

$$\approx X + \int_0^1 \left[Y + \frac{1}{2}[X, Y] + \frac{1}{4}[X,[X,Y]] - \frac{1}{6}[X,[X,Y]] - \frac{t}{6}[Y,[X,Y]]\right] dt$$

$$\approx X + Y + \frac{1}{2}[X, Y] + \frac{1}{12}[X,[X,Y]] - \frac{1}{12}[Y,[X,Y]],$$

which is the expression in (5.3).

5.7   Group Versus Lie Algebra Homomorphisms

Recall Theorem 3.28, which says that given matrix Lie groups G and H and a Lie group homomorphism $\Phi : G \rightarrow H$, we can find a Lie algebra homomorphism $\phi : \mathfrak{g} \rightarrow \mathfrak{h}$ such that $\Phi(e^X) = e^{\phi(X)}$ for all $X \in \mathfrak{g}$. In this section, we prove a converse to this result in the case that G is *simply connected*.

Theorem 5.6. *Let G and H be matrix Lie groups with Lie algebras $\mathfrak{g}$ and $\mathfrak{h}$, respectively, and let $\phi : \mathfrak{g} \rightarrow \mathfrak{h}$ be a Lie algebra homomorphism. If G is simply connected, there exists a unique Lie group homomorphism $\Phi : G \rightarrow H$ such that $\Phi(e^X) = e^{\phi(X)}$ for all $X \in \mathfrak{g}$.*

This result has the following corollary.

Corollary 5.7. *Suppose G and H are simply connected matrix Lie groups with Lie algebras $\mathfrak{g}$ and $\mathfrak{h}$, respectively. If $\mathfrak{g}$ is isomorphic to $\mathfrak{h}$, then G is isomorphic to H.*

Proof. Let $\phi : \mathfrak{g} \rightarrow \mathfrak{h}$ be a Lie algebra isomorphism. By Theorem 5.6, there exists an associated Lie group homomorphism $\Phi : G \rightarrow H$. Since $\psi := \phi^{-1}$ is also a Lie algebra homomorphism, there is a corresponding Lie group homomorphism $\Psi : H \rightarrow G$. We want to show that Φ and Ψ are inverses of each other.

Now, the Lie algebra homomorphism associated to $\Phi \circ \Psi$ is, by Proposition 3.30, equal to $\phi \circ \psi = I$, and similarly for $\Psi \circ \Phi$. Thus, by Corollary 3.49, both $\Phi \circ \Psi$ and $\Psi \circ \Phi$ are equal to the identity maps on H and G, respectively.   □

We now proceed with the proof of Theorem 5.6. The first step is to construct a "local homomorphism" from ϕ. This step is the only place in the argument in which we use the BCH formula.

Definition 5.8. If G and H are matrix Lie groups, a **local homomorphism** of G to H is pair (U, f) where U is a path-connected neighborhood of the identity in G and $f : U \to H$ is a continuous map such that $f(AB) = f(A)f(B)$ whenever A, B, and AB all belong to U.

The definition says that f is as much of a homomorphism as it makes sense to be, given that U is not necessarily a subgroup of G.

Proposition 5.9. *Let G and H be matrix Lie groups with Lie algebras $\mathfrak{g}$ and $\mathfrak{h}$, respectively, and let $\phi : \mathfrak{g} \to \mathfrak{h}$ be a Lie algebra homomorphism. Define $U_\varepsilon \subset G$ by*

$$U_\varepsilon = \{ A \in G \,|\, \|A - I\| < 1 \text{ and } \|\log A\| < \varepsilon \}.$$

Then there exists some $\varepsilon > 0$ such that the map $f : U_\varepsilon \to H$ given by

$$f(A) = e^{\phi(\log A)}$$

is a local homomorphism.

Note that by Theorem 3.42, if ε is small enough, $\log A$ will be in $\mathfrak{g}$ for all $A \in U_\varepsilon$, so that Φ makes sense.

Proof. Choose ε small enough that Theorem 3.42 applies and small enough that for all $A, B \in U_\varepsilon$, the BCH formula applies to $X := \log A$ and $Y := \log B$ and also to $\phi(X)$ and $\phi(Y)$. Then if AB happens to be in U_ε, we have

$$f(AB) = f(e^X e^Y) = e^{\phi(\log(e^X e^Y))}.$$

We now compute $\log(e^X e^Y)$ by the BCH formula and then apply ϕ. Since ϕ is a Lie algebra homomorphism, it will change all the Lie-algebraic quantities involving X and Y in the BCH formula into the analogous quantities involving $\phi(X)$ and $\phi(Y)$. Thus, as in (5.4), we have

$$\phi[\log(e^X e^Y)] = \phi(X) + \int_0^1 \sum_{m=0}^\infty a_m (e^{\mathrm{ad}_{\phi(X)}} e^{t\,\mathrm{ad}_{\phi(Y)}} - I)^m (\phi(Y))\, dt$$

$$= \log(e^{\phi(X)} e^{\phi(Y)}),$$

We obtain, then,

$$f(AB) = \exp \{ \log(e^{\phi(X)} e^{\phi(Y)}) \}$$

$$= e^{\phi(X)} e^{\phi(Y)}$$

$$= f(A)f(B),$$

as claimed. $\square$

Theorem 5.10. *Let G and H be matrix Lie groups, with G simply connected. If (U, f) is a local homomorphism of G into H, there exists a unique (global) Lie group homomorphism $\Phi : G \to H$ such that Φ agrees with f on U.*

Proof.   *Step 1: Define Φ along a path.* Since G is simply connected and thus connected, for any $A \in G$, there exists a path $A(t) \in G$ with $A(0) = I$ and $A(1) = A$. Let us call a partition $0 = t_0 < t_1 < t_2 \cdots < t_m = 1$ of $[0, 1]$ a **good partition** if for all s and t belonging to the same subinterval of the partition, we have

$$A(t)A(s)^{-1} \in U. \tag{5.19}$$

Lemma 3.48 guarantees that good partitions exist. If a partition is good, then, in particular, since $t_0 = 0$ and $A(0) = I$, we have $A(t_1) \in U$. Choose a good partition and write A as

$$A = [A(1)A(t_{m-1})^{-1}][A(t_{m-1})A(t_{m-2})^{-1}] \cdots [A(t_2)A(t_1)^{-1}]A(t_1).$$

Since Φ is supposed to be a homomorphism and is supposed to agree with f near the identity it is reasonable to "define" $\Phi(A)$ by

$$\Phi(A) = f(A(1)A(t_{m-1})^{-1}) \cdots f(A(t_2)A(t_1)^{-1}) f(A(t_1)). \tag{5.20}$$

In the next two steps, we will prove that $\Phi(A)$ is independent of the choice of partition for a fixed path and independent of the choice of path.

Step 2: Prove independence of the partition. For any good partition, if we insert an extra partition point s between t_j and t_{j+1}, the result is easily seen to be another good partition. This change in the partition has the effect of replacing the factor $f(A(t_{j+1})A(t_j)^{-1})$ in (5.20) by

$$f(A(t_{j+1})A(s)^{-1}) f(A(s)A(t_j)^{-1}).$$

Since s is between t_j and t_{j+1}, the condition (5.19) on the original partition guarantees that $A(t_{j+1})A(s)^{-1}$, $A(s)A(t_j)^{-1}$ and $A(t_{j+1})A(t_j)^{-1}$ are all in U. Thus, since f is a local homomorphism, we have

$$f(A(t_{j+1})A(t_j)^{-1}) = f(A(t_{j+1})A(s)^{-1}) f(A(s)A(t_j)^{-1}),$$

showing that the value of $\Phi(A)$ is unchanged by the addition of the extra partition point.

By repeating this argument, we see that the value of $\Phi(A)$ does not change by the addition of any finite number of points to the partition. Now, any two good partitions have a common refinement, namely their union, which is also a good partition. The above argument shows that the value of $\Phi(A)$ computed from the first partition is the same as for the common refinement, which is the same as for the second partition.

Step 3: Prove independence of the path. It is in this step that we use the simple connectedness of G. Suppose $A_0(t)$ and $A_1(t)$ are two paths joining the identity to some $A \in G$. Then, since G is simply connected, a standard topological argument (e.g., Proposition 1.6 in [Hat]) shows that A_0 and A_1 are homotopic with endpoints fixed. This means that there exists a continuous map $A : [0,1] \times [0,1] \rightarrow G$ with

$$A(0,t) = A_0(t), \quad A(1,t) = A_1(t)$$

for all $t \in [0,1]$ and also

$$A(s,0) = I, \quad A(s,1) = A$$

for all $s \in [0,1]$.

As in the proof of Lemma 3.48, there exists an integer N such that for all (s,t) and (s',t') in $[0,1] \times [0,1]$ with $|s - s'| < 2/N$ and $|t - t'| < 2/N$, we have

$$A(s,t)A(s',t')^{-1} \in U.$$

We now deform A_0 "a little bit at a time" into A_1. This means that we define a sequence $B_{k,l}$ of paths, with $k = 0, \ldots, N - 1$ and $l = 0, \ldots, N$. We define these paths so that $B_{k,l}(t)$ coincides with $A((k + 1)/N, t)$ for t between 0 and $(l - 1)/N$, and $B_{k,l}(t)$ coincides with $A(k/N, t)$ for t between l/N and 1. For t between $(l - 1)/N$ and l/N, we define $B_{k,l}(t)$ to coincide with the values of $A(\cdot, \cdot)$ on the path that goes "diagonally" in the (s,t)-plane, as indicated in Figure 5.1. When $l = 0$, there are no t-values between 0 and $(l - 1)/N$, so $B_{k,0}(t) = A(k/N, t)$ for all $t \in [0,1]$. In particular, $B_{0,0}(t) = A_0(t)$.

Fig. 5.1 The path in the (s,t) plane defining $B_{k,l}(t)$

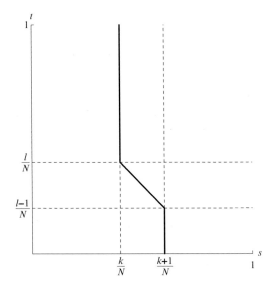

We now deform $A_0 = B_{0,0}$ into $B_{0,1}$ and then into $B_{0,2}$, $B_{0,3}$, and so on until we reach $B_{0,N}$, which we then deform into $B_{1,0}$ and so on until we reach $B_{N-1,N}$, which we finally deform into A_1. We claim that the value of $\Phi(A)$ is the same at each stage of this deformation. Note that for $k < l$, $B_{k,l}(t)$ and $B_{k,l+1}(t)$ are the same except for t's in the interval

$$[(l-1)/N, (l+1)/N].$$

Furthermore, by Step 2, we are free to choose any good partition we like to compute $\Phi(A)$. For both $B_{k,l}$ and $B_{k,l+1}$, we choose the partition points to be

$$0, \frac{1}{N}, \ldots, \frac{l-1}{N}, \frac{l+1}{N}, \frac{l+2}{N}, \ldots, 1,$$

which gives a good partition by the way N was chosen.

Now, from (5.20), the value of $\Phi(A)$ depends only on the values of the path at the partition points. Since we have chosen our partition in such a way that the values of $B_{k,l}$ and $B_{k,l+1}$ are identical at all the partition points, the value of $\Phi(A)$ is the same for these two paths. (See Figure 5.2.) A similar argument shows that the value of $\Phi(A)$ computed along $B_{k,N}$ is the same as along $B_{k+1,0}$. Thus, the value of $\Phi(A)$ is the same for each path from $A_0 = B_{0,0}$ all the way to $B_{N-1,N}$ and then the same as A_1.

Fig. 5.2 The paths $B_{k,l}$ and $B_{k,l+1}$ agree at each partition point. In the figure, s increases as we move from the *top* toward the *bottom*

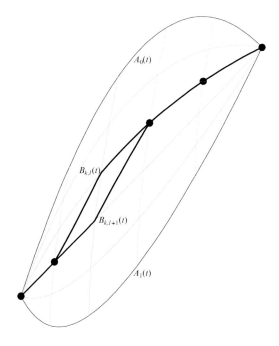

Step 4: Prove that Φ is a homomorphism and agrees with f on U. The proof that Φ is a homomorphism is a straightforward unpacking of the definition of Φ and is left to the reader; see Exercise 7. To show that Φ agrees with f on U, choose $A \in U$. Since U is path connected, we can find a path $A(t)$, $0 \leq t \leq 1$, lying in U joining I to A. Choose a good partition $\{t_j\}_{j=1}^m$ for $A(t)$, and we then claim that for all j, we have $\Phi(A(t_j)) = f(A(t_j))$.

Note that $A(t)$, $0 \leq t \leq t_j$ is a path joining I to $A(t_j)$ and that $\{t_0, t_1, \ldots, t_j\}$ is a good partition of this path. (Technically, we should reparameterize this path so that the time interval is $[0, 1]$.) Hence,

$$\Phi(A(t_j)) = f(A(t_j)A(t_{j-1})^{-1}) \cdots f(A(t_2)A(t_1)^{-1}) f(A(t_1)),$$

for all j. In particular,

$$\Phi(A(t_1)) = f(A(t_1)).$$

Now assume that $\Phi(A(t_j)) = f(A(t_j))$, and compute that

$$\Phi(A(t_{j+1})) = f(A(t_{j+1})A(t_j)^{-1}) f(A(t_j)A(t_{j-1})^{-1}) \cdots f(A(t_1))$$
$$= f(A(t_{j+1})A(t_j)^{-1})\Phi(A(t_j))$$
$$= f(A(t_{j+1})A(t_j)^{-1}) f(A(t_j))$$
$$= f(A(t_j)).$$

The last equality holds because f is a local homomorphism and because $A(t_{j+1})A(t_j)^{-1}$, $A(t_j)$, and their product all lie in U. Thus, by induction, $\Phi(A(t_j)) = f(A(t_j))$ for all j; when $j = m$, we obtain $\Phi(A) = f(A)$. $\square$

It is important to note that when proving independence of the path (Step 3 of the proof), it is essential to know already that the value of $\Phi(A)$ is independent of the choice of good partition (Step 2 of the proof). Specifically, when we move from $B_{k,l}$ to $B_{k,l+1}$, we use one partition for $B_{k,l+1}$, but when we move from $B_{k,l+1}$ to $B_{k,l+2}$, we use a different partition for $B_{k,l+1}$. Essentially, the proof proceeds by deforming the path between partition points—which clearly does not change the value of $\Phi(A)$—then picking a new partition and doing the same thing again. Note also that it is in proving independence of the partition that we use the assumption that f is a local homomorphism.

Proof of Theorem 5.6. For the existence part of the proof, let f be the local homomorphism in Proposition 5.9 and let Φ be the global homomorphism in Theorem 5.10. Then for any $X \in \mathfrak{g}$, the element $e^{X/m}$ will be in U for all sufficiently large m, showing that

$$\Phi(e^{X/m}) = f(e^{X/m}) = e^{\phi(X)/m}.$$

Since Φ is a homomorphism, we have

$$\Phi(e^X) = \Phi(e^{X/m})^m = e^{\phi(X)},$$

as required.

For the uniqueness part of the proof, suppose Φ_1 and Φ_2 are two homomorphisms related in the desired way to ϕ. Then for any $A \in G$, we express A as $e^{X_1} \cdots e^{X_N}$ with $X_j \in \mathfrak{g}$, as in Corollary 3.47, and observe that

$$\Phi_1(A) = \Phi_2(A) = e^{\phi(X_1)} \cdots e^{\phi(X_N)},$$

so that Φ_1 agrees with Φ_2.   □

We conclude this section with a typical application of Theorem 5.6.

Theorem 5.11. *Suppose that G is a simply connected matrix Lie group and that the Lie algebra $\mathfrak{g}$ of G decomposes as a Lie algebra direct sum $\mathfrak{g} = \mathfrak{h}_1 \oplus \mathfrak{h}_2$, for two subalgebras $\mathfrak{h}_1$ and $\mathfrak{h}_2$ of $\mathfrak{g}$. Then there exists closed, simply connected subgroups H_1 and H_2 of G whose Lie algebras are $\mathfrak{h}_1$ and $\mathfrak{h}_2$, respectively. Furthermore, G is isomorphic to the direct product of H_1 and H_2.*

Proof. Consider the Lie algebra homomorphism $\phi : \mathfrak{g} \to \mathfrak{g}$ that sends $X + Y$ to X, where $X \in \mathfrak{h}_1$ and $Y \in \mathfrak{h}_2$. Since G is simply connected, there is a Lie group homomorphism $\Phi : G \to G$ associated to ϕ, and the Lie algebra of the kernel of Φ is the kernel of ϕ (Proposition 3.31), which is $\mathfrak{h}_2$. Let H_2 be the identity component of $\ker \Phi$. Since Φ is continuous, $\ker \Phi$ is closed, and so is its identity component H_2 (Corollary 3.52). Thus, H_2 is a closed, connected subgroup of G with Lie algebra $\mathfrak{h}_2$. By a similar argument, we may construct a closed, connected subgroup H_1 of G whose Lie algebra is $\mathfrak{h}_1$.

Suppose now that $A(t)$ is a loop in H_1. Since G is simply connected, there is a homotopy $A(s,t)$ shrinking $A(t)$ to a point in G. Now, ϕ is the identity on $\mathfrak{h}_1$, from which it follows that Φ is the identity on H_1. Thus, if we define $B(s,t) = \Phi(A(s,t))$, we see that

$$B(0,t) = \Phi(A(t)) = A(t).$$

Furthermore, since ϕ maps G into $\mathfrak{h}_1$, we see that Φ maps G into H_1. We conclude that B is a homotopy of $A(t)$ to a point in H_1. Thus, H_1 is simply connected, and, by a similar argument, so is H_2.

Finally, since $\mathfrak{g}$ is the Lie algebra direct sum of $\mathfrak{h}_1$ and $\mathfrak{h}_2$, elements of $\mathfrak{h}_1$ commutes with elements of $\mathfrak{h}_2$. It follows that elements of H_1 (which are all product of exponentials of elements of $\mathfrak{h}_1$) commute with elements of H_2. Thus, we have a Lie group homomorphism $\Psi : H_1 \times H_2 \to G$ given by $\Psi(A, B) = AB$. The associated Lie algebra homomorphism ψ is then just the original isomorphism of $\mathfrak{h}_1 \oplus \mathfrak{h}_2$ with $\mathfrak{g}$. Since G is simply connected, there is a homomorphism $\Gamma : G \to H_1 \times H_2$ for which the associated Lie algebra homomorphism is ψ^{-1}. By the proof of Corollary 5.7, Γ and Ψ are inverses of each other, showing that G is isomorphic to $H_1 \times H_2$.   □

5.8   Universal Covers

Theorem 5.6 says that if G is simply connected, every homomorphism of the Lie algebra $\mathfrak{g}$ of G can be exponentiated to a homomorphism of G. If G is not simply connected, we may look for another group $\tilde{G}$ that has the same Lie algebra as G but such that $\tilde{G}$ is simply connected.

Definition 5.12. Let G be a connected matrix Lie group. Then a **universal cover** of G is a simply connected matrix Lie group H together with a Lie group homomorphism $\Phi : H \to G$ such that the associated Lie algebra homomorphism $\phi : \mathfrak{h} \to \mathfrak{g}$ is a Lie algebra isomorphism. The homomorphism Φ is called the **covering map.**

If a universal cover of G exists, it is unique up to "canonical isomorphism," as follows.

Proposition 5.13. *If G is a connected matrix Lie group and (H_1, Φ_1) and (H_2, Φ_2) are universal covers of G, then there exists a Lie group isomorphism $\Psi : H_1 \dashrightarrow H_2$ such that $\Phi_2 \circ \Psi = \Phi_1$.*

Proof. See Exercise 9.   □

Since a connected matrix Lie group has at most one universal cover (up to canonical isomorphism), it is reasonable to speak of *the* universal cover $(\tilde{G}, \Phi)$ of G. Furthermore, if H is a simply connected Lie group and $\phi : \mathfrak{h} \to \mathfrak{g}$ is a Lie algebra isomorphism, then by Theorem 5.6, we can construct an associated Lie group homomorphism $\Phi : H \to G$, so that (H, Φ) is a universal cover of G. Since ϕ is an isomorphism, we can use ϕ to identify $\tilde{\mathfrak{g}}$ with $\mathfrak{g}$. Thus, in slightly less formal terms, we may define the notion of universal cover as follows: *The universal cover of a matrix Lie group G is a simply connected matrix Lie group $\tilde{G}$ such that the Lie algebra of $\tilde{G}$ is equal to the Lie algebra of G.* With this perspective, we have the following immediate corollary of Theorem 5.6.

Corollary 5.14. *Let G be a connected matrix Lie group and let $\tilde{G}$ be the universal cover of G, where we think of G and $\tilde{G}$ as having the same Lie algebra $\mathfrak{g}$. If H is a matrix Lie group with Lie algebra $\mathfrak{h}$ and $\phi : \mathfrak{g} \to \mathfrak{h}$ is a Lie algebra homomorphism, there exists a unique homomorphism $\Phi : \tilde{G} \to H$ such that $\Phi(e^X) = e^{\phi(X)}$ for all $X \in \mathfrak{g}$.*

An example of importance in physics is the universal cover of $\mathsf{SO}(3)$.

Example 5.15. The universal cover of $\mathsf{SO}(3)$ is $\mathsf{SU}(2)$.

Proof. The group $\mathsf{SU}(2)$ is simply connected by Proposition 1.15. Proposition 1.19 and Example 3.29 then provide the desired covering map.   □

The topic of universal covers is one place where we pay a price for our decision to consider only *matrix* Lie groups: a matrix Lie group may not have a universal cover that is a matrix Lie group. It is not hard to show that every Lie group G

has a universal cover in the class of (not necessarily matrix) Lie groups. Indeed, G has a universal cover in the topological sense (Definition 13.1), and this cover can be given a group structure in such a way that the covering map is a Lie group homomorphism. It turns out, however, that the universal cover of a *matrix* Lie group may not be a matrix group.

We now show that the group $\mathsf{SL}(2; \mathbb{R})$ does not have a universal cover in the class of matrix Lie groups. We begin by showing that $\mathsf{SL}(2; \mathbb{R})$ is not simply connected. By Theorem 2.17 and Proposition 2.19, $\mathsf{SL}(2; \mathbb{R})$, as a manifold, is homeomorphic to $\mathsf{SO}(2) \times V$, where V is the space of 2×2 real, symmetric matrices with trace zero. Now, V, being a vector space, is certainly simply connected, but $\mathsf{SO}(2)$, which is homeomorphic to the unit circle S^1, is not. Thus, $\mathsf{SL}(2; \mathbb{R})$ itself is not simply connected. By contrast, the group $\mathsf{SL}(2; \mathbb{C})$ decomposes as $\mathsf{SU}(2) \times W$, where W is the space of 2×2 self-adjoint, complex matrices with trace zero. Since both $\mathsf{SU}(2)$ and W are simply connected, $\mathsf{SL}(2; \mathbb{C})$ is simply connected. (See Appendix 13.3 for more information on this type of calculation.)

Proposition 5.16. *Let $G \subset \mathsf{GL}(n; \mathbb{C})$ be a connected matrix Lie group with Lie algebra $\mathfrak{g}$. Suppose $\Phi : G \to \mathsf{SL}(2; \mathbb{R})$ is a Lie group homomorphism for which the associated Lie algebra $\phi : \mathfrak{g} \to \mathsf{sl}(2; \mathbb{R})$ is a Lie algebra isomorphism. Then Φ is a Lie group isomorphism and, therefore, G cannot be simply connected. Thus, $\mathsf{SL}(2; \mathbb{R})$ has no universal cover in the class of matrix Lie groups.*

The result relies essentially on the assumption that G is a *matrix* Lie group—or, more precisely, on the assumption that G is contained in a group whose Lie algebra is complex.

Lemma 5.17. *Suppose $\psi : \mathsf{sl}(2; \mathbb{R}) \to \mathfrak{gl}(n; \mathbb{C})$ is a Lie algebra homomorphism. Then there exists a Lie group homomorphism $\Phi : \mathsf{SL}(2; \mathbb{R}) \to \mathsf{GL}(n; \mathbb{C})$ such that $\Phi(e^X) = e^{\phi(X)}$ for all $X \in \mathsf{sl}(2; \mathbb{R})$.*

The significance of the lemma is that the result holds even though $\mathsf{SL}(2; \mathbb{R})$ is not simply connected.

Proof. Let $\psi_{\mathbb{C}} : \mathsf{sl}(2; \mathbb{C}) \to \mathfrak{gl}(n; \mathbb{C})$ be the complex-linear extension of ψ to $\mathsf{sl}(2; \mathbb{C}) \cong \mathsf{sl}(2; \mathbb{R})_{\mathbb{C}}$, which is a Lie algebra homomorphism (Proposition 3.39). Since $\mathsf{SL}(2; \mathbb{C})$ is simply connected, there exists a Lie group homomorphism $\Psi_{\mathbb{C}} : \mathsf{SL}(2; \mathbb{C}) \to \mathsf{GL}(n; \mathbb{C})$ such that $\Psi_{\mathbb{C}}(e^X) = e^{\psi_{\mathbb{C}}(X)}$ for all $X \in \mathsf{sl}(2; \mathbb{C})$. If we let Ψ be the restriction of $\Psi_{\mathbb{C}}$ to $\mathsf{SL}(2; \mathbb{R})$, then Ψ is a Lie group homomorphism which satisfies $\Psi(e^X) = e^{\psi(X)}$ for $X \in \mathsf{sl}(2; \mathbb{R})$. $\square$

Proof of Proposition 5.16. Since ϕ is a Lie algebra isomorphism, the inverse map $\psi : \mathsf{sl}(2; \mathbb{R}) \to \mathfrak{g}$ is a Lie algebra homomorphism. Thus, by the lemma, there is a Lie group homomorphism $\Psi : \mathsf{SL}(2; \mathbb{R}) \to G$ corresponding to ψ. Since ϕ and ψ are inverses of each other, it follows from Proposition 3.30 and Corollary 3.49 that Φ and Ψ are also inverses of each other. $\square$

5.9   Subgroups and Subalgebras

In this section, we address Question 3 from Sect. 5.1: If G is a matrix Lie group with Lie algebra $\mathfrak{g}$ and $\mathfrak{h}$ is a subalgebra of $\mathfrak{g}$, does there exist a matrix Lie group $H \subset G$ whose Lie algebra is H? *If* the exponential map for G were a homeomorphism between $\mathfrak{g}$ and G and *if* the BCH formula worked globally instead of locally, the answer would be yes, since we could simply define H to be the set of elements of the form e^X, $X \in \mathfrak{g}$, and the BCH formula would show that H is a subgroup.

In reality, the answer to Question 3, as stated, is no. Suppose, for example, that $G = \mathsf{GL}\,(2; \mathbb{C})$ and

$$\mathfrak{h} = \left\{ \begin{pmatrix} it & 0 \\ 0 & ita \end{pmatrix} \middle| t \in \mathbb{R} \right\}, \tag{5.21}$$

where a is irrational. If there is going to be a matrix Lie group H with Lie algebra $\mathfrak{h}$, then H would have to contain the closure of the group

$$H_0 = \left\{ \begin{pmatrix} e^{it} & 0 \\ 0 & e^{ita} \end{pmatrix} \middle| t \in \mathbb{R} \right\}, \tag{5.22}$$

which is (Exercise 10 in Chapter 1) is the group

$$H_1 = \left\{ \begin{pmatrix} e^{i\theta} & 0 \\ 0 & e^{i\phi} \end{pmatrix} \middle| \theta, \phi \in \mathbb{R} \right\}.$$

But then the Lie algebra of H would have to contain the Lie algebra of H_1, which is two dimensional!

Fortunately, all is not lost. We can still get a subgroup H for each subalgebra $\mathfrak{h}$ if we weaken the condition that H be a matrix Lie group. In the above example, the subgroup we want is H_0, despite the fact that H_0 is not closed.

Definition 5.18. If H is any subgroup of $\mathsf{GL}\,(n; \mathbb{C})$, the **Lie algebra** $\mathfrak{h}$ of H is the set of all matrices X such that

$$e^{tX} \in H$$

for all real t.

It is possible to prove that for *any* subgroup H of $\mathsf{GL}(n; \mathbb{C})$, the Lie algebra $\mathfrak{h}$ of H is actually a Lie algebra, that is, a real vector space—possibly zero dimensional— and closed under brackets. (See Proposition 1 and Corollary 7 in Chapter 2 of [Ross].) This result is not, however, directly relevant to our goal in this section, which is to construct, for each subalgebra $\mathfrak{h}$ of $\mathsf{gl}(n; \mathbb{C})$ a subgroup with Lie algebra $\mathfrak{h}$. Note, however, that if $\mathfrak{h}$ is at least a real subspace of $\mathsf{gl}(n; \mathbb{C})$, then the proof of Point 4 of Theorem 3.20 shows that $\mathfrak{h}$ is also closed under brackets.

Definition 5.19. If G is a matrix Lie group with Lie algebra $\mathfrak{g}$, then $H \subset G$ is a **connected Lie subgroup** of G if the following conditions are satisfied:

1. H is a subgroup of G.
2. The Lie algebra $\mathfrak{h}$ of H is a Lie subalgebra of $\mathfrak{g}$.
3. Every element of H can be written in the form $e^{X_1} e^{X_2} \cdots e^{X_m}$, with $X_1, \ldots, X_m \in \mathfrak{h}$.

Connected Lie subgroups are also called **analytic subgroups**. Note that any group H as in the definition is path connected, since each element of H can be connected to the identity in H by a path of the form

$$t \mapsto e^{(1-t)X_1} e^{(1-t)X_2} \cdots e^{(1-t)X_m}.$$

The group H_0 in (5.22) is a connected Lie subgroup of $\mathsf{GL}(2; \mathbb{C})$ whose Lie algebra is the algebra $\mathfrak{h}$ in (5.21).

We are now ready to state the main result of this section, which is our second major application of the Baker–Campbell–Hausdorff formula.

Theorem 5.20. *Let G be a matrix Lie group with Lie algebra $\mathfrak{g}$ and let $\mathfrak{h}$ be a Lie subalgebra of $\mathfrak{g}$. Then there exists a unique connected Lie subgroup H of G with Lie algebra $\mathfrak{h}$.*

If $\mathfrak{h}$ is the subalgebra of $\mathsf{gl}(2; \mathbb{C})$ in (5.21), then the connected Lie subgroup H is the group H_0 in (5.22), which is not closed. In practice, Theorem 5.20 is most useful in those cases where the connected Lie subgroup H turns out to be closed. See Proposition 5.24 and Exercises 10, 13, and 14 for conditions under which this is the case.

We now begin working toward the proof of Theorem 5.20. Since G is assumed to be a matrix Lie group, we may as well assume that $G = \mathsf{GL}(n; \mathbb{C})$. After all, if G is a closed subgroup of $\mathsf{GL}(n; \mathbb{C})$ and H is a connected Lie subgroup of $\mathsf{GL}(n; \mathbb{C})$ whose Lie algebra $\mathfrak{h}$ is contained in $\mathfrak{g}$, then H is also a connected Lie subgroup of G. We now let

$$H = \{ e^{X_1} e^{X_2} \cdots e^{X_N} \, \big| \, X_1, \ldots, X_N \in \mathfrak{h} \}, \tag{5.23}$$

which is a subgroup of G. The key issue is to prove that the Lie algebra of H, in the sense of Definition 5.18, is $\mathfrak{h}$. Once we know that $\mathrm{Lie}(H) = \mathfrak{h}$, we will immediately conclude that H is a connected Lie subgroup with Lie algebra $\mathfrak{h}$, the remaining properties in Definition 5.19 being true by definition. Note that for the claim $\mathrm{Lie}(H) = \mathfrak{h}$ to be true, it essential that $\mathfrak{h}$ be a sub*algebra* of $\mathsf{gl}(n; \mathbb{C})$, and not merely a sub*space*; compare Exercise 11.

As in the proof of Theorem 3.42, we think of $\mathsf{gl}(n; \mathbb{C})$ as $\mathbb{R}^{2n^2}$ and we decompose $\mathsf{gl}(n; \mathbb{C})$ as the direct sum of $\mathfrak{h}$ and D, where D is the orthogonal complement of $\mathfrak{h}$ with respect to the usual inner product on $\mathbb{R}^{2n^2}$. Then, as shown in the proof of Theorem 3.42, there exist neighborhoods U and V of the origin in $\mathfrak{h}$ and D,

respectively, and a neighborhood W of I in $\mathsf{GL}(n;\mathbb{C})$ with the following properties:
Each $A \in W$ can be written uniquely as

$$A = e^X e^Y, \quad X \in U, \, Y \in V, \tag{5.24}$$

in such a way that X and Y depend continuously on A. We think of the
decomposition in (5.24) as our local coordinates in a neighborhood of the identity
in $\mathsf{GL}(n;\mathbb{C})$.

If X is a small element of $\mathfrak{h}$, the decomposition of e^X is just $e^X e^0$. If we take the
product of two elements of the form $e^{X_1} e^{X_2}$, with X_1 and X_2 small elements of $\mathfrak{h}$,
then since $\mathfrak{h}$ is a subalgebra, if we combine the exponentials as $e^{X_1} e^{X_2} = e^{X_3}$ by
means of the Baker–Campbell–Hausdorff formula, X_3 will again be in $\mathfrak{h}$. Thus, if
we take a small number of products as in (5.23) with the X_j's being small elements
of $\mathfrak{h}$, we will move from the identity in the X-direction in the decomposition (5.24).
Globally, however, H may wind around and come back to points in W of the
form (5.24) with $Y \neq 0$. (See Figure 5.3.) Indeed, as the example of the "irrational
line" in (5.22) shows, there may be elements of H in W with arbitrarily small
nonzero values of Y. Nevertheless, we will see that the set of Y values that occurs
is at most countable.

Lemma 5.21. *Decompose* $\mathsf{gl}(n;\mathbb{C})$ *as* $\mathfrak{h} \oplus D$ *and let* V *be a neighborhood of the
origin in* D *as in (5.24). If* $E \subset V$ *is defined by*

$$E = \{Y \in V \,|\, e^Y \in H\},$$

then E *is at most countable.*

Assuming the lemma, we may now prove Theorem 5.20.

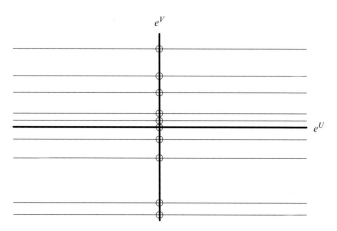

Fig. 5.3 The *black lines* indicate the portion of H in the set W. The group H intersects e^V in at
most countably many points

Proof of Theorem 5.20. As we have already observed, it suffices to show that the Lie algebra of H is $\mathfrak{h}$. Let $\mathfrak{h}'$ be the Lie algebra of H, which clearly contains $\mathfrak{h}$. For $Z \in \mathfrak{h}'$, we may write, for all sufficiently small t,

$$e^{tZ} = e^{X(t)} e^{Y(t)},$$

where $X(t) \in U \subset \mathfrak{h}$ and $Y(t) \in V \subset D$ and where $X(t)$ and $Y(t)$ are continuous functions of t. Since Z is in the Lie algebra of H, we have $e^{tZ} \in H$ for all t. Since, also, $e^{X(t)}$ is in the group H, we conclude that $e^{Y(t)}$ is in H for all sufficiently small t. If $Y(t)$ were not constant, then it would take on uncountably many values, which would mean that E is uncountable, violating Lemma 5.21. So, $Y(t)$ must be constant, and since $Y(0) = 0$, this means that $Y(t)$ is identically equal to zero. Thus, for small t, we have $e^{tZ} = e^{X(t)}$ and, therefore, $tZ = X(t) \in \mathfrak{h}$. This means $Z \in \mathfrak{h}$ and we conclude that $\mathfrak{h}' \subset \mathfrak{h}$. $\square$

Before proving Lemma 5.21, we prove another lemma.

Lemma 5.22. *Pick a basis for $\mathfrak{h}$ and call an element of $\mathfrak{h}$ **rational** if its coefficients with respect to this basis are rational. Then for every $\delta > 0$ and every $A \in H$, there exist rational elements $R_1, \ldots, R_m$ of $\mathfrak{h}$ such that*

$$A = e^{R_1} e^{R_2} \cdots e^{R_m} e^X,$$

where X is in $\mathfrak{h}$ and $\|X\| < \delta$.

Suppose we take δ small enough that the ball of radius δ in $\mathfrak{h}$ is contained in U. Then since there are only countably many m-tuples of the form $(R_1, \ldots, R_m)$ with R_j rational, the lemma tells us that H can be covered by countably many translates of the set e^U.

Proof. Choose $\varepsilon > 0$ so that for all $X, Y \in \mathfrak{h}$ with $\|X\| < \varepsilon$ and $\|Y\| < \varepsilon$, the Baker–Campbell–Hausdorff holds for X and Y. Let $C(\cdot, \cdot)$ denote the right-hand side of the formula, so that

$$e^X e^Y = e^{C(X,Y)}$$

whenever $\|X\|, \|Y\| < \varepsilon$. It is not hard to see that $C(\cdot, \cdot)$ is a continuous. Now, if the lemma holds for some δ, it also holds for any $\delta' > \delta$. Thus, it is harmless to assume δ is less than ε and small enough that if $\|X\|, \|Y\| < \delta$, we have $\|C(X, Y)\| < \varepsilon$.

Since $e^X = (e^{X/k})^k$, every element A of H can be written as

$$A = e^{X_1} \cdots e^{X_N} \tag{5.25}$$

with $X_j \in \mathfrak{h}$ and $\|X_j\| < \delta$. We now proceed by induction on N. If $N = 0$, then $A = I = e^0$, and there is nothing to prove. Assume the lemma for A's that can be expressed as in (5.25) for some integer N, and consider A of the form

$$A = e^{X_1} \cdots e^{X_N} e^{X_{N+1}} \tag{5.26}$$

with $X_j \in \mathfrak{h}$ and $\|X_j\| < \delta$. Applying our induction hypothesis to $e^{X_1} \cdots e^{X_N}$, we obtain

$$A = e^{R_1} \cdots e^{R_m} e^{X} e^{X_{N+1}}$$

$$= e^{R_1} \cdots e^{R_m} e^{C(X, X_{N+1})}.$$

where the R_j's are rational and $\|C(X, X_{N+1})\| < \varepsilon$. Since $\mathfrak{h}$ is a subalgebra of $\mathfrak{gl}(n; \mathbb{C})$, the element $C(X, X_{N+1})$ is again in $\mathfrak{h}$, but may not have norm less than δ.

Now choose a rational element R_{m+1} of $\mathfrak{h}$ that is very close to $C(X, X_{N+1})$ and such that $\|R_{m+1}\| < \varepsilon$. We then have

$$A = e^{R_1} \cdots e^{R_m} e^{R_{m+1}} e^{-R_{m+1}} e^{C(X, X_{N+1})}$$

$$= e^{R_1} \cdots e^{R_m} e^{R_{m+1}} e^{X'},$$

where

$$X' = C(-R_{m+1}, C(X, X_{N+1})).$$

Then X' will be in $\mathfrak{h}$, and by choosing R_{m+1} sufficiently close to $C(X, X_{N+1})$, we can make $\|X'\| < \delta$. After all, since $C(-Z, Z) = \log(e^{-Z} e^{Z}) = 0$ for all small Z, if Z' is close to Z, then $C(-Z', Z)$ will be small. $\qquad\square$

We now supply the proof of Lemma 5.21.

Proof of Lemma 5.21. Fix δ so that for all X and Y with $\|X\|, \|Y\| < \delta$, the quantity $C(X, Y)$ is defined and contained in U. We then claim that for each sequence $R_1, \ldots, R_m$ of rational elements in $\mathfrak{h}$, there is at most one $X \in \mathfrak{h}$ with $\|X\| < \delta$ such that the element

$$e^{R_1} e^{R_2} \cdots e^{R_m} e^{X} \tag{5.27}$$

belongs to e^V. After all, if we have

$$e^{R_1} e^{R_2} \cdots e^{R_m} e^{X_1} = e^{Y_1}, \tag{5.28}$$

$$e^{R_1} e^{R_2} \cdots e^{R_m} e^{X_2} = e^{Y_2} \tag{5.29}$$

with $Y_1, Y_2 \in V$, then

$$e^{-X_1} e^{X_2} = e^{-Y_1} e^{Y_2}$$

and so

$$e^{-Y_1} = e^{-X_1} e^{X_2} e^{-Y_2} = e^{C(-X_1, X_2)} e^{-Y_2},$$

with $C(-X_1, X_2) \in U$. However, each element of $e^U e^V$ has a *unique* representation as $e^Y e^X$ with $X \in U$ and $Y \in V$. Thus, we must have $-Y_2 = -Y_1$ and, by (5.28) and (5.29), $e^{X_1} = e^{X_2}$ and $X_1 = X_2$.

By Lemma 5.22, every element of H can be expressed in the form (5.27) with $\|X\| < \delta$. Now, there are only countably many rational elements in $\mathfrak{h}$ and thus only countably many expressions of the form $e^{R_1} \cdots e^{R_m}$, each of which produces at most one element of the form (5.27) that belongs to e^V. Thus, the set E in Lemma 5.21 is at most countable. $\square$

This completes the proof of Theorem 5.20.

If a connected Lie subgroup H of $\mathsf{GL}(n; \mathbb{C})$ is not closed, the topology H inherits from $\mathsf{GL}(n; \mathbb{C})$ may be pathological, e.g., not locally connected. (Compare Figure 1.1.) Nevertheless, we can give H a new topology that is much nicer.

Theorem 5.23. *Let H be a connected Lie subgroup of $\mathsf{GL}(n; \mathbb{C})$ with Lie algebra $\mathfrak{h}$. Then H can be given the structure of a smooth manifold in such a way that the group operations on H are smooth and the inclusion map of H into $\mathsf{GL}(n; \mathbb{C})$ is smooth.*

Thus, every connected Lie subgroup of $\mathsf{GL}(n; \mathbb{C})$ can be made into a Lie group. In the case of the group H_0 in (5.22), the new topology on H_0 is obtained by identifying H_0 with $\mathbb{R}$ by means of the parameter t in the definition of H_0.

Proof. For any $A \in H$ and any $\varepsilon > 0$, define

$$U_{A,\varepsilon} = \{Ae^X \,|\, X \in \mathfrak{h} \text{ and } \|X\| < \varepsilon\}.$$

Now define a topology on H as follows: A set $U \subset H$ is open if for each $A \in U$ there exists $\varepsilon > 0$ such that $U_{A,\varepsilon} \subset U$. (See Figure 5.4.) In this topology, two elements A and B of H are "close" if we can express B as $B = Ae^X$ with $X \in \mathfrak{h}$ and $\|X\|$ small. This topology is finer than the topology H inherits from G; that is, if A and B are close in this new topology, they are certainly close in the ordinary sense in G, but not vice versa.

It is easy to check that this topology is Hausdorff, and using Lemma 5.22, it is not hard to see that the topology is second countable. Furthermore, in this topology, H is locally homeomorphic to $\mathbb{R}^N$, where $N = \dim \mathfrak{h}$, by identifying each $U_{A,\varepsilon}$ with the ball of radius ε in $\mathfrak{h}$.

We may define a smooth structure on H by using the $U_{A,\varepsilon}$'s, with ε less than some small number ε_0, as our "atlas." If two of these sets overlap, then some element C of H can be written as $C = Ae^X = Be^Y$ for some $A, B \in H$ and $X, Y \in \mathfrak{h}$. It follows that $B = Ae^X e^{-Y}$, which means (since $\|X\|$ and $\|Y\|$ are less than ε_0) that A and B are close. The change-of-coordinates map is then $Y = \log(B^{-1}Ae^X)$. Since A and B are close and $\|X\|$ is small, we will have that $\|B^{-1}Ae^X - I\| < 1$, so that $B^{-1}Ae^X$ is in the domain where the matrix logarithm is defined and smooth. Thus, the change-of-coordinates map is smooth as function of X. Finally, in any of the coordinate neighborhoods $U_{A,\varepsilon}$, the inclusion of H into G is given by $X \mapsto Ae^X$, which is smooth as a function of X. $\square$

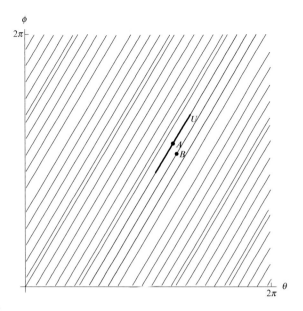

Fig. 5.4 The set U in H is open in the new topology but not in the topology inherited from $\mathsf{GL}(2; \mathbb{C})$. The element B is close to A in $\mathsf{GL}(2; \mathbb{C})$ but not in the new topology on H

As we have already noted, Theorem 5.20 is most useful in cases where the connected Lie subgroup H is actually closed. The following result gives one condition under which this is guaranteed to be the case. See also Exercises 10, 13, and 14.

Proposition 5.24. *Suppose $G \subset \mathsf{GL}(n; \mathbb{C})$ is a matrix Lie group with Lie algebra $\mathfrak{g}$ and that $\mathfrak{h}$ is a* maximal commutative *subalgebra of $\mathfrak{g}$, meaning that $\mathfrak{h}$ is commutative and $\mathfrak{h}$ is not contained in any larger commutative subalgebra of $\mathfrak{g}$. Then the connected Lie subgroup H of G with Lie algebra $\mathfrak{h}$ is closed.*

Proof. Since $\mathfrak{h}$ is commutative, H is also commutative, since every element of H is a product of exponentials of elements of $\mathfrak{h}$. It easily follows that the closure $\bar{H}$ of H in $\mathsf{GL}(n; \mathbb{C})$ is also commutative. We now claim that $\bar{H}$ is connected. To see this, take $A \in \bar{H}$, so that A is in G (since G is closed) and A is the limit of a sequence A_m in H. Since $\bar{H}$ is closed, Theorem 3.42 applies. Thus, for all sufficiently large m, the element AA_m^{-1} is expressible as $AA_m^{-1} = e^X$, for some X in the Lie algebra $\mathfrak{h}'$ of $\bar{H}$. Thus, $A = e^X A_m$, which means that A can be connected to A_m by the path $A(t) = e^{(1-t)X} A_m, 0 \le t \le 1$, in $\bar{H}$. Since A_m can be connected to the identity in $H \subset \bar{H}$, we see that A can be connected to the identity in $\bar{H}$.

Now, since $\bar{H}$ is commutative, its Lie algebra $\mathfrak{h}'$ is also commutative. But since $\mathfrak{h}$ was maximal commutative, we must have $\mathfrak{h}' = \mathfrak{h}$. Since, also, $\bar{H}$ is connected, we conclude that $\bar{H} = H$, showing that H is closed. $\qquad\square$

5.10 Lie's Third Theorem

Lie's third theorem (in its modern, global form) says that for every finite-dimensional, real Lie algebra $\mathfrak{g}$, there exists a Lie group G with Lie algebra $\mathfrak{g}$. We will construct G as a connected Lie subgroup of $\mathsf{GL}(n;\mathbb{C})$.

Theorem 5.25. *If $\mathfrak{g}$ is any finite-dimensional, real Lie algebra, there exists a connected Lie subgroup G of $\mathsf{GL}(n;\mathbb{C})$ whose Lie algebra is isomorphic to $\mathfrak{g}$.*

Our proof assumes Ado's theorem, which asserts that every finite-dimensional real or complex Lie algebra is isomorphic to an algebra of matrices. (See, for example, Theorem 3.17.7 in [Var].)

Proof. By Ado's theorem, we may identify $\mathfrak{g}$ with a real subalgebra of $\mathfrak{gl}(n;\mathbb{C})$. Then by Theorem 5.20, there is a connected Lie subgroup of $\mathsf{GL}(n;\mathbb{C})$ with Lie algebra $\mathfrak{g}$. □

It is actually possible to choose the subgroup G in Theorem 5.25 to be closed. Indeed, according to Theorem 9 on p. 105 of [Got], if a connected Lie group G can be embedded into some $\mathsf{GL}(n;\mathbb{C})$ as a connected Lie subgroup, then G can be embedded into some other $\mathsf{GL}(n';\mathbb{C})$ as a closed subgroup. Assuming this result, we may reach the following conclusion.

Conclusion 5.26. *Every finite-dimensional, real Lie algebra is isomorphic to the Lie algebra of some matrix Lie group.*

This result does not, however, mean that every Lie group is isomorphic to a matrix Lie group, since there can be several nonisomorphic Lie groups with the same Lie algebra. See, for example, Sect. 4.8.

5.11 Exercises

1. Let X be a linear transformation on a finite-dimensional real or complex vector space. Show that

$$\frac{I - e^{-X}}{X}$$

   is invertible if and only if none of the eigenvalues of X (over $\mathbb{C}$) is of the form $2\pi in$, with n an nonzero integer.

   Remark. This exercise, combined with the formula in Theorem 5.4, gives the following result (in the language of differentiable manifolds): The exponential map $\exp : \mathfrak{g} \to G$ is a local diffeomorphism near $X \in \mathfrak{g}$ if and only if $\mathrm{ad}_X : \mathfrak{g} \to \mathfrak{g}$ has no eigenvalue of the form $2\pi in$, with n a nonzero integer.

2. Show that for any X and Y in $M_n(\mathbb{C})$, *even if X and Y do not commute,*

$$\left. \frac{d}{dt} \operatorname{trace}(e^{X+tY}) \right|_{t=0} = \operatorname{trace}(e^X Y).$$

3. Compute $\log(e^X e^Y)$ through third order in X and Y by calculating directly with the power series for the exponential and the logarithm. Show this gives the same answer as the Baker–Campbell–Hausdorff formula.

4. Suppose that X and Y are upper triangular matrices with zeros on the diagonal. Show that the power series for $\log(e^X e^Y)$ is convergent. What happens to the series form of the Baker–Campbell–Hausdorff formula in this case?

5. Suppose X and Y are $n \times n$ complex matrices satisfying $[X, Y] = \alpha Y$ for some complex number α. Suppose further that there is no nonzero integer n such that $\alpha = 2\pi i n$. Show that

$$e^X e^Y = \exp\left\{ X + \frac{\alpha}{1 - e^{-\alpha}} Y \right\}.$$

Hint: Let $A(t) = e^X e^{tY}$ and let

$$B(t) = \exp\left\{ X + \frac{\alpha}{1 - e^{-\alpha}} tY \right\}.$$

Using Theorem 5.4, show that $A(t)$ and $B(t)$ satisfy the same differential equation with the same value at $t = 0$.

6. Give an example of matrices X and Y in $\mathsf{sl}(2; \mathbb{C})$ such that $[X, Y] = 2\pi i Y$ but such that there does not exist any Z in $\mathsf{sl}(2; \mathbb{C})$ with $e^X e^Y = e^Z$. Use Example 3.41 and compare Exercise 5.

7. Complete Step 4 in the proof of Theorem 5.6 by showing that Φ is a homomorphism. For all $A, B \in G$, choose a path $A(t)$ connecting I to A and a path $B(t)$ connecting I to B. Then define a path C connecting I to AB by setting $C(t) = B(2t)$ for $0 \le t \le 1/2$ and setting $C(t) = A(2t - 1)B$ for $1/2 \le t \le 1$. If $t_0, \ldots, t_m$ is a good partition for $A(t)$ and $s_0, \ldots, s_M$ is a good partition for $B(t)$, show that

$$\frac{s_0}{2}, \ldots, \frac{s_M}{2}, \frac{1}{2} + \frac{t_0}{2}, \ldots, \frac{1}{2} + \frac{t_m}{2}$$

is a good partition for $C(t)$. Now, compute $\Phi(A)$, $\Phi(B)$, and $\Phi(AB)$ using these paths and partitions and show that $\Phi(AB) = \Phi(A)\Phi(B)$.

8. If $\tilde{G}$ is a universal cover of a connected group G with projection map Φ, show that Φ maps $\tilde{G}$ onto G.

9. Prove the uniqueness of the universal cover, as stated in Proposition 5.13.

10. Let $\mathfrak{a}$ be a subalgebra of the Lie algebra of the Heisenberg group. Show that $\exp(\mathfrak{a})$ is a connected Lie subgroup of the Heisenberg group and that this subgroup is closed.

11. Consider the Lie algebra $\mathfrak{h}$ of the Heisenberg group H, as computed in Proposition 3.26. Let X, Y, and Z be the basis elements for $\mathfrak{h}$ in (4.18), which satisfy $[X, Y] = Z$ and $[X, Z] = [Y, Z] = 0$. Let V be the subspace of $\mathfrak{h}$ spanned by X and Y (which is *not* a subalgebra of $\mathfrak{h}$) and let K denote the subgroup of H consisting of products of exponential of elements of V. Show that $K = H$ and, thus, that the Lie algebra of K is *not* equal to V.
    Hint: Use Theorem 5.1 and the surjectivity of the exponential map for H (Exercise 18 in Chapter 3).

12. Show that every connected Lie subgroup of $\mathsf{SU}(2)$ is closed. Show that this is not the case for $\mathsf{SU}(3)$.

13. Let G be a matrix Lie group with Lie algebra $\mathfrak{g}$, let $\mathfrak{h}$ be a subalgebra of $\mathfrak{g}$, and let H be the unique connected Lie subgroup of G with Lie algebra $\mathfrak{h}$. Suppose that there exists a simply connected, *compact* matrix Lie group K such that the Lie algebra of K is isomorphic to $\mathfrak{h}$. Show that H is closed. Is H necessarily isomorphic to K?

14. This exercise asks you to prove, assuming Ado's theorem (Sect. 5.10), the following result: If G is a simply connected matrix Lie group with Lie algebra $\mathfrak{g}$ and $\mathfrak{h}$ is an *ideal* in $\mathfrak{g}$, then the connected Lie subgroup H with Lie algebra $\mathfrak{h}$ is closed.

    (a) Show that there exists a Lie algebra homomorphism $\phi : \mathfrak{g} \to \mathsf{gl}(N; \mathbb{C})$ with $\ker(\phi) = \mathfrak{h}$.
        Hint: Since $\mathfrak{h}$ is an ideal in $\mathfrak{g}$, the quotient space $\mathfrak{g}/\mathfrak{h}$ has a natural Lie algebra structure.

    (b) Since G is simply connected, there exists a Lie group homomorphism $\Phi : G \to \mathsf{gl}(N; \mathbb{C})$ for which the associated Lie algebra homomorphism is ϕ. Show that the identity component of the kernel of Φ is a closed subgroup of G whose Lie algebra is $\mathfrak{h}$.

    (c) Show that the result fails if the assumption that G be simply connected is omitted.

Part II
Semisimple Lie Algebras

Chapter 6
The Representations of $\mathsf{sl}(3; \mathbb{C})$

6.1 Preliminaries

In this chapter, we investigate the representations of the Lie algebra $\mathsf{sl}(3; \mathbb{C})$, which is the complexification of the Lie algebra of the group $\mathsf{SU}(3)$. The main result of this chapter is Theorem 6.7, which states that an irreducible finite-dimensional representation of $\mathsf{sl}(3; \mathbb{C})$ can be classified in terms of its "highest weight." This result is analogous to the results of Sect. 4.6, in which we classify the irreducible representations by the largest eigenvalue of $\pi(H)$, namely the non-negative integer m.

The results of this chapter are special cases of the general theory of representations of semisimple Lie algebras (Chapters 7 and 9) and of the theory of representations of compact Lie groups (Chapters 11 and 12). It is nevertheless useful to consider this case separately, in part because of the importance of $\mathsf{SU}(3)$ in physical applications but mainly because seeing roots, weights, and the Weyl group "in action" in a simple example motivates the introduction of these structures later in a more general setting.

Every finite-dimensional representation of $\mathsf{SU}(3)$ (over a complex vector space) gives rise to a representation of $\mathsf{su}(3)$, which can then be extended by complex linearity to $\mathsf{sl}(3; \mathbb{C}) \cong \mathsf{su}(3)_{\mathbb{C}}$. Since $\mathsf{SU}(3)$ is simply connected, we can go in the opposite direction by restricting any representation of $\mathsf{sl}(3; \mathbb{C})$ to $\mathsf{su}(3)$ and then applying Theorem 5.6 to obtain a representation of $\mathsf{SU}(3)$. Propositions 4.5 and 4.6 tell us that a representation of $\mathsf{SU}(3)$ is irreducible if and only if the associated representation of $\mathsf{sl}(3; \mathbb{C})$ is irreducible, thus establishing a one-to-one correspondence

A previous version of this book was inadvertently published without the middle initial of the author's name as "Brian Hall". For this reason an erratum has been published, correcting the mistake in the previous version and showing the correct name as Brian C. Hall (see DOI http://dx.doi.org/10.1007/978-3-319-13467-3_14). The version readers currently see is the corrected version. The Publisher would like to apologize for the earlier mistake.

© Springer International Publishing Switzerland 2015
B.C. Hall, *Lie Groups, Lie Algebras, and Representations*, Graduate
Texts in Mathematics 222, DOI 10.1007/978-3-319-13467-3_6

between the irreducible representations of $SU(3)$ and the irreducible representations of $sl(3; \mathbb{C})$. Furthermore, since $SU(3)$ is compact, Theorem 4.28 then tells us that all finite-dimensional representations of $SU(3)$—and thus, also, of $sl(3; \mathbb{C})$—are completely reducible.

It is desirable, however, to avoid relying unnecessarily on Theorem 5.6, which in turn relies on the Baker–Campbell–Hausdorff formula. If we look the representations from the Lie algebra point of view, we can classify the *irreducible* representations of $sl(3; \mathbb{C})$ without knowing that they come from representations of $SU(3)$. Of course, classifying the irreducible representations of $sl(3; \mathbb{C})$ does not tell one what a general representation of $sl(3; \mathbb{C})$ looks like, unless one knows complete reducibility. Nevertheless, it is possible to give an algebraic proof of complete reducibility, without referring to the group $SU(3)$. This proof is given in the setting of general semisimple Lie algebras in Sect. 10.3, but it should be fairly easy to specialize the argument to the $sl(3; \mathbb{C})$ case.

Meanwhile, if we look at the representations from the group point of view, we can construct the irreducible representations of $SU(3)$ without knowing that every representation of $sl(3; \mathbb{C})$ gives rise to a representation of $SU(3)$. Indeed, the irreducible representations of $SU(3)$ are constructed as subspaces of tensor products of several copies of the standard representation with several copies of the dual of the standard representation. Since the standard representation and its dual are defined directly at the level of the group $SU(3)$, there is no need to appeal to Theorem 5.6.

In short, this chapter provides a self-contained classification of the irreducible representations of both $SU(3)$ and $sl(3; \mathbb{C})$, without needing to know the results of Chapter 5. We establish results for $sl(3; \mathbb{C})$ first, and then pass to $SU(3)$ (Theorem 6.8).

6.2   Weights and Roots

We will use the following basis for $sl(3; \mathbb{C})$:

$$H_1 = \begin{pmatrix} 1 & 0 & 0 \\ 0 & -1 & 0 \\ 0 & 0 & 0 \end{pmatrix}, \quad H_2 = \begin{pmatrix} 0 & 0 & 0 \\ 0 & 1 & 0 \\ 0 & 0 & -1 \end{pmatrix},$$

$$X_1 = \begin{pmatrix} 0 & 1 & 0 \\ 0 & 0 & 0 \\ 0 & 0 & 0 \end{pmatrix}, \quad X_2 = \begin{pmatrix} 0 & 0 & 0 \\ 0 & 0 & 1 \\ 0 & 0 & 0 \end{pmatrix}, \quad X_3 = \begin{pmatrix} 0 & 0 & 1 \\ 0 & 0 & 0 \\ 0 & 0 & 0 \end{pmatrix},$$

$$Y_1 = \begin{pmatrix} 0 & 0 & 0 \\ 1 & 0 & 0 \\ 0 & 0 & 0 \end{pmatrix}, \quad Y_2 = \begin{pmatrix} 0 & 0 & 0 \\ 0 & 0 & 0 \\ 0 & 1 & 0 \end{pmatrix}, \quad Y_3 = \begin{pmatrix} 0 & 0 & 0 \\ 0 & 0 & 0 \\ 1 & 0 & 0 \end{pmatrix}.$$

Note that the span $\langle H_1, X_1, Y_1 \rangle$ of H_1, X_1, and Y_1 is a subalgebra of $\mathsf{sl}(3;\mathbb{C})$ isomorphic to $\mathsf{sl}(2;\mathbb{C})$, as can be seen by ignoring the third row and the third column in each matrix. The subalgebra $\langle H_2, X_2, Y_2 \rangle$ is also, similarly, isomorphic to $\mathsf{sl}(2;\mathbb{C})$. Thus, we have the following commutation relations:

$$\begin{aligned}
[H_1, X_1] &= 2X_1, & [H_2, X_2] &= 2X_2, \\
[H_1, Y_1] &= -2Y_1, & [H_2, Y_2] &= -2Y_2, \\
[X_1, Y_1] &= H_1, & [X_2, Y_2] &= H_2.
\end{aligned}$$

We now list all of the commutation relations among the basis elements which involve at least one of H_1 and H_2. (This includes some repetitions of the above commutation relations.)

$$[H_1, H_2] = 0;$$

$$\begin{aligned}
[H_1, X_1] &= 2X_1, & [H_1, Y_1] &= -2Y_1, \\
[H_2, X_1] &= -X_1, & [H_2, Y_1] &= Y_1;
\end{aligned}$$

$$\begin{aligned}
[H_1, X_2] &= -X_2, & [H_1, Y_2] &= Y_2, \\
[H_2, X_2] &= 2X_2, & [H_2, Y_2] &= -2Y_2;
\end{aligned}$$

$$\begin{aligned}
[H_1, X_3] &= X_3, & [H_1, Y_3] &= -Y_3, \\
[H_2, X_3] &= X_3, & [H_2, Y_3] &= -Y_3.
\end{aligned}$$

(6.1)

Finally, we list all of the remaining commutation relations.

$$\begin{aligned}
[X_1, Y_1] &= H_1, \\
[X_2, Y_2] &= H_2, \\
[X_3, Y_3] &= H_1 + H_2;
\end{aligned}$$

$$\begin{aligned}
[X_1, X_2] &= X_3, & [Y_1, Y_2] &= -Y_3, \\
[X_1, Y_2] &= 0, & [X_2, Y_1] &= 0;
\end{aligned}$$

$$\begin{aligned}
[X_1, X_3] &= 0, & [Y_1, Y_3] &= 0, \\
[X_2, X_3] &= 0, & [Y_2, Y_3] &= 0;
\end{aligned}$$

$$\begin{aligned}
[X_2, Y_3] &= Y_1, & [X_3, Y_2] &= X_1, \\
[X_1, Y_3] &= -Y_2, & [X_3, Y_1] &= -X_2.
\end{aligned}$$

All of our analysis of the representations of $\mathsf{sl}(3;\mathbb{C})$ will be in terms of the above basis. From now on, all representations of $\mathsf{sl}(3;\mathbb{C})$ will be assumed to be finite dimensional and complex linear.

Our basic strategy in classifying the representations of $\mathsf{sl}(3;\mathbb{C})$ is to simultaneously diagonalize $\pi(H_1)$ and $\pi(H_2)$. (See Sect. A.8 for information on simultaneous diagonalization.) Since H_1 and H_2 commute, $\pi(H_1)$ and $\pi(H_2)$ will also commute (for any representation π) and so there is at least a chance that $\pi(H_1)$ and $\pi(H_2)$ can be simultaneously diagonalized. (Compare Proposition A.16.)

Definition 6.1. If (π, V) is a representation of sl(3; $\mathbb{C}$), then an ordered pair $\mu = (m_1, m_2) \in \mathbb{C}^2$ is called a **weight** for π if there exists $v \neq 0$ in V such that

$$\pi(H_1)v = m_1 v,$$
$$\pi(H_2)v = m_2 v. \tag{6.2}$$

A nonzero vector v satisfying (6.2) is called a **weight vector** corresponding to the weight μ. If $\mu = (m_1, m_2)$ is a weight, then the space of all vectors v satisfying (6.2) is the **weight space** corresponding to the weight μ. The **multiplicity** of a weight is the dimension of the corresponding weight space.

Thus, a weight is simply a pair of simultaneous eigenvalues for $\pi(H_1)$ and $\pi(H_2)$. It is easily shown that isomorphic representations have the same weights and multiplicities.

Proposition 6.2. *Every representation of* sl(3; $\mathbb{C}$) *has at least one weight.*

Proof. Since we are working over the complex numbers, $\pi(H_1)$ has at least one eigenvalue $m_1 \in \mathbb{C}$. Let $W \subset V$ be the eigenspace for $\pi(H_1)$ with eigenvalue m_1. Since $[H_1, H_2] = 0$, $\pi(H_2)$ commutes with $\pi(H_1)$, and, so, by Proposition A.2, $\pi(H_2)$ must map W into itself. Then the restriction of $\pi(H_2)$ to W must have at least one eigenvector w with eigenvalue $m_2 \in \mathbb{C}$, which means that w is a simultaneous eigenvector for $\pi(H_1)$ and $\pi(H_2)$ with eigenvalues m_1 and m_2. $\square$

Every representation π of sl(3; $\mathbb{C}$) can be viewed, by restriction, as a representation of the subalgebras $\langle H_1, X_1, Y_1 \rangle$ and $\langle H_2, X_2, Y_2 \rangle$, both of which are isomorphic to sl(2; $\mathbb{C}$).

Proposition 6.3. *If* (π, V) *is a representation of* sl(3; $\mathbb{C}$) *and* $\mu = (m_1, m_2)$ *is a weight of* V, *then both* m_1 *and* m_2 *are integers.*

Proof. Apply Point 1 of Theorem 4.34 to the restriction of π to $\langle H_1, X_1, Y_1 \rangle$ and to the restriction of π to $\langle H_2, X_2, Y_2 \rangle$. $\square$

Our strategy now is to begin with one simultaneous eigenvector for $\pi(H_1)$ and $\pi(H_2)$ and then to apply $\pi(X_j)$ or $\pi(Y_j)$ and see what the effect is. The following definition is relevant in this context.

Definition 6.4. An ordered pair $\alpha = (a_1, a_2) \in \mathbb{C}^2$ is called a **root** if

1. a_1 and a_2 are not both zero, and
2. there exists a nonzero $Z \in$ sl(3; $\mathbb{C}$) such that

$$[H_1, Z] = a_1 Z,$$
$$[H_2, Z] = a_2 Z.$$

The element Z is called a **root vector** corresponding to the root α.

Condition 2 in the definition says that Z is a simultaneous eigenvector for ad_{H_1} and ad_{H_2}. This means that Z is a weight vector for the adjoint representation with weight (a_1, a_2). Thus, taking into account Condition 1, we may say that the roots are precisely the *nonzero weights of the adjoint representation*. The commutation relations (6.1) tell us that we have the following six roots for $\mathsf{sl}(3; \mathbb{C})$:

$$
\begin{array}{cc@{\qquad}cc}
\alpha & Z & \alpha & Z \\
\hline
(2, -1) & X_1 & (-2, 1) & Y_1 \\
(-1, 2) & X_2 & (1, -2) & Y_2 \\
(1, 1) & X_3 & (-1, -1) & Y_3
\end{array}
\tag{6.3}
$$

Note that H_1 and H_2 are also simultaneous eigenvectors for ad_{H_1} and ad_{H_2}, but they are not root vectors because the simultaneous eigenvalues are both zero. Since the vectors in (6.3), together with H_1 and H_2, form a basis for $\mathsf{sl}(3; \mathbb{C})$, it is not hard to show that the roots listed in (6.3) are the only roots (Exercise 1). These six roots form a "root system," conventionally called A_2. (For much more information about root systems, see Chapter 8.)

It is convenient to single out the two roots corresponding to X_1 and X_2:

$$
\alpha_1 = (2, -1); \quad \alpha_2 = (-1, 2),
\tag{6.4}
$$

which we call the **positive simple roots**. They have the property that all of the roots can be expressed as linear combinations of α_1 and α_2 with *integer* coefficients, and these coefficients are (for each root) either all greater than or equal to zero or all less than or equal to zero. This is verified by direct computation:

$$
(2, -1) = \alpha_1; \quad (-1, 2) = \alpha_2; \quad (1, 1) = \alpha_1 + \alpha_2,
$$

with the remaining three roots being the negatives of the ones above. The decision to designate α_1 and α_2 as the positive simple roots is arbitrary; any other pair of roots with similar properties would do just as well.

The significance of the roots for the representation theory of $\mathsf{sl}(3; \mathbb{C})$ is contained in the following lemma, which is the analog of Lemma 4.33 in the $\mathsf{sl}(2; \mathbb{C})$ case.

Lemma 6.5. *Let* $\alpha = (a_1, a_2)$ *be a root and let* $Z_\alpha \in \mathsf{sl}(3; \mathbb{C})$ *be a corresponding root vector. Let* π *be a representation of* $\mathsf{sl}(3; \mathbb{C})$, *let* $\mu = (m_1, m_2)$ *be a weight for* π, *and let* $v \neq 0$ *be a corresponding weight vector. Then we have*

$$
\pi(H_1)\pi(Z_\alpha)v = (m_1 + a_1)\pi(Z_\alpha)v,
$$
$$
\pi(H_2)\pi(Z_\alpha)v = (m_2 + a_2)\pi(Z_\alpha)v.
$$

Thus, either $\pi(Z_\alpha)v = 0$ *or* $\pi(Z_\alpha)v$ *is a new weight vector with weight*

$$
\mu + \alpha = (m_1 + a_1, m_2 + a_2).
$$

Proof. By the definition of a root, we have the commutation relation $[H_1, Z_\alpha] = a_1 Z_\alpha$. Thus,

$$\pi(H_1)\pi(Z_\alpha)v = (\pi(Z_\alpha)\pi(H_1) + a_1\pi(Z_\alpha)) v$$
$$= \pi(Z_\alpha)(m_1 v) + a_1\pi(Z_\alpha)v$$
$$= (m_1 + a_1)\pi(Z_\alpha)v.$$

A similar argument allows us to compute $\pi(H_2)\pi(Z_\alpha)v$.   □

6.3   The Theorem of the Highest Weight

If we have a representation with a weight $\mu = (m_1, m_2)$, then by applying the root vectors X_1, X_2, X_3, Y_1, Y_2, and Y_3, we obtain new weights of the form $\mu + \alpha$, where α is the root. Of course, some of the time, $\pi(Z_\alpha)v$ will be zero, in which case $\mu + \alpha$ is not necessarily a weight. In fact, since our representation is finite dimensional, there can be only finitely many weights, so we must get zero quite often. By analogy to the classification of the representations of sl(2; ℂ), we would like to single out in each representation a "highest" weight and then work from there. The following definition gives the "right" notion of highest.

Definition 6.6. Let $\alpha_1 = (2, -1)$ and $\alpha_2 = (-1, 2)$ be the roots introduced in (6.4). Let μ_1 and μ_2 be two weights. Then μ_1 is **higher** than μ_2 (or, equivalently, μ_2 is **lower** than μ_1) if $\mu_1 - \mu_2$ can be written in the form

$$\mu_1 - \mu_2 = a\alpha_1 + b\alpha_2 \tag{6.5}$$

with $a \geq 0$ and $b \geq 0$. This relationship is written as $\mu_1 \succeq \mu_2$ or $\mu_2 \preceq \mu_1$.

If π is a representation of sl(3; ℂ), then a weight μ_0 for π is said to be a **highest weight** if for all weights μ of π, $\mu \preceq \mu_0$.

Note that the relation of "higher" is only a *partial* ordering; for example, $\alpha_1 - \alpha_2$ is neither higher nor lower than 0. In particular, a finite set of weights need not have a highest element. Note also that the coefficients a and b in (6.5) do not have to be integers, even if both μ_1 and μ_2 have integer entries. For example, $(1,0)$ is higher than $(0,0)$ since $(1,0) = \frac{2}{3}\alpha_1 + \frac{1}{3}\alpha_2$.

We are now ready to state the main theorem regarding the irreducible representations of sl(3; ℂ), the *theorem of the highest weight*. The proof of the theorem is found in Sect. 6.4.

Theorem 6.7. *1. Every irreducible representation π of sl(3; ℂ) is the direct sum of its weight spaces.*
2. Every irreducible representation of sl(3; ℂ) has a unique highest weight μ.

3. *Two irreducible representations of* $\mathsf{sl}(3;\mathbb{C})$ *with the same highest weight are isomorphic.*
4. *The highest weight* μ *of an irreducible representation must be of the form*

$$\mu = (m_1, m_2),$$

    where m_1 *and* m_2 *are* non-negative *integers.*
5. *For every pair* (m_1, m_2) *of non-negative integers, there exists an irreducible representation of* $\mathsf{sl}(3;\mathbb{C})$ *with highest weight* (m_1, m_2).

We will also prove (without appealing to Theorem 5.6) a similar result for the group $\mathsf{SU}(3)$. Since every irreducible representation of $\mathsf{SU}(3)$ gives rise to an irreducible representation of $\mathsf{sl}(3;\mathbb{C}) \cong \mathsf{su}(3)_\mathbb{C}$, the only nontrivial matter is to prove Point 5 for $\mathsf{SU}(3)$.

Theorem 6.8. *For every pair* (m_1, m_2) *of non-negative integers, there exists an irreducible representation* Π *of* $\mathsf{SU}(3)$ *such that the associated representation* π *of* $\mathsf{sl}(3;\mathbb{C})$ *has highest weight* (m_1, m_2).

One might naturally attempt to construct representations of $\mathsf{SU}(3)$ by a method similar to that used in Example 4.10, acting on spaces of homogeneous polynomials on $\mathbb{C}^3$. This is, indeed, possible and the resulting representations of $\mathsf{SU}(3)$ turn out to be irreducible. Not every irreducible representation of $\mathsf{SU}(3)$, however, arises in this way, but only those with highest weight of the form $(0, m)$. See Exercise 8.

For $\lambda = (m_1, m_2) \in \mathbb{C}^2$, we may say that λ is an **integral element** if m_1 and m_2 are integers and that λ is **dominant** if m_1 and m_2 are real and non-negative. Thus, the set of possible highest weights in Theorem 6.7 are the *dominant integral elements*. Figure 6.1 shows the roots and dominant integral elements for $\mathsf{sl}(3;\mathbb{C})$.

Fig. 6.1 The roots (*arrows*) and dominant integral elements (*black dots*), shown in the obvious basis

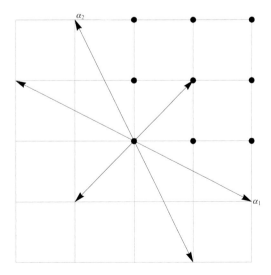

This picture is made using the obvious basis for the space of weights; that is, the x-coordinate is the eigenvalue of H_1 and the y-coordinate is the eigenvalue of H_2. Once we have introduced the Weyl group (Sect. 6.6), we will see the same picture rendered using a Weyl-invariant inner product, which will give a more symmetric view of the situation.

Note the parallels between this result and the classification of the irreducible representations of sl(2; ℂ): In each irreducible representation of sl(2; ℂ), $\pi(H)$ is diagonalizable, and there is a largest eigenvalue of $\pi(H)$. Two irreducible representations of sl(2; ℂ) with the same largest eigenvalue are isomorphic. The highest eigenvalue is always a non-negative integer and every non-negative integer is the highest weight of some irreducible representation.

6.4   Proof of the Theorem

The proof consists of a series of propositions.

Proposition 6.9. *In every irreducible representation* (π, V) *of* sl(3; ℂ), *the operators* $\pi(H_1)$ *and* $\pi(H_2)$ *can be simultaneously diagonalized; that is,* V *is the direct sum of its weight spaces.*

Proof. Let W be the sum of the weight spaces in V. Equivalently, W is the space of all vectors $w \in V$ such that w can be written as a linear combination of simultaneous eigenvectors for $\pi(H_1)$ and $\pi(H_2)$. Since (Proposition 6.2) π always has at least one weight, $W \neq \{0\}$.

On the other hand, Lemma 6.5 tells us that if Z_α is a root vector corresponding to the root α, then $\pi(Z_\alpha)$ maps the weight space corresponding to μ into the weight space corresponding to $\mu + \alpha$. Thus, W is invariant under the action of each of the root vectors, $X_1, X_2, X_3, Y_1, Y_2,$ and Y_3. Since W is certainly also invariant under the action of H_1 and H_2, W is invariant under all of sl(3; ℂ). Thus, by irreducibility, $W = V$. Finally, since, by Proposition A.17, weight vectors with distinct weights are independent, V is actually the direct sum of its weight spaces.   □

Definition 6.10. A representation (π, V) of sl(3; ℂ) is said to be a **highest weight cyclic representation with weight** $\mu = (m_1, m_2)$ if there exists $v \neq 0$ in V such that

1. v is a weight vector with weight μ,
2. $\pi(X_j)v = 0$, for $j = 1, 2, 3$,
3. the smallest invariant subspace of V containing v is all of V.

Proposition 6.11. *Let* (π, V) *be a highest weight cyclic representation of* sl(3; ℂ) *with weight* μ. *Then the following results hold.*

1. *The representation* π *has highest weight* μ.
2. *The weight space corresponding to the weight* μ *is one dimensional.*

Before turning to the proof of this proposition, let us record a simple lemma that applies to arbitrary Lie algebras and which will be useful also in the setting of general semisimple Lie algebras.

Lemma 6.12 (Reordering Lemma). *Suppose that $\mathfrak{g}$ is any Lie algebra and that π is a representation of $\mathfrak{g}$. Suppose that $X_1, \ldots, X_m$ is an ordered basis for $\mathfrak{g}$ as a vector space. Then any expression of the form*

$$\pi(X_{j_1})\pi(X_{j_2}) \cdots \pi(X_{j_N}), \tag{6.6}$$

can be expressed as a linear combination of terms of the form

$$\pi(X_m)^{k_m} \pi(X_{m-1})^{k_{m-1}} \cdots \pi(X_1)^{k_1} \tag{6.7}$$

where each k_l is a non-negative integer and where $k_1 + k_2 + \cdots + k_m \leq N$.

Proof. The idea is to use the commutation relations of $\mathfrak{g}$ to re-order the factors into the desired order, at the expense of generating terms with one fewer factors, which then be handled by the same method. To be more formal, we use induction on N. If $N = 1$, there is nothing to do: Any expression of the form $\pi(X_j)$ is of the form (6.7) with $k_j = 1$ and all the other k_l's equal to zero. Assume, then, that the result holds for a product of at most N factors, and consider an expression of the form (6.6) with $N + 1$ factors. By induction, we can assume that the last N factors are in the desired form, giving an expression of the form

$$\pi(X_j)\pi(X_m)^{k_m} \pi(X_{m-1})^{k_{m-1}} \cdots \pi(X_1)^{k_1}$$

with $k_1 + \cdots + k_m = N$.

We now move the factor of $\pi(X_j)$ to the right one step at a time until it is in the right spot. Each time we have $\pi(X_j)\pi(X_k)$ somewhere in the expression we use the relation

$$\pi(X_j)\pi(X_k) = \pi(X_k)\pi(X_j) + \pi([X_j, X_k])$$

$$= \pi(X_k)\pi(X_j) + \sum_l c_{jkl}\pi(X_l),$$

where the constants c_{jkl} are the structure constants for the basis $\{X_j\}$ (Definition 3.10). Each commutator term has at most at most N factors. Thus, we ultimately obtain several terms with N factors, which can be handled by induction, and one term with N factors that is of the desired form (once $\pi(X_j)$ finally gets to the right spot). $\qquad\square$

We now proceed with the proof of Proposition 6.11.

Proof. Let v be as in the definition. Consider the subspace W of V spanned by elements of the form

$$w = \pi(Y_{j_1})\pi(Y_{j_2})\cdots\pi(Y_{j_N})v \tag{6.8}$$

with each j_l equal to 1, 2, or 3 and $N \geq 0$. (If $N = 0$, then $w = v$.) We now claim that W is invariant. We take as our basis for sl(3; ℂ) the elements X_1, X_2, X_3, H_1, H_2, Y_1, Y_2, and Y_3, in that order. If we apply a basis element to w, the lemma tells us that we can rewrite the resulting vector as a linear combination of terms in which the $\pi(X_j)$'s act first, the $\pi(H_j)$'s act second, and the $\pi(Y_j)$'s act last, and all of these are applied to the vector v. Since v is annihilated by each $\pi(X_j)$, any term having a positive power of any X_j is simply zero. Since v is an eigenvector for each $\pi(H_j)$, any factors of $\pi(H_j)$ acting on v can be replaced by constants. That leaves only factors of $\pi(Y_j)$ applied to v, which means that we have a linear combination of vectors of the form (6.8). Thus, W is invariant and contains v, so $W = V$.

Now, Y_1, Y_2, and Y_3 are root vectors with roots $-\alpha_1$, $-\alpha_2$, and $-\alpha_1 - \alpha_2$, respectively. Thus, by Lemma 6.5, each element of the form (6.8) with $N > 0$ is a weight vector with weight lower than μ. Thus, the only weight vectors with weight μ are multiples of v.   □

Proposition 6.13. *Every irreducible representation of* sl(3; ℂ) *is a highest weight cyclic representation, with a unique highest weight* μ.

Proof. We have already shown that every irreducible representation π is the direct sum of its weight spaces. Since the representation is finite dimensional, there can be only finitely many weights, so there must be a *maximal* weight μ, that is, such that there is no weight strictly higher than μ. Thus, for any nonzero weight vector v with weight μ, we must have

$$\pi(X_j)v = 0, \quad j = 1, 2, 3.$$

Since π is irreducible, the smallest invariant subspace containing v must be the whole space; therefore, the representation is highest weight cyclic.   □

Proposition 6.14. *Suppose* (π, V) *is a* completely reducible *representation of* sl(3; ℂ) *that is also highest weight cyclic. Then* π *is irreducible.*

As it turns out, *every* finite-dimensional representation of sl(3; ℂ) is completely reducible. This claim can be verified analytically (by passing to the simply connected group SU(3) and using Theorem 4.28) or algebraically (as in Sect. 10.3). We do not, however, require this result here, since we will only apply Proposition 6.14 to representations that are manifestly completely reducible.

Meanwhile, it is tempting to think that *any* representation with a cyclic vector (that is, a vector satisfying Point 3 of Definition 6.10) must be irreducible, but this is false. (What is true is that if *every* nonzero vector in a representation is cyclic, then the representation is irreducible.) Thus, Proposition 6.14 relies on the special form of the cyclic vector in Definition 6.10.

Proof. Let (π, V) be a highest weight cyclic representation with highest weight μ and let v be a weight vector with weight μ. By assumption, V decomposes as a direct sum of irreducible representations

$$V \cong \bigoplus_j V_j. \tag{6.9}$$

By Proposition 6.9, each of the V_j's is the direct sum of its weight spaces. Since the weight μ occurs in V, it must occur in some V_j (compare the last part of Proposition A.17). But by Proposition 6.11, v is (up to a constant) the *only* vector in V with weight μ. Thus, V_j is an invariant subspace containing v, which means that $V_j = V$. There is, therefore, only one term in the sum (6.9), and V is irreducible. $\square$

Proposition 6.15. *Two irreducible representations of* $\mathsf{sl}(3; \mathbb{C})$ *with the same highest weight are isomorphic.*

Proof. Suppose (π, V) and (σ, W) are irreducible representations with the same highest weight μ and let v and w be the highest weight vectors for V and W, respectively. Consider the representation $V \oplus W$ and let U be smallest invariant subspace of $V \oplus W$ which contains the vector (v, w). Then U is a highest weight cyclic representation. Furthermore, since $V \oplus W$ is, by definition, completely reducible, it follows from Proposition 4.26 that U is completely reducible. Thus, by Proposition 6.14, U is irreducible.

Consider now the two "projection" maps P_1 and P_2, mapping $V \oplus W$ to V and W, respectively, and given by

$$P_1(v, w) = v; \quad P_2(v, w) = w.$$

Since P_1 and P_2 are easily seen to be intertwining maps, their restrictions to $U \subset V \oplus W$ are also intertwining maps. Now, neither $P_1|_U$ nor $P_2|_U$ is the zero map, since both are nonzero on (v, w). Moreover, U, V, and W are all irreducible. Therefore, by Schur's lemma, $P_1|_U$ is an isomorphism of U with V and $P_2|_U$ is an isomorphism of U with W, showing that $V \cong U \cong W$. $\square$

Proposition 6.16. *If* π *is an irreducible representation of* $\mathsf{sl}(3; \mathbb{C})$ *with highest weight* $\mu = (m_1, m_2)$, *then* m_1 *and* m_2 *non-negative integers.*

Proof. By Proposition 6.3, m_1 and m_2 are integers. If v is a weight vector with weight μ, then $\pi(X_1)v$ and $\pi(X_2)v$ must be zero, or μ would not be the highest weight for π. Thus, if we then apply Point 1 of Theorem 4.34 to the restrictions of π to $\langle H_1, X_1, Y_1 \rangle$ and $\langle H_2, X_2, Y_2 \rangle$, we conclude that m_1 and m_2 are non-negative. $\square$

Proposition 6.17. *If* m_1 *and* m_2 *are non-negative integers, then there exists an irreducible representation of* $\mathsf{sl}(3; \mathbb{C})$ *with highest weight* $\mu = (m_1, m_2)$.

Proof. Since the trivial representation is an irreducible representation with highest weight $(0,0)$, we need only construct representations with at least one of m_1 and m_2 positive.

First, we construct two irreducible representations, with highest weights $(1,0)$ and $(0,1)$, which we call the **fundamental representations**. The standard representation of sl(3; $\mathbb{C}$), acting on $\mathbb{C}^3$ in the obvious say, is easily seen to be irreducible. It has weight vectors e_1, e_2, and e_3, with corresponding weights $(1,0)$, $(-1,1)$, and $(0,-1)$, and with highest weight is $(1,0)$. The dual of the standard representation, given by

$$\pi(Z) = -Z^{tr} \tag{6.10}$$

for all $Z \in$ sl(3; $\mathbb{C}$), is also irreducible. It also has weight vectors e_1, e_2, and e_3, with corresponding weights $(-1,0)$, $(1,-1)$, and $(0,1)$ and with highest weight $(0,1)$.

Let (π_1, V_1) and (π_2, V_2) be the standard representation and its dual, respectively, and let $v_1 = e_1$ and $v_2 = e_3$ be the respective highest weight vectors. Now, consider the representation π_{m_1,m_2} given by

$$(V_1 \otimes \cdots \otimes V_1) \otimes (V_2 \otimes \cdots \otimes V_2), \tag{6.11}$$

where V_1 occurs m_1 times and V_2 occurs m_2 times. The action of sl(3; $\mathbb{C}$) on this space is given by the obvious extension of Definition 4.20 to multiple factors. It then easy to check that the vector

$$v_{m_1,m_2} = v_1 \otimes v_1 \cdots \otimes v_1 \otimes v_2 \otimes v_2 \cdots \otimes v_2$$

is a weight vector with weight (m_1, m_2) and that v_{m_1,m_2} is annihilated by $\pi_{m_1,m_2}(X_j)$, $j = 1, 2, 3$.

Now let W be the smallest invariant subspace containing v_{m_1,m_2}. Assuming that π_{m_1,m_2} is completely reducible, W will also be completely reducible and Proposition 6.14 will tell us that W is the desired irreducible representation with highest weight (m_1, m_2).

It remains only to establish complete reducibility. Note first that both the standard representation and its dual are "unitary" for the action of su(3), meaning that $\pi(X)^* = -\pi(X)$ for all $X \in$ su(3). Meanwhile, it is easy to verify (Exercise 5) that if V and W are inner product spaces, then there is a unique inner product on $V \otimes W$ for which

$$\langle v_1 \otimes w_1, v_2 \otimes w_2 \rangle = \langle v_1, v_2 \rangle \langle w_1, w_2 \rangle$$

for all $v_1, v_2 \in V$ and $w_1, w_2 \in W$. Extending this construction to tensor products of several vector spaces, use the standard inner product on $\mathbb{C}^3$ to construct an inner product on the space in (6.11). It is then easy to check that π_{m_1,m_2} is also unitary for the action of su(3). Thus, by Proposition 4.27, π_{m_1,m_2} is completely reducible under the action of su(3) and thus, also, under the action of sl(3; $\mathbb{C}$) $\cong$ su(3)$_{\mathbb{C}}$. $\square$

We have now completed the proof of Theorem 6.7.

Proof of Theorem 6.8. The standard representation π_1 of $\mathsf{sl}(3; \mathbb{C})$ comes from the standard representation Π_1 of $\mathsf{SU}(3)$, and similarly for the dual of the standard representation. By taking tensor products, we see that there is a representation Π_{m_1, m_2} corresponding to the representation π_{m_1, m_2} of $\mathsf{sl}(3; \mathbb{C})$. The irreducible invariant subspace W in the proof of Proposition 6.17 is then also invariant under the action of $\mathsf{SU}(3)$, so that the restriction of Π_{m_1, m_2} to W is the desired representation of $\mathsf{SU}(3)$.   □

6.5   An Example: Highest Weight $(1, 1)$

To obtain the irreducible representation with highest weight $(1, 1)$, we take the tensor product of the standard representation and its dual, take the highest weight vector in the tensor product, and then consider the space obtained by repeated applications of the operators $\pi_{1,1}(Y_j)$, $j = 1, 2, 3$. Since, however, $Y_3 = -[Y_1, Y_2]$, it suffices to apply only $\pi_{1,1}(Y_1)$ and $\pi_{1,1}(Y_2)$.

Now, the standard representation has highest weight e_1 and the action of the operators $\pi(Y_1) = Y_1$ and $\pi(Y_2) = Y_2$ is given by

$$Y_1 e_1 = e_2 \quad Y_1 e_2 = 0 \quad Y_1 e_3 = 0$$
$$Y_2 e_1 = 0 \quad Y_2 e_2 = e_3 \quad Y_2 e_3 = 0 .$$

For the dual of the standard representation, let use the notation $\overline{Z} = -Z^{tr}$, so that $\pi(Z) = \overline{Z}$. If we introduce the new basis

$$f_1 = e_3; \quad f_2 = -e_2; \quad f_3 = e_1,$$

then the highest weight is f_1 and we have

$$\overline{Y_1} f_1 = 0 \quad \overline{Y_1} f_2 = f_3 \quad \overline{Y_1} f_3 = 0$$
$$\overline{Y_2} f_1 = f_2 \quad \overline{Y_2} f_2 = 0 \quad \overline{Y_2} f_3 = 0 .$$

We must now repeatedly apply the operators

$$\pi_{1,1}(Y_1) = Y_1 \otimes I + I \otimes \overline{Y_1}$$
$$\pi_{1,1}(Y_2) = Y_2 \otimes I + I \otimes \overline{Y_2} \tag{6.12}$$

until we get zero. This calculation is contained in the following chart. Here, there are two arrows coming out of each vector. Of these, the left arrow indicates the action of $Y_1 \otimes I + I \otimes \overline{Y_1}$ and the right arrow indicates the action of $Y_2 \otimes I + I \otimes \overline{Y_2}$. To save space, we omit the tensor product symbol, writing, for example, $e_2 f_2$ instead of $e_2 \otimes f_2$.

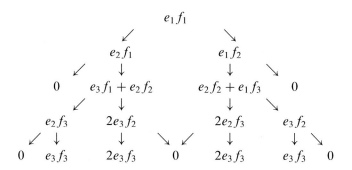

A basis for the space spanned by these vectors is $e_1 f_1$, $e_2 f_1$, $e_1 f_2$, $e_3 f_1 + e_2 f_2$, $e_2 f_2 + e_1 f_3$, $e_2 f_3$, $e_3 f_2$, and $e_3 f_3$. Thus, the dimension of this representation is 8; it is (isomorphic to) the adjoint representation. Now, e_1, e_2, and e_3 have weights $(1, 0)$, $(-1, 1)$, and $(0, -1)$, respectively, whereas f_1, f_2, and f_3 have weights $(0, 1)$, $(1, -1)$, and $(-1, 0)$, respectively. From (6.12), we can see that the weight for $e_j \otimes f_k$ is just the sum of the weight for e_j and the weight for f_k. Thus, the weights for the basis elements listed above are $(1, 1)$, $(-1, 2)$, $(2, -1)$, $(0, 0)$ (twice), $(1, -2)$, $(-2, 1)$, and $(-1, -1)$. Each weight has multiplicity 1 except for $(0, 0)$, which has multiplicity 2. See the first image in Figure 6.4.

6.6   The Weyl Group

This section describes an important symmetry of the representations of SU(3), involving something called the Weyl group. Our discussion follows the compact-group approach to the Weyl group. See Sect. 7.4 for the Lie algebra approach, in the context of general semisimple Lie algebras.

Definition 6.18. Let $\mathfrak{h}$ be the two-dimensional subspace of sl(3; $\mathbb{C}$) spanned by H_1 and H_2. Let N be the subgroup of SU(3) consisting of those $A \in$ SU(3) such that $\text{Ad}_A(H)$ is an element of $\mathfrak{h}$ for all H in $\mathfrak{h}$. Let Z be the subgroup of SU(3) consisting of those $A \in$ SU(3) such that $\text{Ad}_A(H) = H$ for all $H \in \mathfrak{h}$.

The space $\mathfrak{h}$ is a **Cartan subalgebra** of sl(3; $\mathbb{C}$). It is a straightforward exercise (Exercise 9) to verify that Z and N are subgroups of SU(3) and that Z is a normal subgroup of N. This leads us to the definition of the Weyl group.

Definition 6.19. The **Weyl group** of SU(3), denoted W, is the quotient group N/Z.

The primary significance of W for the representation theory of SU(3) is that it gives rise to a symmetry of the weights occurring in a fixed representation; see Theorem 6.22. We can define an action of W on $\mathfrak{h}$ as follows. For each element w of W, choose an element A of the corresponding coset in N. Then for H in $\mathfrak{h}$ we define the action $w \cdot H$ of w on H by

$$w \cdot H = \text{Ad}_A(H).$$

To see that this action is well defined, suppose B is an element of the same coset as A. Then $B = AC$ with $C \in Z$ and, thus,

$$\mathrm{Ad}_B(H) = \mathrm{Ad}_A(\mathrm{Ad}_C(H)) = \mathrm{Ad}_A(H),$$

by the definition of Z. Note that by definition, if $w \cdot H = H$ for all $H \in \mathfrak{h}$, then w is the identity element of W (that is, the associated $A \in N$ is actually in Z). Thus, we may identify W with the group of linear transformations of $\mathfrak{h}$ that can be expressed in the form $H \mapsto w \cdot H$ for some $w \in W$.

Proposition 6.20. *The group Z consists precisely of the diagonal matrices inside* $\mathsf{SU}(3)$*, namely the diagonal matrices with diagonal entries* $(e^{i\theta}, e^{i\phi}, e^{-i(\theta+\phi)})$ *with* $\theta, \phi \in \mathbb{R}$*. The group N consists of precisely those matrices $A \in \mathsf{SU}(3)$ such that for each $j = 1, 2, 3$, there exist $k_j \in \{1, 2, 3\}$ and $\theta_j \in \mathbb{R}$ such that $Ae_j = e^{i\theta_j} e_{k_j}$. Here, e_1, e_2, e_3 is the standard basis for $\mathbb{C}^3$.*

The Weyl group $W = N/Z$ is isomorphic to the permutation group on three elements.

Proof. Suppose A is in Z, which means that A commutes with all elements of $\mathfrak{h}$, including H_1, which has eigenvectors e_1, e_2, and e_3, with corresponding eigenvalues $1, -1$, and 0. Since A commutes with H_1, it must preserve each of these eigenspaces (Proposition A.2). Thus, Ae_j must be a multiple of e_j for each j, meaning that A is diagonal. Conversely, any diagonal matrix in $\mathsf{SU}(3)$ does indeed commute not only with H_1 but also with H_2 and, thus, with every element of $\mathfrak{h}$.

Suppose, now, that A is in N. Then $AH_1 A^{-1}$ must be in $\mathfrak{h}$ and therefore must be diagonal, meaning that e_1, e_2, and e_3 are eigenvectors for $AH_1 A^{-1}$, with the same eigenvalues $1, -1, 0$ as H_1, but not necessarily in the same order. On the other hand, the eigenvectors of $AH_1 A^{-1}$ must be Ae_1, Ae_2, and Ae_3. Thus, Ae_j must be a multiple of some e_{k_j}, and the constant must have absolute value 1 if A is unitary. Conversely, if Ae_j is a multiple of e_{k_j} for each j, then for any (diagonal) matrix H in $\mathfrak{h}$, the matrix AHA^{-1} will again be diagonal and thus in $\mathfrak{h}$.

Finally, if A maps each e_j to a multiple of e_{k_j}, for some k_j depending on j, then for each diagonal matrix H, the matrix AHA^{-1} will be diagonal with diagonal entries rearranged by the permutation $j \mapsto k_j$. For any permutation, we can choose the constants to that the map taking e_j to $e^{i\theta_j} e_{k_j}$ has determinant 1, showing that every permutation actually arises in this way. Thus, W—which we think of as the group of linear transformations of $\mathfrak{h}$ of the form Ad_A, $A \in N$—is isomorphic to the permutation group on three elements. □

We want to show that the Weyl group is a symmetry of the weights of any finite-dimensional representation of $\mathsf{sl}(3; \mathbb{C})$. To understand this, we need to adopt a less basis-dependent view of the weights. We have defined a weight as a pair (m_1, m_2) of simultaneous eigenvalues for $\pi(H_1)$ and $\pi(H_2)$. However, if a vector v is an eigenvector for $\pi(H_1)$ and $\pi(H_2)$ then it is also an eigenvector for $\pi(H)$ for any element H of the space $\mathfrak{h}$ spanned by H_1 and H_2, and the eigenvalues will depend linearly on H in $\mathfrak{h}$. Thus, we may think of a weight not as a pair of numbers but as a linear functional on $\mathfrak{h}$.

It is then convenient to use an inner product on $\mathfrak{h}$ to identity linear functionals on $\mathfrak{h}$ with elements of $\mathfrak{h}$ itself. We define the inner product of H and H' in $\mathfrak{h}$ by

$$\langle H, H' \rangle = \text{trace}(H^* H'), \qquad (6.13)$$

or, explicitly,

$$\langle \text{diag}(a, b, c), \text{diag}(d, e, f) \rangle = \bar{a}d + \bar{b}e + \bar{c}f,$$

where $\text{diag}(\cdot, \cdot, \cdot)$ is the diagonal matrix with the indicated diagonal entries. If ϕ is a linear functional on $\mathfrak{h}$, there is (Proposition A.11) a unique vector λ in $\mathfrak{h}$ such that ϕ may be represented as $\phi(H) = \langle \lambda, H \rangle$ for all $H \in \mathfrak{h}$. If we represent the linear functional in the previous paragraph in this way, we arrive at a new, basis-independent notion of a weight.

Definition 6.21. Let $\mathfrak{h}$ be the subspace of $sl(3; \mathbb{C})$ spanned by H_1 and H_2 and let (π, V) be a representation of $sl(3; \mathbb{C})$. An element λ of $\mathfrak{h}$ is called a **weight** for π if there exists a nonzero vector v in V such that

$$\pi(H)v = \langle \lambda, H \rangle v$$

for all H in $\mathfrak{h}$. Such a vector v is called a **weight vector** with weight λ.

If λ is a weight in our new sense, the ordered pair (m_1, m_2) in Definition 6.1 is given by

$$m_1 = \langle \lambda, H_1 \rangle; \quad m_2 = \langle \lambda, H_2 \rangle.$$

It is easy to check that for all $U \in N$, the adjoint action of U on $\mathfrak{h}$ preserves the inner product in (6.13). Thus, the action of the Weyl group on $\mathfrak{h}$ is unitary: $\langle w \cdot H, w \cdot H' \rangle = \langle H, H' \rangle$. Since the roots are just the nonzero weights of the adjoint representation, we now also think of the roots as elements of $\mathfrak{h}$.

Theorem 6.22. *Suppose that* (Π, V) *is a finite-dimensional representation of* $SU(3)$ *with associated representation* (π, V) *of* $sl(3; \mathbb{C})$. *If* $\lambda \in \mathfrak{h}$ *is a weight for* V *then* $w \cdot \lambda$ *is also a weight of* V *with the same multiplicity. In particular, the roots are invariant under the action of the Weyl group.*

Proof. Suppose that λ is a weight for V with weight vector v. Then for all $U \in N$ and $H \in \mathfrak{h}$, we have

$$\pi(H)\Pi(U)v = \Pi(U)(\Pi(U)^{-1}\pi(H)\Pi(U))v$$

$$= \Pi(U)\pi(U^{-1}HU)v$$

$$= \langle \lambda, U^{-1}HU \rangle \Pi(U)v.$$

Here, we have used that U is in N, which guarantees that $U^{-1}HU$ is, again, in $\mathfrak{h}$. Thus, if w is the Weyl group element represented by U, we have

$$\pi(H)\Pi(U)v = \left\langle \lambda, w^{-1} \cdot H \right\rangle \Pi(U)v = \langle w \cdot \lambda, H \rangle \, \Pi(U)v.$$

We conclude that $\Pi(U)v$ is a weight vector with weight $w \cdot \lambda$.

The same sort of reasoning shows that $\Pi(U)$ is an invertible map of the weight space with weight λ onto the weight space with weight $w \cdot \lambda$, whose inverse is $\Pi(U)^{-1}$. This means that the two weights have the same multiplicity. $\qquad\square$

To represent the basic weights, $(1,0)$ and $(0,1)$, in our new approach, we look for diagonal, trace-zero matrices μ_1 and μ_2 such that

$$\langle \mu_1, H_1 \rangle = 1, \quad \langle \mu_1, H_2 \rangle = 0$$
$$\langle \mu_2, H_1 \rangle = 0, \quad \langle \mu_2, H_2 \rangle = 1.$$

These are easily found as

$$\mu_1 = \operatorname{diag}(2/3, -1/3, -1/3); \quad \mu_2 = \operatorname{diag}(1/3, 1/3, -2/3).$$

The positive simple roots $(2, -1)$ and $(-1, 2)$ are then represented as

$$\alpha_1 = 2\mu_1 - \mu_2 = \operatorname{diag}(1, -1, 0);$$
$$\alpha_2 = -\mu_1 + 2\mu_2 = \operatorname{diag}(0, 1, -1). \tag{6.14}$$

Note that both α_1 and α_2 have length $\sqrt{2}$ and $\langle \alpha_1, \alpha_2 \rangle = -1$. Thus, the angle θ between them satisfies $\cos\theta = -1/2$, so that $\theta = 2\pi/3$.

Figure 6.2 shows the same information as Figure 6.1, namely, the roots and the dominant integral elements, but now drawn relative to the Weyl-invariant inner product in (6.13). We draw only the two-dimensional *real* subspace of $\mathfrak{h}$ consisting

Fig. 6.2 The roots and dominant integral elements for $\mathsf{sl}(3;\mathbb{C})$, computed relative to a Weyl-invariant inner product

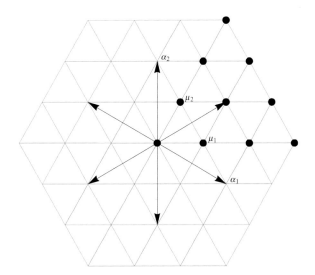

Fig. 6.3 The Weyl group is
the symmetry group of the
indicated equilateral triangle

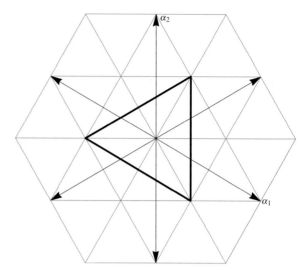

of those elements μ such that $\langle \mu, H_1 \rangle$ and $\langle \mu, H_2 \rangle$ are real, since all the roots and
weights have this property. Let $w_{(1,2,3)}$ denote the Weyl group element that acts by
cyclically permuting the diagonal entries of each $H \in \mathfrak{h}$. Then $w_{(1,2,3)}$ takes α_1 to α_2
and α_2 to $-(\alpha_1 + \alpha_2)$, which is a counterclockwise rotation by $2\pi/3$ in Figure 6.2.
Similarly, if $w_{(1,2)}$ the element that interchanges the first two diagonal entries of
$H \in \mathfrak{h}$, then $w_{(1,2)}$ maps α_1 to $-\alpha_1$ and α_2 to $\alpha_1 + \alpha_2$. Thus, $w_{(1,2)}$ is the reflection
across the line perpendicular to α_1. The reader is invited to compute the action of the
remaining elements of the Weyl group and to verify that it is the symmetry group of
the equilateral triangle in Figure 6.3.

We previously defined a pair (m_1, m_2) to be integral if m_1 and m_2 are integers and
dominant if $m_1 \geq 0$ and $m_2 \geq 0$. These concepts translate into our new language
as follows. If $\lambda \in \mathfrak{h}$, then λ is **integral** if $\langle \lambda, H_1 \rangle$ and $\langle \lambda, H_2 \rangle$ are integers and λ
is **dominant** if $\langle \lambda, H_1 \rangle \geq 0$ and $\langle \lambda, H_2 \rangle \geq 0$. Geometrically, the set of dominant
elements is a sector spanning an angle of $\pi/3$.

6.7   Weight Diagrams

In this section, we display the weights and multiplicities for several irreducible
representations of sl(3; ℂ). Figure 6.4 covers the irreducible representations with
highest weighs $(1, 1)$, $(1, 2)$, $(0, 4)$, and $(2, 2)$. The first of these examples was
analyzed in Sect. 6.5, and the other examples can be analyzed by the same method.
In each part of the figure, the arrows indicate the roots, the two black lines indicate
the boundary of the set of dominant elements, and the dashed lines indicate the
boundary of the set of points lower than the highest weight. Each weight of

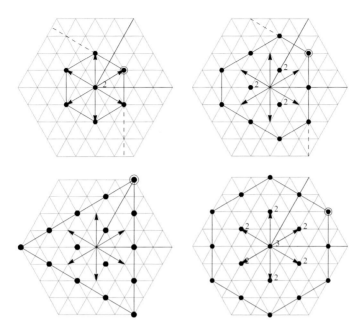

Fig. 6.4 Weight diagrams for representations with highest weights $(1, 1)$, $(1, 2)$, $(0, 4)$, and $(2, 2)$

a particular representation is indicated by a black dot, with a number next to a dot indicating its multiplicity. A dot without a number indicates a weight of multiplicity 1.

Our last example is the representation with highest weight $(9, 2)$ (Figure 6.5), which cannot feasibly be analyzed using the method of Sect. 6.5. Instead, the weights are determined by the results of Sect. 6.8 and the multiplicities are computed using the Kostant multiplicity formula. (See Figure 10.8 in Sect. 10.6.) See also Exercises 11 and 12 for another approach to computing multiplicities.

6.8   Further Properties of the Representations

Although we now have a classification of the irreducible representations of $\mathsf{sl}(3; \mathbb{C})$ by means of their highest weights, there are other things we might like to know about the representations, such as (1) the other weights that occur, besides the highest weight, (2) the multiplicities of those weights, and (3) the dimension of the representation. In this section, we establish which weights occur and state without proof the formula for the dimension. A formula for the multiplicities and a proof of the dimension formula are given in Chapter 10 in the setting of general semisimple Lie algebras.

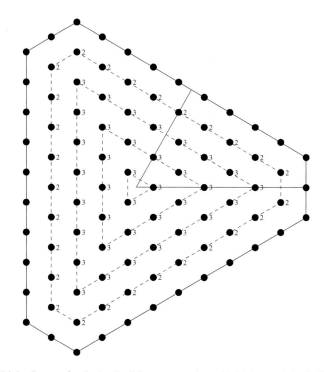

Fig. 6.5 Weight diagram for the irreducible representation with highest weight $(9, 2)$

Definition 6.23. If $v_1, \ldots, v_N$ are elements of a real or complex vector space, the **convex hull** of $v_1, \ldots, v_N$ is the set of all vectors of the form

$$c_1 v_1 + c_2 v_2 + \cdots + c_N v_N$$

where the c_j's are non-negative real numbers satisfying $c_1 + c_2 + \cdots + c_N = 1$.

Equivalently, the convex hull of $v_1, \ldots, v_N$ is the smallest convex set that contains all of the v_j's.

Theorem 6.24. *Let μ be a dominant integral element and let V_μ be the irreducible representation with highest weight μ. If λ is a weight of V_μ, then λ satisfies the following two conditions: (1) $\mu - \lambda$ can be expressed as an integer combination of roots, and (2) λ belongs to the convex hull of $W \cdot \mu$, the orbit of μ under the action of W.*

Proof. According to the proof of Proposition 6.11, V_μ is spanned by vectors of the form in (6.8). These vectors are weight vectors with weights of the form $\lambda :=$ $\mu - \alpha_{j_1} - \cdots - \alpha_{j_N}$. Thus, every weight of V_μ satisfies the first property in the theorem.

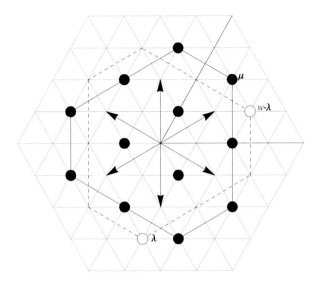

Fig. 6.6 The integral element λ is outside the convex hull of the orbit of μ, and the element $w \cdot \lambda$ is not lower than μ

The second property in the theorem is based on the following idea: If λ is a weight of V_μ, then $w \cdot \lambda$ is also a weight for all $w \in W$, which means that $w \cdot \lambda$ is lower than μ. We can now argue "pictorially" that if λ were not in the convex hull of $W \cdot \mu$, there would be some $w \in W$ for which $w \cdot \lambda$ is not lower than μ, so that λ could not be a weight of V_μ. See Figure 6.6.

We can give a more formal argument as follows. For any weight λ of V_μ, we can, by Exercise 10, find some $w \in W$ so that $\lambda' := w \cdot \lambda$ is dominant. Since λ' is also a weight of V_μ, we must have $\lambda' \preceq \mu$. Thus, λ' is in the quadrilateral Q_μ consisting of dominant elements that are lower than μ (Figure 6.7). We now argue that the vertices of Q_μ are all in the convex hull. First, it is easy to see that for any μ, the average of $w \cdot \mu$ over all $w \in W$ is zero, which means that 0 is in E_μ. Second, the vertices marked v_1 and v_2 in the figure are expressible as follows:

$$v_1 = \frac{1}{2}\mu + \frac{1}{2}s_{\alpha_1} \cdot \mu$$

$$v_2 = \frac{1}{2}\mu + \frac{1}{2}s_{\alpha_2} \cdot \mu,$$

where s_{α_1} and s_{α_2} are the Weyl group elements given by reflecting about the lines orthogonal to α_1 and α_2. Thus, all the vertices of Q_μ are in E_μ, from which it follows that Q_μ itself is contained in E_μ.

Now, $W \cdot \mu$ is clearly W-invariant, which means that E_μ is also W-invariant. Since $\lambda' \in Q_\mu \subset E_\mu$, we have $\lambda = w^{-1}\lambda' \in E_\mu$ as well. $\square$

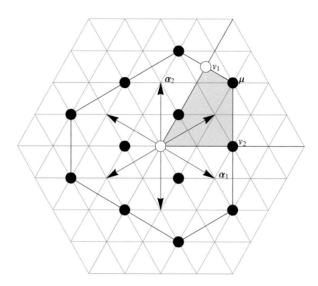

Fig. 6.7 The shaded quadrilateral is the set of all points that are dominant and lower than μ

Theorem 6.25. *Suppose V_μ is an irreducible representation with highest weight μ and that λ is an integral element satisfying the two conditions in Theorem 6.24. Then λ is a weight of V_μ.*

Theorem 6.25 says, in effect, that there are no unexpected holes in the set of weights of V_μ. The key to the proof is the "no holes" result (Point 4 of Theorem 4.34) we previously established for sl(2; ℂ).

Lemma 6.26. *Let γ be a weight of V_μ, let α be a root, and let $s_\alpha \in W$ be the reflection about the line orthogonal to α. Suppose λ is a point on the line segment joining γ to $s_\alpha \cdot \gamma$ with the property that $\gamma - \lambda$ is an integer multiple of α. Then λ is also a weight of V_μ.*

See Figure 6.8 for an example. Note from Figure 6.3 that for each root α, the reflection s_α is an element of the Weyl group.

Proof. Since the reflections associated to α and $-\alpha$ are the same, it suffices to consider the roots α_1, α_2, and $\alpha_3 := \alpha_1 + \alpha_2$. If we let $H_3 = H_1 + H_2$, then for $j = 1, 2, 3$ we have a subalgebra $\mathfrak{s}_j = \langle X_j, Y_j, H_j \rangle$ isomorphic to sl(2; ℂ) such that X_j is a root vector with root α_j and Y_j is a root vector with root $-\alpha_j$. Since

$$[H_j, X_j] = 2X_j = \langle \alpha_j, H_j \rangle X_j,$$

we have $\langle \alpha_j, H_j \rangle = 2$ for each j.

Let us now fix a weight γ of V_μ and let U be the span of all the weight vectors in V_μ whose weights are of the form $\gamma + k\alpha_j$ for some real number k. (These weights are circled in Figure 6.8.) Since, by Lemma 6.5, $\pi(X_j)$ and $\pi(Y_j)$ shift weights

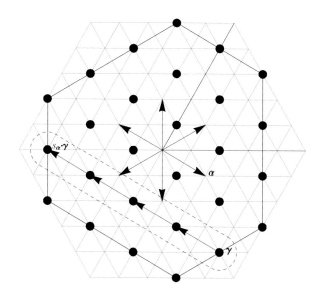

Fig. 6.8 Since γ is a weight of V_μ, each of the elements $\gamma - \alpha, \gamma - 2\alpha, \dots, s_\alpha \cdot \gamma$ must also be a weight of V_μ

by $\pm\alpha_j$, we see that U is invariant under $\mathfrak{s}_j$ and thus constitutes a representation of $\mathfrak{s}_j$ (not necessarily irreducible). With our new perspective that roots are elements of $\mathfrak{h}$, we can verify from (6.14) that for each j, we have $\alpha_j = H_j$, from which it follows that $s_{\alpha_j} \cdot H_j = -H_j$. Thus, if u and v are weight vectors with weights γ and $s_\alpha \cdot \gamma$, respectively, u and v are in U and are eigenvectors for $\pi(H_j)$ with eigenvalues $\langle \gamma, H_j \rangle$ and

$$\langle s_\alpha \cdot \gamma, H_j \rangle = \langle \gamma, s_\alpha \cdot H_j \rangle = -\langle \gamma, H_j \rangle,$$

respectively.

If λ is on the line segment joining γ to $s_\alpha \cdot \gamma$, we see that $\langle \lambda, H_j \rangle$ is between $\langle \gamma, H_j \rangle$ and $\langle s_\alpha \cdot \gamma, H_j \rangle = -\langle \gamma, H_j \rangle$. If, in addition, λ differs from γ by an integer multiple of α_j, then $\langle \gamma, H_j \rangle$ differs from $\langle \lambda, H_j \rangle$ by an integer multiple of $\langle \alpha_j, H_j \rangle = 2$. Thus, by applying Point 4 of Theorem 4.34 to the action of $\mathfrak{s}_j$ on U, there must be an eigenvector w for $\pi(H_j)$ in U with eigenvalue $l = \langle \lambda, H_j \rangle$. Since the unique weight of the form $\gamma + k\alpha_j$ for which $\langle \gamma + k\alpha_j, H_j \rangle = \langle \lambda, H_j \rangle$ is the one where $\gamma + k\alpha_j = \lambda$, we conclude that λ is a weight of V_μ.   □

Proof of Theorem 6.25. Suppose that λ satisfies the two conditions in the theorem, and write $\lambda = \mu - n_1\alpha_1 - n_2\alpha_2$. Consider first the case $n_1 \geq n_2$, so that

$$\lambda = \mu - (n_1 - n_2)\alpha_1 - n_2(\alpha_1 + \alpha_2)$$
$$= \mu - (n_1 - n_2)\alpha_1 - n_2\alpha_3,$$

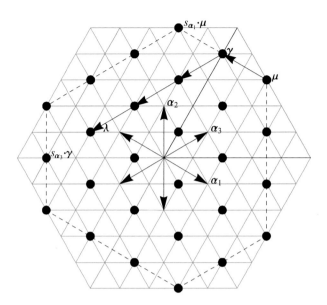

Fig. 6.9 By applying Lemma 6.26 twice, we can see that γ and λ must be weights of V_μ

where $\alpha_3 = \alpha_1 + \alpha_2$. If we start at λ and travel in the direction of α_3, we will hit the boundary of E_μ at the point

$$\gamma := \mu - (n_1 - n_2)\alpha_1.$$

(See Figure 6.9.) Thus, γ is in E_μ and must therefore be between μ and $s_{\alpha_1} \cdot \mu$. Since also γ differs from μ by an integer multiple of α_1 (namely $n_1 - n_2$) Lemma 6.26 says that γ is a weight of V. Meanwhile, λ is between γ and $s_{\alpha_3} \cdot \gamma$ (see, again, Figure 6.9) and differs from γ by an integer multiple of α_3 (namely n_2). Thus, the lemma tells us that λ must be a weight of V, as claimed. If $n_1 \leq n_2$, we can use a similar argument with the roles of α_1 and α_2 reversed.   □

We close this section by stating the formula for the dimension of an irreducible representation of sl(3; ℂ). We will prove the result in Chapter 10 as a special case of the Weyl dimension formula.

Theorem 6.27. *The dimension of the irreducible representation with highest weight* (m_1, m_2) *is*

$$\frac{1}{2}(m_1 + 1)(m_2 + 1)(m_1 + m_2 + 2).$$

The reader is invited to verify this formula by direct computation in the representations depicted in Figure 6.4.

6.9   Exercises

1. Show that the roots listed in (6.3) are the only roots.
2. Let π be an irreducible finite-dimensional representation of $\mathsf{sl}(3;\mathbb{C})$ acting on a space V and let π^* be the dual representation to π, acting on V^*, as defined in Sect. 4.3.3. Show that the weights of π^* are the negatives of the weights of π.
   Hint: Choose a basis for V in which both $\pi(H_1)$ and $\pi(H_2)$ are diagonal.
3. Let π be an irreducible representation of $\mathsf{sl}(3;\mathbb{C})$ with highest weight μ.

   (a) Let $\alpha_3 = \alpha_1 + \alpha_2$ and let s_{α_3} denote the reflection about the line orthogonal to α_3. Show the lowest weight for π is $s_{\alpha_3} \cdot \mu$.
   (b) Show that the highest weight for the dual representation π^* to π is the weight

   $$\mu' := -s_{\alpha_3} \cdot \mu.$$

   (c) Let μ_1 and μ_2 be the fundamental weights, as in Figure 6.2. If $\mu = m_1\mu_1 + m_2\mu_2$, show that $\mu' = m_2\mu_1 + m_1\mu_2$. That is to say, the dual to the representation with highest weight (m_1, m_2) has highest weight (m_2, m_1).

4. Consider the adjoint representation of $\mathsf{sl}(3;\mathbb{C})$ as a representation of $\mathsf{sl}(2;\mathbb{C})$ by restricting the adjoint representation to the subalgebra spanned by X_1, Y_1, and H_1. Decompose this representation as a direct sum of irreducible representations of $\mathsf{sl}(2;\mathbb{C})$. Which representations occur and with what multiplicity?
5. Suppose that V and W are finite-dimensional inner product spaces over $\mathbb{C}$. Show that there exists a unique inner product on $V \otimes W$ such that

   $$\langle v \otimes w, v' \otimes w' \rangle = \langle v, v' \rangle \langle w, w' \rangle$$

   for all $v, v' \in V$ and $w, w' \in W$.
   Hint: Let $\{e_j\}$ and $\{f_k\}$ be orthonormal bases for V and W, respectively. Take the inner product on $V \otimes W$ for which $\{e_j \otimes f_k\}$ is an orthonormal basis.
6. Following the method of Sect. 6.5, work out the representation of $\mathsf{sl}(3;\mathbb{C})$ with highest weight $(2,0)$, acting on a subspace of $\mathbb{C}^3 \otimes \mathbb{C}^3$. Determine all the weights of this representation and their multiplicity (i.e., the dimension of the corresponding weight space). Verify that the dimension formula (Theorem 6.27) holds in this case.
7. Consider the nine-dimensional representation of $\mathsf{sl}(3;\mathbb{C})$ considered in Sect. 6.5, namely the tensor product of the representations with highest weights $(1,0)$ and $(0,1)$. Decompose this representation as a direct sum of irreducibles. Do the same for the tensor product of two copies of the irreducible representation with highest weight $(1,0)$. (Compare Exercise 6.)

8. Let W_m denote the space of homogeneous polynomials on $\mathbb{C}^3$ of degree m. Let SU(3) act on W_m by the obvious generalization of the action in Example 4.10.

   (a) Show that the associated representation of sl(3; $\mathbb{C}$) contains a highest weight cyclic representation with highest weight $(0, m)$ and highest weight vector z_3^m.
   (b) By imitating the proof of Proposition 4.11, show that any nonzero invariant subspace of W_m must contain z_3^m.
   (c) Conclude that W_m is irreducible with highest weight $(0, m)$.

9. Show that Z and N (defined in Definition 6.18) are subgroups of SU(3). Show that Z is a normal subgroup of N.

10. Suppose λ is an integral element, that is, one of the triangular lattice points in Figure 6.2. Show that there is an element w of the Weyl group such that $w \cdot \lambda$ is dominant integral, that is, one of the black dots in Figure 6.2.

   Hint: Recall that the Weyl group is the symmetry group of the triangle in Figure 6.3.

   (a) Regard the Weyl group as a group of linear transformations of $\mathfrak{h}$. Show that $-I$ is not an element of the Weyl group.
   (b) Which irreducible representations of sl(3; $\mathbb{C}$) have the property that their weights are invariant under $-I$?

11. Suppose (π, V) is an irreducible representation of sl(3; $\mathbb{C}$) with highest weight μ and highest weight vector v_0. Show that the weight space with weight $\mu - \alpha_1 - \alpha_2$ has multiplicity at most 2 and is spanned by the vectors

$$\pi(Y_1)\pi(Y_2)v_0, \quad \pi(Y_2)\pi(Y_1)v_0.$$

12. Let (π, V) be the irreducible representation with highest weight (m_1, m_2). As in the proof of Proposition 6.17, choose an inner product on V such that $\pi(X)^* = -\pi(X)$ for all $X \in$ su(3) $\subset$ sl(3; $\mathbb{C}$). Let v_0 be a highest weight vector for V, normalized to be a unit vector, and define vectors u_1 and u_2 in V as

$$u_1 = \pi(Y_1)\pi(Y_2)v_0; \quad u_2 = \pi(Y_2)\pi(Y_1)v_0.$$

Each of these vectors is either zero or a weight vector with weight $\mu - \alpha_1 - \alpha_2$.

   (a) Using and the commutation relations among the basis elements of sl(3; $\mathbb{C}$), show that

$$\langle u_1, u_1 \rangle = m_2(m_1 + 1)$$
$$\langle u_2, u_2 \rangle = m_1(m_2 + 1)$$
$$\langle u_1, u_2 \rangle = m_1 m_2$$

   Hint: Show that $\pi(X_j)^* = \pi(Y_j)$ for $j = 1, 2, 3$.

(b) Show that if $m_1 \geq 1$ and $m_2 \geq 1$ then u_1 and u_2 are linearly independent. Conclude that the weight $\mu - \alpha_1 - \alpha_2$ has multiplicity 2.
(c) Show that if $m_1 = 0$ and $m_2 \geq 1$ or $m_1 \geq 1$ and $m_2 = 0$, then the weight $\mu - \alpha_1 - \alpha_2$ has multiplicity 1.

Note: The reader may verify the results of this exercise in the representations depicted in Figure 6.4.

Chapter 7
Semisimple Lie Algebras

In this chapter we introduce a class of Lie algebras, the *semisimple* algebras, for which we can classify the irreducible representations using a strategy similar to the one we used for $\mathsf{sl}(3;\mathbb{C})$. In this chapter, we develop the relevant structures of semisimple Lie algebras. In Chapter 8, we look into the properties of the set of roots. Then in Chapter 9, we construct and classify the irreducible, finite-dimensional representations of semisimple Lie algebras. Finally, in Chapter 10, we consider several additional properties of the representations constructed in Chapter 9. Meanwhile, in Chapters 11 and 12, we consider representation theory from the closely related viewpoint of compact Lie groups.

7.1 Semisimple and Reductive Lie Algebras

We begin by defining the term semisimple. There are many equivalent characterizations of semisimple Lie algebras. It is not, however, always easy to *prove* that two of these various characterizations are equivalent. We will use an atypical definition, which allows for a rapid development of the structure of semisimple Lie algebras. Recall from Sect. 3.6 the notion of the complexification of a real Lie algebra.

Definition 7.1. A complex Lie algebra $\mathfrak{g}$ is **reductive** if there exists a compact matrix Lie group K such that

$$\mathfrak{g} \cong \mathfrak{k}_{\mathbb{C}}.$$

A previous version of this book was inadvertently published without the middle initial of the author's name as "Brian Hall". For this reason an erratum has been published, correcting the mistake in the previous version and showing the correct name as Brian C. Hall (see DOI http://dx.doi.org/10.1007/978-3-319-13467-3_14). The version readers currently see is the corrected version. The Publisher would like to apologize for the earlier mistake.

© Springer International Publishing Switzerland 2015
B.C. Hall, *Lie Groups, Lie Algebras, and Representations*, Graduate
Texts in Mathematics 222, DOI 10.1007/978-3-319-13467-3_7

A complex Lie algebra $\mathfrak{g}$ is **semisimple** if it is reductive and the center of $\mathfrak{g}$ is trivial.

Definition 7.2. If $\mathfrak{g}$ is a semisimple Lie algebra, a real subalgebra $\mathfrak{k}$ of $\mathfrak{g}$ is a **compact real form** of $\mathfrak{g}$ if $\mathfrak{k}$ is isomorphic to the Lie algebra of some compact matrix Lie group and every element Z of $\mathfrak{g}$ can be expressed uniquely as $Z = X + iY$, with $X, Y \in \mathfrak{k}$.

On the one hand, using Definition 7.1 gives an easy method of constructing Cartan subalgebras and fits naturally with our study of compact Lie groups in Part III. On the other hand, this definition covers an apparently smaller class of Lie algebras than some of the more standard definitions. That is to say, we will prove (Theorem 7.8 and Exercise 6) that the condition in Definition 7.1 implies two of the standard definitions of "semisimple," but we will not prove the reverse implications. These reverse implications are, in fact, true, so that our definition of semisimplicity is ultimately equivalent to any other definition. But it is not possible to prove the reverse implications without giving up the gains in efficiency that go with Definition 7.1. The reader who wishes to see a development of the theory starting from a more traditional definition of semisimplicity may consult Chapter II (along with the first several sections of Chapter I) of [Kna2].

The only time we use the compact group in Definition 7.1 is to construct the inner product in Proposition 7.4. In the standard treatment of semisimple Lie algebras, the Killing form (Exercise 6) is used in place of this inner product. Our use of an inner product in place of the bilinear Killing form substantially simplifies some of the arguments. Notably, in our construction of Cartan subalgebras (Proposition 7.11), we use that a skew self-adjoint operator is always diagonalizable. By contrast, an operator that is skew symmetric with respect to a nondegenerate bilinear form need not be diagonalizable. Thus, the construction of Cartan subalgebras in the conventional approach is substantially more involved than in our approach.

For a complex semisimple Lie algebra $\mathfrak{g}$, we will always assume we have chosen a compact real form $\mathfrak{k}$ of $\mathfrak{g}$, so that $\mathfrak{g} = \mathfrak{k}_{\mathbb{C}}$.

Example 7.3. The following Lie algebras are semisimple:

$$\mathsf{sl}(n; \mathbb{C}), \quad n \geq 2$$
$$\mathsf{so}(n; \mathbb{C}), \quad n \geq 3$$
$$\mathsf{sp}(n; \mathbb{C}), \quad n \geq 1.$$

The Lie algebras $\mathsf{gl}(n; \mathbb{C})$ and $\mathsf{so}(2; \mathbb{C})$ are reductive but not semisimple.

Proof. It is easy to see that the listed Lie algebras are reductive, with the corresponding compact groups K being $\mathsf{SU}(n)$, $\mathsf{SO}(n)$, $\mathsf{Sp}(n)$, $\mathsf{U}(n)$, and $\mathsf{SO}(2)$, respectively. (Compare (3.17) in Sect. 3.6.) The Lie algebra $\mathsf{gl}(n; \mathbb{C})$ has a nontrivial center, consisting of scalar multiples of the identity, while the Lie algebra $\mathsf{so}(2; \mathbb{C})$ is commutative. It remains only to show that the centers of $\mathsf{sl}(n; \mathbb{C})$, $\mathsf{so}(n; \mathbb{C})$, and $\mathsf{sp}(n; \mathbb{C})$ are trivial for the indicated values of n.

Consider first the case of $\mathsf{sl}(n;\mathbb{C})$ and let X be an element of the center of $\mathsf{sl}(n;\mathbb{C})$. For any $1 \leq j,k \leq n$, let E_{jk} be the matrix with a 1 in the (j,k) spot and zeros elsewhere. Consider the matrix $H_{jk} \in \mathsf{sl}(n;\mathbb{C})$ given by

$$H_{jk} = E_{jj} - E_{kk},$$

for $j < k$. Then we may easily calculate that

$$0 = [H_{jk}, X] = 2X_{jk}E_{jk} - 2X_{kj}E_{kj}.$$

Since E_{jk} and E_{kj} are linearly independent for $j < k$, we conclude that $X_{jk} = X_{kj} = 0$. Since this holds for all $j < k$, we see that X must be diagonal.
   Once X is known to be diagonal, we may compute that for $j \neq k$,

$$0 = [X, E_{jk}] = (X_{jj} - X_{kk})E_{jk}.$$

Thus, all the diagonal entries of X must be equal. But since, also, $\operatorname{trace}(X) = 0$, we conclude that X must be zero.
   For the remaining semisimple Lie algebras in Example 7.3, the calculations in Sect. 7.7 will allow us to carry out a similar argument. It is proved there that the center of $\mathsf{so}(2n;\mathbb{C})$ is trivial for $n \geq 2$, and a similar analysis shows that the centers of $\mathsf{so}(2n+1;\mathbb{C})$ and $\mathsf{sp}(n;\mathbb{C})$ are also trivial. □

Proposition 7.4. *Let* $\mathfrak{g} := \mathfrak{k}_{\mathbb{C}}$ *be a reductive Lie algebra. Then there exists an inner product on* $\mathfrak{g}$ *that is real valued on* $\mathfrak{k}$ *and such that the adjoint action of* $\mathfrak{k}$ *on* $\mathfrak{g}$ *is "unitary," meaning that*

$$\langle \operatorname{ad}_X(Y), Z \rangle = -\langle Y, \operatorname{ad}_X(Z) \rangle \tag{7.1}$$

for all $X \in \mathfrak{k}$ *and all* $X, Y \in \mathfrak{g}$. *If we define a operation* $X \mapsto X^*$ *on* $\mathfrak{g}$ *by the formula*

$$(X_1 + iX_2)^* = -X_1 + iX_2 \tag{7.2}$$

for $X_1, X_2 \in \mathfrak{k}$, *then any inner product satisfying (7.1) also satisfies*

$$\langle \operatorname{ad}_X(Y), Z \rangle = \langle Y, \operatorname{ad}_{X^*}(Z) \rangle \tag{7.3}$$

for all X, Y, *and* Z *in* $\mathfrak{g}$.

The motivation for the definition of X^* is that $\mathfrak{g} = \mathsf{gl}(n;\mathbb{C})$ and $\mathfrak{k} = \mathsf{u}(n)$, then X^* is the usual matrix adjoint of X.

Proof. By the proof of Theorem 4.28, there is a (real-valued) inner product on $\mathfrak{k}$ that is invariant under the adjoint action of K. This inner product then extends to a complex inner product on $\mathfrak{g} = \mathfrak{k}_{\mathbb{C}}$ for which the adjoint action of K is unitary. Thus, by Proposition 4.8, the adjoint action of $\mathfrak{k}$ on $\mathfrak{g}$ is unitary in the sense of (7.1). It is then a simple matter to check the relation (7.3).   □

Recall from Definition 3.9 what it means for a Lie algebra to decompose as a Lie algebra direct sum of subalgebras.

Proposition 7.5. *Suppose* $\mathfrak{g} = \mathfrak{k}_{\mathbb{C}}$ *is a reductive Lie algebra. Choose an inner product on* $\mathfrak{g}$ *as in Proposition 7.4. Then if* $\mathfrak{h}$ *is an ideal in* $\mathfrak{g}$, *the orthogonal complement of* $\mathfrak{h}$ *is also an ideal, and* $\mathfrak{g}$ *decomposes as the Lie algebra direct sum of* $\mathfrak{h}$ *and* $\mathfrak{h}^{\perp}$.

Proof. An ideal in $\mathfrak{g}$ is nothing but an invariant subspace for the adjoint action of $\mathfrak{g}$ on itself. If $\mathfrak{h}$ is a complex subspace of $\mathfrak{g}$ and $\mathfrak{h}$ is invariant under the adjoint action of $\mathfrak{k}$, it will also be invariant under adjoint action of $\mathfrak{g} = \mathfrak{k}_{\mathbb{C}}$. Thus, if $\mathfrak{h}$ is an ideal in $\mathfrak{g}$, then by Proposition 4.27, $\mathfrak{h}^{\perp}$ is invariant under the adjoint action of $\mathfrak{k}$ and is, therefore, also an ideal.

Now, $\mathfrak{g}$ decomposes as a vector space as $\mathfrak{g} = \mathfrak{h} \oplus \mathfrak{h}^{\perp}$. But since both $\mathfrak{h}$ and $\mathfrak{h}^{\perp}$ are ideals, for all $X \in \mathfrak{h}$ and $Y \in \mathfrak{h}^{\perp}$, we have

$$[X, Y] \in \mathfrak{h} \cap \mathfrak{h}^{\perp} = \{0\}.$$

Thus, $\mathfrak{g}$ is actually the Lie algebra direct sum of $\mathfrak{h}$ and $\mathfrak{h}^{\perp}$.   □

Proposition 7.6. *Every reductive Lie algebra* $\mathfrak{g}$ *over* $\mathbb{C}$ *decomposes as a Lie algebra direct sum* $\mathfrak{g} = \mathfrak{g}_1 \oplus \mathfrak{z}$, *where* $\mathfrak{z}$ *is the center of* $\mathfrak{g}$ *and where* $\mathfrak{g}_1$ *is semisimple.*

Proof. Since $\mathfrak{z}$ is an ideal in $\mathfrak{g}$, Proposition 7.5 shows that $\mathfrak{g}_1 := \mathfrak{z}^{\perp}$ is also an ideal in $\mathfrak{g}$ and that $\mathfrak{g}$ decomposes as a Lie algebra direct sum $\mathfrak{g} = \mathfrak{g}_1 \oplus \mathfrak{z}$. It remains only to show that $\mathfrak{g}_1$ is semisimple. It is apparent that $\mathfrak{g}_1$ has trivial center, since any central element Z of $\mathfrak{g}_1$ would also be central in $\mathfrak{g} = \mathfrak{g}_1 \oplus \mathfrak{z}$, in which case, Z must be in $\mathfrak{z} \cap \mathfrak{g}_1 = \{0\}$. Thus, we must only construct a compact real form of $\mathfrak{g}_1$.

It is easy to see that $Z \in \mathfrak{g}$ belongs to $\mathfrak{z}$ if and only if Z commutes with every element of $\mathfrak{k}$. From this, it is easily seen that $\mathfrak{z}$ is invariant under "conjugation," that is, under the map $X + iY \mapsto X - iY$, where $X, Y \in \mathfrak{k}$. Since $\mathfrak{z}$ is invariant under conjugation, its orthogonal complement $\mathfrak{g}_1$ is also closed under conjugation. It follows that $\mathfrak{z}$ is the complexification of $\mathfrak{z}' := \mathfrak{z} \cap \mathfrak{k}$ and that $\mathfrak{g}_1$ is the complexification of $\mathfrak{k}_1 := \mathfrak{g}_1 \cap \mathfrak{k}$.

We will now show that $\mathfrak{k}_1$ is a compact real form of $\mathfrak{g}_1$. Let K' be the adjoint group of K, that is, the image in $\mathsf{GL}(\mathfrak{k})$ of the adjoint map. Since K is compact and the adjoint map is continuous, K' is compact and thus closed. Now, by Proposition 3.34, the Lie algebra map associated to the map $A \mapsto \mathrm{Ad}_A$ is the map $X \mapsto \mathrm{ad}_X$, the kernel of which is the center $\mathfrak{z}'$ of $\mathfrak{k}$. Thus, the image of ad is isomorphic to $\mathfrak{k}/\mathfrak{z} \cong \mathfrak{k}_1$, showing that $\mathfrak{k}_1$ is isomorphic to the Lie algebra of the image of Ad, namely K'. □

Proposition 7.7. *If* K *is a* simply connected *compact matrix Lie group with Lie algebra* $\mathfrak{k}$, *then* $\mathfrak{g} := \mathfrak{k}_{\mathbb{C}}$ *is semisimple.*

Proof. As in the proof of Proposition 7.6, $\mathfrak{k}$ decomposes as a Lie algebra direct sum $\mathfrak{k} = \mathfrak{k}_1 \oplus \mathfrak{z}'$, where $\mathfrak{z}'$ is the center of $\mathfrak{k}$ and where $\mathfrak{g}_1 := (\mathfrak{k}_1)_{\mathbb{C}}$ is semisimple. Then by Theorem 5.11, K decomposes as a direct product of closed, simply connected subgroups K_1 and Z'. Now, since $\mathfrak{z}'$ is commutative, it is isomorphic to the Lie algebra of the commutative Lie group $\mathbb{R}^n$, for some n. Since both Z' and $\mathbb{R}^n$ are simply connected, Corollary 5.7 tells us that $Z' \cong \mathbb{R}^n$. On the other hand, since Z' is a closed subgroup of K, we see that $Z' \cong \mathbb{R}^n$ is compact, which is possible only if $n = 0$. Thus, $\mathfrak{z}'$ and $\mathfrak{z} = \mathfrak{z}' + i\mathfrak{z}'$ are zero dimensional, meaning that $\mathfrak{g} = \mathfrak{g}_1$ is semisimple.   □

Recall (Definition 3.11) that a Lie algebra $\mathfrak{g}$ is said to be simple if $\mathfrak{g}$ has no nontrivial ideals *and* dim $\mathfrak{g}$ is at least 2. That is to say, a one-dimensional Lie algebra is, by decree, not simple, even though it clearly has no nontrivial ideals.

Theorem 7.8. *Suppose that $\mathfrak{g}$ is semisimple in the sense of Definition 7.1. Then $\mathfrak{g}$ decomposes as a Lie algebra direct sum*

$$\mathfrak{g} = \bigoplus_{j=1}^{m} \mathfrak{g}_j, \tag{7.4}$$

where each $\mathfrak{g}_j \subset \mathfrak{g}$ is a simple Lie algebra.

We will see in Sect. 7.6 that most of our examples of semisimple Lie algebras are actually simple, meaning that there is only one term in the decomposition in (7.4). The converse of Theorem 7.8 is also true; if a complex Lie algebra $\mathfrak{g}$ decomposes as a direct sum of simple algebras, then $\mathfrak{g}$ is semisimple in the sense of Definition 7.1. (See Theorem 6.11 in [Kna2].)

Proof. If $\mathfrak{g}$ has a nontrivial ideal $\mathfrak{h}$, then by Proposition 7.5, $\mathfrak{g}$ decomposes as the Lie algebra direct sum $\mathfrak{g} = \mathfrak{h} \oplus \mathfrak{h}^\perp$, where $\mathfrak{h}^\perp$ is also an ideal in $\mathfrak{g}$. Suppose that, say, $\mathfrak{h}$ has a nontrivial ideal $\mathfrak{h}'$. Since every element of $\mathfrak{h}$ commutes with every element of $\mathfrak{h}^\perp$, we see that $\mathfrak{h}'$ is actually an ideal in $\mathfrak{g}$. Thus, $\mathfrak{h}'' := (\mathfrak{h}')^\perp \cap \mathfrak{h}$ is an ideal in $\mathfrak{g}$ and we can decompose $\mathfrak{g}$ as $\mathfrak{h}' \oplus \mathfrak{h}'' \oplus \mathfrak{h}^\perp$. Proceeding on in the same way, we can decompose $\mathfrak{g}$ as a direct sum of ideals $\mathfrak{g}_j$, where each $\mathfrak{g}_j$ has no nontrivial ideals. It remains only to show that each $\mathfrak{g}_j$ has dimension at least 2. Suppose, toward a contradiction, that dim $\mathfrak{g}_j = 1$ for some j. Then $\mathfrak{g}_j$ is necessarily commutative, which means (since elements of $\mathfrak{g}_j$ commute with elements of $\mathfrak{g}_k$ for $j \neq k$) that $\mathfrak{g}_j$ is contained in the center $Z(\mathfrak{g})$ of $\mathfrak{g}$, contradicting the assumption that the center of $\mathfrak{g}$ is trivial.   □

Proposition 7.9. *If $\mathfrak{g}$ is a complex semisimple Lie algebra, then the subalgebras $\mathfrak{g}_j$ appearing in the decomposition (7.8) are unique up to order.*

To be more precise, suppose $\mathfrak{g}$ is isomorphic to a Lie algebra direct sum of simple subalgebras $\mathfrak{g}_1, \ldots, \mathfrak{g}_l$ and also to a Lie algebra direct sum of simple subalgebras $\mathfrak{h}_1, \ldots, \mathfrak{h}_m$. Then the proposition asserts that each $\mathfrak{h}_j$ is actually *equal to* (not just isomorphic to) $\mathfrak{g}_k$ for some k. The proposition depends crucially on the fact that the summands $\mathfrak{g}_j$ in (7.8) are simple (hence dim $\mathfrak{g}_j \geq 2$) and not just that $\mathfrak{g}_j$ has no nontrivial ideals. Indeed, if $\mathfrak{g}$ is a two-dimensional commutative Lie algebra, then $\mathfrak{g}$ can be decomposed as a direct sum of one-dimensional commutative algebras in many different ways.

Proof. If $\mathfrak{h}$ is an ideal in $\mathfrak{g}$, then $\mathfrak{h}$ can be viewed as a representation of $\mathfrak{g}$ by means of the map $X \mapsto (\mathrm{ad}_X)|_{\mathfrak{h}}$. If we decompose $\mathfrak{g}$ as the direct sum of simple subalgebra $\mathfrak{g}_1, \ldots, \mathfrak{g}_m$, then each $\mathfrak{g}_j$ is irreducible as a representation of $\mathfrak{g}$, since any invariant subspace would be an ideal in $\mathfrak{g}$ and thus an ideal in $\mathfrak{g}_j$. Furthermore, these representations are nonisomorphic, since action of $\mathfrak{g}_j$ on $\mathfrak{g}_j$ is nontrivial (since $\mathfrak{g}_j$ is not commutative), whereas the action of each $\mathfrak{g}_k$, $k \neq j$, on $\mathfrak{g}_j$ is trivial. Suppose now that $\mathfrak{h}$ is an ideal in $\mathfrak{g}$ that is simple as a Lie algebra and, thus, irreducible under the adjoint action of $\mathfrak{g}$. Now, for any j, the projection map $\pi_j : \mathfrak{g} \to \mathfrak{g}_j$ is an intertwining map of representations of $\mathfrak{g}$. Thus, for each j, the restriction of π_j to $\mathfrak{h}$ must be either 0 or an isomorphism. There must be some j for which $\pi_j|_{\mathfrak{h}}$ is nonzero and hence an isomorphism. But since the various $\mathfrak{g}_j$'s are nonisomorphic representations of $\mathfrak{g}$, we must have $\pi_k|_{\mathfrak{h}} = 0$ for all $k \neq j$. Thus, actually, $\mathfrak{h} = \mathfrak{g}_j$. $\qquad\square$

Before getting into the details of semisimple Lie algebras, let us briefly outline what our strategy will be in classifying their representations and what structures we will need to carry out this strategy. We will look for commuting elements $H_1, \ldots, H_r$ in our Lie algebra that we will try to simultaneously diagonalize in each representation. We should find as many such elements as possible, and if they are going to be simultaneously diagonalizable in every representation, they must certainly be diagonalizable in the adjoint representation. This leads (in basis-independent language) to the definition of a **Cartan subalgebra**. The nonzero sets of simultaneous eigenvalues for $\mathrm{ad}_{H_1}, \ldots, \mathrm{ad}_{H_r}$ are called **roots** and the corresponding simultaneous eigenvectors are called **root vectors**. The root vectors will serve to raise and lower the eigenvalues of $\pi(H_1), \ldots, \pi(H_r)$ in each representation π. We will also have the **Weyl group**, which is an important symmetry of the roots and also of the weights in each representation.

One crucial part of the structure of semisimple Lie algebras is the existence of certain special subalgebras isomorphic to $\mathsf{sl}(2; \mathbb{C})$. Several times over the course of this chapter and the subsequent ones, we will make use of our knowledge of the representations of $\mathsf{sl}(2; \mathbb{C})$, applied to these subalgebras. If X, Y, and H are the usual basis elements for $\mathsf{sl}(2; \mathbb{C})$, then of particular importance is the fact that in any finite-dimensional representation π of $\mathsf{sl}(2; \mathbb{C})$ (not necessarily irreducible), every eigenvalue of $\pi(H)$ must be an integer. (Compare Point 1 of Theorem 4.34 and Exercise 13 in Chapter 4.)

7.2 Cartan Subalgebras

Our first task is to identify certain special sorts of commutative subalgebras, called Cartan subalgebras.

Definition 7.10. If $\mathfrak{g}$ is a complex semisimple Lie algebra, then a **Cartan subalgebra** of $\mathfrak{g}$ is a complex subspace $\mathfrak{h}$ of $\mathfrak{g}$ with the following properties:

1. For all H_1 and H_2 in $\mathfrak{h}$, $[H_1, H_2] = 0$.
2. If, for some $X \in \mathfrak{g}$, we have $[H, X] = 0$ for all H in $\mathfrak{h}$, then X is in $\mathfrak{h}$.
3. For all H in $\mathfrak{h}$, ad_H is diagonalizable.

Condition 1 says that $\mathfrak{h}$ is a commutative subalgebra of $\mathfrak{g}$. Condition 2 says that $\mathfrak{h}$ is a *maximal* commutative subalgebra (i.e., not contained in any larger commutative subalgebra). Condition 3 says that each ad_H $(H \in \mathfrak{h})$ is diagonalizable. Since the H's in $\mathfrak{h}$ commute, the ad_H's also commute, and thus they are *simultaneously* diagonalizable (Proposition A.16).

Of course, the definition of a Cartan subalgebra makes sense in any Lie algebra, semisimple or not. However, if $\mathfrak{g}$ is not semisimple, then $\mathfrak{g}$ may not have any Cartan subalgebras in the sense of Definition 7.10; see Exercise 1. (Sometimes a different definition of "Cartan subalgebra" is used, one that allows every complex Lie algebra to have a Cartan subalgebra. This other definition is equivalent to Definition 7.10 when $\mathfrak{g}$ is semisimple but not in general.) Even in the semisimple case we must prove that a Cartan subalgebra exists.

Proposition 7.11. *Let* $\mathfrak{g} = \mathfrak{k}_\mathbb{C}$ *be a complex semisimple Lie algebra and let* $\mathfrak{t}$ *be any maximal commutative subalgebra of* $\mathfrak{k}$. *Define* $\mathfrak{h} \subset \mathfrak{g}$ *by*

$$\mathfrak{h} = \mathfrak{t}_\mathbb{C} = \mathfrak{t} + i\,\mathfrak{t}.$$

Then $\mathfrak{h}$ *is a Cartan subalgebra of* $\mathfrak{g}$.

Note that $\mathfrak{k}$ (or any other finite-dimensional Lie algebra) contains a maximal commutative subalgebra. After all, any one-dimensional subalgebra $\mathfrak{t}_1$ of $\mathfrak{k}$ is commutative. If $\mathfrak{t}_1$ is maximal, then we are done; if not, then we choose some commutative subalgebra $\mathfrak{t}_2$ properly containing $\mathfrak{t}_1$, and so on.

Proof of Proposition 7.11. It is clear that $\mathfrak{h}$ is a commutative subalgebra of $\mathfrak{g}$. We must first show that $\mathfrak{h}$ is *maximal* commutative. So, suppose that $X \in \mathfrak{g}$ commutes with every element of $\mathfrak{h}$, which certainly means that it commutes with every element of $\mathfrak{t}$. If we write $X = X_1 + iX_2$ with X_1 and X_2 in $\mathfrak{k}$, then for H in $\mathfrak{t}$, we have

$$[H, X_1 + iX_2] = [H, X_1] + i[H, X_2] = 0,$$

where $[H, X_1]$ and $[H, X_2]$ are in $\mathfrak{k}$. However, since every element of $\mathfrak{g}$ has a *unique* decomposition as an element of $\mathfrak{k}$ plus an element of $i\mathfrak{k}$, we see that $[H, X_1]$ and $[H, X_2]$ must separately be zero. Since this holds for all H in $\mathfrak{t}$ and since $\mathfrak{t}$ is maximal commutative, we must have X_1 and X_2 in $\mathfrak{t}$, which means that $X = X_1 + iX_2$ is in $\mathfrak{h}$. This shows that $\mathfrak{h}$ is maximal commutative.

If $\langle \cdot, \cdot \rangle$ is an inner product on $\mathfrak{g}$ as in Proposition 7.4, then for each X in $\mathfrak{k}$, the operator $\mathrm{ad}_X : \mathfrak{g} \to \mathfrak{g}$ is skew self-adjoint. In particular, each ad_H, $H \in \mathfrak{t}$, is skew self-adjoint and thus diagonalizable (Theorem A.3). Finally, if H is any element of $\mathfrak{h} = \mathfrak{t} + i\,\mathfrak{t}$, then $H = H_1 + iH_2$, with H_1 and H_2 in $\mathfrak{t}$. Since H_1 and H_2 commute, ad_{H_1} and ad_{H_2} also commute and, thus, by Proposition A.16, ad_{H_1} and ad_{H_2} are simultaneously diagonalizable. It follows that ad_H is diagonalizable. $\qquad\square$

Throughout this chapter and the subsequent chapters, we consider only Cartan subalgebras of the form $\mathfrak{h} = \mathfrak{t}_{\mathbb{C}}$, where $\mathfrak{t}$ is a maximal commutative subalgebra of some compact real form $\mathfrak{k}$ of $\mathfrak{g}$. It is true, but not obvious, that every Cartan subalgebra arises in this way. Indeed, for any two Cartan subalgebras $\mathfrak{h}_1$ and $\mathfrak{h}_2$ of $\mathfrak{g}$, there exists an automorphism of $\mathfrak{g}$ mapping $\mathfrak{h}_1$ to $\mathfrak{h}_2$. (See Proposition 2.13 and Theorem 2.15 in Chapter II of [Kna2].) While we will not prove this result, we will prove in Chapter 11 that for a fixed $\mathfrak{k}$, the maximal commutative subalgebra $\mathfrak{t} \subset \mathfrak{k}$ is unique up to an automorphism of $\mathfrak{k}$. (See Proposition 11.7 and Theorem 11.9.) In light of the uniqueness of Cartan subalgebras up to automorphism, the following definition is meaningful.

Definition 7.12. If $\mathfrak{g}$ is a complex semisimple Lie algebra, the **rank** of $\mathfrak{g}$ is the dimension of any Cartan subalgebra.

7.3 Roots and Root Spaces

For the rest of this chapter, we assume that we have chosen a compact real form $\mathfrak{k}$ of $\mathfrak{g}$ and a maximal commutative subalgebra $\mathfrak{t}$ of $\mathfrak{k}$, and we consider the Cartan subalgebra $\mathfrak{h} = \mathfrak{t} + i\mathfrak{t}$. We assume also that we have chosen an inner product on $\mathfrak{g}$ that is real on $\mathfrak{k}$ and invariant under the adjoint action of K (Proposition 7.4).

Since the operators ad_H, $H \in \mathfrak{h}$, commute, and each such ad_H is diagonalizable, Proposition A.16 tell us that the ad_H's with $H \in \mathfrak{h}$ are *simultaneously* diagonalizable. If $X \in \mathfrak{g}$ is a simultaneous eigenvalue for each ad_H, $H \in \mathfrak{h}$, then the corresponding eigenvalues depend linearly on $H \in \mathfrak{h}$. If this linear functional is nonzero, we call it a *root*. As in Sect. 6.6, it is convenient to express this linear functional as $H \mapsto \langle \alpha, H \rangle$ for some $\alpha \in \mathfrak{h}$. The preceding discussion leads to the following definition.

Definition 7.13. A nonzero element α of $\mathfrak{h}$ is a **root** (for $\mathfrak{g}$ relative to $\mathfrak{h}$) if there exists a nonzero $X \in \mathfrak{g}$ such that

$$[H, X] = \langle \alpha, H \rangle X$$

for all H in $\mathfrak{h}$. The set of all roots is denoted R.

Note that we follow the convention that an inner product be linear in the *second* factor, so that $\langle \alpha, H \rangle$ depends linearly on H for a fixed α.

Proposition 7.14. *Each root α belongs to $i\mathfrak{t} \subset \mathfrak{h}$.*

Proof. As we have already noted, each ad_H, $H \in \mathfrak{t}$, is a skew self-adjoint operator on $\mathfrak{h}$, which means that ad_H has pure imaginary eigenvalues. It follows that if α is a root, $\langle \alpha, H \rangle$ must be pure imaginary for $H \in \mathfrak{t}$, which (since our inner product is real on $\mathfrak{t} \subset \mathfrak{k}$) can only happen if α is in $i\mathfrak{t}$. $\qquad\square$

Definition 7.15. If α is a root, then the **root space** $\mathfrak{g}_\alpha$ is the space of all X in $\mathfrak{g}$ for which $[H, X] = \langle \alpha, H \rangle X$ for all H in $\mathfrak{h}$. A nonzero element of $\mathfrak{g}_\alpha$ is called a **root vector** for α.

More generally, if α is any element of $\mathfrak{h}$, we define $\mathfrak{g}_\alpha$ to be the space of all X in $\mathfrak{g}$ for which $[H, X] = \langle \alpha, H \rangle X$ for all H in $\mathfrak{h}$ (but we do not call $\mathfrak{g}_\alpha$ a root space unless α is actually a root).

Taking $\alpha = 0$, we see that $\mathfrak{g}_0$ is the set of all elements of $\mathfrak{g}$ that commute with every element of $\mathfrak{h}$. Since $\mathfrak{h}$ is a maximal commutative subalgebra, we conclude that $\mathfrak{g}_0 = \mathfrak{h}$. If α is not zero and not a root, then $\mathfrak{g}_\alpha = \{0\}$.

As we have noted, the operators ad_H, $H \in \mathfrak{h}$, are simultaneously diagonalizable. As a result, $\mathfrak{g}$ can be decomposed as the sum of $\mathfrak{h}$ and the root spaces $\mathfrak{g}_\alpha$. Actually, by Proposition A.17, the sum is direct and we have established the following result.

Proposition 7.16. *The Lie algebra $\mathfrak{g}$ can be decomposed as a direct sum of vector spaces as follows:*

$$\mathfrak{g} = \mathfrak{h} \oplus \bigoplus_{\alpha \in R} \mathfrak{g}_\alpha.$$

That is to say, every element of $\mathfrak{g}$ can be written uniquely as a sum of an element of $\mathfrak{h}$ and one element from each root space $\mathfrak{g}_\alpha$. Note that the decomposition is not a Lie algebra direct sum, since, for example, elements of $\mathfrak{h}$ do not, in general, commute with elements of $\mathfrak{g}_\alpha$.

Proposition 7.17. *For any α and β in $\mathfrak{h}$, we have*

$$[\mathfrak{g}_\alpha, \mathfrak{g}_\beta] \subset \mathfrak{g}_{\alpha+\beta}.$$

In particular, if X is in $\mathfrak{g}_\alpha$ and Y is in $\mathfrak{g}_{-\alpha}$, then $[X, Y]$ is in $\mathfrak{h}$. Furthermore, if X is in $\mathfrak{g}_\alpha$, Y is in $\mathfrak{g}_\beta$, and $\alpha + \beta$ is neither zero nor a root, then $[X, Y] = 0$.

Proof. It follows from the Jacobi identity that ad_H is a derivation:

$$[H, [X, Y]] = [[H, X], Y] + [X, [H, Y]].$$

Thus, if X is in $\mathfrak{g}_\alpha$ and Y is in $\mathfrak{g}_\beta$, we have

$$[H, [X, Y]] = [\langle \alpha, H \rangle X, Y] + [X, \langle \beta, H \rangle Y]$$
$$= \langle \alpha + \beta, H \rangle [X, Y].$$

for all $H \in \mathfrak{h}$, showing that $[X, Y]$ is in $\mathfrak{g}_{\alpha+\beta}$. $\qquad\square$

Proposition 7.18. *1. If $\alpha \in \mathfrak{h}$ is a root, so is $-\alpha$. Specifically, if X is in $\mathfrak{g}_\alpha$, then X^* is in $\mathfrak{g}_{-\alpha}$, where X^* is defined by (7.2) in Proposition 7.4.*
2. The roots span $\mathfrak{h}$.

Proof. For Point 1, if $X = X_1 + iX_2$ with $X_1, X_2 \in \mathfrak{k}$, let

$$\bar{X} = X_1 - iX_2. \tag{7.5}$$

Since $\mathfrak{k}$ is closed under brackets, if $H \in \mathfrak{t} \subset \mathfrak{k}$ and $X \in \mathfrak{g}$, we have

$$\overline{[H, X]} = [H, X_1] - i[H, X_2] = [H, \bar{X}].$$

In particular, if X is a root vector with root $\alpha \in i\mathfrak{t}$, then for all $H \in \mathfrak{h}$, we have

$$[H, \bar{X}] = \overline{[H, X]} = \overline{\langle \alpha, H \rangle X} = -\langle \alpha, H \rangle \bar{X}, \tag{7.6}$$

since $\langle \alpha, H \rangle$ is pure imaginary for $H \in \mathfrak{t}$. Extending (7.6) by linearity in H, we see that $[H, \bar{X}] = -\langle \alpha, H \rangle \bar{X}$ for all $H \in \mathfrak{h}$. Thus, $\bar{X}$ is a root vector corresponding to the root $-\alpha$. It follows that $X^* = -\bar{X}$ belongs to $\mathfrak{g}_{-\alpha}$.

For Point 2, suppose that the roots did not span $\mathfrak{h}$. Then there would exist a nonzero $H \in \mathfrak{h}$ such that $\langle \alpha, H \rangle = 0$ for all $\alpha \in R$. Then we would have $[H, H'] = 0$ for all H' in $\mathfrak{h}$, and also

$$[H, X] = \langle \alpha, H \rangle X = 0$$

for X in $\mathfrak{g}_\alpha$. Thus, by Proposition 7.16, H would be in the center of $\mathfrak{g}$, contradicting the definition of a semisimple Lie algebra. $\square$

We now develop a key tool in the study of a semisimple Lie algebra $\mathfrak{g}$, the existence of certain subalgebras of $\mathfrak{g}$ isomorphic to $\mathsf{sl}(2; \mathbb{C})$.

Theorem 7.19. *For each root α, we can find linearly independent elements X_α in $\mathfrak{g}_\alpha$, Y_α in $\mathfrak{g}_{-\alpha}$, and H_α in $\mathfrak{h}$ such that H_α is a multiple of α and such that*

$$[H_\alpha, X_\alpha] = 2X_\alpha,$$
$$[H_\alpha, Y_\alpha] = -2Y_\alpha,$$
$$[X_\alpha, Y_\alpha] = H_\alpha. \tag{7.7}$$

Furthermore, Y_α can be chosen to equal X_α^*.

If $X_\alpha, Y_\alpha, H_\alpha$ are as in the theorem, then on the one hand, $[H_\alpha, X_\alpha] = 2X_\alpha$, while on the other hand, $[H_\alpha, X_\alpha] = \langle \alpha, H_\alpha \rangle X_\alpha$. Thus,

$$\langle \alpha, X_\alpha \rangle = 2. \tag{7.8}$$

Meanwhile, H_α is a multiple of α and the unique such multiple compatible with (7.8) is

$$H_\alpha = 2\frac{\alpha}{\langle \alpha, \alpha \rangle}. \tag{7.9}$$

For use in Part III, we record the following consequence of Theorem 7.19.

Corollary 7.20. *For each root* α, *let* X_α, Y_α, *and* H_α *be as in Theorem 7.19, with* $Y_\alpha = X_\alpha^*$. *Then the elements*

$$E_1^\alpha := \frac{i}{2}H_\alpha; \quad E_2^\alpha := \frac{i}{2}(X_\alpha + Y_\alpha); \quad E_3^\alpha := \frac{1}{2}(Y_\alpha - X_\alpha)$$

are linearly independent elements of $\mathfrak{k}$ *and satisfy the commutation relations*

$$[E_1^\alpha, E_2^\alpha] = E_3^\alpha; \quad [E_2^\alpha, E_3^\alpha] = E_1^\alpha; \quad [E_3^\alpha, E_1^\alpha] = E_2^\alpha.$$

Thus, by Example 3.27, the span of E_1^α, E_2^α, *and* E_3^α *is a subalgebra of* $\mathfrak{k}$ *isomorphic to* $\mathsf{su}(2)$.

Proof. Since α belongs to $i\mathfrak{t}$ and H_α is, by (7.9), a real multiple of α, the element $(i/2)H_\alpha$ will belong to $\mathfrak{t} \subset \mathfrak{k}$. Meanwhile, we may check that $(E_2^\alpha)^* = -E_2^\alpha$ and $(E_3^\alpha)^* = -E_3^\alpha$. From the way the map $X \mapsto X^*$ was defined (Proposition 7.4), it follows that E_2^α and E_3^α also belong to $\mathfrak{k}$. Furthermore, since X_α, Y_α, and H_α are linearly independent, E_1^α, E_2^α, and E_3^α are also independent. Finally, direct computation with the commutation relations in (7.7) confirms the claimed commutation relations for the E_j^α's. $\square$

Definition 7.21. The element $H_\alpha = 2\alpha/\langle \alpha, \alpha \rangle$ in Theorem 7.19 is the **coroot** associated to the root α.

We begin the proof of Theorem 7.19 with a lemma.

Lemma 7.22. *Suppose that X is in $\mathfrak{g}_\alpha$, that Y is in $\mathfrak{g}_{-\alpha}$, and that H is in $\mathfrak{h}$. Then $[X, Y]$ is in $\mathfrak{h}$ and*

$$\langle [X, Y], H \rangle = \langle \alpha, H \rangle \langle Y, X^* \rangle, \tag{7.10}$$

where X^* is as in Proposition 7.4.

In the proof of Theorem 7.19, we will need to know not just that $[X, Y] \in \mathfrak{h}$, but *where* in $\mathfrak{h}$ the element $[X, Y]$ lies. This information is obtained by computing the inner product of $[X, Y]$ with each element of H, as we have done in (7.10).

Proof. That $[X, Y]$ is in $\mathfrak{h}$ follows from Proposition 7.17. Then, using Proposition 7.4, we compute that

$$\langle [X, Y], H \rangle = \langle \mathrm{ad}_X(Y), H \rangle = \langle Y, \mathrm{ad}_{X^*}(H) \rangle = -\langle Y, [H, X^*] \rangle. \tag{7.11}$$

Since X is in $\mathfrak{g}_\alpha$, Proposition 7.18 tells us that the element X^* is in $\mathfrak{g}_{-\alpha}$, so that $[H, X^*] = -\langle \alpha, H \rangle X^*$ and (7.11) reduces to the desired result. (Recall that we take inner products to be linear in the second factor.) $\square$

Proof of Theorem 7.19. Choose a nonzero X in $\mathfrak{g}_\alpha$, so that $X^* = -\bar{X}$ is a nonzero element of $\mathfrak{g}_{-\alpha}$. Applying Lemma 7.22 with $Y = X^*$ gives

$$\langle [X, X^*], H \rangle = \langle \alpha, H \rangle \langle X^*, X^* \rangle. \tag{7.12}$$

From (7.12), we see that $[X, X^*] \in \mathfrak{h}$ is orthogonal to every $H \in \mathfrak{h}$ that is orthogonal to α, which happens only if $[X, X^*]$ is a multiple of α. Furthermore, if we choose H so that $\langle \alpha, H \rangle \neq 0$, we see that $\langle [X, X^*], H \rangle \neq 0$, so that $[X, X^*] \neq 0$. Now, if we evaluate (7.12) with $H = [X, X^*]$, we obtain

$$\langle [X, X^*], [X, X^*] \rangle = \langle \alpha, [X, X^*] \rangle \langle X^*, X^* \rangle.$$

Since $[X, X^*] \neq 0$, we conclude that $\langle \alpha, [X, X^*] \rangle$ is real and strictly positive.

Let $H = [X, X^*]$ and define elements of $\mathfrak{g}$ as follows:

$$H_\alpha = \frac{2}{\langle \alpha, H \rangle} H,$$

$$X_\alpha = \sqrt{\frac{2}{\langle \alpha, H \rangle}} X,$$

$$Y_\alpha = \sqrt{\frac{2}{\langle \alpha, H \rangle}} Y.$$

Then $\langle \alpha, H_\alpha \rangle = 2$, from which it follows that $[H_\alpha, X_\alpha] = 2X_\alpha$ and $[H_\alpha, Y_\alpha] = -2Y_\alpha$. Furthermore, $[X_\alpha, Y_\alpha] = 2[X, Y]/\langle \alpha, H \rangle = H_\alpha$. Thus, these elements satisfy the relations claimed in the theorem, and $Y_\alpha = X_\alpha^*$. Furthermore, since these elements are eigenvectors for ad_H with distinct eigenvalues 2, -2, and 0, they must be linearly independent (Proposition A.1). $\square$

We now make use of the subalgebras in Theorem 7.19 to obtain results about roots and root spaces. Note that

$$\mathfrak{s}^\alpha := \langle X_\alpha, Y_\alpha, H_\alpha \rangle \tag{7.13}$$

acts on $\mathfrak{g}$ by (the restriction to $\mathfrak{s}^\alpha$ of) the adjoint representation.

Theorem 7.23. *1. For each root α, the only multiples of α that are roots are α and $-\alpha$.*
2. For each root α, the root space $\mathfrak{g}_\alpha$ is one dimensional.

Point 1 of Theorem 7.23 should be contrasted with the results of Exercise 8 in the case of a Lie algebra that is not semisimple.

Lemma 7.24. *If α and $c\alpha$ are both roots with $|c| > 1$, then $c = \pm 2$.*

Proof. Let $\mathfrak{s}^\alpha$ be as in (7.13). If $\beta = c\alpha$ is also a root and X is a nonzero element of $\mathfrak{g}_\beta$, then by (7.8), we have

$$[H_\alpha, X] = \langle \beta, H_\alpha \rangle X = \bar{c} \langle \alpha, H_\alpha \rangle X = 2\bar{c} X.$$

Thus, $2\bar{c}$ is an eigenvalue for the adjoint action of $H_\alpha \in \mathfrak{s}^\alpha$ on $\mathfrak{g}$. By Point 1 of Theorem 4.34, any such eigenvalue must be an integer, meaning that c is an integer multiple of $1/2$. But by reversing the roles of α and β in the argument, we see that $1/c$ must also be an integer multiple of $1/2$.

Fig. 7.1 If $\beta = 2\alpha$ were a root, the orthogonal complement of $\mathfrak{s}^\alpha$ in V^α would contain an element of $\mathfrak{h}$ orthogonal to H_α

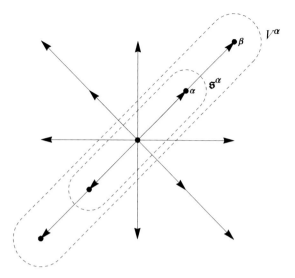

Now, suppose $c = n/2$ for some integer n and that $1/c = 2/n$ is an integer multiple of $1/2$, so that $2/c = 4/n$ is an integer. Then $n = \pm 1, \pm 2$, or ± 4. Thus, $c = \pm 1/2, \pm 1$, or ± 2, of which only ± 2 have $|c| > 1$.   □

Proof of Theorem 7.23. Since there are only finitely many roots, there is some smallest positive multiple of α that is a root. Replacing α by this smallest multiple, we can assume that if $c\alpha$ is a root and $c\alpha \neq \pm \alpha$, then $|c| > 1$. By Lemma 7.24, we must then have $c = \pm 2$.

Let $\mathfrak{s}^\alpha$ be as in (7.13), with Y_α chosen to equal X_α^*. Let V^α be the subspace of $\mathfrak{g}$ spanned by H_α and all the root spaces $\mathfrak{g}_\beta$ for which β is a multiple of α. (See Figure 7.1.) We claim that V^α is a subalgebra of $\mathfrak{g}$. To verify this claim, observe first that if a root β is a multiple of α then by Lemma 7.22, every element of $[\mathfrak{g}_\beta, \mathfrak{g}_{-\beta}]$ is in $\mathfrak{h}$ and is orthogonal to every element of $\mathfrak{h}$ that is orthogonal to β. Thus, every element of $[\mathfrak{g}_\beta, \mathfrak{g}_{-\beta}]$ is a multiple of β, which is a multiple of α, which is a multiple of H_α. Observe next that if $X \in \mathfrak{g}_\beta$, then $[H_\alpha, X]$ is a multiple of X. Observe, finally, that if β and β' are roots that are multiples of α with $\beta' \neq -\beta$, then $[\mathfrak{g}_\beta, \mathfrak{g}_{\beta'}] \subset \mathfrak{g}_{\beta+\beta'}$, where $\beta + \beta' \neq 0$ is again a multiple of α.

Since V^α is a subalgebra of $\mathfrak{g}$, it is certainly invariant under the adjoint action of $\mathfrak{s}^\alpha \subset V^\alpha$. Note also that $\mathfrak{s}^\alpha$ itself is an invariant subspace for the adjoint action of $\mathfrak{s}^\alpha$ on V^α. Now, since $Y_\alpha = X_\alpha^*$ and H_α is a positive multiple of $\alpha \in i\mathfrak{t} \subset i\mathfrak{k}$, we see that $X^* \in \mathfrak{s}^\alpha$ for every $X \in \mathfrak{s}^\alpha$. Thus, by Propositions 7.4 and 4.27, the orthogonal complement U^α of $\mathfrak{s}^\alpha$ in V^α will also be an invariant under the adjoint action of $\mathfrak{s}^\alpha$.

Now, $\langle \alpha, H_\alpha \rangle = 2$ and, by Lemma 7.24, any multiples β of α that are roots are of the form $\beta = \pm \alpha$ or $\beta = \pm 2\alpha$. Thus, the eigenvalues of ad_{H_α} in V^α are $0, \pm 2$, and, possibly, ± 4. If $U^\alpha \neq \{0\}$, then U^α will contain an eigenvector for ad_{H_α}, with an eigenvalue $\lambda \in \{0, \pm 2, \pm 4\}$. Since λ is even, it follows from Point 4 of Theorem 4.34 that 0 must also be an eigenvalue for ad_{H_α} in U^α. But this is impossible, since H_α is the only eigenvector for ad_{H_α} in V^α with eigenvalue 0, and U^α is orthogonal to $H_\alpha \in \mathfrak{s}^\alpha$. Thus, actually, $U^\alpha = \{0\}$, which means that $V^\alpha = \mathfrak{s}^\alpha$. Thus, the only multiples of α that are roots are $\pm\alpha$ and the root spaces with roots $\pm\alpha$ are one dimensional. $\qquad\qquad\qquad\qquad\qquad\qquad\qquad\qquad\qquad\qquad\qquad\qquad\square$

Figure 7.1 shows a configuration of "roots" consistent with Lemma 7.24, but which cannot actually be the root system of a semisimple Lie algebra.

7.4 The Weyl Group

We now introduce an important symmetry of the set R of roots, known as the Weyl group. In this section, we follow the Lie algebra approach to the Weyl group, as opposed to the Lie group approach we followed in Sect. 6.6 in the SU(3) case. We will return to the Lie group approach to the Weyl group in Chapter 11 and show (Sect. 11.7) that the two approaches are equivalent.

Definition 7.25. For each root $\alpha \in R$, define a linear map $s_\alpha : \mathfrak{h} \to \mathfrak{h}$ by the formula

$$s_\alpha \cdot H = H - 2\frac{\langle \alpha, H \rangle}{\langle \alpha, \alpha \rangle}\alpha. \tag{7.14}$$

The **Weyl group** of R, denoted W, is then the subgroup of $\mathsf{GL}(\mathfrak{h})$ generated by all the s_α's with $\alpha \in R$.

Note that since each root α is in $i\mathfrak{t}$ and our inner product is real on $\mathfrak{t}$, if H is in $i\mathfrak{t}$, then $s_\alpha \cdot H$ is also in $i\mathfrak{t}$. As a map of $i\mathfrak{t}$ to itself, s_α is the *reflection about the hyperplane orthogonal to α*. That is to say, $s_\alpha \cdot H = H$ whenever H is orthogonal to α, and $s_\alpha \cdot \alpha = -\alpha$. Since each reflection is an orthogonal linear transformation, we see that W is a subgroup of the orthogonal group $\mathsf{O}(i\mathfrak{t})$.

Theorem 7.26. *The action of W on $i\mathfrak{t}$ preserves R. That is to say, if α is a root, then $w \cdot \alpha$ is a root for all $w \in W$.*

Proof. For each $\alpha \in R$, consider the invertible linear operator S_α on $\mathfrak{g}$ given by

$$S_\alpha = e^{\mathrm{ad}_{X_\alpha}} e^{-\mathrm{ad}_{Y_\alpha}} e^{\mathrm{ad}_{X_\alpha}}.$$

Now, if $H \in \mathfrak{h}$ satisfies $\langle \alpha, H \rangle = 0$, then $[H, X_\alpha] = \langle \alpha, H \rangle X_\alpha = 0$. Thus, H and X_α commute, which means that ad_H and ad_{X_α} also commute, and similarly for ad_H and ad_{Y_α}. Thus, if $\langle \alpha, H \rangle = 0$, the operator S_α will commute with ad_H, so that

$$S_\alpha \mathrm{ad}_H S_\alpha^{-1} = \mathrm{ad}_H, \quad \langle \alpha, H \rangle = 0. \tag{7.15}$$

On the other hand, if we apply Point 3 of Theorem 4.34 to the adjoint action of $\mathfrak{s}^\alpha \cong \mathsf{sl}(2; \mathbb{C})$ on $\mathfrak{g}$, we see that

$$S_\alpha \mathrm{ad}_{H_\alpha} S_\alpha^{-1} = -\mathrm{ad}_{H_\alpha}. \tag{7.16}$$

By combining (7.15) and (7.16), we see that for all $H \in \mathfrak{h}$, we have

$$S_\alpha \mathrm{ad}_H S_\alpha^{-1} = \mathrm{ad}_{s_\alpha \cdot H}. \tag{7.17}$$

Now if β is any root and X is an associated root vector, consider the vector $S_\alpha^{-1}(X) \in \mathfrak{g}$. We compute that

$$\mathrm{ad}_H(S_\alpha^{-1}(X)) = S_\alpha^{-1}(S_\alpha \mathrm{ad}_H S_\alpha^{-1})(X)$$
$$= S_\alpha^{-1} \mathrm{ad}_{s_\alpha \cdot H}(X)$$
$$= \langle \beta, s_\alpha \cdot H \rangle S_\alpha^{-1}(X)$$
$$= \langle s_\alpha^{-1} \cdot \beta, H \rangle S_\alpha^{-1}(X).$$

Thus, $S_\alpha^{-1}(X)$ is a root vector with root $s_\alpha^{-1} \cdot \beta = s_\alpha \cdot \beta$. This shows that the set of roots is invariant under each s_α and, thus, under W. □

Actually, since $s_\alpha \cdot s_\alpha \cdot \beta = \beta$, each reflection maps R onto R. It follows that each $w \in W$ also maps R onto R.

Corollary 7.27. *The Weyl group is finite.*

Proof. Since the roots span $\mathfrak{h}$, each $w \in W$ is determined by its action on R. Since, also, w maps R onto R, we see that W may be thought of as a subgroup of the permutation group on the roots. □

7.5 Root Systems

In this section, we record several important properties of the roots, using results from the two previous sections. Recall that for each root α, we have an element H_α of $\mathfrak{h}$ contained in $[\mathfrak{g}_\alpha, \mathfrak{g}_{-\alpha}]$ as in Theorem 7.19. As we saw in (7.8) and (7.9), H_α satisfies $\langle \alpha, H_\alpha \rangle = 2$ and is related to α by the formula $H_\alpha = 2\alpha / \langle \alpha, \alpha \rangle$. In particular, the element H_α is independent of the choice of X_α and Y_α in Theorem 7.19.

Definition 7.28. For each root α, the element $H_\alpha \in \mathfrak{h}$ given by

$$H_\alpha = 2\frac{\alpha}{\langle \alpha, \alpha \rangle}$$

is the **coroot** associated to the root α.

Proposition 7.29. *For all roots α and β, we have that*

$$\langle \beta, H_\alpha \rangle = 2 \frac{\langle \alpha, \beta \rangle}{\langle \alpha, \alpha \rangle} \tag{7.18}$$

is an integer.

We have actually already made use of this result in the proof of Lemma 7.24.

Proof. If $\mathfrak{s}^\alpha = \langle X_\alpha, Y_\alpha, H_\alpha \rangle$ is as in Theorem 7.19 and X is a root vector associated to the root β, then $[H_\alpha, X] = \langle \beta, H_\alpha \rangle X$. Thus, $\langle \beta, H_\alpha \rangle$ is an eigenvalue for the adjoint action of $\mathfrak{s}^\alpha \cong \mathsf{sl}(2; \mathbb{C})$ on $\mathfrak{g}$. Point 1 of Theorem 4.34 then shows that $\langle \beta, H_\alpha \rangle$ must be an integer. $\qquad\qquad\qquad\square$

Recall from elementary linear algebra that if α and β are elements of an inner product space, the orthogonal projection of β onto α is given by

$$\frac{\langle \alpha, \beta \rangle}{\langle \alpha, \alpha \rangle} \alpha.$$

The quantity on the right-hand side of (7.18) is thus twice the coefficient of α in the projection of β onto α. We may therefore interpret the integrality result in Proposition 7.29 in the following geometric way:

> If α and β are roots, the orthogonal projection of α onto β must be an integer or half-integer multiple of β.

Alternatively, we may think about Proposition 7.29 as saying that β and $s_\alpha \cdot \beta$ must differ by an integer multiple of α [compare (7.14)].

If we think of the set R of roots as a subset of the real inner product space $E := i\mathfrak{t}$, we may summarize the properties of R as follows.

Theorem 7.30. *The set R of roots is a finite set of nonzero elements of a real inner product space E, and R has the following additional properties.*

1. *The roots span E.*
2. *If $\alpha \in R$, then $-\alpha \in R$ and the only multiples of α in R are α and $-\alpha$.*
3. *If α and β are in R, so is $s_\alpha \cdot \beta$, where s_α is the reflection defined by (7.14).*
4. *For all α and β in R, the quantity*

$$2 \frac{\langle \alpha, \beta \rangle}{\langle \alpha, \alpha \rangle}$$

is an integer.

Any such collection of vectors is called a **root system**. We will look in detail at the properties of root systems in Chapter 8.

7.6   Simple Lie Algebras

Every semisimple Lie algebra decomposes as a direct sum of simple algebras (Theorem 7.8). In this section, we give a criterion for a semisimple Lie algebra to be simple. We will eventually see (Sect. 8.11) that most of the familiar examples of semisimple Lie algebras are actually simple.

Proposition 7.31. *Suppose $\mathfrak{g}$ is a real Lie algebra and that the complexification $\mathfrak{g}_{\mathbb{C}}$ of $\mathfrak{g}$ is simple. Then $\mathfrak{g}$ is also simple.*

Proof. Since $\mathfrak{g}_{\mathbb{C}}$ is simple, the dimension of $\mathfrak{g}_{\mathbb{C}}$ over $\mathbb{C}$ is at least 2, so that the dimension of $\mathfrak{g}$ over $\mathbb{R}$ is also at least 2. If $\mathfrak{g}$ had a nontrivial ideal $\mathfrak{h}$, then the complexification $\mathfrak{h}_{\mathbb{C}}$ of $\mathfrak{h}$ would be a nontrivial ideal in $\mathfrak{g}$. ☐

The converse of Proposition 7.31 is false in general. The Lie algebra $\mathsf{so}(3;1)$, for example, is simple as a real Lie algebra, and yet its complexification is isomorphic to $\mathsf{so}(4;\mathbb{C})$, which in turn is isomorphic to $\mathsf{sl}(2;\mathbb{C}) \oplus \mathsf{sl}(2;\mathbb{C})$. See Exercise 14.

Theorem 7.32. *Suppose K is a* compact *matrix Lie group whose Lie algebra $\mathfrak{k}$ is simple as a real Lie algebra. Then the complexification $\mathfrak{g} := \mathfrak{k}_{\mathbb{C}}$ of $\mathfrak{k}$ is simple as a complex Lie algebra.*

For more results about simple algebras over $\mathbb{R}$, see Exercises 12 and 13. Before proving Theorem 7.32 result, we introduce a definition.

Definition 7.33. *If $\mathfrak{g}$ is a real Lie algebra, $\mathfrak{g}$* **admits a complex structure** *if there exists a "multiplication by i" map $J : \mathfrak{g} \to \mathfrak{g}$ that makes $\mathfrak{g}$ into a complex vector space in such a way that the bracket map $[\cdot, \cdot] : \mathfrak{g} \times \mathfrak{g} \to \mathfrak{g}$ is complex bilinear.*

Here, bilinearity of the bracket means, in particular, that $[JX, Y] = J[X, Y]$ for all $X, Y \in \mathfrak{k}$. Equivalently, $\mathfrak{g}$ admits a complex structure if there exists a complex Lie algebra $\mathfrak{h}$ and a real-linear map $\phi : \mathfrak{h} \to \mathfrak{g}$ that is one-to-one and onto and satisfies $\phi([X, Y]) = [\phi(X), \phi(Y)]$ for all $X, Y \in \mathfrak{h}$.

Lemma 7.34. *Suppose that K is a compact matrix Lie group whose Lie algebra $\mathfrak{k}$ is noncommutative. Then $\mathfrak{k}$ does not admit a complex structure.*

Proof. Suppose, toward a contradiction, that $\mathfrak{k}$ does admit a complex structure J. By Proposition 7.4, there exists a real inner product on $\mathfrak{k}$ with respect to which ad_H is skew symmetric for all $H \in \mathfrak{k}$. If we choose H not in the center of $\mathfrak{k}$ then ad_H is nonzero and skew symmetric, hence diagonalizable in $\mathfrak{k}_{\mathbb{C}}$ with pure-imaginary eigenvalues, not all of which are zero. In particular, ad_H is not nilpotent.

On the other hand, since J is complex bilinear, if we view $\mathfrak{k}$ as a complex vector space with respect to the map J, then ad_H is complex linear. Since ad_H is not nilpotent, it has a nonzero eigenvalue $\lambda = a + ib \in \mathbb{C}$. Thus, there is a nonzero $X \in \mathfrak{k}$ such that

$$[H, X] = (a + ib) \cdot X = aX + bJX.$$

If we then consider element

$$H' := \bar{\lambda} \cdot H = aH - bJH,$$

we may compute, using the linearity of the bracket with respect to J and the identity $J^2 = -I$, that

$$[H', X] = |\lambda|^2 X = (a^2 + b^2)X.$$

But $\mathrm{ad}_{H'}$ is also skew symmetric, and, thus,

$$|\lambda|^2 \langle X, X \rangle = \langle \mathrm{ad}_{H'}(X), X \rangle = - \langle X, \mathrm{ad}_{H'}(X) \rangle = - |\lambda|^2 \langle X, X \rangle,$$

which is impossible if λ and X are both nonzero.   □

Proof of Theorem 7.32. Suppose to the contrary that $\mathfrak{g}$ were not simple, so that it decomposes as a sum of at least two simple algebras $\mathfrak{g}_j$. By Proposition 7.9, the decomposition of a semisimple algebra Lie algebra into a sum of simple algebras is unique up to ordering of the summands. On the other hand, if $\mathfrak{g}$ decomposes as the sum of the $\mathfrak{g}_j$'s, it also decomposes as the sum of the $\overline{\mathfrak{g}_j}$'s, where the map $X \mapsto \bar{X}$ is as in (7.5). Thus, for each j, we must have $\overline{\mathfrak{g}_j} = \mathfrak{g}_k$ for some k.

Suppose there is some j for which $\overline{\mathfrak{g}_j} = \mathfrak{g}_j$. Then for all $X \in \mathfrak{g}_j$, the element $X + \bar{X}$ is in $\mathfrak{g}_j \cap \mathfrak{k}$, from which it follows that $\mathfrak{g}_j \cap \mathfrak{k}$ is a nonzero ideal in $\mathfrak{k}$. But $\mathfrak{g}_j \cap \mathfrak{k}$ cannot be all of $\mathfrak{k}$, or else $\mathfrak{g}_j$ would be all of $\mathfrak{g}$. Thus, $\mathfrak{g}_j \cap \mathfrak{k}$ would be a nontrivial ideal in $\mathfrak{k}$, contradicting our assumption that $\mathfrak{k}$ is simple. On the other hand, if we pick some j with $\overline{\mathfrak{g}_j} = \mathfrak{g}_k, k \neq j$, then $(\mathfrak{g}_j \oplus \mathfrak{g}_k) \cap \mathfrak{k}$ is a nonzero ideal in $\mathfrak{k}$, which must be all of $\mathfrak{k}$. We conclude, then, that there must be exactly two summands $\mathfrak{g}_1$ and $\mathfrak{g}_2$ in the decomposition of $\mathfrak{g}$, satisfying $\overline{\mathfrak{g}_1} = \mathfrak{g}_2$.

Let us then define a real-linear map $\phi : \mathfrak{g}_1 \to \mathfrak{k}$ by $\phi(X) = X + \bar{X}$. Note that for any $X \in \mathfrak{g}_1$, the element $\bar{X}$ is in $\mathfrak{g}_2$, so that $[X, \bar{X}] = 0$. From this, we can easily check that ϕ satisfies $\phi([X, Y]) = [\phi(X), \phi(Y)]$ for all $X, Y \in \mathfrak{g}_1$. Furthermore, ϕ is injective because, for any $X \in \mathfrak{g}_1$, we have $\bar{X} \in \mathfrak{g}_2$ and thus $\bar{X}$ cannot equal $-X$ unless $X = 0$. Finally, by counting real dimensions, we see that ϕ is also surjective. Since $\mathfrak{g}_1$ is a complex Lie algebra and ϕ is a real Lie algebra isomorphism, we see that $\mathfrak{k}$ admits a complex structure, contradicting Lemma 7.34.   □

Theorem 7.35. *Let $\mathfrak{g} \cong \mathfrak{k}_{\mathbb{C}}$ be a complex semisimple Lie algebra, let $\mathfrak{t}$ be a maximal commutative subalgebra of $\mathfrak{k}$, and let $\mathfrak{h} = \mathfrak{t}_{\mathbb{C}}$ be the associated Cartan subalgebra of $\mathfrak{g}$. Let $R \subset \mathfrak{h}$ be the root system for $\mathfrak{g}$ relative to $\mathfrak{h}$. If $\mathfrak{g}$ is not simple then $\mathfrak{h}$ decomposes as an orthogonal direct sum of nonzero subspaces $\mathfrak{h}_1$ and $\mathfrak{h}_2$ in such a way that every element of R is either in $\mathfrak{h}_1$ or in $\mathfrak{h}_2$. Conversely, if $\mathfrak{h}$ decomposes in this way, then $\mathfrak{g}$ is not simple.*

We may restate the first part of the theorem in contrapositive form: If there *does not* exist an orthogonal decomposition of $\mathfrak{h}$ as $\mathfrak{h} = \mathfrak{h}_1 \oplus \mathfrak{h}_2$ (with dim $\mathfrak{h}_1 > 0$ and dim $\mathfrak{h}_2 > 0$) such that every root is either in $\mathfrak{h}_1$ or $\mathfrak{h}_2$, then $\mathfrak{g}$ is simple. In Sect. 8.6, we will show that if the "Dynkin diagram" of the root system is connected, then no such decomposition of $\mathfrak{h}$ exists. We will then be able to check that most of our examples of semisimple Lie algebras are actually simple.

Proof of Theorem 7.35, Part 1. Assume first that $\mathfrak{g}$ is not simple, so that, by Theorem 7.32, $\mathfrak{k}$ is not simple either. Thus, $\mathfrak{k}$ has a nontrivial ideal $\mathfrak{k}_1$. Now, ideals in $\mathfrak{k}$ are precisely invariant subspaces for the adjoint action of $\mathfrak{k}$ on itself, which are the same as the invariant subspaces for the adjoint action of K on $\mathfrak{k}$. If we choose an inner product on $\mathfrak{k}$ is Ad-K-invariant, the orthogonal complement of $\mathfrak{k}_1$ is also an ideal. Thus, $\mathfrak{k}$ decomposes as the Lie algebra direct sum $\mathfrak{k}_1 \oplus \mathfrak{k}_2$, where $\mathfrak{k}_2 = (\mathfrak{k}_1)^\perp$, in which case $\mathfrak{g}$ decomposes as $\mathfrak{g}_1 \oplus \mathfrak{g}_2$, where $\mathfrak{g}_1 = (\mathfrak{k}_1)_\mathbb{C}$ and $\mathfrak{g}_2 = (\mathfrak{k}_2)_\mathbb{C}$.

Let $\mathfrak{t}$ be any maximal commutative subalgebra of $\mathfrak{k}$ and $\mathfrak{h} = \mathfrak{t}_\mathbb{C}$ the associated Cartan subalgebra of $\mathfrak{g}$. We claim that $\mathfrak{t}$ must decompose as $\mathfrak{t}_1 \oplus \mathfrak{t}_2$, where $\mathfrak{t}_1 \subset \mathfrak{k}_1$ and $\mathfrak{t}_2 \subset \mathfrak{k}_2$. Suppose H and H' are two elements of $\mathfrak{t}$, with $H = X_1 + X_2$ and $H' = Y_1 + Y_2$, with $X_1, Y_1 \in \mathfrak{k}_1$ and $X_2, Y_2 \in \mathfrak{k}_2$. Then

$$0 = [H, H'] = [X_1, Y_1] + [X_2, Y_2],$$

with $[X_1, Y_1] \in \mathfrak{k}_1$ and $[X_2, Y_2] \in \mathfrak{k}_2$, which can happen only if $[X_1, Y_1] = [X_2, Y_2] = 0$. Thus,

$$[X_1, H'] = [X_1, Y_1] = 0,$$

showing that X_1 commutes with every element of $\mathfrak{h}$. Since $\mathfrak{h}$ is maximal commutative, X_1 must actually be in $\mathfrak{h}$, and similarly for X_2. That is to say, for every $X \in \mathfrak{t}$, the $\mathfrak{k}_1$- and $\mathfrak{k}_2$-components of X also belong to $\mathfrak{t}$. From this observation, it follows easily that $\mathfrak{h}$ is the direct sum of the subalgebras

$$\mathfrak{t}_1 := \mathfrak{t} \cap \mathfrak{k}_1, \quad \mathfrak{t}_2 := \mathfrak{t} \cap \mathfrak{k}_2.$$

The algebra $\mathfrak{h}$ then splits as $\mathfrak{h}_1 \oplus \mathfrak{h}_2$, where $\mathfrak{h}_1 = (\mathfrak{t}_1)_\mathbb{C}$ and $\mathfrak{h}_2 = (\mathfrak{t}_2)_\mathbb{C}$. Let R_1 and R_2 be the roots for $\mathfrak{g}_1$ relative to $\mathfrak{h}_1$ and $\mathfrak{g}_2$ relative to $\mathfrak{h}_2$, respectively. If α is an element of R_1 and $X \in \mathfrak{g}_1$ is an associated root vector, then consider any element $H = H_1 + H_2$ of $\mathfrak{h}$, where $H_1 \in \mathfrak{h}_1$ and $H_2 \in \mathfrak{h}_2$. We have

$$[H, X] = [H_1, X] + [H_2, X] = \langle \alpha, H_1 \rangle X,$$

since $H_2 \in \mathfrak{g}_2$ and $X \in \mathfrak{g}_1$. Thus, X is also a root vector for $\mathfrak{g}$ relative to $\mathfrak{h}$, with the associated root being α. Similarly, every root $\beta \in R_2$ is also a root for $\mathfrak{g}$ relative to $\mathfrak{h}$.

We now claim that *every* root α for $\mathfrak{g}$ relative to $\mathfrak{h}$ is either an element of R_1 or an element of R_2. If $X = X_1 + X_2$ is a root vector associated to the root α, then for $H_1 \in \mathfrak{h}_1$ we have

$$[H_1, X] = [H_1, X_1 + X_2] = [H_1, X_1] = \langle \alpha, H_1 \rangle X_1.$$

On the other hand, $[H_1, X]$ must be a multiple of X. Thus, either $X_2 = 0$ or α is orthogonal to $\mathfrak{h}_1$. If α is orthogonal to $\mathfrak{h}_1$ then it belongs to $\mathfrak{h}_2$ and so α is a root for $\mathfrak{g}_2$ (i.e., $\alpha \in R_2$). If, on the other hand, $X_2 = 0$, then $0 = [H_2, X] = \langle \alpha, H_2 \rangle X$ for all $H_2 \in \mathfrak{h}_2$, so that α is orthogonal to $\mathfrak{h}_2$. In that case, $\alpha \in \mathfrak{h}_1$ and α must belong to R_1. $\square$

Proof of Theorem 7.35, Part 2. Suppose now $\mathfrak{h}$ splits as $\mathfrak{h} = \mathfrak{h}_1 \oplus \mathfrak{h}_2$, with $\mathfrak{h}_1$ and $\mathfrak{h}_2$ being nonzero, orthogonal subspaces of $\mathfrak{h}$, in such a way that every root is either in $\mathfrak{h}_1$ or $\mathfrak{h}_2$. Let $R_j = R \cap \mathfrak{h}_j$, $j = 1, 2$, and define subspaces $\mathfrak{g}_j$ of $\mathfrak{g}$ as

$$\mathfrak{g}_j = \mathfrak{h}_j \oplus \bigoplus_{\alpha \in R_j} \mathfrak{g}_\alpha, \quad j = 1, 2.$$

Then $\mathfrak{g}$ decomposes as a vector space as $\mathfrak{g}_1 \oplus \mathfrak{g}_2$. But $\mathfrak{h}_1$ commutes with each $\mathfrak{g}_\alpha$ with $\alpha \in R_2$, because α is in $\mathfrak{h}_2$, which is orthogonal to $\mathfrak{h}_1$. Similarly, $\mathfrak{h}_2$ commutes with $\mathfrak{g}_\alpha$, $\alpha \in R_1$, and, of course, with $\mathfrak{h}_1$. Finally, if $\alpha \in R_1$ and $\beta \in R_2$, then $[\mathfrak{g}_\alpha, \mathfrak{g}_\beta] = \{0\}$, because $\alpha + \beta$ is not a root. Thus, actually, $\mathfrak{g}$ is the Lie algebra direct sum of $\mathfrak{g}_1$ and $\mathfrak{g}_2$, showing that $\mathfrak{g}$ is not simple. $\square$

7.7 The Root Systems of the Classical Lie Algebras

In this section, we look at how the structures described in this chapter work out in the case of the "classical" semisimple Lie algebras, that is, the special linear, orthogonal, and symplectic algebras over $\mathbb{C}$. We label each of our Lie algebras in such a way that the rank is n, and we split our analysis of the orthogonal algebras into the even and odd cases. For each of the classical Lie algebras, we use a constant multiple of the Hilbert–Schmidt inner product $\langle X, Y \rangle = \operatorname{trace}(X^* Y)$, which is invariant under the adjoint action of the corresponding compact group.

7.7.1 The Special Linear Algebras $\mathsf{sl}(n + 1; \mathbb{C})$

We work with the compact real form $\mathfrak{k} = \mathsf{su}(n + 1)$ and the commutative subalgebra $\mathfrak{t}$ which is the intersection of the set of diagonal matrices with $\mathsf{su}(n + 1)$; that is,

$$\mathfrak{t} = \left\{ \begin{pmatrix} ia_1 & & \\ & \ddots & \\ & & ia_{n+1} \end{pmatrix} \middle| a_j \in \mathbb{R}, \quad a_1 + \cdots + a_{n+1} = 0 \right\}. \tag{7.19}$$

We also consider $\mathfrak{h} := \mathfrak{t}_\mathbb{C}$, which is described as follows:

$$\mathfrak{h} = \left\{ \begin{pmatrix} \lambda_1 & & \\ & \ddots & \\ & & \lambda_{n+1} \end{pmatrix} \middle| \lambda_j \in \mathbb{C}, \quad \lambda_1 + \cdots + \lambda_{n+1} = 0 \right\}. \tag{7.20}$$

If a matrix X commutes with each element of $\mathfrak{t}$, it will also commute with each element of $\mathfrak{h}$. But then, as in the proof of Example 7.3, X would have to be diagonal, and if $X \in \mathsf{su}(n+1)$, then X would have to be in $\mathfrak{t}$. Thus, $\mathfrak{t}$ is actually a *maximal* commutative subalgebra of $\mathsf{su}(n+1)$.

Now, let E_{jk} denote the matrix that has a one in the jth row and kth column and has zeros elsewhere. A simple calculation shows that if $H \in \mathfrak{h}$ is as in (7.20), then $HE_{jk} = \lambda_j E_{jk}$ and $E_{jk}H = \lambda_k E_{jk}$. Thus,

$$[H, E_{jk}] = (\lambda_j - \lambda_k)E_{jk}. \tag{7.21}$$

If $j \neq k$, then E_{jk} is in $\mathsf{sl}(n+1;\mathbb{C})$ and (7.21) shows that E_{jk} is a simultaneous eigenvector for each ad_H with H in $\mathfrak{h}$, with eigenvalue $\lambda_j - \lambda_k$. Note that every element X of $\mathsf{sl}(n+1;\mathbb{C})$ can be written uniquely as an element of the Cartan subalgebra (the diagonal entries of X) plus a linear combination of the E_{jk}'s with $j \neq k$ (the off-diagonal entries of X).

If we think at first of the roots as elements of $\mathfrak{h}^*$, then [according to (7.21)] the roots are the linear functionals α_{jk} that associate to each $H \in \mathfrak{h}$, as in (7.20), the quantity $\lambda_j - \lambda_k$. We identify $\mathfrak{h}$ with the subspace of $\mathbb{C}^{n+1}$ consisting vectors whose components sum to zero. The inner product $\langle X, Y \rangle = \mathrm{trace}(X^*Y)$ on $\mathfrak{h}$ is just the restriction to this subspace of the usual inner product on $\mathbb{C}^{n+1}$. If we use this inner product to transfer the roots from $\mathfrak{h}^*$ to $\mathfrak{h}$, we obtain the vectors

$$\alpha_{jk} = e_j - e_k \quad (j \neq k).$$

The roots of $\mathsf{sl}(n+1;\mathbb{C})$ form a root system that is conventionally called A_n, with the subscript n indicating that the rank of $\mathsf{sl}(n+1;\mathbb{C})$ (i.e., the dimension of $\mathfrak{h}$) is n. We see that each root has length $\sqrt{2}$ and

$$\langle \alpha_{jk}, \alpha_{j'k'} \rangle$$

has the value $0, \pm 1$, or ± 2, depending on whether $\{j, k\}$ and $\{j', k'\}$ have zero, one, or two elements in common. Thus

$$2\frac{\langle \alpha_{jk}, \alpha_{j'k'} \rangle}{\langle \alpha_{jk}, \alpha_{jk} \rangle} \in \{0, \pm 1, \pm 2\}.$$

If α and β are roots and $\alpha \neq \beta$ and $\alpha \neq -\beta$, then the angle between α and β is either $\pi/3, \pi/2$, or $2\pi/3$, depending on whether $\langle \alpha, \beta \rangle$ has the value $1, 0$, or -1.

It is easy to see that for any j and k, the reflection $s_{\alpha_{jk}}$ acts on $\mathbb{C}^{n+1}$ by interchanging the jth and kth entries of each vector. It follows that the Weyl group of the A_n root system is the permutation group on $n+1$ elements.

7.7.2 The Orthogonal Algebras so(2n; ℂ)

The root system for $so(2n; \mathbb{C})$ is denoted D_n. We consider the compact real form $so(2n)$ of $so(2n; \mathbb{C})$, and we consider in $so(2n)$ the commutative subalgebra $\mathfrak{t}$ consisting of 2×2 block-diagonal matrices in which the jth diagonal block is of the form

$$\begin{pmatrix} 0 & a_j \\ -a_j & 0 \end{pmatrix}, \tag{7.22}$$

for some $a_j \in \mathbb{R}$. We then consider $\mathfrak{h} = \mathfrak{t} + i\mathfrak{t}$ of $so(2n; \mathbb{C})$, which consists of 2×2 block-diagonal matrices in which the jth diagonal block is of the form (7.22) with $a_j \in \mathbb{C}$. As we will show below, $\mathfrak{t}$ is actually a maximal commutative subalgebra of $so(2n)$, so that $\mathfrak{h}$ is a Cartan subalgebra of $so(2n; \mathbb{C})$.

The root vectors are now 2×2 block matrices having a 2×2 matrix C in the (j, k) block $(j < k)$, the matrix $-C^{tr}$ in the (k, j) block, and zero in all other blocks, where C is one of the four matrices

$$C_1 = \begin{pmatrix} 1 & i \\ i & -1 \end{pmatrix}, \quad C_2 = \begin{pmatrix} 1 & -i \\ -i & -1 \end{pmatrix},$$

$$C_3 = \begin{pmatrix} 1 & -i \\ i & 1 \end{pmatrix}, \quad C_4 = \begin{pmatrix} 1 & i \\ -i & 1 \end{pmatrix}.$$

Direct calculation shows that these are, indeed, root vectors and that the corresponding roots are the linear functionals on $\mathfrak{h}$ given by $i(a_j + a_k), -i(a_j + a_k), i(a_j - a_k)$, and $-i(a_j - a_k)$, respectively.

Let us use on $\mathfrak{h}$ the inner product $\langle X, Y \rangle = \text{trace}(X^*Y)/2$. If we then identify $\mathfrak{h}$ with $\mathbb{C}^n$ by means of the map

$$H \mapsto i(a_1, \ldots, a_n),$$

our inner product on $\mathfrak{h}$ will correspond to the standard inner product on $\mathbb{C}^n$. The roots as elements of $\mathbb{C}^n$ are then the vectors

$$\pm e_j \pm e_k, \quad j < k, \tag{7.23}$$

where $\{e_j\}$ is the standard basis for $\mathbb{R}^n$.

We now demonstrate that $\mathfrak{t}$ is a maximal commutative subalgebra of $\mathfrak{k}$, and also that the center of $so(2n; \mathbb{C})$ is trivial for $n \geq 2$, as claimed in Example 7.3. It is easy to check that every $X \in \mathfrak{g}$ can be expresses as an element of $\mathfrak{h}$ plus a linear combination of the root vectors described above. If X commutes with every element of $\mathfrak{t}$, then X also commutes with every element of $\mathfrak{h}$. Since each of the linear functionals $i(\pm a_j \pm a_k), j < k$, is nonzero on $\mathfrak{h}$, the coefficients of the root vectors in the expansion of X must be zero; that is, X must be in $\mathfrak{h}$. If, also, X is in $\mathfrak{k}$, then X must be in $\mathfrak{h} \cap \mathfrak{k} = \mathfrak{t}$. This shows that $\mathfrak{t}$ is maximal commutative in $\mathfrak{k}$.

Meanwhile, if X is in the center of $\mathsf{so}(2n;\mathbb{C})$, then as shown in the previous paragraph, X must belong to $\mathfrak{h}$. But then for each root vector X_α with root α, we must have

$$0 = [X, X_\alpha] = \langle \alpha, X \rangle X_\alpha,$$

so that $\langle \alpha, X \rangle = 0$. It is then easy to check that if $n \geq 2$, the roots in (7.23) span $\mathfrak{h} \cong \mathbb{C}^n$ and we conclude that X must be zero. (If, on the other hand, $n = 1$, then there are no roots and $\mathsf{so}(2;\mathbb{C}) = \mathfrak{h}$ is commutative.)

7.7.3 The Orthogonal Algebras $\mathsf{so}(2n + 1; \mathbb{C})$

The root system for $\mathsf{so}(2n + 1; \mathbb{C})$ is denoted B_n. We consider the compact real form $\mathsf{so}(2n + 1)$ of $\mathsf{so}(2n + 1; \mathbb{C})$, and we consider in $\mathsf{so}(2n + 1)$ the commutative subalgebra $\mathfrak{t}$ consisting of block diagonal matrices with n blocks of size 2×2 followed by one block of size 1×1. We take the 2×2 blocks to be of the same form as in $\mathsf{so}(2n)$ and we take the 1×1 block to be zero. Then $\mathfrak{h} := \mathfrak{t}_\mathbb{C}$ consists of those matrices in $\mathsf{so}(2n + 1; \mathbb{C})$ of the same form as in $\mathfrak{t}$ except that the off-diagonal elements of the 2×2 blocks are permitted to be complex. The same argument as in the case of $\mathsf{so}(2n;\mathbb{C})$, based on the calculations below, shows that $\mathfrak{t}$ is maximal commutative, so that $\mathfrak{h}$ is a Cartan subalgebra.

The Cartan subalgebra in $\mathsf{so}(2n + 1; \mathbb{C})$ is identifiable in an obvious way with the Cartan subalgebra in $\mathsf{so}(2n;\mathbb{C})$. In particular, both $\mathsf{so}(2n;\mathbb{C})$ and $\mathsf{so}(2n+1;\mathbb{C})$ have rank n. With this identification of the Cartan subalgebras, every root for $\mathsf{so}(2n;\mathbb{C})$ is also a root for $\mathsf{so}(2n + 1;\mathbb{C})$. There are $2n$ additional roots for $\mathsf{so}(2n + 1;\mathbb{C})$. These have the matrices having the following 2×1 block in entries $(2k, 2n + 1)$ and $(2k + 1, 2n + 1)$ as their root vectors:

$$B_1 = \begin{pmatrix} 1 \\ i \end{pmatrix}$$

and having $-B_1^{tr}$ in entries $(2n + 1, 2k)$ and $(2n + 1, 2k + 1)$, together with the matrices having

$$B_2 = \begin{pmatrix} 1 \\ -i \end{pmatrix}$$

in entries $(2k, 2n + 1)$ and $(2k + 1, 2n + 1)$ and $-B_2^{tr}$ in entries $(2n + 1, 2k)$ and $(2n + 1, 2k + 1)$. The corresponding roots, viewed as elements of $\mathfrak{h}^*$, are given by ia_k and $-ia_k$.

If we use the inner product to identify the roots with elements of $\mathfrak{h}$ and then identify $\mathfrak{h}$ with $\mathbb{C}^n$ in the same way as in the previous subsection, the roots for $\mathsf{so}(2n+1;\mathbb{C})$ consist of the roots $\pm e_j \pm e_k$, $j < k$, for $\mathsf{so}(2n;\mathbb{C})$, together with additional roots of the form

$$\pm e_j, \quad j = 1, \ldots, n.$$

These additional roots are shorter by a factor of $\sqrt{2}$ than the roots $\pm e_j \pm e_k$ for $\mathsf{so}(2n;\mathbb{C})$.

7.7.4 The Symplectic Algebras $\mathsf{sp}(n;\mathbb{C})$

The root system for $\mathsf{sp}(n;\mathbb{C})$ is denoted C_n. We consider $\mathsf{sp}(n;\mathbb{C})$, the space of $2n \times 2n$ complex matrices X satisfying $\Omega X^{tr}\Omega = X$, where Ω is the $2n \times 2n$ matrix

$$\Omega = \begin{pmatrix} 0 & I \\ -I & 0 \end{pmatrix}.$$

(Compare Proposition 3.25.) Explicitly, the elements of $\mathsf{sp}(n;\mathbb{C})$ are matrices of the form

$$X = \begin{pmatrix} A & B \\ C & -A^{tr} \end{pmatrix},$$

where A is an arbitrary $n \times n$ matrix and B and C are arbitrary symmetric matrices. We consider the compact real form $\mathsf{sp}(n) = \mathsf{sp}(n;\mathbb{C}) \cap \mathsf{u}(2n)$. (Compare Sect. 1.2.8.)

We consider the commutative subalgebra $\mathfrak{t}$ of $\mathsf{sp}(n)$ consisting of matrices of the form

$$\begin{pmatrix} a_1 & & & & & \\ & \ddots & & & & \\ & & a_n & & & \\ & & & -a_1 & & \\ & & & & \ddots & \\ & & & & & -a_n \end{pmatrix},$$

where each a_j is pure imaginary. We then consider the subalgebra $\mathfrak{h} = \mathfrak{t} + i\mathfrak{t}$ of $\mathsf{sp}(n;\mathbb{C})$, which consists of matrices of the same form but where each a_j is now an arbitrary complex number. As in previous subsections, the calculations below will show that $\mathfrak{t}$ is maximal commutative, so that $\mathfrak{h}$ is a Cartan subalgebra.

The $2n \times 2n$ matrices of the block form

$$\begin{pmatrix} 0 & E_{jk} + E_{kj} \\ 0 & 0 \end{pmatrix}, \quad \begin{pmatrix} 0 & 0 \\ E_{jk} + E_{kj} & 0 \end{pmatrix} \tag{7.24}$$

$(j \neq k)$ are root vectors for which the corresponding roots are $(a_j + a_k)$ and $-(a_j + a_k)$, respectively. Next, matrices of the block form

$$\begin{pmatrix} E_{jk} & 0 \\ 0 & -E_{kj} \end{pmatrix}, \tag{7.25}$$

$(j \neq k)$ are root vectors for which the corresponding roots are $(a_k - a_l)$. Finally, matrices of the block form

$$\begin{pmatrix} 0 & E_{jj} \\ 0 & 0 \end{pmatrix}, \quad \begin{pmatrix} 0 & 0 \\ E_{jj} & 0 \end{pmatrix} \tag{7.26}$$

are root vectors for which the corresponding roots are $2a_j$ and $-2a_j$.

We use on $\mathfrak{h}$ the inner product $\langle X, Y \rangle = \mathrm{trace}(X^*Y)/2$. If we then identify $\mathfrak{h}$ with $\mathbb{C}^n$ by means of the map

$$H \mapsto (a_1, \dots, a_n),$$

our inner product on $\mathfrak{h}$ will correspond to the standard inner product on $\mathbb{C}^n$. The roots are then the vectors of the form

$$\pm e_j \pm e_k, \quad j < k$$

and of the form

$$\pm 2e_j, \quad j = 1, \dots, n.$$

This root system is the same as that for $\mathsf{so}(2n + 1; \mathbb{C})$, except that instead of $\pm e_j$ we now have $\pm 2e_j$.

7.8 Exercises

1. Let $\mathfrak{h}_{\mathbb{C}}$ denote the complexification of the Lie algebra of the Heisenberg group, namely the space of all complex 3×3 upper triangular matrices with zeros on the diagonal.

   (a) Show the maximal commutative subalgebras of $\mathfrak{h}_{\mathbb{C}}$ are precisely the two-dimensional subalgebras of $\mathfrak{h}_{\mathbb{C}}$ that contain the center of $\mathfrak{h}_{\mathbb{C}}$.
   (b) Show that $\mathfrak{h}_{\mathbb{C}}$ does not have any Cartan subalgebras (in the sense of Definition 7.10).

2. Give an example of a maximal commutative subalgebra of $\mathsf{sl}(2;\mathbb{C})$ that is not a Cartan subalgebra.
3. Show that the Hilbert–Schmidt inner product on $\mathsf{sl}(n;\mathbb{C})$, given by $\langle X, Y \rangle = \mathrm{trace}(X^*Y)$, is invariant under the adjoint action of $\mathsf{SU}(n)$ and is real on $\mathsf{su}(n)$.
4. Using the root space decomposition in Sect. 7.7.2, show that the Lie algebra $\mathsf{so}(4;\mathbb{C})$ is isomorphic to the Lie algebra direct sum $\mathsf{sl}(2;\mathbb{C}) \oplus \mathsf{sl}(2;\mathbb{C})$. Then show that $\mathsf{so}(4)$ is isomorphic to $\mathsf{su}(2) \oplus \mathsf{su}(2)$.
5. Suppose $\mathfrak{g}$ is a Lie subalgebra of $M_n(\mathbb{C})$ and assume that for all $X \in \mathfrak{g}$, we have $X^* \in \mathfrak{g}$, where X^* is the usual matrix adjoint of X. Let $\mathfrak{k} = \mathfrak{g} \cap \mathsf{u}(n)$.

   (a) Show that $\mathfrak{k}$ is a real subalgebra of $\mathfrak{g}$ and that $\mathfrak{g} \cong \mathfrak{k}_\mathbb{C}$.
   (b) Define an inner product on $\mathfrak{g}$ by

$$\langle X, Y \rangle = \mathrm{trace}(X^*Y).$$

   Show that this inner product satisfies all the properties in Proposition 7.4.

6. For any Lie algebra $\mathfrak{g}$, let the **Killing form** be the symmetric bilinear form B on $\mathfrak{g}$ defined by

$$B(X, Y) = \mathrm{trace}(\mathrm{ad}_X \, \mathrm{ad}_Y).$$

   (a) Show that for any $X \in \mathfrak{g}$, the operator $\mathrm{ad}_X : \mathfrak{g} \to \mathfrak{g}$ is skew-symmetric with respect to B, meaning that

$$B(\mathrm{ad}_X(Y), Z) = -B(Y, \mathrm{ad}_X(Z))$$

   for all $Y, Z \in \mathfrak{g}$.
   (b) Suppose $\mathfrak{g} = \mathfrak{k}_\mathbb{C}$ is semisimple. Show that B is nondegenerate, meaning that for all nonzero $X \in \mathfrak{g}$, there is some $Y \in \mathfrak{g}$ for which $B(X, Y) \neq 0$.

   Hint: Use Proposition 7.4.

7. Let $\mathfrak{g}$ be a Lie algebra and let B be the Killing form on $\mathfrak{g}$ (Exercise 6). Show that if $\mathfrak{g}$ is a nilpotent Lie algebra (Definition 3.15), then $B(X, Y) = 0$ for all $X, Y \in \mathfrak{g}$.
8. Let $\mathfrak{g}$ denote the vector space of 3×3 complex matrices of the form

$$\begin{pmatrix} A & B \\ 0 & 0 \end{pmatrix},$$

   where A is a 2×2 matrix with trace zero and B is an arbitrary 2×1 matrix.

   (a) Show that $\mathfrak{g}$ is a subalgebra of $M_3(\mathbb{C})$.
   (b) Let X, Y, H, e_1, and e_2 be the following basis for $\mathfrak{g}$. We let X, Y, and H be the usual $\mathsf{sl}(2;\mathbb{C})$ basis in the "A" slot, with $B = 0$. We let e_1 and

e_2 be the matrices with $A = 0$ and with $(1 \ 0)^{tr}$ and $(0 \ 1)^{tr}$ in the "B"
slot, respectively. Compute the commutation relations among these basis
elements.

(c) Show that $\mathfrak{g}$ has precisely one nontrivial ideal, namely the span of e_1 and
e_2. Conclude that $\mathfrak{g}$ is *not* semisimple.

Hint: First, determine the subspaces of $\mathfrak{g}$ that are invariant under the adjoint
action of the $\mathsf{sl}(2; \mathbb{C})$ algebra spanned by X, Y, and H, and then determine
which of these subspaces are also invariant under the adjoint action of e_1
and e_2. In determining the $\mathsf{sl}(2; \mathbb{C})$-invariant subspaces, use Exercise 10 of
Chapter 4.

(d) Let $\mathfrak{h}$ be the one-dimensional subspace of $\mathfrak{g}$ spanned by the element H.
Show that $\mathfrak{h}$ is a maximal commutative subalgebra of $\mathfrak{g}$ and that ad_H is
diagonalizable. Show that the eigenvalues of ad_H are 0, ± 1, and ± 2.

Note: The "roots" for $\mathfrak{h}$ are thus the numbers ± 1 *and* ± 2, which would
not be possible if $\mathfrak{h}$ were a Cartan subalgebra of the form $\mathfrak{h} = t_{\mathbb{C}}$ in a
semisimple Lie algebra.

(e) Let $\mathfrak{g}_1$ and $\mathfrak{g}_{-1}$ denote the eigenspaces of ad_H with eigenvalues 1 and -1,
respectively. Show that $[\mathfrak{g}_1, \mathfrak{g}_{-1}] = \{0\}$.

Note: This result should be contrasted with the semisimple case, where
$[\mathfrak{g}_\alpha, \mathfrak{g}_{-\alpha}]$ is always a one-dimensional subspace of $\mathfrak{h}$, so that $\mathfrak{g}_\alpha$, $\mathfrak{g}_{-\alpha}$, and
$[\mathfrak{g}_\alpha, \mathfrak{g}_{-\alpha}]$ span a three-dimensional subalgebra of $\mathfrak{g}$ isomorphic to $\mathsf{sl}(2; \mathbb{C})$.

9. Using Theorem 7.35 and the calculations in Sect. 7.7.3, show that the Lie
algebra $\mathsf{so}(5; \mathbb{C})$ is simple.

10. Let $E = \mathbb{R}^n$, $n \geq 2$, and consider the D_n root system, consisting of the vectors
of the form $\pm e_j \pm e_k$, with $j < k$. Show that the Weyl group of D_n is the group
of transformations of $\mathbb{R}^n$ expressible as a composition of a permutation of the
entries and an *even* number of sign changes.

11. Let $E = \mathbb{R}^n$ and consider the B_n root system, consisting of the vectors of
the form $\pm e_j \pm e_k$, with $j < k$, together with the vectors of the form $\pm e_j$,
$j = 1, \ldots, n$. Show that the Weyl group of B_n is the group of transformations
expressible as a composition of a permutation of the entries and an *arbitrary*
number of sign changes.

Note: Since each root in C_n is a multiple of a root in B_n, and vice versa, the
reflections for C_n are the same as the reflections for B_n. Thus, the Weyl group
of C_n is the same as that of B_n.

12. Let $\mathfrak{g}$ be a complex simple Lie algebra, with complex structure denoted by
J. Let $\mathfrak{g}_{\mathbb{R}}$ denote the Lie algebra $\mathfrak{g}$ viewed as a real Lie algebra of twice the
dimension. Now let $\mathfrak{g}'$ be the complexification of $\mathfrak{g}_{\mathbb{R}}$, with the complex structure
on $\mathfrak{g}'$ denoted by i.

(a) Show that $\mathfrak{g}'$ decomposes as a Lie algebra direct sum $\mathfrak{g}' = \mathfrak{g}_1 \oplus \mathfrak{g}_2$, where
both $\mathfrak{g}_1$ and $\mathfrak{g}_2$ are isomorphic to $\mathfrak{g}$.

Hint: Consider element of $\mathfrak{g}'$ of the form $X + iJX$ and of the form $X - iJX$,
with $X \in \mathfrak{g}$.

(b) Show that $\mathfrak{g}_{\mathbb{R}}$ is simple as a real Lie algebra. (That is to say, there is no nontrivial *real* subspace $\mathfrak{h}$ of $\mathfrak{g}$ such that $[X, Y] \in \mathfrak{h}$ for all $X \in \mathfrak{g}$ and $Y \in \mathfrak{h}$.)

13. Let $\mathfrak{h}$ be a real simple Lie algebra, and assume that $\mathfrak{h}_{\mathbb{C}}$ is not simple. Show that there is a complex simple Lie algebra $\mathfrak{g}$ such that $\mathfrak{h} \cong \mathfrak{g}_{\mathbb{R}}$, where the notation is as in Exercise 12.
    Hint: Imitate the proof of Theorem 7.32.

14. Show that the real Lie algebra $\mathsf{so}(3; 1)$ is isomorphic to $\mathsf{sl}(2; \mathbb{C})_{\mathbb{R}}$, where the notation is as in Exercise 12. Conclude that $\mathsf{so}(3; 1)$ is simple as a real Lie algebra, but that $\mathsf{so}(3; 1)_{\mathbb{C}}$ is not simple and is isomorphic to $\mathsf{sl}(2; \mathbb{C}) \oplus \mathsf{sl}(2; \mathbb{C})$.
    Hint: First show that $\mathsf{so}(3; 1)_{\mathbb{C}} \cong \mathsf{sl}(2; \mathbb{C}) \oplus \mathsf{sl}(2; \mathbb{C})$ and then show that the two copies of $\mathsf{sl}(2; \mathbb{C})$ are conjugates of each other.

Chapter 8
Root Systems

In this chapter, we consider root systems apart from their origins in semisimple Lie algebras. We establish numerous "factoids" about root systems, which will be used extensively in subsequent chapters. Here is one example of how results about root systems will be used. In Chapter 9, we construct each finite-dimensional, irreducible representation of a semisimple Lie algebra as a quotient of an *infinite-dimensional* representation known as a Verma module. To prove that the quotient representation is finite-dimensional, we prove that the weights of the quotient are invariant under the action of the Weyl group, that is, the group generated by the reflections about the hyperplanes orthogonal to the roots. It is not possible, however, to prove directly that the weights are invariant under *all* reflections, but only under reflections coming from a special subset of the root system known as the base. To complete the argument, then, we need to know that the Weyl group is actually generated by the reflections associated to the roots in the base. This claim is the content of Proposition 8.24.

8.1 Abstract Root Systems

A root system is any collection of vectors having the properties satisfied by the roots (viewed as a subset of $i\mathfrak{t} \subset \mathfrak{h}$) of a semisimple Lie algebra, as encoded in the following definition.

A previous version of this book was inadvertently published without the middle initial of the author's name as "Brian Hall". For this reason an erratum has been published, correcting the mistake in the previous version and showing the correct name as Brian C. Hall (see DOI http://dx.doi.org/10.1007/978-3-319-13467-3_14). The version readers currently see is the corrected version. The Publisher would like to apologize for the earlier mistake.

© Springer International Publishing Switzerland 2015   197
B.C. Hall, *Lie Groups, Lie Algebras, and Representations*, Graduate
Texts in Mathematics 222, DOI 10.1007/978-3-319-13467-3_8

Definition 8.1. A root system (E, R) is a finite-dimensional real vector space E with an inner product $\langle \cdot, \cdot \rangle$, together with a finite collection R of nonzero vectors in E satisfying the following properties:

1. The vectors in R span E.
2. If α is in R and $c \in \mathbb{R}$, then $c\alpha$ is in R only if $c = \pm 1$.
3. If α and β are in R, then so is $s_\alpha \cdot \beta$, where s_α is the linear transformation of E defined by

$$s_\alpha \cdot \beta = \beta - 2\frac{\langle \beta, \alpha \rangle}{\langle \alpha, \alpha \rangle}\alpha, \quad \beta \in E.$$

4. For all α and β in R, the quantity

$$2\frac{\langle \beta, \alpha \rangle}{\langle \alpha, \alpha \rangle}$$

is an integer.

The dimension of E is called the **rank** of the root system and the elements of R are called **roots**.

Note that since $s_\alpha \cdot \alpha = -\alpha$, we have that $-\alpha \in R$ whenever $\alpha \in R$. In the theory of symmetric spaces, there arise systems satisfying Conditions 1, 3, and 4, but not Condition 2. These are called "nonreduced" root systems. In the theory of Coxeter groups, there arise systems satisfying Conditions 1, 2, and 3, but not Property 4. These are called "noncrystallographic" or "nonintegral" root systems. In this book, we consider only root systems satisfying *all* of the conditions in Definition 8.1.

The map s_α is the *reflection* about the hyperplane orthogonal to α; that is, $s_\alpha \cdot \alpha = -\alpha$ and $s_\alpha \cdot \beta = \beta$ for all β that are orthogonal to α, as is easily verified from the formula for s_α. From this description, it should be evident that s_α is an orthogonal transformation of E with determinant -1.

We can interpret Property 4 geometrically in one of two ways. In light of the formula for s_α, Property 4 is equivalent to saying that $s_\alpha \cdot \beta$ should differ from β by an *integer* multiple of α. Alternatively, if we recall that the orthogonal projection of β onto α is given by $(\langle \beta, \alpha \rangle / \langle \alpha, \alpha \rangle)\alpha$, we note that the quantity in Property 4 is twice the coefficient of α in this projection. Thus, Property 4 is equivalent to saying that *the projection of β onto α is an integer or half-integer multiple of α*.

We have shown that one can associate a root system to every complex semisimple Lie algebra. It turns out that *every* root system arises in this way, although this is far from obvious—see Sect. 8.11.

Definition 8.2. If (E, R) is a root system, the **Weyl group** W of R is the subgroup of the orthogonal group of E generated by the reflections s_α, $\alpha \in R$.

By assumption, each s_α maps R into itself, indeed *onto* itself, since each $\beta \in R$ satisfies $\beta = s_\alpha \cdot (s_\alpha \cdot \beta)$. It follows that every element of W maps R onto itself. Since the roots span E, a linear transformation of E is determined by its action on R. Thus, the Weyl group is a finite subgroup of $O(E)$ and may be regarded as a subgroup of the permutation group on R. We denote the action of $w \in W$ on $H \in E$ by $w \cdot H$.

Proposition 8.3. *Suppose (E, R) and (F, S) are root systems. Consider the vector space $E \oplus F$, with the natural inner product determined by the inner products on E and F. Then $R \cup S$ is a root system in $E \oplus F$, called the **direct sum** of R and S.*

Here, we are identifying E with the subspace of $E \oplus F$ consisting of all vectors of the form $(e, 0)$ with e in E, and similarly for F. Thus, more precisely, $R \cup S$ means the elements of the form $(\alpha, 0)$ with α in R together with elements of the form $(0, \beta)$ with β in S. (Elements of the form (α, β) with $\alpha \in R$ and $\beta \in S$ are *not* in $R \cup S$.)

Proof. If R spans E and S spans F, then $R \cup S$ spans $E \oplus F$, so Condition 1 is satisfied. Condition 2 holds because R and S are root systems in E and F, respectively. For Condition 3, if α and β are both in R or both in S, then $s_\alpha \cdot \beta \in R \cup S$ because R and S are root systems. If $\alpha \in R$ and $\beta \in S$ or vice versa, then $\langle \alpha, \beta \rangle = 0$, so that

$$s_\alpha \cdot \beta = \beta \in R \cup S.$$

Similarly, if α and β are both in R or both in S, then $2 \langle \alpha, \beta \rangle / \langle \alpha, \alpha \rangle$ is an integer because R and S are root systems, and if $\alpha \in R$ and $\beta \in S$ or vice versa, then $2 \langle \alpha, \beta \rangle / \langle \alpha, \alpha \rangle = 0$. Thus, Condition 4 holds for $R \cup S$. $\square$

Definition 8.4. A root system (E, R) is called **reducible** if there exists an orthogonal decomposition $E = E_1 \oplus E_2$ with dim $E_1 > 0$ and dim $E_2 > 0$ such that every element of R is either in E_1 or in E_2. If no such decomposition exists, (E, R) is called **irreducible**.

If (E, R) is reducible, then it is not hard to see that the part of R in E_1 is a root system in E_1 and the part of R in E_2 is a root system in E_2. Thus, a root system is reducible precisely if it can be realized as a direct sum of two other root systems. In the Lie algebra setting, the root system associated to a complex semisimple Lie algebra $\mathfrak{g}$ is irreducible precisely if $\mathfrak{g}$ is simple (Theorem 7.35).

Definition 8.5. Two root systems (E, R) and (F, S) are said to be **isomorphic** if there exists an invertible linear transformation $A : E \to F$ such that A maps R onto S and such that for all $\alpha \in R$ and $\beta \in E$, we have

$$A \left(s_\alpha \cdot \beta \right) = s_{A\alpha} \cdot (A\beta).$$

A map A with this property is called an **isomorphism**.

Note that the linear map A is *not* required to preserve inner products, but only to preserve the reflections about the roots. If, for example, $F = E$ and S consists of elements of the form $c\alpha$ with $\alpha \in R$, then (F, S) is isomorphic to (E, R), with the isomorphism being the map $A = cI$.

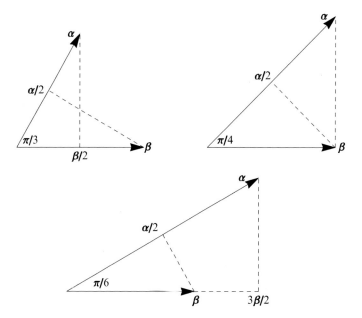

Fig. 8.1 The basic acute angles and length ratios

We now establish a basic result limiting the possible angles and length ratios occurring in a root system.

Proposition 8.6. *Suppose* α *and* β *are roots,* α *is not a multiple of* β, *and* $\langle \alpha, \alpha \rangle \geq \langle \beta, \beta \rangle$. *Then one of the following holds:*

1. $\langle \alpha, \beta \rangle = 0$
2. $\langle \alpha, \alpha \rangle = \langle \beta, \beta \rangle$ *and the angle between* α *and* β *is* $\pi/3$ *or* $2\pi/3$
3. $\langle \alpha, \alpha \rangle = 2\langle \beta, \beta \rangle$ *and the angle between* α *and* β *is* $\pi/4$ *or* $3\pi/4$
4. $\langle \alpha, \alpha \rangle = 3\langle \beta, \beta \rangle$ *and the angle between* α *and* β *is* $\pi/6$ *or* $5\pi/6$

Figure 8.1 shows the allowed angles and length ratios, for the case of an acute angle. In each case, $2\langle \alpha, \beta \rangle / \langle \alpha, \alpha \rangle = 1$, whereas $2\langle \beta, \alpha \rangle / \langle \beta, \beta \rangle$ takes the values 1, 2, and 3 in the three successive cases. Section 8.2 shows that each of the angles and length ratios permitted by Proposition 8.6 actually occurs in some root system.

Proof. Suppose that α and β are roots and let $m_1 = 2\langle \alpha, \beta \rangle / \langle \alpha, \alpha \rangle$ and $m_2 = 2\langle \beta, \alpha \rangle / \langle \beta, \beta \rangle$, so that m_1 and m_2 are integers. Assume $\langle \alpha, \alpha \rangle \geq \langle \beta, \beta \rangle$ and note that

$$m_1 m_2 = 4\frac{\langle \alpha, \beta \rangle^2}{\langle \alpha, \alpha \rangle \langle \beta, \beta \rangle} = 4\cos^2 \theta, \tag{8.1}$$

where θ is the angle between α and β, and that

$$\frac{m_2}{m_1} = \frac{\langle \alpha, \alpha \rangle}{\langle \beta, \beta \rangle} \geq 1 \tag{8.2}$$

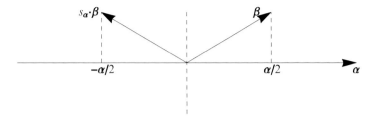

Fig. 8.2   The projection of β onto α equals $\alpha/2$ and $s_\alpha \cdot \beta$ equals $\beta - \alpha$

whenever $\langle \alpha, \beta \rangle \neq 0$. From (8.1), we conclude that $0 \leq m_1 m_2 \leq 4$. If $m_1 m_2 = 0$, then $\cos \theta = 0$, so α and β are orthogonal. If $m_1 m_2 = 4$, then $\cos^2 \theta = 1$, which means that α and β are multiples of one another.

The remaining possible values for $m_1 m_2$ are 1, 2, and 3. If $m_1 m_2 = 1$, then $\cos^2 \theta = 1/4$, so θ is $\pi/3$ or $2\pi/3$. Since m_1 and m_2 are both integers, we must have $m_1 = 1$ and $m_2 = 1$ or $m_1 = -1$ and $m_2 = -1$. In the first case, $\langle \alpha, \beta \rangle > 0$ and we have $\theta = \pi/3$ and in the second case, $\langle \alpha, \beta \rangle < 0$ and we have $\theta = 2\pi/3$. In either case, (8.2) tells us that α and β have the same length, establishing Case 2 of the proposition.

If $m_1 m_2 = 2$, then $\cos^2 \theta = 1/2$, so that θ is $\pi/4$ or $3\pi/4$. Since m_1 and m_2 are integers and $|m_2| \geq |m_1|$ by (8.2), we must have $m_1 = 1$ and $m_2 = 2$ or $m_1 = -1$ and $m_2 = -2$. In the first case, we have $\theta = \pi/4$ and in the second case, $\theta = 3\pi/4$. In either case, (8.2) tells us that α is longer than β by a factor of $\sqrt{2}$. The analysis of the case $m_1 m_2 = 3$ is similar. $\square$

Corollary 8.7. *Suppose α and β are roots. If the angle between α and β is strictly obtuse (i.e., strictly between $\pi/2$ and π), then $\alpha + \beta$ is a root. If the angle between α and β is strictly acute (i.e., strictly between 0 and $\pi/2$), then $\alpha - \beta$ and $\beta - \alpha$ are also roots.*

See Figure 8.2.

Proof. The proof is by examining each of the three obtuse angles and each of the three acute angles allowed by Proposition 8.6. Consider first the acute case and adjust the labeling so that $\langle \alpha, \alpha \rangle \geq \langle \beta, \beta \rangle$. An examination of Cases 2, 3, and 4 in the proposition (see Figure 8.1) shows that in each of these cases, the projection of β onto α is equal to $\alpha/2$. Thus, $s_\alpha \cdot \beta = \beta - \alpha$ is again a root—and so, therefore, is $-(\beta - \alpha) = \alpha - \beta$. In the obtuse case (with $\langle \alpha, \alpha \rangle \geq \langle \beta, \beta \rangle$), the projection of β onto α equals $-\alpha/2$, and, thus, $s_\alpha \cdot \beta = \alpha + \beta$ is again a root. $\square$

8.2   Examples in Rank Two

If the rank of the root system is one, then there is only one possibility: R must consist of a pair $\{\alpha, -\alpha\}$, where α is a nonzero element of E. In rank two, there are four possibilities, pictured in Figure 8.3 with their conventional names. In the case

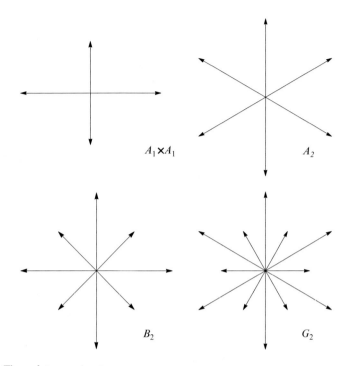

Fig. 8.3 The rank-two root systems

of $A_1 \times A_1$, the lengths of the horizontal roots are unrelated to the lengths of the vertical roots. In A_2, all roots have the same length; in B_2, the length of the longer roots is $\sqrt{2}$ times the length of the shorter roots; in G_2, the length of the longer roots is $\sqrt{3}$ times the length of the shorter roots. The angle between successive roots is $\pi/2$ for $A_1 \times A_1$, $\pi/3$ for A_2, $\pi/4$ for B_2, and $\pi/6$ for G_2. It is easy to check that each of these systems is actually a root system; in particular, Proposition 8.6 is satisfied for each pair of roots.

Proposition 8.8. *Every rank-two root system is isomorphic to one of the systems in Figure 8.3.*

Proof. It is harmless to assume $E = \mathbb{R}^2$; thus, let $R \subset \mathbb{R}^2$ be a root system. Let θ be the smallest angle occurring between any two vectors in R. Since the elements of R span $\mathbb{R}^2$, we can find two linearly independent vectors α and β in R. If the angle between α and β is greater than $\pi/2$, then the angle between α and $-\beta$ is less than $\pi/2$; thus, the minimum angle θ is at most $\pi/2$. Then, according to Proposition 8.6, θ must be one of the following: $\pi/2, \pi/3, \pi/4, \pi/6$.

Let α and β be two elements of R such that the angle between them is the minimum angle θ. Then the vector $-s_\beta \cdot \alpha$ will be a vector that is at angle θ to β but on the opposite side of β from α, as shown in Figure 8.4. Thus, $-s_\beta \cdot \alpha$ is at

Fig. 8.4 The root $-s_\beta \cdot \alpha$ is at angle 2θ from α

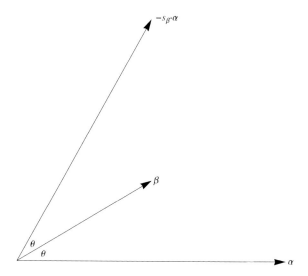

angle 2θ to α. Similarly, $-s_{s_\beta \cdot \alpha} \cdot \beta$ is at angle 3θ to α. Continuing in the same way, we can obtain vectors at angle $n\theta$ to α for all n. These vectors are unique since a nontrivial positive multiple of a root is not allowed to be a root. Now, since each of the allowed values of θ evenly divides 2π, we will eventually come around to α again. Furthermore, there cannot be any other vectors besides those at angles $n\theta$ to α, or else there would be an angle smaller than θ.

Thus, R must consist of n equally spaced vectors, with consecutive vectors separated by angle θ, where θ is one of the acute angles in Proposition 8.6. If, say, $\theta = \pi/4$, then in order to satisfy the length requirement in Proposition 8.6, the roots must alternate between a shorter length and a second length that is greater by a factor of $\sqrt{2}$. Thus, our root system must be isomorphic to B_2. Similar reasoning shows that all remaining values of θ yield one of the root systems in Figure 8.3. $\quad\square$

Using the results of Sect. 7.7, we can see that the root systems $A_1 \times A_1$, A_2, and B_2 arise as root systems of classical Lie algebras as follows. The root system $A_1 \times A_1$ is the root system of $\mathsf{so}(4; \mathbb{C})$, which is isomorphic to $\mathsf{sl}(2; \mathbb{C}) \oplus \mathsf{sl}(2; \mathbb{C})$; A_2 is the root system of $\mathsf{sl}(3; \mathbb{C})$; and B_2 is the root system of $\mathsf{so}(5; \mathbb{C})$, which is isomorphic to the root system of $\mathsf{sp}(2; \mathbb{C})$. The root system G_2, meanwhile, is the root system of an "exceptional" Lie algebra, which is also referred to as G_2 (Sect. 8.11).

Proposition 8.9. *If R is a rank-two root system with minimum angle $\theta = 2\pi/n$, $n = 4, 6, 8, 12$, then the Weyl group of R is the symmetry group of a regular $n/2$-gon.*

Proof. The group W will contain at least $n/2$ reflections, one for each pair $\pm\alpha$ of roots. If α and β are roots with some angle ϕ between them, then the composition of the reflections s_α and s_β will be a rotation by angle $\pm 2\phi$, with the direction of the rotation depending on the order of the composition. To see this, note that s_α

and s_β both have determinant -1 and so $s_\alpha s_\beta$ has determinant 1 and is, therefore, a rotation by some angle. To determine the angle, it suffices to apply $s_\beta s_\alpha$ to any nonzero vector, for example, α. However, $s_\alpha \cdot \alpha = -\alpha$ and $s_\beta \cdot (-\alpha)$ is at angle 2ϕ to α, as in Figure 8.4.

Now, since ϕ can be any integer multiple of θ, we obtain all rotations by integer multiples of 2θ. Meanwhile, the composition of a reflection s_α and a rotation by angle 2ϕ will be another reflection s_β, where β is a root at angle ϕ to α, as the reader may verify. Thus, the set of $n/2$ reflections together with rotations by integer multiples of 2θ form a group; this is the Weyl group of the rank-two root system. But this group is also the symmetry group of a regular $n/2$-gon, also known as the dihedral group on $n/2$ elements. $\qquad\Box$

Note that in the case of A_2, the Weyl group consists of three reflections together with three rotations (by multiples of $2\pi/3$). In this case, the Weyl group is not the full symmetry group of the root system: Rotations by $\pi/3$ map R onto itself but are not elements of the Weyl group.

8.3 Duality

In this section, we introduce an important duality operation on root systems.

Definition 8.10. If (E, R) is a root system, then for each root $\alpha \in R$, the **coroot** H_α is the vector given by

$$H_\alpha = 2\frac{\alpha}{\langle \alpha, \alpha\rangle}.$$

The set of all coroots is denoted $R^\vee$ and is called the **dual root system** to R.

This definition is consistent with the use of the term "coroot" in Chapter 7; see Definition 7.28. Point 4 in the definition of a root system may be restated as saying that $\langle \beta, H_\alpha\rangle$ should be an integer for all roots α and β.

Proposition 8.11. *If R is a root system, then $R^\vee$ is also a root system and the Weyl group for $R^\vee$ is the same as the Weyl group for R. Furthermore, $(R^\vee)^\vee = R$.*

Proof. We compute that

$$\langle H_\alpha, H_\alpha\rangle = 4\frac{\langle \alpha, \alpha\rangle}{\langle \alpha, \alpha\rangle^2} = \frac{4}{\langle \alpha, \alpha\rangle}$$

and, therefore,

$$2\frac{H_\alpha}{\langle H_\alpha, H_\alpha\rangle} = 2\left(\frac{2\alpha}{\langle \alpha, \alpha\rangle}\right)\frac{\langle \alpha, \alpha\rangle}{4} = \alpha. \qquad (8.3)$$

If we take the inner product of (8.3) with H_β, we see that

$$2\frac{\langle H_\alpha, H_\beta\rangle}{\langle H_\alpha, H_\alpha\rangle} = \langle \alpha, H_\beta\rangle = 2\frac{\langle \alpha, \beta\rangle}{\langle \beta, \beta\rangle}, \tag{8.4}$$

which means that the left-hand side of (8.4) is an integer.

Furthermore, since H_α is a multiple of α, it is evident that $s_\alpha = s_{H_\alpha}$. Since, also, s_α is an orthogonal transformation, we have

$$s_\alpha \cdot H_\beta = 2\frac{s_\alpha \cdot \beta}{\langle \beta, \beta\rangle} = 2\frac{s_\alpha \cdot \beta}{\langle s_\alpha \cdot \beta, s_\alpha \cdot \beta\rangle} = H_{s_\alpha\cdot\beta}.$$

Thus, the set of coroots is invariant under each reflection s_α ($= s_{H_\alpha}$). This observation, together with (8.4), shows that $R^\vee$ is again a root system, with the remaining properties of root systems for $R^\vee$ following immediately from the corresponding properties of R.

Since $s_\alpha = s_{H_\alpha}$, we see that R and $R^\vee$ have the same Weyl group. Finally, (8.3) says that the formula for α in terms of H_α is the same as the formula for H_α in terms of α. Thus, $H_{H_\alpha} = \alpha$ and $(R^\vee)^\vee = R$.   □

Note from (8.4) that the integer associated to the pair (H_α, H_β) in $R^\vee$ is the same as the integer associated to the pair (β, α) (*not* (α, β)) in R. If all the roots in R have the same length, $R^\vee$ is isomorphic to R. Even if not all the roots have the same length, R and $R^\vee$ could be isomorphic. In the case of B_2, for example, the dual root system $R^\vee$ can be converted back to R by a $\pi/4$ rotation and a scaling. (See Figure 8.5.) On the other hand, the rank-three root systems B_3 and C_3 (Sect. 8.9) are dual to each other but not isomorphic to each other.

Figure 8.5 shows the dual root system for the root system B_2. On the left-hand side of the figure, the long roots have been normalized to have length $\sqrt{2}$. Thus, for each long root we have $H_\alpha = \alpha$ and for each short root we have $H_\alpha = 2\alpha$, yielding the root system on the right-hand side of the figure.

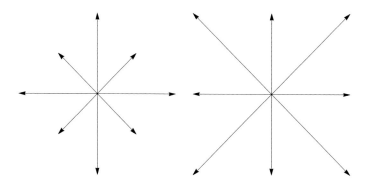

Fig. 8.5 The B_2 root system and its dual

8.4   Bases and Weyl Chambers

We now introduce the notion of a base, or a system of positive simple roots, for a root system.

Definition 8.12. If (E, R) is a root system, a subset Δ of R is called a **base** if the following conditions hold:

1. Δ is a basis for E as a vector space.
2. Each root $\alpha \in R$ can be expressed as a linear combination of elements of Δ with integer coefficients and in such a way that the coefficients are either all non-negative or all nonpositive.

The roots for which the coefficients are non-negative are called **positive roots** and the others are called **negative roots** (relative to the base Δ). The set of positive roots relative to a fixed base Δ is denoted R^+ and the set of negative roots is denoted R^-. The elements of Δ are called the **positive simple roots**.

Note that since Δ is a basis for E, each α can be expressed *uniquely* as a linear combination the elements of Δ. We require that Δ be such that the coefficients in the expansion of each $\alpha \in R$ be integers and such that all the nonzero coefficients have the same sign.

Proposition 8.13. *If α and β are distinct elements of a base Δ for R, then* $\langle \alpha, \beta \rangle \leq 0.$

Geometrically, this means that either α and β are orthogonal or the angle between them is obtuse.

Proof. Since $\alpha \neq \beta$, if we had $\langle \alpha, \beta \rangle > 0$, then the angle between α and β would be strictly between 0 and $\pi/2$. Then, by Corollary 8.7, $\alpha - \beta$ would be an element of R. Since the elements of Δ form a basis for E as a vector space, each element of R has a *unique* expansion in terms of elements of Δ, and the coefficients of that expansion are supposed to be either all non-negative or all nonpositive. The expansion of $\alpha - \beta$, however, has one positive and one negative coefficient. Thus, $\alpha - \beta$ cannot be a root, which means that $\langle \alpha, \beta \rangle \leq 0$. □

The reader is invited to find a base for each of the rank-two root systems in Figure 8.3. We now show that every root system has a base.

Proposition 8.14. *There exists a hyperplane V through the origin in E that does not contain any element of R.*

Proof. For each $\alpha \in R$, let V_α denote the hyperplane

$$V_\alpha = \{ H \in E \mid \langle \alpha, H \rangle = 0 \}.$$

Since there are only finitely many of these hyperplanes, there exists $H \in E$ not in any V_α. (See Exercise 2.) Let V be the hyperplane through the origin orthogonal to

H. Since H is not in any V_α, we see that H is not orthogonal to any α, which means that no α is in V. □

Definition 8.15. Let (E, R) be a root system. Let V be a hyperplane through the origin in E such that V does not contain any root. Choose one "side" of V and let R^+ denote the set of roots on this side of V. An element α of R^+ is **decomposable** if there exist β and γ in R^+ such that $\alpha = \beta + \gamma$; if no such elements exist, α is **indecomposable**.

The "sides" of V can be defined as the connected components of the set $E \setminus V$. Alternatively, if H is a nonzero vector orthogonal to V, then V is the set of $\mu \in E$ for which $\langle \mu, H \rangle = 0$. The two "sides" of V are then the sets $\{ \mu \in E \mid \langle \mu, H \rangle > 0 \}$ and $\{ \mu \in E \mid \langle \mu, H \rangle < 0 \}$.

Theorem 8.16. *Suppose (E, R) is a root system, V is a hyperplane through the origin in E not containing any element of R, and R^+ is the set of roots lying on a fixed side of V. Then the set of indecomposable elements of R^+ is a base for R.*

This construction of a base motivates the term "positive simple root" for the elements of a base. We first define the positive roots (R^+) as the roots on one side of V and then define the positive simple roots (the base) as the set of indecomposable (or "simple") elements of R^+.

Proof. Let Δ denote the set of indecomposable elements in R^+. Choose a nonzero vector H orthogonal to V so that the chosen side of V is the set of $\mu \in E$ for which $\langle H, \mu \rangle > 0$.

Step 1: Every $\alpha \in R^+$ can be expressed as a linear combination of elements of Δ with non-negative integer coefficients. If not, then among all of the elements of R^+ that cannot be expressed in this way, choose α so that $\langle H, \alpha \rangle$ is as small as possible. Certainly α cannot be an element of Δ, so α must be decomposable, $\alpha = \beta_1 + \beta_2$, with $\beta_1, \beta_2 \in R^+$. Now, β_1 and β_2 cannot both be expressible as linear combinations of elements of Δ with non-negative integer coefficients, or else α would be expressible in this way. However, $\langle H, \alpha \rangle = \langle H, \beta_1 \rangle + \langle H, \beta_2 \rangle$, and since the numbers $\langle H, \beta_1 \rangle$ and $\langle H, \beta_2 \rangle$ are both positive, they must be smaller than $\langle H, \alpha \rangle$, contradicting the minimality of α.

Step 2: If α and β are distinct elements of Δ, then $\langle \alpha, \beta \rangle \leq 0$. Note that since Δ is not yet known to be a base, Proposition 8.13 does not apply. Nevertheless, if we had $\langle \alpha, \beta \rangle > 0$, then by Corollary 8.7, $\alpha - \beta$ and $\beta - \alpha$ would both be roots, one of which would have to be in R^+. If $\alpha - \beta$ were in R^+, we would have $\alpha = (\alpha - \beta) + \beta$ and α would be decomposable. If, on the other hand, $\beta - \alpha$ were in R^+, we would have $\beta = (\beta - \alpha) + \alpha$, and α would, again, be decomposable. Since α and β are assumed indecomposable, we must have $\langle \alpha, \beta \rangle \leq 0$.

Step 3: The elements of Δ are linearly independent. Suppose we have

$$\sum_{\alpha \in \Delta} c_\alpha \alpha = 0 \tag{8.5}$$

for some collection of constants c_α. Separate the sum into those terms where $c_\alpha \geq 0$ and those where $c_\alpha = -d_\alpha < 0$, so that

$$\sum c_\alpha \alpha = \sum d_\beta \beta \tag{8.6}$$

where the sums range over disjoint subsets of Δ. If u denotes the vector in (8.6), we have

$$\langle u, u \rangle = \left\langle \sum c_\alpha \alpha, \sum d_\beta \beta \right\rangle$$
$$= \sum \sum c_\alpha d_\beta \langle \alpha, \beta \rangle.$$

However, c_α and d_α are non-negative and (by Step 2) $\langle \alpha, \beta \rangle \leq 0$. Thus, $\langle u, u \rangle \leq 0$ and u must be zero.

Now, if $u = 0$, then $\langle H, u \rangle = \sum c_\alpha \langle H, \alpha \rangle = 0$, which implies that all the c_α's are zero since $c_\alpha \geq 0$ and $\langle H, \alpha \rangle > 0$. The same argument then shows that all the d_α's are zero as well.

Step 4: Δ is a base. We have shown that Δ is linearly independent and that all of the elements of R^+ can be expressed as linear combinations of elements of Δ with non-negative integer coefficients. The remaining elements of R, namely the elements of R^-, are simply the negatives of the elements of R^+, and so they can be expressed as linear combinations of elements of Δ with nonpositive integer coefficients. Since the elements of R span E, then Δ must also span E and it is a base. $\square$

Figure 8.6 illustrates Theorem 8.16 in the case of the G_2 root system, with α_1 and α_2 being the indecomposable roots on one side of the dashed line.

Theorem 8.17. *For any base Δ for R, there exists a hyperplane V and a side of V such that Δ arises as in Theorem 8.16.*

Proof. If $\Delta = \{\alpha_1, \ldots, \alpha_r\}$ is a base for R, then Δ is a basis for E in the vector space sense. Then, by elementary linear algebra, for any sequence of numbers $c_1, \ldots, c_r$ there exists a unique $\gamma \in E$ with $\langle \gamma, \alpha_j \rangle = c_j, j = 1, \ldots, r$. In particular, we can choose γ so that $\langle \gamma, \alpha_j \rangle > 0$ for all j. Then if R^+ denotes the positive roots with respect to Δ, we will have $\langle \gamma, \alpha \rangle > 0$ for all $\alpha \in R^+$, since α is a linear combination of elements of Δ with non-negative coefficients. Thus, all of the elements of R^+ lie on the same side of the hyperplane $V = \{H \in E \,|\, \langle \gamma, H \rangle = 0\}$.

Suppose now that α is an element of Δ and that α were expressible as a sum of at least two elements of R^+. Then at least one of these elements would be distinct from α and, thus, not a multiple of α. Thus, α would be expressible as a linear combination of the elements of Δ with non-negative coefficients, where the coefficient of some $\beta \neq \alpha$ would have to be nonzero. This would contradict the independence of the elements of Δ. We conclude, then, that every $\alpha \in \Delta$ is an indecomposable element of R^+. Thus, Δ is contained in the base associated to V as

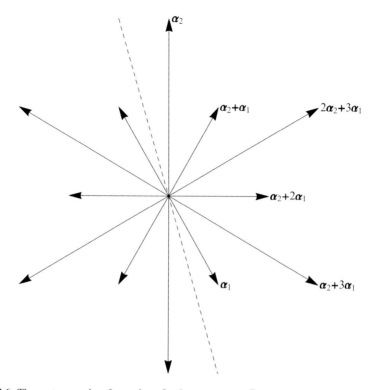

Fig. 8.6 The roots α_1 and α_2 form a base for the root system G_2

in Theorem 8.16. However, the number of elements in a base must equal dim E, so, actually, Δ *is* the base associated to V.   □

Proposition 8.18. *If Δ is a base for R, then the set of all coroots H_α, $\alpha \in \Delta$, is a base for the dual root system $R^\vee$.*

Figure 8.7 illustrates Proposition 8.18 in the case of the root system B_2.

Lemma 8.19. *Let Δ be a base, R^+ the associated set of positive roots, and α an element of Δ. Then α cannot be expressed as a linear combination of elements of $R^+ \setminus \{\alpha\}$ with non-negative real coefficients.*

Proof. Let $\alpha_1 = \alpha$ and let $\alpha_2, \dots, \alpha_r$ be the remaining elements of Δ. Suppose α_1 is a linear combination elements $\beta \neq \alpha_1$ in R^+ with non-negative coefficients. Each such β can then be expanded in terms of $\alpha_1, \dots, \alpha_r$ with non-negative (integer) coefficients. Thus, we end up with

$$\alpha_1 = c_1\alpha_1 + c_2\alpha_2 + \cdots + c_r\alpha_r,$$

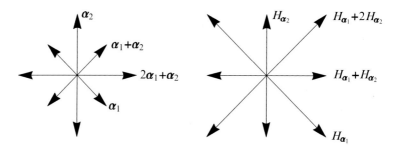

Fig. 8.7 Bases for B_2 and its dual

with each $c_j \geq 0$. Since $\alpha_1, \ldots, \alpha_r$ are independent, we must have $c_1 = 1$ and all other c_j's equal 0. But since each β in the expansion of α has non-negative coefficients in the basis $\alpha_1, \ldots, \alpha_r$, each β would have to be a multiple of α_1, and thus actually equal to α_1, since the only multiple of α_1 in R^+ is α_1. But this contradicts the assumption that β was different from α_1. $\qquad\qquad\qquad\square$

Proof of Proposition 8.18. Choose a hyperplane V such that the base Δ for R arises as in Theorem 8.16, and call the side of V on which Δ lies the positive side. Let R^+ denote the set of positive roots in R relative to the base Δ. Then the coroots H_α, $\alpha \in R^+$, also lie on the positive side of V, and all the remaining coroots lie on the negative side of V. Thus, applying Theorem 8.16 to $R^\vee$, there exists a base $\Delta^\vee$ for $R^\vee$ such that the positive roots associated to Δ are precisely the H_α's with $\alpha \in R^+$.

Now, if $\alpha \in R^+$ but $\alpha \notin \Delta$, then α is a linear combination of $\alpha_1, \ldots, \alpha_r$ with non-negative integer coefficients, at least two of which are nonzero. Thus, H_α is a linear combination of $H_{\alpha_1}, \ldots, H_{\alpha_r}$ with non-negative real coefficients, at least two of which are nonzero. Since, by Lemma 8.19, such an H_α cannot be in $\Delta^\vee$, the r elements of $\Delta^\vee$ must be precisely $H_{\alpha_1}, \ldots, H_{\alpha_r}$. $\qquad\qquad\square$

Definition 8.20. The **open Weyl chambers** for a root system (E, R) are the connected components of

$$E - \bigcup_{\alpha \in R} V_\alpha,$$

where V_α is the hyperplane through the origin orthogonal to α. If $\Delta = \{\alpha_1, \ldots, \alpha_r\}$ is a base for R, then the **open fundamental Weyl chamber** in E (relative to Δ) is the set of all H in E such that $\langle \alpha_j, H \rangle > 0$ for all $j = 1, \ldots, r$.

Figure 8.8 shows that open fundamental Weyl chamber C associated to a particular base for the A_2 root system. Note that the boundary of C is made up of portions of the lines *orthogonal to* the roots (not the lines through the roots).

Since the elements of a base Δ form a basis for E as a vector space, elementary linear algebra shows that the open fundamental Weyl chamber is convex, hence connected, and nonempty. Since the only way one can exit a fundamental Weyl

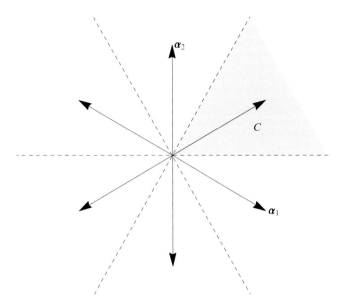

Fig. 8.8 The shaded region C is the open fundamental Weyl chamber associated to the base $\{\alpha_1, \alpha_2\}$ for A_2

chamber is by passing through a point H where $\langle \alpha_j, H \rangle = 0$, the open fundamental Weyl chamber is, indeed, an open Weyl chamber. Note, also, that if $\langle \alpha_j, H \rangle > 0$ for all $j = 1, \ldots, r$, then $\langle \alpha, H \rangle > 0$ for all $\alpha \in R^+$, since α is a linear combination of $\alpha_1, \ldots, \alpha_r$ with non-negative coefficients.

Each $w \in W$ is an orthogonal linear transformation that maps R to R and, thus, maps the set of hyperplanes orthogonal to the roots to itself. It then easily follows that for each open Weyl chamber C, the set $w \cdot C$ is another open Weyl chamber.

For any base Δ and the associated set R^+ of positive roots, we have defined the fundamental Weyl chamber to be the set of those elements having positive inner product with each element of Δ, and therefore also with each element of R^+. Our next result says that we can reverse this process.

Proposition 8.21. *For each open Weyl chamber C, there exists a unique base Δ_C for R such that C is the open fundamental Weyl chamber associated to Δ_C. The positive roots with respect to Δ_C are precisely those elements α of R such that α has positive inner product with each element of C.*

Thus, there is a one-to-one correspondence between bases and Weyl chambers.

Proof. Let H be any element of C and let V be the hyperplane orthogonal to H. Since H is contained in an open chamber, H is not orthogonal to any root, and, thus, V does not contain any root. Thus, by Theorem 8.16, there exists a base $\Delta = \{\alpha_1, \ldots, \alpha_r\}$ lying on the same side of V as H. Since $\langle \alpha_j, H \rangle$ has constant sign on C, we see that every element of C has positive inner product with each element of Δ and thus with every element of the associated set R^+ of positive roots. Thus, C must be the fundamental Weyl chamber associated to Δ. We have just said that every $\alpha \in R^+$ has positive inner product with every element of C, which means that each $\alpha \in R^-$ has negative inner product with each element of C. Thus, R^+ consists precisely of those roots having positive inner product with C.

Finally, if Δ' is any base whose fundamental chamber is C, then each element of Δ' has positive inner product with $H \in C$, meaning that Δ' lies entirely on the same side of V as Δ. Thus, Δ' has the same positive roots as Δ and therefore also the same set of positive simple (i.e., indecomposable) roots as Δ. That is to say, $\Delta' = \Delta$. $\qquad\square$

Proposition 8.22. *Every root is an element of some base.*

Since C is the set of $H \in E$ for which $\langle \alpha_j, H \rangle > 0$ for $j = 1, \ldots, r$, the codimension-one pieces of the boundary of C will consist of portions of the hyperplanes V_{α_j} orthogonal to the elements of Δ. Thus, to prove the proposition, we merely need to show that for every root α, the hyperplane V_α contains a codimension-one piece of the boundary of some C.

Proof. Let α be a root and V_α the hyperplane orthogonal to α. If we apply Exercise 2 to V_α we see that there exists some $H \in V_\alpha$ such that H does not belong to V_β for any root β other than $\beta = \pm\alpha$. Then for small positive ε, the element $H + \varepsilon\alpha$ will be in a open Weyl chamber C.

We now claim that α must be a member of the base $\Delta_C = \{\alpha_1, \ldots, \alpha_r\}$ in Proposition 8.21. Since α has positive inner product with $H + \varepsilon\alpha \in C$, we must at least have $\alpha \in R^+$, the set of positive roots associated to Δ_C. Write $\alpha = \sum_{j=1}^{r} c_j \alpha_j$, with $c_j \geq 0$, so that

$$0 = \langle \alpha, H \rangle = \sum_{j=1}^{r} c_j \langle \alpha_j, H \rangle.$$

Since H is clearly in the closure of C, we must have $\langle H, \alpha_j \rangle \geq 0$, which means that $\langle \alpha_j, H \rangle$ must be zero whenever $c_j \neq 0$. If more than one of the c_j's were nonzero, H would be orthogonal to two distinct roots in Δ, contradicting our choice of H. Thus, actually, α is in Δ. $\qquad\square$

8.5   Weyl Chambers and the Weyl Group

In this section, we establish several results about how the action of the Weyl group relates to the Weyl chambers.

Proposition 8.23. *The Weyl group acts transitively on the set of open Weyl chambers.*

Proposition 8.24. *If Δ is a base, then W is generated by the reflections s_α with $\alpha \in \Delta$.*

Proof of Propositions 8.23 and 8.24. Fix a base Δ, let C be the fundamental chamber associated to Δ, and let W' be the subgroup of W generated by the s_α's with $\alpha \in \Delta$. Let D be any other chamber and let H and H' be fixed elements of C and D, respectively. We wish to show that there exists $w \in W'$ so that $w \cdot H'$ belongs to C. To this end, choose $w \in W'$ so that $|w \cdot H' - H|$ is minimized, which is possible because $W' \subset W$ is finite. If $w \cdot H'$ were not in C, there would be some $\alpha \in \Delta$ such that $\langle \alpha, w \cdot H' \rangle < 0$. In that case, direct calculation would show that

$$\left| w \cdot H' - H \right|^2 - \left| s_\alpha \cdot w \cdot H' - H \right|^2 = -4 \frac{\langle \alpha, w \cdot H' \rangle}{\langle \alpha, \alpha \rangle} \langle \alpha, H \rangle > 0,$$

which means that $s_\alpha \cdot w \cdot H'$ is closer to H than $w \cdot H'$ is (Figure 8.9). Since $s_\alpha \in W'$, this situation would contradict the minimality of w. Thus, actually, $w \cdot H' \in C$, showing that D can be mapped to C by an element of W'. Since any chamber can be mapped to C by an element of W', any chamber can be mapped to any other chamber by an element of W'. We conclude that W', and thus also W, acts transitively on the chambers, proving Proposition 8.23.

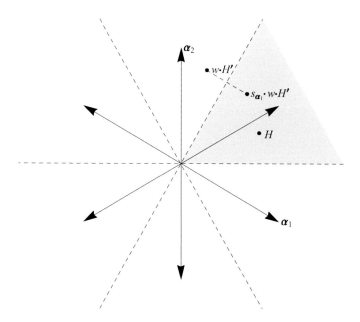

Fig. 8.9 If $w \cdot H'$ and H are on opposite sides of the line orthogonal to α_1, then $s_{\alpha_1} \cdot w \cdot H'$ is closer to H than $w \cdot H'$ is

To prove Proposition 8.24, we must show that $W' = W$. Let s_α be the reflection associated to an arbitrary root α. By Proposition 8.22, α belongs to Δ_D for some chamber D. If $w \in W'$ is chosen so that $w \cdot D = C$, then $w \cdot \alpha$ will belong to Δ. Now, it is easily seen that

$$s_{w \cdot \alpha} = w s_\alpha w^{-1},$$

so that $s_\alpha = w^{-1} s_{w \cdot \alpha} w$. But since $w \cdot \alpha \in \Delta$, both w and $s_{w \cdot \alpha}$ belong to W'. Thus, s_α belongs to W' for every root α, which means that W (the group generated by the s_α's) equals W'. $\square$

Proposition 8.25. *Let C be a Weyl chamber and let H and H' be elements of $\bar{C}$, the closure of C. If $w \cdot H = H'$ for some $w \in W$, then $H = H'$.*

That is to say, two distinct elements of $\bar{C}$ cannot be in the same orbit of W.

By Proposition 8.24, each $w \in W$ can be written as a product of reflections from Δ. If $k \geq 0$ is the smallest number of reflections from Δ needed to express w, then any expression for w as the product of k reflections from Δ is called a **minimal expression** for w. Such a minimal expression need not be unique.

The following technical lemma is the key to the proof of Proposition 8.25.

Lemma 8.26. *Let Δ be a base for R and let C be the associated fundamental Weyl chamber. Let w be an element of W with $w \neq I$ and let $w = s_{\alpha_1} s_{\alpha_2} \cdots s_{\alpha_k}$, with $\alpha_j \in \Delta$, be a minimal expression for w. Then C and $w \cdot C$ lie on opposite sides of the hyperplane V_{α_1} orthogonal to α_1.*

Proof. Since $w \neq I$, we must have $k \geq 1$. If $k = 1$, then $w = s_{\alpha_1}$, so that $w \cdot C = s_{\alpha_1} \cdot C$ is on the opposite side of V_{α_1} from C. Assume, inductively, that the result holds for $u \in W$ where the minimal number of reflections needed to express u is $k - 1$. Then consider $w \in W$ having a minimal expression of the form $w = s_{\alpha_1} \cdots s_{\alpha_k}$. If we let $u = s_{\alpha_1} \cdots s_{\alpha_{k-1}}$, then $s_{\alpha_1} \cdots s_{\alpha_{k-1}}$ must be a minimal expression for u, since any shorter expression for u would result in a shorter expression for $w = u s_{\alpha_k}$. Thus, by induction, $u \cdot C$ and C must lie on opposite sides of V_{α_1}. Suppose, toward a contradiction, that $w \cdot C$ lies on the same side of V_{α_1} as C. Then $u \cdot C$ and $w \cdot C = u s_{\alpha_k} \cdot C$ lie on opposite sides of V_{α_1}, which implies that C and $s_{\alpha_k} \cdot C$ lie on opposite sides of $u^{-1} V_{\alpha_1} = V_{u^{-1} \cdot \alpha_1}$.

We now claim that there is only *one* hyperplane V_β, with $\beta \in R$, such that C and $s_{\alpha_k} \cdot C$ lie on opposite sides of V_β, namely, $V_\beta = V_{\alpha_k}$. After all, as in the proof of Proposition 8.22, we can choose H in the boundary of C so that H lies in V_{α_k} but in no other hyperplane orthogonal to a root. We may then pass from C to $s_{\alpha_k} \cdot C$ along a line segment of the form $H + t \alpha_k$, $-\varepsilon < t < \varepsilon$, and we will pass through no hyperplane orthogonal to a root, other than V_{α_k}. We conclude, then, that $V_{u^{-1} \cdot \alpha_1} = V_{\alpha_k}$.

Now, if $V_{u^{-1} \cdot \alpha_1} = V_{\alpha_k}$, it follows that $s_{u^{-1} \cdot \alpha_1} = s_{\alpha_k}$, so that

$$s_{\alpha_k} = s_{u^{-1} \cdot \alpha_1} = u^{-1} s_{\alpha_1} u. \tag{8.7}$$

Substituting (8.7) into the formula $w = us_{\alpha_k}$ gives

$$w = s_{\alpha_1} u = s_{\alpha_1}^2 s_{\alpha_2} \cdots s_{\alpha_{k-1}}.$$

Since $s_\alpha^2 = I$, we conclude that $w = s_{\alpha_2} \cdots s_{\alpha_{k-1}}$, which contradicts the assumption that $s_{\alpha_1} \cdots s_{\alpha_k}$ was a minimal expression for w.   □

Proof of Proposition 8.25. We proceed by induction on the minimal number of reflections from Δ_C needed to express w. If the minimal number is zero, then $w = I$ and the result holds. If the minimal number is greater than zero, let $w = s_{\alpha_1} \cdots s_{\alpha_k}$ be a minimal expression for w. By Lemma 8.26, C and $w \cdot C$ lie on opposite sides of the hyperplane V_{α_1} orthogonal to α_1. Thus,

$$(w \cdot \bar{C}) \cap \bar{C} \subset V_{\alpha_1},$$

which means that $w \cdot H = H'$ must lie in V_{α_1}. It follows that

$$s_{\alpha_1} \cdot w \cdot H = s_{\alpha_1} \cdot H' = H'.$$

That is to say, the Weyl group element $w' := s_{\alpha_1} w$ *also* maps H to H'. But

$$w' = s_{\alpha_1} w = s_{\alpha_1}^2 s_{\alpha_2} \cdots s_{\alpha_k} = s_{\alpha_2} \cdots s_{\alpha_k}$$

is a product of fewer than k reflections from Δ_C. Thus, by induction, we have $H' = H$.   □

Proposition 8.27. *The Weyl group acts freely on the set of open Weyl chambers. If H belongs to an open chamber C and $w \cdot H = H$ for some $w \in W$, then $w = I$.*

Proof. Let C be any chamber and let Δ be the associated base (Proposition 8.21). If $w \in W$ and $W \neq I$, Lemma 8.26 tells us that $w \cdot C$ and C lie on opposite sides of a hyperplane, so that $w \cdot C$ cannot equal C.

Meanwhile, if H belongs to an open chamber C and $w \cdot H = H$, then $w \cdot C$ must equal C, so that $w = I$.   □

Proposition 8.28. *For any two bases Δ_1 and Δ_2 for R, there exists a unique $w \in W$ such that $w \cdot \Delta_1 = \Delta_2$.*

Proof. By Proposition 8.21, there is a bijective correspondence between bases and open Weyl chambers, and this correspondence is easily seen to respect the action of the Weyl group. Since, by Propositions 8.23 and 8.27, W acts freely and transitively on the chambers, the same is true for bases.   □

Proposition 8.29. *Let C be a Weyl chamber and H an element of E. Then there exists exactly one point in the W-orbit of H that lies in the closure $\bar{C}$ of C.*

We are *not* saying that there is a unique w such that $w \cdot H \in \bar{C}$, but rather that there exists a unique point $H' \in \bar{C}$ such that H' can be expressed (not necessarily uniquely) as $H' = w \cdot H$.

Proof. If U is any neighborhood of H, then by the argument in Exercise 2, the hyperplanes V_α, $\alpha \in R$, cannot fill up U, which means U contains points in some open Weyl chamber. It follows that H belongs to $\bar{D}$ for some chamber D. By Proposition 8.23, there exists $w \in W$ such that $w \cdot D = C$ and, thus, $w \cdot \bar{D} = \bar{C}$. Thus, $H' := w \cdot H$ is in $\bar{C}$. Meanwhile, if H'' is a point in the W-orbit of H such that $H'' \in \bar{C}$, then H' and H'' lie in the same W-orbit, which means (Proposition 8.25) that $H' = H''$. $\square$

Proposition 8.30. *Let Δ be a base for R, let R^+ be the associated set of positive roots, and let α be an element of Δ. If $\beta \in R^+$ and $\beta \neq \alpha$, then $s_\alpha \cdot \beta \in R^+$. That is to say, if $\alpha \in \Delta$, then s_α permutes the positive roots different from α.*

Proof. Write $\beta = \sum_{\gamma \in \Delta} c_\gamma \alpha_\gamma$ with $c_\gamma \geq 0$. Since $\beta \neq \alpha$, there is some c_γ with $\gamma \neq \alpha$ and $c_\gamma > 0$. Now, since $s_\alpha \cdot \beta = \beta - n\alpha$ for some integer n, in the expansion of $s_\alpha \cdot \beta$, only the coefficient of α has changed compared to the expansion of β. Thus, the coefficient of γ in the expansion of $s_\alpha \cdot \beta$ remains positive. But if one coefficient of $s_\alpha \cdot \beta$ is positive, all the other coefficients must be non-negative, showing that $s_\alpha \cdot \beta$ is a positive root. $\square$

8.6   Dynkin Diagrams

A Dynkin diagram is a convenient graphical way of encoding the structure of a base for a root system R, and thus also of R itself.

Definition 8.31. If $\Delta = \{\alpha_1, \ldots, \alpha_r\}$ is a base for a root system R, the **Dynkin diagram** for R is a graph having vertices $v_1, \ldots, v_r$. Between two distinct vertices v_j and v_k, we place zero, one, two, or three edges according to whether the angle between α_j and α_k is $\pi/2$, $2\pi/3$, $3\pi/4$, or $5\pi/6$. In addition, if α_j and α_k are not orthogonal and have different lengths, we decorate the edges between v_j and v_k with an arrow pointing from the vertex associated to the longer root to the vertex associated to the shorter root.

Note that by Proposition 8.6, angles of $2\pi/3$, $3\pi/4$, or $5\pi/6$ correspond to length ratios of $1, \sqrt{2}, \sqrt{3}$, respectively. Thus, the number of edges between vertices corresponding to two nonorthogonal roots is 1, 2, or 3 according to whether the length ratio (of the longer to the shorter) is $1, \sqrt{2}$, or $\sqrt{3}$. Thinking of the arrow decorating the edges as a "greater than" sign helps one to recall which way the arrow should go.

Two Dynkin diagrams are said to be **isomorphic** if there is a one-to-one, onto map of the vertices of one to the vertices of the other that preserves the number of bonds and the direction of the arrows. By Proposition 8.28, any two bases Δ_1 and

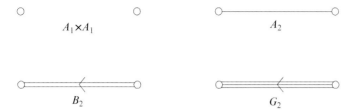

Fig. 8.10 The Dynkin diagrams for the rank-two root systems

Δ_2 for a fixed root system are related by the action of a unique Weyl group element w. Since w preserves angles and lengths, the Dynkin diagrams associated to two different bases for the same root system are isomorphic.

In the case of the root system G_2, for example, a base consists of two roots at angle of $5\pi/6$, with a length ratio of $\sqrt{3}$ (Figure 8.6). Thus, the Dynkin diagram consists of two vertices connected by three edges, with an arrow pointing from the longer root (α_2) to the shorter (α_1). One can similarly read off the Dynkin diagram for A_2 from Figure 6.2 and the diagram for B_2 from Figure 8.7. Finally, for $A_1 \times A_1$, the two elements of the base are orthogonal, yielding the results in Figure 8.10.

Proposition 8.32. *1. A root system is irreducible if and only if its Dynkin diagram is connected.*
2. If the Dynkin diagrams of two root systems R_1 and R_2 are isomorphic, then R_1 and R_2 themselves are isomorphic.

Proof. For Point 1, if a root system R decomposes as the direct sum of two root systems R_1 and R_2, then we can obtain a base Δ for R as $\Delta = \Delta_1 \cup \Delta_2$, where Δ_1 and Δ_2 are bases for R_1 and R_2, respectively. Since elements of R_1 are orthogonal to elements R_2, each element of Δ_1 is orthogonal to each element of Δ_2. Thus, the Dynkin diagram associated to Δ is disconnected.

Conversely, suppose the Dynkin diagram of R is disconnected. Then the base Δ decomposes into two pieces Δ_1 and Δ_2 where (since there are no edges connecting the pieces), each element of Δ_1 is orthogonal to each element of Δ_2. Thus, E is the orthogonal direct sum of $E_1 := \text{span}(\Delta_1)$ and $E_2 := \text{span}(\Delta_2)$. If $R_1 = R \cap E_1$ and $R_2 = R \cap E_2$, then it easy to check that R_1 and R_2 are root systems in E_1 and E_2, respectively, and that Δ_1 is a base for R_1 and Δ_2 is a base for R_2.

Now, for each $\alpha \in \Delta_1$, the reflection s_α will act as the identity on E_2 and similarly for $\alpha \in \Delta_2$. Since the Weyl groups of R, R_1, and R_2 are generated by the reflections from Δ, Δ_1, and Δ_2, respectively, we see that the Weyl group W of R is the direct product of the Weyl groups W_1 and W_2 of R_1 and R_2, with W_1 acting only on E_1 and W_2 acting only on E_2. Since W acts transitively on the bases of R and every element of R is part of some base, we see that $W \cdot \Delta = R$. But since $W = W_1 \times W_2$, we have $W \cdot \Delta = (W_1 \cdot \Delta_1) \cup (W_2 \cdot \Delta_2)$. Thus, every element of R is either in E_1 or in E_2, meaning that R is the direct sum of R_1 and R_2.

For Point 2, using Point 1, we can reduce the problem to the case in which R_1 and R_2 are irreducible and the Dynkin diagrams of R_1 and R_2 are connected. Let $\Delta_1 = \{\alpha_1, \ldots, \alpha_r\}$ and $\Delta_2 = \{\beta_1, \ldots, \beta_r\}$ be bases for R_1 and R_2, respectively, ordered so that the isomorphism of the Dynkin diagrams maps the vertex associated to α_j to the vertex associated to β_j. We may rescale the inner product on R_1 so that $\|\alpha_1\| = \|\beta_1\|$. Since the Dynkin diagrams are connected and the diagram determines the length ratios between vertices joined by an edge, it follows that $\langle \alpha_j, \alpha_k \rangle = \langle \beta_j, \beta_k \rangle$ for all j and k. It is then easy to check that the unique linear map $A : E_1 \rightarrow E_2$ such that $A\alpha_j = \beta_j$, $j = 1, \ldots, r$ is an isometry. We then have that $As_{\alpha_j} = s_{\beta_j} A$ for all j.

Now, if α is any element of R_1, then, since W_1 is generated by the reflections from Δ_1 and $W_1 \cdot \Delta = R_1$, we see that

$$\alpha = s_{\alpha_{j_1}} \cdots s_{\alpha_{j_N}} \alpha_k$$

for some indices $j_1, \ldots, j_N$ and k. Thus,

$$A\alpha = s_{\beta_{j_1}} \cdots s_{\beta_{j_N}} \beta_k$$

is an element of R_2. The same reasoning shows that $A^{-1}\beta \in R_1$ for all $\beta \in R_2$. Thus, A is an isometry mapping R_1 onto R_2, which implies that A is an isomorphism of R_1 with R_2. $\qquad\square$

Corollary 8.33. *Let $\mathfrak{g} = \mathfrak{k}_{\mathbb{C}}$ be a semisimple Lie algebra, let $\mathfrak{h} = \mathfrak{t}_{\mathbb{C}}$ be a Cartan subalgebra of $\mathfrak{g}$, and let $R \subset$ it be the root system of $\mathfrak{g}$ relative to $\mathfrak{h}$. Then $\mathfrak{g}$ is simple if and only if the Dynkin diagram of R is connected.*

Proof. According to Theorem 7.35, $\mathfrak{g}$ is simple if and only if R is irreducible. But according to Point 1 of Proposition 8.32, R is irreducible if and only if the Dynkin diagram of R is connected. $\qquad\square$

8.7 Integral and Dominant Integral Elements

We now introduce a notion of integrality for elements of E. In the setting of the representations of a semisimple Lie algebra $\mathfrak{g}$, the weights of a finite-dimensional representation of $\mathfrak{g}$ are always integral elements. Recall from Definition 8.10 the notion of the coroot H_α associated to a root α.

Definition 8.34. An element μ of E is an **integral element** if for all α in R, the quantity

$$\langle \mu, H_\alpha \rangle = 2\frac{\langle \mu, \alpha \rangle}{\langle \alpha, \alpha \rangle}$$

is an integer. If Δ is a base for R, an element μ of E is **dominant** (relative to Δ) if

$$\langle \alpha, \mu \rangle \geq 0$$

for all $\alpha \in \Delta$ and **strictly dominant** if

$$\langle \alpha, \mu \rangle > 0$$

for all $\alpha \in \Delta$.

See Figure 6.2 for the integral and dominant integral elements in the case of A_2. Additional examples will be given shortly.

A point $\mu \in E$ is strictly dominant relative to Δ if and only if μ is contained in the open fundamental Weyl chamber associated to Δ, and μ is dominant if and only if μ is contained in the closure of the open fundamental Weyl chamber. Proposition 8.29 therefore implies the following result: *For all $\mu \in E$, there exists $w \in W$ such that $w \cdot \mu$ is dominant.*

Note that by the definition of a root system, every root is an integral element. Thus, every integer linear combination of roots is also an integral element. In most cases, however, there exist integral elements that are not expressible as an integer combination of roots. In the case of A_2, for example, the elements labeled μ_1 and μ_2 in Figure 6.2 are integral, but their expansions in terms of α_1 and α_2 are $\mu_1 = 2\alpha_1/3 + \alpha_2/3$ and $\mu_2 = \alpha_1/3 + 2\alpha_2/3$. Since α_1 and α_2 form a base for A_2, if μ_1 or μ_2 were an integer combination of roots, it would also be an integer combination of α_1 and α_2.

Proposition 8.35. *If $\mu \in E$ has the property that*

$$2 \frac{\langle \mu, \alpha \rangle}{\langle \alpha, \alpha \rangle}$$

is an integer for all $\alpha \in \Delta$, then the same holds for all $\alpha \in R$ and, thus, μ is an integral element.

Proof. An element μ is integral if and only if $\langle \mu, H_\alpha \rangle$ is an integer for all $\alpha \in R$. By Proposition 8.18, each H_α with $\alpha \in R$ can be expressed as a linear combination of the H_α's with $\alpha \in \Delta$, with *integer* coefficients. Thus, if $\langle \mu, H_\alpha \rangle \in \mathbb{Z}$ for $\alpha \in \Delta$, then same is true for $\alpha \in R$, showing that μ is integral. $\square$

Definition 8.36. Let $\Delta = \{\alpha_1, \ldots, \alpha_r\}$ be a base. Then the **fundamental weights** (relative to Δ) are the elements $\mu_1, \ldots, \mu_r$ with the property that

$$2 \frac{\langle \mu_j, \alpha_k \rangle}{\langle \alpha_k, \alpha_k \rangle} = \delta_{jk}, \quad j, k = 1, \ldots, r. \tag{8.8}$$

Elementary linear algebra shows there exists a unique set of integral elements satisfying (8.8). Geometrically, the jth fundamental weight is the unique element of E that is orthogonal to each α_j, $j \neq k$, and whose orthogonal projection onto α_j is one-half of α_j. Note that the set of dominant integral elements is precisely the set of linear combinations of the fundamental weights with non-negative integer coefficients; the set of all integral elements is the set of linear combinations of fundamental weights with arbitrary integer coefficients.

Definition 8.37. Let Δ be a base for R and R^+ the associated set of positive roots. We then let δ denote **half the sum of the positive roots**:

$$\delta = \frac{1}{2} \sum_{\alpha \in R^+} \alpha.$$

The element δ plays a key role in many of the developments in subsequent chapters. It appears, for example, in the statement of the Weyl character formula and the Weyl integral formula. Figure 8.11 shows the integral and dominant integral elements for B_2 and G_2. In each case, the base $\{\alpha_1, \alpha_2\}$ and the element δ are labeled, and the fundamental weights are circled. The background square lattice (B_2 case) or triangular lattice (G_2 case) indicates the set of all integral elements. The root system G_2 is unusual in that the fundamental weights are roots, which means that every integral element is expressible as an integer combination of roots.

Proposition 8.38. *The element δ is a strictly dominant integral element; indeed,*

$$2\frac{\langle \beta, \delta \rangle}{\langle \beta, \beta \rangle} = 1$$

for each $\beta \in \Delta$.

Proof. If $\beta \in \Delta$, then by Proposition 8.30, s_β permutes the elements of R^+ different from β. Thus, we can decompose $R^+ \setminus \{\beta\}$ as $E_1 \cup E_2$, where elements of E_1 are orthogonal to β and elements of E_2 are not. Then the elements of E_2 split up into pairs $\{\alpha, s_\beta \cdot \alpha\}$, where

$$\langle s_\beta \cdot \alpha, \beta \rangle = \langle \alpha, s_\beta \cdot \beta \rangle = -\langle \alpha, \beta \rangle.$$

Thus, when we compute the inner product of β with half the sum of the positive roots, the roots in E_1 do not contribute and the contributions from roots in E_2 cancel in pairs. Thus, only the contribution from β itself remains:

$$2\frac{\langle \beta, \delta \rangle}{\langle \beta, \beta \rangle} = 2\frac{\langle \beta, \frac{1}{2}\beta \rangle}{\langle \beta, \beta \rangle} = 1,$$

as claimed.   □

Fig. 8.11 Integral and dominant integral elements for B_2 (*top*) and G_2 (*bottom*). The *black dots* indicate dominant integral elements, while the *background lattice* indicates the set of all integral elements

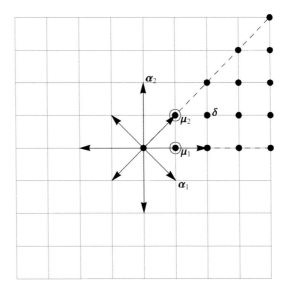

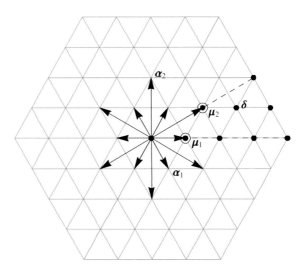

8.8   The Partial Ordering

We now introduce a partial ordering on the set of integral elements, which will be used to formulate the theorem of the highest weight for representations of a semisimple Lie algebra.

Fig. 8.12 Points that are
higher than zero (*light and
dark gray*) and points that are
dominant (*dark gray*) for B_2

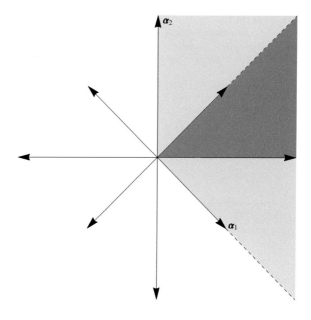

Definition 8.39. If $\Delta = \{\alpha_1, \ldots, \alpha_r\}$ is a base, an element $\mu \in E$ is said to be **higher** than $\lambda \in E$ (relative to Δ) if $\mu - \lambda$ can be expressed as

$$\mu - \lambda = c_1\alpha_1 + \cdots + c_r\alpha_r,$$

where each c_j is a non-negative real number. We equivalently say that λ is **lower** than μ and we write this relation as $\mu \succeq \lambda$ or $\lambda \preceq \mu$.

The relation $\succeq$ defines a partial ordering on E. We now develop various useful properties of this relation. For the rest of the section, we assume a base Δ for R has been chosen, and that the notions of higher, lower, and dominant are defined relative to Δ.

Proposition 8.40. *If $\mu \in E$ is dominant, then $\mu \succeq 0$.*

See Figure 8.12 for an example of the proposition.

For any basis $\{v_1, \ldots, v_r\}$ of E, we can form a basis for the dual space E^* by considering the linear functionals ξ_j given by $\xi_j(v_k) = \delta_{jk}$. We can then find unique vectors $v_j^* \in E$ such that $\xi_j(u) = \left\langle v_j^*, u \right\rangle$, so that

$$\left\langle v_j^*, v_k \right\rangle = \delta_{jk}.$$

The basis $\{v_1^*, \ldots, v_r^*\}$ for E is called the **dual basis** to $\{v_1, \ldots, v_r\}$.

Lemma 8.41. *Suppose* $\{v_1, \ldots, v_r\}$ *is an obtuse basis for* E, *meaning that* $\langle v_j, v_k \rangle \leq 0$ *for all* $j \neq k$. *Then* $\{v_1^*, \ldots, v_r^*\}$ *is an acute basis for* E, *meaning that* $\langle v_j^*, v_k^* \rangle \geq 0$ *for all* j, k.

The proof of this elementary lemma is deferred until the end of this section.

Proof of Proposition 8.40. Any vector u can be expanded as $u = \sum_j c_j \alpha_j$ and the coefficients may be computed as $c_j = \langle \alpha_j^*, u \rangle$, where $\{\alpha_j^*\}$ is the dual basis to $\{\alpha_j\}$. Applying this with $u = \alpha_j^*$ gives

$$\alpha_j^* = \sum_{k=1}^r \langle \alpha_k^*, \alpha_j^* \rangle \alpha_k.$$

Now, if μ is dominant, the coefficients in the expansion $\mu = \sum_j c_j \alpha_j$ are given by

$$c_j = \langle \alpha_j^*, \mu \rangle = \sum_k \langle \alpha_k^*, \alpha_j^* \rangle \langle \alpha_k, \mu \rangle.$$

Since μ is dominant, $\langle \alpha_k, \mu \rangle \geq 0$. Furthermore, the α_j's form an obtuse basis for E (Proposition 8.13) and thus by Lemma 8.41, we have $\langle \alpha_k^*, \alpha_j^* \rangle \geq 0$ for all j, k. Thus, $c_j \geq 0$ for all j, which shows that μ is higher than zero. $\qquad\square$

Proposition 8.42. *If* μ *is dominant, then* $w \cdot \mu \preceq \mu$ *for all* $w \in W$.

Proof. Let O be the Weyl-group orbit of μ. Since O is a finite set, it contains a maximal element λ, i.e., one such that there is no $\lambda' \neq \lambda$ in O that is higher than λ. Then for all $\alpha \in \Delta$, we must have $\langle \alpha, \lambda \rangle \geq 0$, since if $\langle \alpha, \lambda \rangle$ were negative, then

$$s_\alpha \cdot \lambda = \lambda - 2 \frac{\langle \lambda, \alpha \rangle}{\langle \alpha, \alpha \rangle} \alpha$$

would be higher than λ. Thus, λ is dominant. But since, by Proposition 8.29, μ is the unique dominant element of O, we must have $\lambda = \mu$. Thus, μ is the *unique* maximal element of O.

We now claim that every element of O is lower than μ. Certainly, no element of O can be higher than μ. Let O' be the set of $\lambda \in O$ that are neither higher nor lower than μ. If O' is not empty, it contains a maximal element λ. We now argue that λ is actually maximal in O. For any $\gamma \in O$, if $\gamma \in O'$, then certainly γ cannot be higher than λ, which is maximal in O'. On the other hand, if $\gamma \in O \setminus O'$, then γ is lower than μ, in which case, γ cannot be higher than λ, or else we would have $\mu \succeq \gamma \succeq \lambda$, so that λ would not be in O'. We conclude that λ is maximal in O, not just in O', which means that λ must equal μ, the unique maximal element of O. But this contradicts the assumption that $\lambda \in O'$. Thus, O' must actually be empty, and μ is the highest element of O. $\qquad\square$

Proposition 8.43. *If μ is a strictly dominant integral element, then $\mu \succeq \delta$, where δ is as in Definition 8.37.*

Proof. Since μ is strictly dominant, $\mu - \delta$ will still be dominant in light of Proposition 8.38. Thus, by Proposition 8.40, $\mu - \delta \succeq 0$, which is equivalent to $\mu \succeq \delta$. □

Recall from Definition 6.23 the notion of the convex hull of a finite collection of vectors in E. We let $W \cdot \mu$ denote the Weyl-group orbit of $\mu \in E$ and we let $\mathrm{Conv}(W \cdot \mu)$ denote the convex hull of $W \cdot \mu$.

Proposition 8.44. *1. If μ and λ are dominant, then λ belongs to $\mathrm{Conv}(W \cdot \mu)$ if and only if $\lambda \preceq \mu$.*
2. Let μ and λ be elements of E with μ dominant. Then λ belongs to $\mathrm{Conv}(W \cdot \mu)$ if and only if $w \cdot \lambda \preceq \mu$ for all $w \in W$.

Figure 8.13 illustrates Point 2 of the proposition in the case of B_2. In the figure, the shaded region represents the set of points that are lower than μ. The point λ_1 is inside $\mathrm{Conv}(W \cdot \mu)$ and $w \cdot \lambda_1$ is lower than μ for all w. By contrast, λ_2 is outside $\mathrm{Conv}(W \cdot \mu)$ and there is some w for which $w \cdot \lambda_2$ is not lower than μ.

Since $\mathrm{Conv}(W \cdot \mu)$ is convex and Weyl invariant, we see that if λ belongs to $\mathrm{Conv}(W \cdot \mu)$, then every point in $\mathrm{Conv}(W \cdot \lambda)$ also belongs to $\mathrm{Conv}(W \cdot \mu)$. Thus, Point 1 of the proposition may be restated as follows:

If μ and λ are dominant, then $\lambda \preceq \mu$ if and only if

$$\mathrm{Conv}(W \cdot \lambda) \subset \mathrm{Conv}(W \cdot \mu).$$

We establish two lemmas that will lead to a proof of Proposition 8.44.

Lemma 8.45. *Suppose K is a compact, convex subset of E and λ is an element of E that it is not in K. Then there is an element γ of E such that for all $\eta \in K$, we have*

$$\langle \gamma, \lambda \rangle > \langle \gamma, \eta \rangle .$$

If we let V be the hyperplane (not necessarily through the origin) given by

$$V = \{\rho \in E \,|\, \langle \gamma, \rho \rangle = \langle \gamma, \eta \rangle - \varepsilon\}$$

for some small ε, then K and λ lie on opposite sides of V. Lemma 8.45 is a special case of the *hyperplane separation theorem* in the theory of convex sets.

Proof. Since K is compact, we can choose an element η_0 of K that minimizes the distance to λ. Set $\gamma = \lambda - \eta_0$, so that

$$\langle \gamma, \lambda - \eta_0 \rangle = \langle \lambda - \eta_0, \lambda - \eta_0 \rangle > 0,$$

and, thus, $\langle \gamma, \lambda \rangle > \langle \gamma, \eta_0 \rangle$.

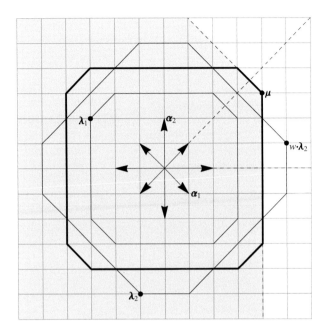

Fig. 8.13 The element $w \cdot \lambda_2$ is not lower than μ

Now, for any $\eta \in K$, the vector $\eta_0 + s(\eta - \eta_0)$ belongs to K for $0 \leq s \leq 1$, and we compute that

$$d(\lambda, \eta_0 + s(\eta - \eta_0))^2 = \langle \lambda - \eta_0, \lambda - \eta_0 \rangle - 2s \langle \lambda - \eta_0, \eta - \eta_0 \rangle$$
$$+ s^2 \langle \eta - \eta_0, \eta - \eta_0 \rangle.$$

The only way this quantity can be greater than or equal to $\langle \lambda - \eta_0, \lambda - \eta_0 \rangle = d(\lambda, \eta_0)^2$ for small positive s is if

$$\langle \lambda - \eta_0, \eta - \eta_0 \rangle = \langle \gamma, \eta - \eta_0 \rangle \leq 0.$$

Thus,

$$\langle \gamma, \eta \rangle \leq \langle \gamma, \eta_0 \rangle < \langle \gamma, \lambda \rangle,$$

which is what we wanted to prove.   □

Lemma 8.46. *If μ and λ are dominant and $\lambda \notin \mathrm{Conv}(W \cdot \mu)$, there exists a dominant element $\gamma \in E$ such that*

$$\langle \gamma, \lambda \rangle > \langle \gamma, w \cdot \mu \rangle \tag{8.9}$$

for all $w \in W$.

Proof. By Lemma 8.45, we can find some γ in E, not necessarily dominant, such that $\langle \gamma, \lambda \rangle > \langle \gamma, \eta \rangle$ for all $\eta \in \mathrm{Conv}(W \cdot \mu)$. In particular,

$$\langle \gamma, \lambda \rangle > \langle \gamma, w \cdot \mu \rangle$$

for all $w \in W$. Choose some w_0 so that $\gamma' := w_0 \cdot \gamma$ is dominant. We will show that replacing γ by γ' makes $\langle \gamma, \lambda \rangle$ bigger while permuting the values of $\langle \gamma, w \cdot \mu \rangle$.

By Proposition 8.42, $\gamma \preceq \gamma'$, meaning that γ' equals γ plus a non-negative linear combination of positive simple roots. But since λ is dominant, it has non-negative inner product with each positive simple root, and we see that $\langle \gamma', \lambda \rangle \geq \langle \gamma, \lambda \rangle$. Thus,

$$\langle \gamma', \lambda \rangle \geq \langle \gamma, \lambda \rangle > \langle \gamma, w \cdot \mu \rangle$$

for all w. But

$$\langle \gamma, w \cdot \mu \rangle = \langle w_0^{-1} \cdot \gamma', w \cdot \mu \rangle = \langle \gamma', (w_0 w) \cdot \mu \rangle.$$

Thus, as w ranges over W, the values of $\langle \gamma, w \cdot \mu \rangle$ and $\langle \gamma', (w_0 w) \cdot \mu \rangle$ range through the same set of real numbers. Thus, $\langle \gamma', \lambda \rangle > \langle \lambda', w \cdot \mu \rangle$ for all w, as claimed. □

The proof of Lemma 8.46 is illustrated in Figure 8.14. The dominant element λ is not in $\mathrm{Conv}(W \cdot \mu)$ and is separated from $\mathrm{Conv}(W \cdot \mu)$ by a line with orthogonal vector γ. The element $\gamma' := s_{\alpha_2} \cdot \gamma$ is dominant and λ is also separated from

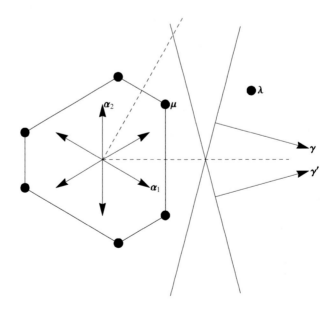

Fig. 8.14 The element λ is separated from $\mathrm{Conv}(W \cdot \mu)$ first by a line orthogonal to γ and then by a line orthogonal to the dominant element γ'

Conv($W \cdot \mu$) by a line with orthogonal vector γ'. The existence of such a line means that λ cannot be lower than μ.

Proof of Proposition 8.44. For Point 1, let μ and λ be dominant. Assume first that λ is in Conv($W \cdot \mu$). By Proposition 8.42, every element of the form $w \cdot \mu$ is lower than μ. But the set E of elements lower than μ is easily seen to be convex, and so E must contain Conv($W \cdot \mu$) and, in particular, λ. Next, assume $\lambda \preceq \mu$ and suppose, toward a contradiction, that $\lambda \notin$ Conv($W \cdot \mu$). Let γ be a dominant element as in Lemma 8.46. Then $\mu - \lambda$ is a non-negative linear combination of positive simple roots, and γ, being dominant, has non-negative inner product with each positive simple root. Thus, $\langle \gamma, \mu - \lambda \rangle \geq 0$ and, hence, $\langle \gamma, \mu \rangle \geq \langle \gamma, \lambda \rangle$, which contradicts (8.9). Thus, λ must actually belong to Conv($W \cdot \mu$).

For Point 2, assume first that $w \cdot \lambda \preceq \mu$ for all $w \in W$, and choose w so that $w \cdot \lambda$ is dominant. Since, $w \cdot \lambda \preceq \mu$, Point 1 tells us that $w \cdot \lambda$ belongs to Conv($W \cdot \mu$), which implies that λ also belongs to Conv($W \cdot \mu$). In the other direction, assume $\lambda \in$ Conv($W \cdot \mu$) so that $w \cdot \lambda \in$ Conv($W \cdot \mu$) for all $w \in W$. Using Proposition 8.42 we can easily see that every element of Conv($W \cdot \mu$) is lower than μ. Thus, $w \cdot \lambda \preceq \mu$ for all w. $\square$

It remains only to supply the proof of Lemma 8.41.

Proof of Lemma 8.41. We proceed by induction on the dimension r of E. When $r = 1$ the result is trivial. When $r = 2$, the result should be geometrically obvious, but we give an algebraic proof. The **Gram matrix** of a basis is the collection of inner products, $G_{jk} := \langle v_j, v_k \rangle$. It is an elementary exercise (Exercise 3) to show that the Gram matrix of the dual basis is the inverse of the Gram matrix of the original basis. Thus, in the $r = 2$ case, we have

$$\begin{pmatrix} \langle v_1^*, v_1^* \rangle & \langle v_1^*, v_2^* \rangle \\ \langle v_1^*, v_2^* \rangle & \langle v_2^*, v_2^* \rangle \end{pmatrix}$$

$$= \frac{1}{(\langle v_1, v_1 \rangle \langle v_2, v_2 \rangle - \langle v_1, v_2 \rangle^2)} \begin{pmatrix} \langle v_2, v_2 \rangle & -\langle v_1, v_2 \rangle \\ -\langle v_1, v_2 \rangle & \langle v_1, v_1 \rangle \end{pmatrix}. \qquad (8.10)$$

Since v_1 is not a multiple of v_2, the Cauchy–Schwarz inequality tells us that the denominator on the right-hand side of (8.10) is positive, which means that $\langle v_1^*, v_2^* \rangle$ has the opposite sign of $\langle v_1, v_2 \rangle$.

Assume now that the result holds in dimension $r \geq 2$ and consider the case of dimension $r + 1$. Fix any index m and let P be the orthogonal projection onto the orthogonal complement of v_m, which is given by

$$P(u) = u - \frac{\langle v_m, u \rangle}{\langle v_m, v_m \rangle} v_m.$$

The operator P is easily seen to be self-adjoint, meaning that $\langle u, Pv \rangle = \langle Pu, v \rangle$ for all u, v.

We now claim that $Pv_1, \ldots, \widehat{Pv_m}, \ldots, Pv_{r+1}$ is an obtuse basis for $\langle v_m \rangle^{\perp}$, where the notation $\widehat{Pv_m}$ indicates that Pv_m is omitted. Indeed, a little algebra shows that

$$\langle Pv_j, Pv_k \rangle = \langle v_j, v_k \rangle - \frac{\langle v_m, v_j \rangle \langle v_m, v_k \rangle}{\langle v_m, v_m \rangle} \le 0,$$

since $\langle v_j, v_k \rangle$, $\langle v_m, v_j \rangle$, and $\langle v_m, v_k \rangle$ are all less than or equal to zero. Furthermore, for j and k different from m, we have

$$\left\langle v_j^*, Pv_k \right\rangle = \left\langle Pv_j^*, v_k \right\rangle = \left\langle v_j^*, v_k \right\rangle = \delta_{jk}$$

since v_j^* is orthogonal to v_m. Thus, the dual basis to $Pv_1, \ldots, \widehat{Pv_m}, \ldots, Pv_{r+1}$ consists simply of the vectors $v_1^*, \ldots, \widehat{v_m^*}, \ldots, v_{r+1}^*$ (all of which are orthogonal to v_m).

Now fix any two distinct indices j and k. Since $r + 1 \ge 3$, we can choose some other index m distinct from both j and k. Applying our induction hypothesis to the basis $Pv_1, \ldots, \widehat{Pv_m}, \ldots, Pv_{r+1}$ for $\langle v_m \rangle^{\perp}$, we conclude that $\left\langle v_j^*, v_k^* \right\rangle \ge 0$, which is what we are trying to prove. $\square$

8.9   Examples in Rank Three

In rank three, we can have a reducible root system, which must be a direct sum of A_1 with one of the rank-two root systems described in the previous section. In this section, we will consider only the *irreducible* root systems of rank three. There are, up to isomorphism, three irreducible root systems in rank three, customarily denoted A_3, B_3, and C_3. They arise from the Lie algebras $\mathsf{sl}(4; \mathbb{C})$, $\mathsf{so}(7; \mathbb{C})$, and $\mathsf{sp}(3; \mathbb{C})$, respectively, as described in Sect. 7.7.

The models in this section can be constructed using the Zometool system, available at www.zometool.com. The reader is encouraged to obtain some Zometool pieces and build the rank-three root systems for him- or herself. The models require the green lines, which are not part of the basic Zometool kits. The models of the C_3 root system use half-length greens, although one can alternatively use whole greens together with double (end to end) blue pieces. The images shown here were rendered in Scott Vorthmann's vZome software, available at vzome.com. For detailed instructions on how to build rank-three root systems using Zometool, click on the "Book" tab of the author's web site: www.nd.edu/~bhall/.

Figure 8.15 shows the A_3 root system, with a base highlighted. The elements of A_3 form the vertices of a polyhedron known as a cuboctahedron, which has six square faces and eight triangular faces, as shown in Figure 8.16. The points in A_3 can also be visualized as the midpoints of the edges of a cube, as in Figure 8.17. Algebraically, we can describe A_3 as the set of 12 vectors in $\mathbb{R}^3$ of the form

Fig. 8.15 The A_3 root system, with the elements of the base in *dark gray*

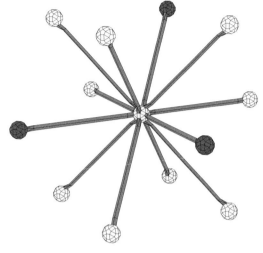

Fig. 8.16 The roots in A_3 make up the vertices of a cuboctahedron

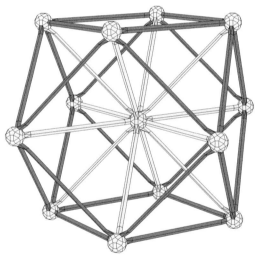

$(\pm 1, \pm 1, 0), (\pm 1, 0, \pm 1)$, and $(0, \pm 1, \pm 1)$. (This set of vectors actually corresponds to the conventional description of the D_3 root system, as in Sect. 7.7.2, which turns out to be isomorphic to A_3.) It is then a simple exercise to check that this collection of vectors is, in fact, a root system. A base for this system is given by the vectors $(1, -1, 0), (0, 1, -1)$, and $(0, 1, 1)$.

The Weyl group W for A_3 is the symmetry group of the tetrahedron pictured in Figure 8.18. The group W is the full permutation group on the four vertices of the tetrahedron. As in the A_2 case, the Weyl group of A_3 is not the full symmetry group of the root system, since $-I$ is not an element of W.

The B_3 root system is obtained from the A_3 root system by adding six additional vectors, consisting of three mutually orthogonal pairs. Each of the new roots

Fig. 8.17 The roots in A_3 lie at the midpoints of the edges of a cube

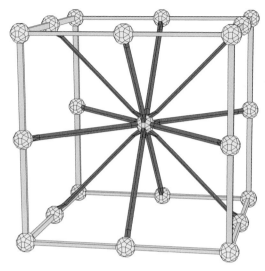

Fig. 8.18 The Weyl group of A_3 is the symmetry group of a regular tetrahedron

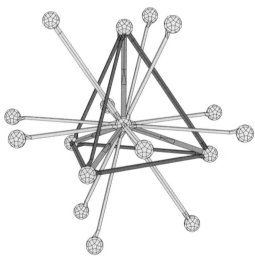

is shorter than the original roots by a factor of $\sqrt{2}$, as shown in Figure 8.19. Algebraically, B_3 consists of the twelve vectors $(\pm 1, \pm 1, 0)$, $(\pm 1, 0, \pm 1)$, and $(0, \pm 1, \pm 1)$ of A_3, together with the six vectors $(\pm 1, 0, 0)$, $(0, \pm 1, 0)$, and $(0, 0, \pm 1)$.

The C_3 root system, meanwhile, is obtained from A_3 by adding six new vectors, as in the case of B_3, except that this time the new roots are longer than the original roots by a factor of $\sqrt{2}$, as in Figure 8.20. That is to say, the new roots are the six vectors $(\pm 2, 0, 0)$, $(0, \pm 2, 0)$, and $(0, 0, \pm 2)$. The C_3 root system is the dual of B_3, in the sense of Definition 8.10. The elements of C_3 make up the vertices of an octahedron, together with the midpoints of the edges of the octahedron, as shown in Figure 8.21.

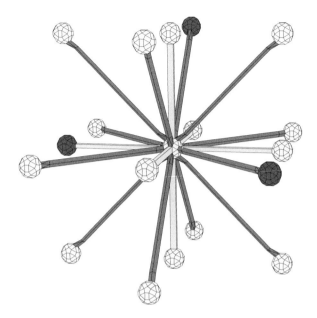

Fig. 8.19   The B_3 root system, with the elements of the base in *dark gray*

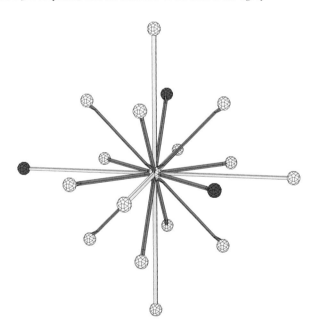

Fig. 8.20   The C_3 root system, with the elements of the base in *dark gray*

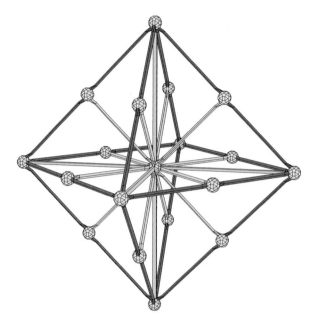

Fig. 8.21 The C_3 root system consists of the vertices of an octahedron, together with the midpoints of the edges of the octahedron

The root systems B_3 and C_3 have the same Weyl group, which is the symmetry group of the cube in Figure 8.17. In both cases, the Weyl group is the full symmetry group of the root system.

8.10 The Classical Root Systems

We now return to the root systems of the classical semisimple Lie algebras, as computed in Sect. 7.7. For each of these root systems, we describe a base and determine the associated Dynkin diagram.

8.10.1 The A_n Root System

The A_n root system is associated to the Lie algebra $\mathsf{sl}(n+1; \mathbb{C})$. For this root system, E is the subspace of $\mathbb{R}^{n+1}$ consisting of vectors whose entries sum to zero. The roots are the vectors of the form

$$e_j - e_k, \quad j \neq k,$$

where $\{e_j\}$ is the standard basis for $\mathbb{R}^{n+1}$. As a base, we may take the vectors

$$e_1 - e_2, \ e_2 - e_3, \ \ldots, \ e_n - e_{n+1}$$

Note that for $j < k$,

$$e_j - e_k = (e_j - e_{j+1}) + (e_{j+1} - e_{j+2}) + \cdots + (e_{k-1} - e_k),$$

so that every root is a sum of elements of the base, or the negative thereof.

All roots in the base have the same length, two consecutive roots are at an angle of $2\pi/3$ to one another, and nonconsecutive roots are orthogonal.

8.10.2   The D_n Root System

The D_n root system is associated to the Lie algebra $\mathsf{so}(2n;\mathbb{C})$, $n \geq 2$. For this root system, $E = \mathbb{R}^n$ and the roots are the vectors of the form

$$\pm e_j \pm e_k, \quad j < k.$$

As a base, we may take the $n - 1$ roots

$$e_1 - e_2, \ e_2 - e_3, \ \cdots, \ ,e_{n-2} - e_{n-1}, \ e_{n-1} - e_n, \tag{8.11}$$

together with the one additional root,

$$e_{n-1} + e_n. \tag{8.12}$$

Note that for $j < k$, we have the following formulas:

$$e_j - e_k = (e_j - e_{j+1}) + (e_{j+1} - e_{j+2}) + \cdots + (e_{k-1} - e_k),$$
$$e_j + e_n = (e_j - e_{n-1}) + (e_{n-1} + e_n),$$
$$e_j + e_k = (e_j + e_n) + (e_k - e_n). \tag{8.13}$$

This shows that every root of the form $e_j - e_k$ or $e_j + e_k$ ($j < k$) can be written as a linear combination of the base in (8.11) and (8.12) with non-negative integer coefficients. The roots of this form are the positive roots, and the remaining roots are the negatives of these roots.

Two consecutive roots in the list (8.11) have an angle of $2\pi/3$ and two nonconsecutive roots in the list (8.11) are orthogonal. The angle between the root in (8.12) and the *second-to-last* element in the list (8.11) is $2\pi/3$; the root in (8.12) is orthogonal to all the other roots in (8.11).

8.10.3   The B_n Root System

The B_n root system is associated to the Lie algebra $\mathsf{so}(2n + 1; \mathbb{C})$. For this root system, $E = \mathbb{R}^n$ and the roots are the vectors of the form

$$\pm e_j \pm e_k, \quad j < k,$$

and of the form

$$\pm e_j, \quad j = 1, \ldots, n.$$

As a base for our root system, we may take the $n - 1$ roots

$$e_1 - e_2, \; e_2 - e_3, \; \ldots, \; e_{n-1} - e_n, \tag{8.14}$$

(exactly as in the $\mathsf{so}(2n; \mathbb{C})$ case) together with the one additional root,

$$e_n. \tag{8.15}$$

The positive roots are those of the form $e_j + e_k$ or $e_j - e_k$ $(j < k)$ and those of the form e_j $(1 \leq j \leq n)$. To expand every positive root in terms of the base, we use the formulas in (8.13), except with the second line replaced by

$$e_j + e_n = (e_j - e_n) + 2e_n, \tag{8.16}$$

and with the additional relation

$$e_j = (e_j - e_n) + e_n. \tag{8.17}$$

As in the $\mathsf{so}(2n; \mathbb{C})$ case, consecutive roots in the list (8.14) have an angle of $2\pi/3$, whereas nonconsecutive roots on the list (8.14) are orthogonal. Meanwhile, the root in (8.15) has an angle of $3\pi/4$ with the *last* root in (8.14) and is orthogonal to the remaining roots in (8.14).

In Sect. 8.2, we have pictured the B_2 root system rotated by $\pi/4$ relative to the $n = 2$ case of the root system described in this subsection. The pictures in Sect. 8.2 actually correspond to the conventional description of the C_2 root system (Sect. 8.10.4), which is isomorphic to B_2.

8.10.4   The C_n Root System

The C_n root system is associated to the Lie algebra $\mathsf{sp}(n; \mathbb{C})$. For this root system, $E = \mathbb{R}^n$ and the roots are the vectors of the form

$$\pm e_j \pm e_k, \quad j < k$$

and of the form

$$\pm 2e_j, \quad j = 1, \ldots, n.$$

As a base, we may take the $n - 1$ roots

$$e_1 - e_2, e_2 - e_3, \ldots, e_{n-1} - e_n \qquad (8.18)$$

(as in the two preceding subsections), together with the root $2e_n$. We use the same formula for expanding roots in terms of the base as in the case of $\mathsf{so}(2n + 1; \mathbb{C})$, except that (8.17) is rewritten as

$$2e_j = 2(e_j - e_n) + (2e_n).$$

The angle between two consecutive roots in (8.18) is $2\pi/3$; nonconsecutive roots in (8.18) are orthogonal. The angle between $2e_n$ and the last root in (8.18) is $3\pi/4$; the root $2e_n$ is orthogonal to the other roots in (8.18).

8.10.5 The Classical Dynkin Diagrams

From the calculations in the previous subsections, we can read off the Dynkin diagram for the root systems A_n, B_n, C_n, and D_n; the results are recorded in Figure 8.22. We can see that certain special things happen for small values of n. First, the Dynkin diagram for D_n does not make sense when $n = 1$, since the diagram always has at least two vertices. This observation reflects the fact that the Lie algebra $\mathsf{so}(2; \mathbb{C})$ is not semisimple. Second, the Dynkin diagram for D_2 is not connected, which means (Corollary 8.33) that the Lie algebra $\mathsf{so}(4; \mathbb{C})$ is semisimple but not simple. (Compare Exercise 4 in Chapter 7.) Third, the Dynkin diagrams for A_1, B_1, and C_1 are isomorphic, reflecting that the rank-one Lie algebras $\mathsf{sl}(2; \mathbb{C})$, $\mathsf{so}(3; \mathbb{C})$, and $\mathsf{sp}(1; \mathbb{C})$ are isomorphic. Last, we have an isomorphism between the diagrams for B_2 and C_2 and an isomorphism between the diagrams for

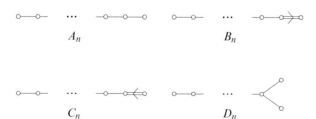

Fig. 8.22 The Dynkin diagrams of the classical Lie algebras

A_3 and D_3, which reflects (Sect. 8.11) an isomorphism between the corresponding Lie algebras.

Corollary 8.47. *The following semisimple Lie algebras are simple: the special linear algebras* $\mathsf{sl}(n + 1; \mathbb{C})$, $n \geq 1$; *the odd orthogonal algebras* $\mathsf{so}(2n + 1; \mathbb{C})$, $n \geq 1$; *the even orthogonal algebras* $\mathsf{so}(2n; \mathbb{C})$, $n \geq 3$; *and the symplectic algebras* $\mathsf{sp}(n; \mathbb{C})$, $n \geq 1$.

Proof. The Dynkin diagrams for A_n, B_n, and C_n are always connected, whereas the Dynkin diagram for D_n is connected for $n \geq 3$. Thus, Corollary 8.33 shows that the claimed Lie algebras are simple.   □

8.11   The Classification

In this section, we describe, without proof, the classification of irreducible root systems and of simple Lie algebras. Recall (Corollary 8.33) that a semisimple Lie algebra is simple if and only if its Dynkin diagram is connected.

Every irreducible root system is either the root system of a classical Lie algebra (types A_n, B_n, C_n, and D_n) or one of five exceptional root systems. We begin by listing the Dynkin diagrams of the exceptional root systems.

Theorem 8.48. *For each of the graphs in Figure 8.23, there exists a root system having that graph as its Dynkin diagram.*

We have already described the root system G_2 in Figure 8.3. Although it is possible to write down the remaining exceptional root systems explicitly, it is not terribly useful to do so, since there is no comparably easy way to construct the Lie algebras associated to these root systems.

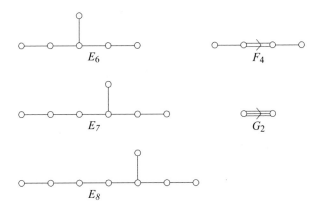

Fig. 8.23 The exceptional Dynkin diagrams

Theorem 8.49. *Every irreducible root system is isomorphic to exactly one of the following:*

- $A_n, n \geq 1$
- $B_n, n \geq 2$
- $C_n, n \geq 3$
- $D_n, n \geq 4$
- *One of the exceptional root systems* G_2, F_4, E_6, E_7, *and* E_8.

The restrictions on n are to avoid the case of D_2, which is not irreducible, and to avoid repetitions. The Dynkin diagram for D_3, for example, is isomorphic to the Dynkin diagram for A_3, which means (Proposition 8.32) that the A_3 and D_3 root systems are also isomorphic. Similarly, all root systems in rank one are isomorphic, and B_2 is isomorphic to C_2.

Theorem 8.50.   *1. If $\mathfrak{g}$ is a complex semisimple Lie algebra and $\mathfrak{h}_1$ and $\mathfrak{h}_2$ are Cartan subalgebra of $\mathfrak{g}$, there exists an automorphism $\phi : \mathfrak{g} \to \mathfrak{g}$ such that $\phi(\mathfrak{h}_1) = \phi(\mathfrak{h}_2)$.*
2. *Suppose $\mathfrak{g}_1$ and $\mathfrak{g}_2$ are semisimple Lie algebras with Cartan subalgebras $\mathfrak{h}_1$ and $\mathfrak{h}_2$, respectively. If the root systems associated to $(\mathfrak{g}_1, \mathfrak{h}_1)$ and $(\mathfrak{g}_2, \mathfrak{h}_2)$ are isomorphic, then $\mathfrak{g}_1$ and $\mathfrak{g}_2$ are isomorphic.*
3. *For every root system R, there exists a semisimple Lie algebra $\mathfrak{g}$ and a Cartan subalgebra $\mathfrak{h}$ of $\mathfrak{g}$ such that the root system of $\mathfrak{g}$ relative to $\mathfrak{h}$ is isomorphic to R.*

Point 1 of the theorem says that there is only one root system associated to each semisimple Lie algebra $\mathfrak{g}$. Since all bases of a fixed root system are equivalent under W, it follows that there is only one Dynkin diagram associated to each $\mathfrak{g}$. Points 2 and 3 then tell us that there is a one-to-one correspondence between isomorphism classes of semisimple Lie algebras and isomorphism classes of root systems. Thus, the classification of irreducible root systems also gives rise to a classification of simple Lie algebras.

For Point 1 of the theorem see Section 16.4 of [Hum] and for Point 2 see Section 14.2 of [Hum]. For Point 3, one can proceed on a case-by-case basis, where the Lie algebras A_n, B_n, C_n, and D_n have already been constructed as classical Lie algebras. The exceptional Lie algebras can then be constructed by special methods, as in [Jac] or [Baez]. Alternatively, one can construct all of the simple Lie algebras by a unified method, as in Section 18 of [Hum].

In particular, the isomorphisms among root systems of small rank translate into isomorphisms among the associated semisimple Lie algebras. In rank one, for example, $\mathsf{sl}(2; \mathbb{C})$, $\mathsf{so}(3; \mathbb{C})$, and $\mathsf{sp}(1; \mathbb{C})$ are isomorphic. In rank two, the isomorphism between B_2 and C_2 reflects an isomorphism of the Lie algebras $\mathsf{so}(5; \mathbb{C})$ and $\mathsf{sp}(2; \mathbb{C})$. In rank three, the isomorphism between A_3 and D_3 reflects an isomorphism of the Lie algebras $\mathsf{sl}(4; \mathbb{C})$ and $\mathsf{so}(6; \mathbb{C})$.

By combining the classification of irreducible root systems in Theorem 8.49 with Proposition 8.32 and Theorem 8.50, we arrive at the following classification of simple Lie algebras over the field of complex numbers.

Theorem 8.51. *Every simple Lie algebra over $\mathbb{C}$ is isomorphic to precisely one algebra from the following list:*

1. $\mathsf{sl}(n+1;\mathbb{C})$, $n \geq 1$
2. $\mathsf{so}(2n+1;\mathbb{C})$, $n \geq 2$
3. $\mathsf{sp}(n;\mathbb{C})$, $n \geq 3$
4. $\mathsf{so}(2n;\mathbb{C})$, $n \geq 4$
5. *The exceptional Lie algebras G_2, F_4, E_6, E_7, and E_8*

A semisimple Lie algebra is then determined up to isomorphism by specifying which simple summands occur and how many times each one occurs; see Proposition 7.9. It is also possible to classify simple Lie algebras over $\mathbb{R}$. As we showed in Sect. 7.6, every such algebra is either a complex simple Lie algebra, viewed as a real Lie algebra, or a real form of a complex simple Lie algebra. Real forms of complex Lie algebras can then be enumerated using the Dynkin diagram of the complex Lie algebra as a starting point. See Section VI.10 and Appendix C of [Kna2] for a description of this enumeration.

8.12   Exercises

Unless otherwise noted, the notation in the exercises is as follows: (E, R) is a root system with Weyl group W, $\Delta = \{\alpha_1, \ldots, \alpha_r\}$ is a fixed base for R, R^+ is the associated set of positive roots, and C is the open fundamental chamber associated to Δ.

1. (a) Suppose that α and β are linearly independent elements of R and that for some positive integer k, the vector $\alpha + k\beta$ belongs to R. Show that $\alpha + l\beta$ also belongs to R for all integers l with $0 < l < k$.
   Hint: If $F = \text{span}(\alpha, \beta)$, then $R \cap F$ is a rank-two root system in F.
   (b) A collection of roots of the form $\alpha, \alpha + \beta, \ldots, \alpha + k\beta$ for which neither $\alpha - \beta$ nor $\alpha + (k+1)\beta$ is a root is called a **root string**. What is the maximum number of roots that can occur in a root string?
2. Let E be a finite-dimensional real inner product space and let $V_1, \ldots, V_k$ be subspaces of E of codimension at least one. Show that the union of the V_k's is not all of E.
   Hint: Show by induction on k that the complement of the union the V_k's is a *nonempty* open subset of E.
3. Let E be a finite-dimensional real inner product space, let $\{v_1, \ldots, v_r\}$ be basis for E, and let $\{v_1^*, \ldots, v_r^*\}$ the dual basis, satisfying $\langle v_j^*, v_k \rangle = \delta_{jk}$ for all j and k. Let G and H be the Gram matrices for these two bases: $G_{jk} = \langle v_j, v_k \rangle$ and $H_{jk} = \langle v_j^*, v_k^* \rangle$. Show that G and H are inverses of each other.

Hint: First show that for any $u \in E$, we have $u = \sum_j \langle v_j^*, u \rangle v_j$. Then apply this result to the vector $u = v_k^*$.

4. Show that Lemma 8.26 fails (even with $u = e$) if β is not assumed to be an element of Δ_C.

5. Show that if R is an irreducible root system, then W acts irreducibly on E.
   Hint: Suppose $V \subset E$ is a W-invariant subspace. Show that every element of R is either in V or in the orthogonal complement of E.

6. Let (E, R) be an irreducible root system and let $\langle \cdot, \cdot \rangle$ be the inner product on E. Using Exercise 5, show that if $\langle \cdot, \cdot \rangle_1$ is a W-invariant inner product on E, there is some constant c such that $\langle H, H' \rangle_1 = c \langle H, H' \rangle$ for all $H, H' \in E$.
   Hint: Consider the unique linear operator $A : E \to E$ that is symmetric with respect to $\langle \cdot, \cdot \rangle$ and that satisfies

$$\langle H, H' \rangle_1 = \langle H, AH' \rangle$$

for all $H, H' \in E$. Then imitate the proof of Schur's lemma, noting that the eigenvalues of A are real.

7. Suppose (E, R) and (F, S) are irreducible root systems and that $A : E \to F$ is an isomorphism of R with S. Show that A is a constant multiple of an isometry.

8. Using the outline below, prove the following result: For all $\mu, \lambda \in E$, we have $\mu \succeq \lambda$ if and only if $\langle \mu - \lambda, H \rangle \geq 0$ for all $H \in C$.

   (a) Show that if $\mu \succeq \lambda$, then $\langle \mu - \lambda, H \rangle \geq 0$ for all $H \in C$.
   (b) Let $\{\alpha_1^*, \ldots, \alpha_r^*\}$ be the dual basis to Δ, satisfying $\langle \alpha_j^*, \alpha_k \rangle = \delta_{jk}$ for all j and k. Show that if $\langle \gamma, H \rangle \geq 0$ for all $H \in C$, then $\langle \gamma, \alpha_j^* \rangle \geq 0$ for all j.
   (c) Show that if $\langle \mu - \lambda, H \rangle \geq 0$ for all $H \in C$, then $\mu \succeq \lambda$.

9. Let $P : E \to \mathbb{R}$ be the function given by

$$P(H) = \prod_{\alpha \in R^+} \langle \alpha, H \rangle .$$

   Show that P satisfies

$$P(w \cdot H) = \det(w) P(H)$$

   for all $w \in W$ and all $H \in E$.

10. Show that if $-I$ is not in the Weyl of R, the Dynkin diagram of R must have a nontrivial automorphism.
    Hint: By Proposition 8.23, there exists an element w of W mapping $-C$ to C. Consider the map $H \mapsto -w \cdot H$, which maps C to itself.

11. For which rank-two root systems is $-I$ an element of the Weyl group?

12. Show that the Weyl group of the A_n root system, described in Sect. 7.7.1, does not contain $-I$, except when $n = 1$.

13. Let $E = \mathbb{R}^n$ and let R denote the collection of vectors of the following three forms:

$$\pm e_j \pm e_k \quad j < k$$
$$\pm e_j \quad j = 1, \ldots, n .$$
$$\pm 2e_j \quad j = 1, \ldots, n$$

Show that R satisfies all the properties of a root system in Definition 8.1 except Condition 2. The collection R is a "nonreduced root system" and is known as BC_n, since it is the union of B_n and C_n. (Compare Figure 7.1 in the $n = 2$ case.)

14. Determine which of the Dynkin diagrams in Figures 8.22 and 8.23 have a nontrivial automorphism. Show that only the Dynkin diagram of D_4 has an automorphism group with more than two elements.

Chapter 9
Representations of Semisimple Lie Algebras

In this chapter, we prove the theorem of the highest weight for irreducible, finite-dimensional representations of a complex semisimple Lie algebra $\mathfrak{g}$. We first prove that every such representation has a highest weight, that two irreducible representations with the same highest weight are isomorphic, and that the highest weight of an irreducible representation must be dominant integral. This part of the theorem is established in precisely the same way as in the case of $\mathsf{sl}(3;\mathbb{C})$ in Chapter 6. It then remains to prove that *every* dominant integral element is, in fact, the highest weight of some irreducible representation. In the $\mathsf{sl}(3;\mathbb{C})$ case, we did this by first constructing the representations whose highest weights were the fundamental weights $(1,0)$ and $(0,1)$, and then taking tensor products of these representations. For a general semisimple Lie algebra $\mathfrak{g}$, however, there is no simple way to construct the representations whose highest weights are the fundamental weights in Definition 8.36. Thus, we require a new method of constructing the irreducible representation of $\mathfrak{g}$ with a given dominant integral highest weight. This construction will be the main topic of the present chapter.

In Chapter 10, we will derive several additional properties of the irreducible representations, including the structure of the set of weights, the multiplicities of the weights, and the dimensions of the representations. In that chapter, we will also prove *complete reducibility* for representations of $\mathfrak{g}$, that is, that every finite-dimensional representation of $\mathfrak{g}$ decomposes as a direct sum of irreducibles.

A previous version of this book was inadvertently published without the middle initial of the author's name as "Brian Hall". For this reason an erratum has been published, correcting the mistake in the previous version and showing the correct name as Brian C. Hall (see DOI http://dx.doi.org/10.1007/978-3-319-13467-3_14). The version readers currently see is the corrected version. The Publisher would like to apologize for the earlier mistake.

© Springer International Publishing Switzerland 2015
B.C. Hall, *Lie Groups, Lie Algebras, and Representations*, Graduate
Texts in Mathematics 222, DOI 10.1007/978-3-319-13467-3_9

9.1   Weights of Representations

Throughout the chapter, we assume that $\mathfrak{g} = \mathfrak{k}_\mathbb{C}$ is a complex semisimple Lie algebra and that $\mathfrak{h} = \mathfrak{t}_\mathbb{C}$ is a fixed Cartan subalgebra of $\mathfrak{g}$ (compare Proposition 7.11). We fix on $\mathfrak{g}$ an inner product that is real on $\mathfrak{k}$ and that is invariant under the adjoint action of $\mathfrak{k}$, as in Proposition 7.4. We let $R \subset i\mathfrak{t}$ denote the set of roots of $\mathfrak{g}$ relative to $\mathfrak{h}$, we let Δ be a fixed base for R, and we let R^+ and R^- be the set of positive and negative roots relative to Δ, respectively. For each root α, we consider the coroot $H_\alpha \in \mathfrak{h}$ given by

$$H_\alpha = 2\frac{\alpha}{\langle \alpha, \alpha \rangle}.$$

We also consider the Weyl group W, that is, the group of linear transformations of $\mathfrak{h}$ generated by the reflections about the hyperplanes orthogonal to the roots. Finally, we consider the notions of integral and dominant integral elements, as in Definition 8.34.

We now introduce the notion of a weight of a representation, as in the $\mathsf{sl}(3; \mathbb{C})$ case.

Definition 9.1. Let (π, V) be a representation of $\mathfrak{g}$, possibly infinite dimensional. An element λ of $\mathfrak{h}$ is a **weight** of π if there exists a nonzero vector $v \in V$ such that

$$\pi(H)v = \langle \lambda, H \rangle v \qquad\qquad (9.1)$$

for all $H \in \mathfrak{h}$. The **weight space** corresponding to λ is the set of all $v \in V$ satisfying (9.1) and the **multiplicity** of λ is the dimension of the corresponding weight space.

Throughout the chapter, we will use, without comment, Proposition A.17, which says that weight vectors with distinct weights are linearly independent.

Proposition 9.2. *If (π, V) is a* finite-dimensional *representation of $\mathfrak{g}$, every weight of π is an integral element.*

Proof. For each root α, let $\mathfrak{s}^\alpha = \langle X_\alpha, Y_\alpha, H_\alpha \rangle \cong \mathsf{sl}(2; \mathbb{C})$ be the subalgebra of $\mathfrak{g}$ in Theorem 7.19. If v is a weight vector with weight λ, then

$$\pi(H_\alpha)v = \langle \lambda, H_\alpha \rangle v.$$

Thus, by applying Point 1 of Theorem 4.34 to the restriction of π to $\mathfrak{s}^\alpha$, we see that $\langle \lambda, H_\alpha \rangle$ must be an integer, showing that λ is integral. $\qquad\qquad\qquad\square$

Theorem 9.3. *If (π, V) is a finite-dimensional representation of $\mathfrak{g}$, the weights of π and their multiplicities are invariant under the action of W on H.*

Proof. For each $\alpha \in R$, we may construct the operator

$$S_\alpha := e^{\pi(X_\alpha)}e^{-\pi(Y_\alpha)}e^{\pi(X_\alpha)}.$$

If $\langle \alpha, H \rangle = 0$, then H will commute with both X_α and Y_α and thus with S_α. On the other hand, by Point 3 of Theorem 4.34, we have $S_\alpha \pi(H_\alpha) S_\alpha^{-1} = -\pi(H_\alpha)$. We see, then, that

$$S_\alpha \pi(H) S_\alpha^{-1} = \pi(s_\alpha \cdot H)$$

for all $H \in \mathfrak{h}$.

Suppose now that v is a weight vector with some weight λ. Then

$$\pi(H) S_\alpha^{-1} v = S_\alpha^{-1} \pi(s_\alpha \cdot H) v$$
$$= S_\alpha^{-1} \langle \lambda, s_\alpha \cdot H \rangle \, v$$
$$= \langle s_\alpha^{-1} \cdot \lambda, H \rangle S_\alpha^{-1} v,$$

showing that $S_\alpha^{-1} v$ is a weight vector with weight $s_\alpha^{-1} \cdot \lambda$. Thus, S_α^{-1} maps the weight space with weight λ into the weight space with weight $s_\alpha^{-1} \cdot \lambda$. Meanwhile, essentially the same argument shows that S_α maps the weight space with weight $s_\alpha^{-1} \cdot \lambda$ into the weight space with weight λ, showing that the two spaces are isomorphic. Thus, $s_\alpha^{-1} \cdot \lambda$ is again a weight with the same multiplicity as λ. Thus, the weights and multiplicities are invariant under each $s_\alpha^{-1} = s_\alpha$ and, thus, under W. $\square$

We now state the "easy" part of the theorem of the highest weight for representations of $\mathfrak{g}$.

Theorem 9.4. *1. Every irreducible, finite-dimensional representation of $\mathfrak{g}$ has a highest weight.*
2. Two irreducible, finite-dimensional representations of $\mathfrak{g}$ with the same highest weight are isomorphic.
3. If (π, V) is an irreducible, finite-dimensional representation of $\mathfrak{g}$ with highest weight μ, then μ is dominant integral.

Proof. Enumerate the positive roots as $\alpha_1, \ldots, \alpha_N$. Choose a basis for $\mathfrak{g}$ consisting of elements $X_1, \ldots, X_N$ with $X_j \in \mathfrak{g}_{\alpha_j}$, elements $Y_1, \ldots, Y_N$ with $Y_j \in \mathfrak{g}_{-\alpha_j}$, and a basis $H_1, \ldots, H_r$ for $\mathfrak{h}$. Then the proof of Proposition 6.11 from Chapter 6 carries over to the present setting, with only the obvious notational changes, showing that every irreducible, finite-dimensional representation of $\mathfrak{g}$ has a highest weight.

The proofs of Propositions 6.14 and 6.15 then also go through without change to show that two irreducible, finite-dimensional representations with the same highest weight are isomorphic. Finally, using the $\mathsf{sl}(2;\mathbb{C})$-subalgebras in Theorem 7.19, we may follow the proof of Proposition 6.16 to show that the highest weight of a finite-dimensional, irreducible representation must be dominant integral. $\square$

We now come to the "hard" part of the theorem of the highest weight.

Theorem 9.5. *If μ is a dominant integral element, there exists an irreducible, finite-dimensional representation of $\mathfrak{g}$ with highest weight μ.*

As we have noted, the method of proof of Proposition 6.17, from the $\mathsf{sl}(3;\mathbb{C})$ case, does not readily extend to general semisimple Lie algebras. The proof of Theorem 9.5 will occupy the remainder of this chapter.

9.2   Introduction to Verma Modules

Our goal is to construct, for each dominant integral element $\mu \in \mathfrak{h}$, a finite-dimensional, irreducible representation of $\mathfrak{g}$ with highest weight μ. Our construction will proceed in two stages. The first stage consists of constructing an *infinite-dimensional* representation V_μ of $\mathfrak{g}$, known as a Verma module. This representation will not be irreducible, but will be a highest weight cyclic representation with highest weight μ. We will construct V_μ as a quotient of the so-called *universal enveloping algebra* $U(\mathfrak{g})$ of $\mathfrak{g}$. In order to show that the highest weight vector in V_μ is nonzero, we will need to develop a structure result for $U(\mathfrak{g})$ known as the Poincaré–Birkhoff–Witt theorem (Theorem 9.10). Unlike the finite-dimensional representations of $\mathfrak{g}$, the weights of the Verma module are *not* invariant under the action of the Weyl group.

The second stage in our construction consists of showing that when μ is dominant integral, V_μ has an invariant subspace W_μ for which the quotient space V_μ / W_μ is finite dimensional and irreducible, but not zero. To establish the finite dimensionality of the quotient, we will show that when μ is dominant integral, the weights of V_μ / W_μ, unlike those of V_μ, are invariant under the action of the Weyl group. Thus, each weight λ of V_μ / W_μ is integral and satisfies $w \cdot \lambda \preceq \mu$ for all w in the Weyl group. It turns out that there are only finitely many λ's with this property. Since each weight λ has finite multiplicity (even in V_μ), it follows that V_μ / W_μ is finite dimensional.

Definition 9.6. A (possibly infinite-dimensional) representation (π, V) of $\mathfrak{g}$ is **highest weight cyclic** with highest weight $\mu \in \mathfrak{h}$ if there exists a nonzero vector $v \in V$ such that (1) $\pi(H)v = \langle \mu, H \rangle v$ for all $H \in \mathfrak{h}$, (2) $\pi(X)v = 0$ for all $X \in \mathfrak{g}_\alpha$ with $\alpha \in R^+$, (3) the smallest invariant subspace containing v is V.

Note that $\mu \in \mathfrak{h}$ is not required to be integral. Although it will turn out that all *finite-dimensional* highest weight cyclic representations are irreducible, this is not the case in infinite dimensions. Furthermore, two highest weight cyclic representations with the same highest weight may not be isomorphic, unless both of them are finite dimensional. In this chapter, we will construct, for any $\mu \in \mathfrak{h}$, a particular highest weight cyclic representation V_μ with highest weight μ, known as a Verma module. The Verma module is the "maximal" highest weight cyclic representation with a particular highest weight, and it is always infinite dimensional, even when μ is dominant integral.

In the $\mathsf{sl}(2; \mathbb{C})$ case, Verma modules can be constructed explicitly as follows. For any complex number μ, construct an infinite-dimensional vector space V_μ with basis $v_0, v_1, \ldots$. (The elements of V_μ are *finite* linear combinations of the v_j's.) We define an action of $\mathsf{sl}(2; \mathbb{C})$ on V_μ by the same formulas as in Sect. 4.6:

$$\pi_\mu(Y)v_j = v_{j+1}$$
$$\pi_\mu(H)v_j = (\mu - 2j)v_j$$

$$\pi_\mu(X)v_0 = 0$$

$$\pi_\mu(X)v_j = j(\mu - (j-1))v_{j-1}.$$

Note that in Sect. 4.6, the vectors v_j equaled zero for large j, whereas here the v_j's are, by definition, linearly independent. Direct calculation shows that these formulas do, in fact, define a representation of $\mathsf{sl}(2;\mathbb{C})$.

When μ is a non-negative integer m, the space W_μ spanned by $v_{m+1}, v_{m+2}, \ldots$ is invariant under the action of $\mathsf{sl}(2;\mathbb{C})$. After all, this space is clearly invariant under the action of $\pi_\mu(Y)$ and $\pi_\mu(H)$, and it is invariant under $\pi_\mu(X)$ because (with $\mu = m$) we have

$$\pi_\mu(X)v_{m+1} = k(m-m)v_m = 0.$$

Since W_μ is invariant, the quotient vector space V_μ/W_μ inherits a natural action of $\mathsf{sl}(2;\mathbb{C})$. This quotient space is then the unique finite-dimensional irreducible representation of $\mathsf{sl}(2;\mathbb{C})$ with highest weight μ.

In the case of general semisimple Lie algebra $\mathfrak{g}$, we would like to do something similar. Pick an basis consisting of vectors

$$Y_1, \ldots, Y_N, \ H_1, \ldots, H_r, \ X_1, \ldots, X_N, \tag{9.2}$$

as in the proof of Theorem 9.4. If V_μ is any highest weight cyclic representation with highest weight μ and highest weight vector v_0, then V_μ is spanned by products of the basis elements applied to v_0. By the reordering lemma (Lemma 6.12), we can reorder any such product as a linear combination of terms in which the elements are in the order listed in (9.2). Once the basis elements are in this order, any term that contains any X_j's will give zero when applied to v_0. Furthermore, in any term that does not have any X_j's, any factors of H_k will simply give $\mu(H_j)$ when applied to v_0. Thus, V_μ must be spanned by elements of the form

$$\pi_\mu(Y_1)^{n_1} \pi_\mu(Y_2)^{n_2} \cdots \pi_\mu(Y_N)^{n_N} v_0. \tag{9.3}$$

The idea of a Verma module is that we should proceed as in the $\mathsf{sl}(2;\mathbb{C})$ case and simply decree that the vectors in (9.3) form a basis for our Verma module. The weights of the Verma module will the consist of all elements of the form

$$\mu - n_1\alpha_1 - \cdots - n_N\alpha_N,$$

where each n_j is a non-negative integer. (See Figure 9.1.) If we do this, then there is only one possible way that the Lie algebra $\mathfrak{g}$ can act. After all, if we apply some Lie algebra element $\pi(Z)$ to a vector as in (9.3), we can reorder the elements as in the previous paragraph until they are in the order of (9.2). Then, as we have already noted, any factors of $\pi_\mu(X_j)$ give zero and any factors of $\pi_\mu(H_k)$ give constants. We will, thus, eventually get back a linear combination of elements of the form (9.3).

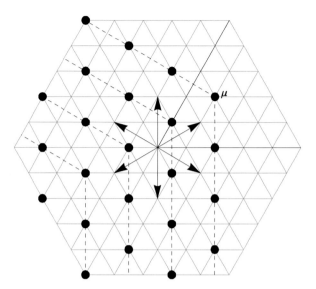

Fig. 9.1 The weights of the Verma module with highest weight μ

The difficulty with the above description of the Verma module is that it does not provide any reasonable method for checking that π_μ *actually constitutes a representation* of $\mathfrak{g}$. After all, unless $\mathfrak{g} = \mathsf{sl}(2; \mathbb{C})$, it is impossible to write down an *explicit* description of how the various basis elements act, and it is thus impossible to verify directly that these elements satisfy the correct commutation relations. Nevertheless, we will eventually prove (Sect. 9.5) that there is a well-defined representation of $\mathfrak{g}$ having the elements (9.3) as a basis and in which $\mathfrak{g}$ acts in the way described above. In the case in which μ is dominant and integral, we will then construct an invariant subspace of the Verma module for which the quotient is finite dimensional and irreducible.

9.3   Universal Enveloping Algebras

In Sect. 9.5, we will construct each Verma module V_μ as a quotient of something called the *universal enveloping algebra* of a Lie algebra $\mathfrak{g}$. If $\mathfrak{g}$ is a Lie algebra, we may try to embed $\mathfrak{g}$ as a subspace of some associative algebra $\mathcal{A}$ in such a way that the bracket on $\mathfrak{g}$ may be computed as $[X, Y] = XY - YX$, where XY and YX are computed in $\mathcal{A}$. If $\mathfrak{g}$ is the Lie algebra of a matrix Lie group $G \subset \mathsf{Gl}(n; \mathbb{C})$, then $\mathfrak{g}$ is a subspace of the associative algebra $M_n(\mathbb{C})$ and the bracket on $\mathfrak{g}$ is indeed given as $[X, Y] = XY - YX$. There may be, however, many other ways to embed $\mathfrak{g}$ into an associative algebra $\mathcal{A}$. For example, if $\mathfrak{g} = \mathsf{sl}(2; \mathbb{C})$, then for each $m \geq 1$, the irreducible representation π_m of $\mathfrak{g}$ of dimension $m + 1$ allows us to embed $\mathfrak{g}$ into $M_{m+1}(\mathbb{C})$.

Let us now give a useful but slightly imprecise definition of the universal enveloping algebra of $\mathfrak{g}$, denoted $U(\mathfrak{g})$ For any Lie algebra $\mathfrak{g}$, the universal enveloping algebra of $\mathfrak{g}$ will be an associative algebra $\mathcal{A}$ with identity with the following properties. (1) The Lie algebra $\mathfrak{g}$ embeds into $\mathcal{A}$ in such a way that $[X, Y] = XY - YX$. (2) The algebra $\mathcal{A}$ is generated by elements of $\mathfrak{g}$, meaning that the smallest subalgebra with identity of $\mathcal{A}$ containing $\mathfrak{g}$ is all of $\mathcal{A}$. (3) The algebra $\mathcal{A}$ is *maximal* among all algebras $\mathcal{A}$ with the two previous properties. The maximality property of $\mathcal{A}$ will be explained more precisely in the discussion following Theorem 9.7.

Consider, for example, the case of a one-dimensional Lie algebra $\mathfrak{g}$ spanned by a single nonzero element X, which of course satisfies $[X, X] = 0$. Then $U(\mathfrak{g})$ should be an associative algebra with identity generated a single element X, in which case, $U(\mathfrak{g})$ must also be commutative. Now, *any* associative algebra $\mathcal{A}$ with identity generated by a single nonzero element X will satisfy Properties 1 and 2 in the definition of the enveloping algebra. But for $\mathcal{A}$ to be maximal, there should be no relations between the different powers of X, meaning that $p(X)$ should be a nonzero element of $\mathcal{A}$ for every nonzero polynomial p. In this case, then, we may take $U(\mathfrak{g})$ to be the algebra of polynomials in a single variable.

Suppose the algebra $\mathfrak{g}$ in the previous paragraph is a matrix algebra, meaning that X is a single $n \times n$ matrix. Contrary to what we might, at first, expect, the enveloping algebra $U(\mathfrak{g})$ will *not* coincide with the associative algebra with identity $\mathcal{A}$ generated by X inside $M_n(\mathbb{C})$. After all, for any $X \in M_n(\mathbb{C})$, the Cayley–Hamilton theorem implies that there exists a nonzero polynomial p (namely, the characteristic polynomial of X) for which $p(X) = 0$. In $U(\mathfrak{g})$, by contrast, we have said that $p(X)$ should be nonzero for all nonzero polynomials p.

Actually, it follows from the PBW theorem (Theorem 9.10) that the universal enveloping algebra of *any* nonzero Lie algebra $\mathfrak{g}$ is infinite dimensional. In fact, if X is any nonzero element of $\mathfrak{g}$, the elements $1, X, X^2, \ldots$ will be linearly independent in $U(\mathfrak{g})$. Thus, even if $\mathfrak{g}$ is an algebra of matrices, $U(\mathfrak{g})$ cannot be isomorphic to a subalgebra of an algebra of matrices.

We now give the formal definition of a universal enveloping algebra.

Theorem 9.7. *For any Lie algebra $\mathfrak{g}$, there exists an associative algebra with identity, denoted $U(\mathfrak{g})$, together with a linear map $i : \mathfrak{g} \to U(\mathfrak{g})$ such that the following properties hold. (1) For all $X, Y \in \mathfrak{g}$, we have*

$$i([X, Y]) = i(X)i(Y) - i(Y)i(X). \tag{9.4}$$

(2) The algebra $U(\mathfrak{g})$ is generated by elements of the form $i(X)$, $X \in \mathfrak{g}$, meaning that the smallest subalgebra with identity of $U(\mathfrak{g})$ containing every $i(X)$ is $U(\mathfrak{g})$. (3) Suppose $\mathcal{A}$ is an associative algebra with identity and $j : \mathfrak{g} \to \mathcal{A}$ is a linear map such that $j([X, Y])$ coincides with $j(X)j(Y) - j(Y)j(X)$ for all $X, Y \in \mathfrak{g}$. Then there exists a unique algebra homomorphism $\phi : U(\mathfrak{g}) \to \mathcal{A}$ such that $\phi(1) = 1$ and such that $\phi(i(X)) = j(X)$ for all $X \in \mathfrak{g}$.

*A pair $(U(\mathfrak{g}), i)$ with the preceding properties is called a **universal enveloping algebra** for $\mathfrak{g}$.*

A simple argument (Exercise 1) shows that any two universal enveloping algebras for a fixed Lie algebra $\mathfrak{g}$ are "canonically" isomorphic.

Let us define *an* enveloping algebra of $\mathfrak{g}$ to be an associative algebra $\mathcal{A}$ together with a linear map $j : \mathfrak{g} \rightarrow \mathcal{A}$ as in Point 3 of the theorem, with the additional property that $\mathcal{A}$ is generated by elements of the form $j(X)$, $X \in \mathfrak{g}$. In this case, the homomorphism $\phi : U(\mathfrak{g}) \rightarrow \mathcal{A}$ as in Theorem 9.7 is surjective and $\mathcal{A}$ is isomorphic to the quotient algebra $U(\mathfrak{g})/\ker(\phi)$. Thus, the *universal* enveloping algebra $U(\mathfrak{g})$ of $\mathfrak{g}$ has the property that every other enveloping algebra of $\mathfrak{g}$ is a quotient of $U(\mathfrak{g})$. This property of $U(\mathfrak{g})$ is a more precise formulation of the maximality condition we discussed in the second paragraph of this subsection.

Example 9.8. Let $\mathfrak{g} = \mathsf{sl}(2; \mathbb{C})$ with the usual basis $\{X, Y, H\}$. The universal enveloping algebra of $\mathfrak{g}$ is then the associative algebra with identity generated by three elements x, y, and h, subject to the relations

$$hx - xh = 2x$$

$$hy - yh = -2y$$

$$xy - yx = h,$$

and *no other relations*. The map $i : \mathfrak{g} \rightarrow U(\mathfrak{g})$ is then the unique linear map such that $i(X) = x$, $i(Y) = y$, and $i(H) = h$ and $\phi : U(\mathfrak{g}) \rightarrow \mathcal{A}$ is the unique homomorphism such that $\phi(x) = j(X)$, $\phi(y) = j(Y)$, and $\phi(h) = j(H)$.

The meaning of the phrase "no other relations" will be made more precise in the proof of Theorem 9.7, in which J is the *smallest* two-sided ideal containing all elements of the form $X \otimes Y - Y \otimes X - [X, Y]$. The fact that $U(\mathfrak{g})$ has no other relations guarantees that the homomorphism ϕ is well defined.

The construction of $U(\mathfrak{g})$ is in some sense easy or "soft." But for $U(\mathfrak{g})$ to be useful in practice (for example, in constructing Verma modules), we need a structure theorem for it known as the Poincaré–Birkhoff–Witt theorem. The Poincaré–Birkhoff–Witt theorem will, in particular, show that the map i in Theorem 9.7 is actually injective. (See also Exercise 2.) Once this is established, we will be able to identify $\mathfrak{g}$ with its image under i and thus think of $\mathfrak{g}$ as embedded into $U(\mathfrak{g})$.

Proof of Theorem 9.7. The operation of tensor product on vector spaces is associative, in the sense that $U \otimes (V \otimes W)$ is canonically isomorphic to $(U \otimes V) \otimes W$, with the isomorphism taking $u \otimes (v \otimes w)$ to $(u \otimes v) \otimes w$ for each $u \in U$, $v \in V$, and $w \in W$. We may thus drop the parentheses and write simply $U \otimes V \otimes W$ and $u \otimes v \otimes w$, and similarly for the tensor product of any finite number of vector spaces. In particular, we will let $V^{\otimes k}$ denote the k-fold tensor product $V \otimes \cdots \otimes V$. The 0-fold tensor product $V^{\otimes 0}$ is defined to be $\mathbb{C}$ (or whatever field we are working over).

For a Lie algebra $\mathfrak{g}$, let us first define the **tensor algebra** $T(\mathfrak{g})$ over $\mathfrak{g}$, which is defined as

$$T(\mathfrak{g}) = \bigoplus_{k=0}^{\infty} \mathfrak{g}^{\otimes k}.$$

In the direct sum, each element of $T(\mathfrak{g})$ is required to be a *finite* linear combinations of elements $\mathfrak{g}^{\otimes k}$ for different values of k. We can make $T(\mathfrak{g})$ into an associative algebra with identity by defining

$$(u_1 \otimes u_2 \otimes \cdots \otimes u_k) \cdot (v_1 \otimes v_2 \otimes \cdots \otimes v_l)$$

$$= u_1 \otimes u_2 \otimes \cdots \otimes u_k \otimes v_1 \otimes v_2 \otimes \cdots \otimes v_l \qquad (9.5)$$

and then extending the product by linearity. That is to say, the product operation is the unique bilinear map of $T(\mathfrak{g}) \times T(\mathfrak{g})$ into $T(\mathfrak{g})$ that coincides with the tensor product (9.5) on $\mathfrak{g}^{\otimes k} \times \mathfrak{g}^{\otimes l}$. Since $\mathbb{C} \otimes \mathfrak{g}^{\otimes k}$ is naturally isomorphic to $\mathfrak{g}^{\otimes k}$, the identity element $1 \in \mathbb{C} = \mathfrak{g}^{\otimes 0}$ is the multiplicative identity for $T(\mathfrak{g})$. The associativity of the tensor product assures that $T(\mathfrak{g})$ is an associative algebra.

We now claim that the algebra $T(\mathfrak{g})$ has the following property: If $\mathcal{A}$ is any associative algebra with identity and $j : \mathfrak{g} \to \mathcal{A}$ is any linear map, there exists an algebra homomorphism $\psi : T(\mathfrak{g}) \to \mathcal{A}$ such that $\psi(1) = 1$ and $\psi(X) = j(X)$ for all $X \in \mathfrak{g} \subset T(\mathfrak{g})$. Note that this property differs from the desired property of $U(\mathfrak{g})$ in that j is an arbitrary linear map and does not have to have any particular relationship to the algebra structure of $\mathcal{A}$. To construct ψ, we require that the restriction of ψ to $\mathfrak{g}^{\otimes k}$ to be the unique linear map of $\mathfrak{g}^{\otimes k}$ into $\mathcal{A}$ such that

$$\psi(X_1 \otimes \cdots \otimes X_k) = j(X_1) \cdots j(X_k) \qquad (9.6)$$

for all $X_1, \ldots, X_k$ in $\mathfrak{g}$. (Here we are using the natural k-fold extension of the universal property of tensor products in Definition 4.13.) It is then simple to check that ψ is an algebra homomorphism. Furthermore, if ψ is to be an algebra homomorphism that agrees with j on $\mathfrak{g}$, then ψ *must* have the form in (9.6).

We now proceed to construct $U(\mathfrak{g})$ as a quotient of $T(\mathfrak{g})$. A **two-sided ideal** in $T(\mathfrak{g})$ is a subspace J of $T(\mathfrak{g})$ such that for all $\alpha \in T(\mathfrak{g})$ and $\beta \in J$, the elements $\alpha\beta$ and $\beta\alpha$ belong to J. We now let J be the smallest two-sided ideal in $T(\mathfrak{g})$ containing all elements of the form

$$X \otimes Y - Y \otimes X - [X, Y], \quad X, Y \in \mathfrak{g}. \qquad (9.7)$$

That is to say, J is the intersection of all two-sided ideals in $T(\mathfrak{g})$ containing all such elements, which is, again, a two-sided ideal containing these elements. More concretely, J can be constructed as the space of elements of the form

$$\sum_{j=1}^{N} \alpha_j (X_j \otimes Y_j - Y_j \otimes X_j - [X_j, Y_j])\beta_j,$$

with X_j and Y_j in $\mathfrak{g}$ and α_j and β_j being arbitrary elements of $T(\mathfrak{g})$.

We now form the quotient vector space $T(\mathfrak{g})/J$, which is an algebra. If $j : \mathfrak{g} \to \mathcal{A}$ is any linear map, we can form the algebra homomorphism $\psi : T(\mathfrak{g}) \to \mathcal{A}$ as above. If j satisfies $j([X, Y]) = j(X)j(Y) - j(Y)j(X)$, then the kernel of ψ will contain all elements of the form $X \otimes Y - Y \otimes X - [X, Y]$.

Furthermore, the kernel of an algebra homomorphism is always a two-sided ideal. Thus, $\ker(\psi)$ contains J. It follows that the map $\psi : T(\mathfrak{g}) \to \mathcal{A}$ factors through $U(\mathfrak{g}) := T(\mathfrak{g})/J$, giving the desired homomorphism ϕ of $U(\mathfrak{g})$ into $\mathcal{A}$. Since $U(\mathfrak{g})$ is spanned by products of elements of $\mathfrak{g}$, there can be at most one map ϕ with the desired property, establishing the claimed uniqueness of ϕ. $\square$

Proposition 9.9. *If* $\pi : \mathfrak{g} \to \mathrm{End}(V)$ *is a representation of a Lie algebra* $\mathfrak{g}$ *(not necessarily finite dimensional), there is a unique algebra homomorphism* $\tilde{\pi} : U(\mathfrak{g}) \to \mathrm{End}(V)$ *such that* $\tilde{\pi}(1) = I$ *and* $\tilde{\pi}(X) = \pi(X)$ *for all* $X \in \mathfrak{g} \subset U(\mathfrak{g})$.

Proof. Apply Theorem 9.7 with $\mathcal{A} = \mathrm{End}(V)$ and $j(X) = \pi(X)$. $\square$

We now state the Poincaré–Birkhoff–Witt theorem—or PBW theorem, for short—which is the key structure result for universal enveloping algebras. Although the result holds even for infinite-dimensional Lie algebras, we state it here in the finite-dimensional case, for notational simplicity. The proof of the PBW theorem is in Sect. 9.4.

Theorem 9.10 (PBW Theorem). *If* $\mathfrak{g}$ *is a finite-dimensional Lie algebra with basis* $X_1, \ldots, X_k$, *then elements of the form*

$$i(X_1)^{n_1} i(X_2)^{n_2} \cdots i(X_k)^{n_k}, \tag{9.8}$$

where each n_k *is a non-negative integer, span* $U(\mathfrak{g})$ *and are linearly independent. In particular, the elements* $i(X_1), \ldots, i(X_k)$ *are linearly independent, meaning that the map* $i : \mathfrak{g} \to U(\mathfrak{g})$ *is injective.*

In (9.8), we interpret $i(X_j)^{n_j}$ as 1 if $n_j = 0$. Since, actually, i is injective, we will henceforth identify $\mathfrak{g}$ with its image under i and thus regard $\mathfrak{g}$ as a subspace of $U(\mathfrak{g})$. Thus, we may now write X in place of $i(X)$. In our new notation, we may write (9.4) as

$$[X, Y] = XY - YX$$

and we may write the basis elements (9.8) as

$$X_1^{n_1} X_2^{n_2} \cdots X_k^{n_k}. \tag{9.9}$$

It is straightforward to show that the elements in (9.8) span $U(\mathfrak{g})$; the hard part is to prove they are linearly independent.

Corollary 9.11. *If* $\mathfrak{g}$ *is a Lie algebra and* $\mathfrak{h}$ *is a subalgebra of* $\mathfrak{g}$, *then there is a natural injection of* $U(\mathfrak{h})$ *into* $U(\mathfrak{g})$ *given by mapping any product* $X_1 X_2 \cdots X_N$ *of elements of* $\mathfrak{h}$ *to the same product in* $U(\mathfrak{g})$.

Proof. The inclusion of $\mathfrak{h}$ into $\mathfrak{g}$ induces an algebra homomorphism of $\phi : U(\mathfrak{h}) \to U(\mathfrak{g})$. Let us choose a basis $X_1, \ldots, X_k$ for $\mathfrak{h}$ and extend it to a basis $X_1, \ldots, X_N$ for $\mathfrak{g}$. By the PBW theorem for $\mathfrak{h}$, the elements $X_1^{n_1} \cdots X_k^{n_k}$ form a basis for $U(\mathfrak{h})$. Then by the PBW theorem for $\mathfrak{g}$, the corresponding elements of $U(\mathfrak{g})$ are linearly independent, showing that ϕ is injective. $\square$

9.4   Proof of the PBW Theorem

It is notationally convenient to write the elements of the claimed basis for $U(\mathfrak{g})$ as

$$i(X_{j_1})i(X_{j_2})\cdots i(X_{j_N}), \tag{9.10}$$

with $j_1 \leq j_2 \leq \cdots \leq j_N$, where we interpret the above expression as 1 if $N = 0$. The easy part of the PBW theorem is to show that these elements span $U(\mathfrak{g})$. The proof of this claim is essentially the same as the proof of the reordering lemma (Lemma 6.12). Every element of the tensor algebra $T(\mathfrak{g})$, and hence also of the universal enveloping algebra $U(\mathfrak{g})$, is a linear combination of products of Lie algebra elements. Expanding each Lie algebra element in our basis shows that every element of $U(\mathfrak{g})$ is a linear combination of products of basis elements, but not (so far) necessarily in nondecreasing order. But using the relation $XY - YX = [X, Y]$, we can reorder any product of basis elements into the desired order, at the expense of introducing several terms that are products of one fewer basis elements. These smaller products can then, inductively, be rewritten as a linear combination of terms that are in the correct order.

It may seem "obvious" that elements of the form (9.10) are linearly independent. Note, however, that any proof of independence of these elements must make use of the Jacobi identity. After all, if $\mathfrak{g}$ is a vector space with *any* skew-symmetric, bilinear "bracket" operation, we can still construct a "universal enveloping algebra" by the construction in Sect. 9.3, and the elements of the form (9.10) will still span this enveloping algebra. If, however, the bracket does not satisfy the Jacobi identity, the elements in (9.10) will *not* be linearly independent. If they were, then, in particular, the map $i : \mathfrak{g} \to U(\mathfrak{g})$ would be injective. We could then identify $\mathfrak{g}$ with its image under i, which means that the bracket on $\mathfrak{g}$ would be given by $[X, Y] = XY - YX$, where XY and YX are computed in the associative algebra $U(\mathfrak{g})$. But any bracket of this form *does* satisfy the Jacobi identity.

We now proceed with the proof of the independence of the elements in (9.10). The reader is encouraged to note the role of the Jacobi identity in our proof. Let D be any vector space having a basis

$$\{v_{(j_1,\cdots,j_N)}\}.$$

indexed by all nondecreasing tuples $(j_1, \ldots, j_N)$. We wish to construct a linear map $\gamma : U(\mathfrak{g}) \to D$ with the property that

$$\gamma(i(X_{j_1})i(X_{j_2})\cdots i(X_{j_N})) = v_{(j_1,\cdots,j_N)}$$

for each nondecreasing tuple $(j_1, \ldots, j_N)$. Since the elements $v_{(j_1,\cdots,j_N)}$ are, by construction, linearly independent, if such a map γ exists, the elements $X_{j_1}X_{j_2}\cdots X_{j_N}$ must be linearly independent as well. (Any linear relation among the $X_{j_1}X_{j_2}\cdots X_{j_N}$'s would translate under γ into a linear relation among the $v_{(j_1,\cdots,j_N)}$'s.)

Instead of directly constructing γ, we will construct a linear map $\delta : T(\mathfrak{g}) \to D$ with the properties (1) that

$$\delta(X_{j_1} \otimes X_{j_2} \otimes \cdots \otimes X_{j_N}) = v_{(j_1,\cdots,j_N)} \tag{9.11}$$

for all nondecreasing tuples $(j_1, \ldots, j_N)$, and (2) that δ is zero on the two-sided ideal J. Since δ is zero on J, it gives rise to a map γ of $U(\mathfrak{g}) := T(\mathfrak{g})/J$ into D with the analogous property.

To keep our notation compact, we will now omit the tensor product symbol for multiplication in $T(\mathfrak{g})$. Since all computations in the remainder of the section are in $T(\mathfrak{g})$, there will be no confusion. Suppose we can construct δ in such a way that (9.11) holds for nondecreasing tuples and that for *all* tuples $(j_1, \ldots, j_N)$, we have

$$\delta(X_{j_1} \cdots X_{j_k} X_{j_{k+1}} \cdots X_{j_N})$$
$$= \delta(X_{j_1} \cdots X_{j_{k+1}} X_{j_k} \cdots X_{j_N}) + \delta(X_{j_1} \cdots [X_{j_k}, X_{j_{k+1}}] \cdots X_{j_N}). \tag{9.12}$$

Then δ will indeed by zero on J. After all, J is spanned by elements of the form

$$\alpha(XY - YX - [X,Y])\beta.$$

After moving all terms in (9.12) to the other side and taking linear combinations, we can see that δ will be zero on every such element.

Define the **degree** of a monomial $X_{j_1} X_{j_2} \cdots X_{j_N}$ to be the number N and the **index** of the monomial to be the number of pairs $l < k$ for which $j_l > j_k$. We will construct δ inductively, first on the degree of the monomial, and then on the index of the monomial for a given degree, verifying (9.12) as we proceed. If $N = 0$, we set $\delta(1) = v_{(0,\ldots,0)}$ and if $N = 1$, we set $\delta(X_j) = v_{(0,\ldots,1,\ldots,0)}$. In both cases, (9.11) holds by construction and (9.12) holds vacuously.

For a fixed $N \geq 2$ and p, we now assume that δ has been defined on the span of all monomials of degree less than N, and also on all monomials of degree N and index less than p. If $p = 0$, this means simply that δ has been defined on the span of all monomials of degree less than N. Our induction hypothesis is that δ, as defined up to this point, satisfies (9.12) whenever all the terms in (9.12) have been defined. That is to say, we assume (9.12) holds whenever *both* the monomials on the left-hand side of (9.12) have degree less than N or degree N and index less than p. (Under these assumptions, the argument of δ on the right-hand side of (9.12) will be a linear combination of monomials of degree less than N, so that the left-hand side of (9.12) has been defined.)

We now need to show that we can extend the definition of δ to monomials of degree N and index p in such a way that (9.12) continues to hold. If $p = 0$, the new monomials we have to consider are the nondecreasing ones, in which case (9.11) requires us to set

$$\delta(X_{j_1} \cdots X_{j_N}) = v_{(j_1,\cdots,j_N)}.$$

Now, the only way *both* the monomials on the left-hand side of (9.12) can have degree N and index zero is if $j_{k+1} = j_k$, in which case, both sides of (9.12) will be zero. It remains, then, to consider the case $p > 0$.

Let us consider an example that illustrates the most important part of the argument. Suppose $N = 3$ and $p = 3$, meaning that we have defined a map δ satisfying (9.12) on all monomials of degree less than 3 and all monomials of degree 3 and index less than 3. We now attempt to define δ on monomials of degree 3 and index 3 and verify that (9.12) still holds. A representative such monomial would be $X_3 X_2 X_1$. Since we want (9.12) to hold, we may attempt to use (9.12) as our definition of $\delta(X_3 X_2 X_1)$. But this strategy gives two possible ways of defining $\delta(X_3 X_2 X_1)$, either

$$\delta(X_3 X_2 X_1) = \delta(X_2 X_3 X_1) + \delta([X_3, X_2]X_1) \tag{9.13}$$

or

$$\delta(X_3 X_2 X_1) = \delta(X_3 X_1 X_2) + \delta(X_3[X_2, X_1]). \tag{9.14}$$

Note that the monomials on the right-hand sides of (9.13) and (9.14) have degree 2 or degree 3 and index 2, so that δ has already been defined on these monomials. We now verify that these two expression for $\delta(X_3 X_2 X_1)$ agree.

Since δ has already been defined for the terms on the right-hand side of (9.13), we may apply our induction hypothesis to these terms. Using induction twice, we may simplify the right-hand side of (9.13) until we obtain a term in the correct PBW order of $X_1 X_2 X_3$, plus commutator terms:

$$\delta(X_2 X_3 X_1) + \delta([X_3, X_2]X_1)$$
$$= \delta(X_2 X_1 X_3) + \delta(X_2[X_3, X_1]) + \delta([X_3, X_2]X_1)$$
$$= \delta(X_1 X_2 X_3) + \delta([X_2, X_1]X_3)$$
$$+ \delta(X_2[X_3, X_1]) + \delta([X_3, X_2]X_1).$$

Similarly, the other candidate (9.14) for $\delta(X_3 X_2 X_1)$ may be computed by our induction hypothesis as

$$\delta(X_3 X_1 X_2) + \delta(X_3[X_2, X_1])$$
$$= \delta(X_1 X_3 X_2) + \delta([X_3, X_1]X_2) + \delta(X_3[X_2, X_1])$$
$$= \delta(X_1 X_2 X_3) + \delta(X_1[X_3, X_2])$$
$$+ \delta([X_3, X_1]X_2) + \delta(X_3[X_2, X_1]).$$

Subtracting the two expressions gives the quantity

$$\delta([X_2, X_1]X_3) - \delta(X_3[X_2, X_1])$$
$$+ \delta(X_2[X_3, X_1]) - \delta([X_3, X_1]X_2)$$
$$+ \delta([X_3, X_2]X_1) - \delta(X_1[X_3, X_2]).$$

Since all terms are of degree 2, we can use our induction hypothesis to reduce this quantity to

$$\delta([[X_2, X_1], X_3] + [X_2, [X_3, X_1]] + [[X_3, X_2], X_1])$$
$$= \delta([X_3, [X_1, X_2]] + [X_2, [X_3, X_1]] + [X_1, [X_2, X_3]])$$
$$= 0,$$

by the Jacobi identity.

Thus, the two apparently different definitions of $\delta(X_3 X_2 X_1)$ in (9.13) and (9.14) agree. Using this result, it should be apparent that (9.12) holds when we extend the domain of definition of δ to include the monomial $X_3 X_2 X_1$ of degree 3 and index 3.

We now proceed with the general induction step in the construction of δ, meaning that we assume δ has been constructed on monomials of degree less than N and on monomials of degree N and index less than p, in such a way that (9.12) holds whenever both monomials on the left-hand side of (9.12) are in the current domain of definition of δ. Since we have already addressed the $p = 0$ case, we assume $p > 0$. We now consider a monomial $X_{j_1} X_{j_2} \cdots X_{j_N}$ of index $p \geq 1$. Since the index of the monomial is at least 1, the monomial is not weakly increasing and there must be some j_k with $j_k > j_{k+1}$. Pick such a k and "define" δ on the monomial by

$$\delta(X_{j_1} \cdots X_{j_k} X_{j_{k+1}} \cdots X_{j_N}) = \delta(X_{j_1} \cdots X_{j_{k+1}} X_{j_k} \cdots X_{j_N})$$
$$+ \delta(X_{j_1} \cdots [X_{j_k}, X_{j_{k+1}}] \cdots X_{j_N}). \qquad (9.15)$$

Note that the first term on the right-hand side of (9.15) has index $p - 1$ and the second term on the right-hand side has degree $N - 1$, which means that both of these terms have been previously defined.

The crux of the matter is to show that the value of δ on a monomial of index p is independent of the choice of k in (9.15). Suppose, then, that there is some $l < k$ such that $j_l > j_{l+1}$ and $j_k > j_{k+1}$. We now proceed to check that the value of the right-hand side of (9.15) is unchanged if we replace k by l.

Case 1: $l \leq k - 2$. In this case, the numbers $l, l + 1, k, k + 1$ are all distinct. Let us consider the two apparently different ways of calculating δ. If we use l, then we have

$$\delta(\cdots X_l X_{l+1} \cdots X_k X_{k+1} \cdots)$$
$$= \delta(\cdots X_{l+1} X_l \cdots X_k X_{k+1} \cdots) + \delta(\cdots [X_l, X_{l+1}] \cdots X_k X_{k+1} \cdots). \qquad (9.16)$$

Now, the second term on the right-hand side of (9.16) has degree $N - 1$. The first term has index $p - 1$, and if we reverse X_k and X_{k+1} we obtain a term of index $p - 2$. Thus, we can apply our induction hypothesis to reverse the order of X_k and X_{k+1} in both terms on the right-hand side of (9.16), giving

$$\delta(\cdots X_l X_{l+1} \cdots X_k X_{k+1} \cdots)$$
$$= \delta(\cdots X_{l+1} X_l \cdots X_{k+1} X_k \cdots) + \delta(\cdots X_{l+1} X_l \cdots [X_k, X_{k+1}] \cdots)$$
$$+ \delta(\cdots [X_l, X_{l+1}] \cdots X_{k+1} X_k \cdots) + \delta(\cdots [X_l, X_{l+1}] \cdots [X_k, X_{k+1}] \cdots).$$
$$(9.17)$$

(Note that on the right-hand side of (9.17), all terms have *both* X_l and X_{l+1} and X_k and X_{k+1} back in their correct PBW order, with X_{l+1} to the left of X_l and X_{k+1} to the left of X_k.) Since the right-hand side of (9.17) is symmetric in k and l, we would get the same result if we started with k instead of l.

Case 2: $l = k - 1$. In this case, the indices j_l, $j_{l+1} = j_k$, and $j_{l+2} = j_{k+1}$ are in the completely wrong order, $j_l > j_{l+1} > j_{l+2}$. Let us use the notation $X = X_l$, $Y = X_{l+1}$, and $Z = X_{l+2}$. We wish to show that the value of $\delta(\cdots XYZ \cdots)$ is the same whether we use l (that is, interchanging X and Y) or we use k (that is, interchanging Y and Z). But the argument is then precisely the same as in the special case of $\delta(X_3 X_2 X_1)$ considered at the beginning of our proof, with some extra factors, indicated by $\cdots$, tagging along for the ride.

Once we have verified that the value of δ is independent of the choice of k in (9.15), it should be clear that (9.12) holds, since we have used (9.12) with respect to a given pair of indices as our "definition" of δ. We have, therefore, completed the construction of δ and the proof of the PBW theorem.

9.5   Construction of Verma Modules

The proof will make use of the following definition: A subspace I of $U(\mathfrak{g})$ is called a **left ideal** if $\alpha\beta \in I$ for all $\alpha \in U(\mathfrak{g})$ and all $\beta \in I$. For any collection of vectors $\{\alpha_j\}$ in $U(\mathfrak{g})$, we may form the left ideal I "generated by" these vectors, that is, the smallest left ideal in $U(\mathfrak{g})$ containing each α_j. The left ideal I is precisely the space of elements of the form

$$\sum_j \beta_j \alpha_j$$

with β_j being arbitrary elements of $U(\mathfrak{g})$.

Let I_μ denote the left ideal in $U(\mathfrak{g})$ generated by elements of the form

$$H - \langle \mu, H \rangle 1, \quad H \in \mathfrak{h} \tag{9.18}$$

and of the form

$$X \in \mathfrak{g}_\alpha, \quad \alpha \in R^+. \tag{9.19}$$

We now let W_μ denote the quotient vector space

$$W_\mu = U(\mathfrak{g})/I_\mu,$$

and we let $[\alpha]$ denote the image of $\alpha \in U(\mathfrak{g})$ in the quotient space.

We may define a representation π_μ of $U(\mathfrak{g})$ acting on W_μ by setting

$$\pi_\mu(\alpha)([\beta]) = [\alpha\beta] \tag{9.20}$$

for all α and β in $U(\mathfrak{g})$. To verify that $\pi_\mu(\alpha)$ is well defined, note that if β' is another representative of the equivalence class $[\beta]$, then $\beta' = \beta + \gamma$ for some γ in I_μ. But then $\alpha\beta' = \alpha\beta + \alpha\gamma$, and $\alpha\gamma$ belongs to I_μ, because I_μ is a left ideal. Thus, $[\alpha\beta'] = [\alpha\beta]$. We may check that π_μ is a homomorphism by noting that $\pi_\mu(\alpha\beta)[\gamma]$ and $\pi_\mu(\alpha)\pi_\mu(\beta)[\gamma]$ both equal $[\alpha\beta\gamma]$, by the associativity of $U(\mathfrak{g})$. The restriction of π_μ to $\mathfrak{g}$ constitutes a representation of $\mathfrak{g}$ acting on W_μ.

Definition 9.12. The **Verma module** with highest weight μ, denoted W_μ, is the quotient space $U(\mathfrak{g})/I_\mu$, where I_μ is the left ideal in $U(\mathfrak{g})$ generated by elements of the form (9.18) and (9.19).

Theorem 9.13. *The vector* $v_0 := [1]$ *is a nonzero element of* W_μ *and* W_μ *is a highest weight cyclic representation with highest weight* μ *and highest weight vector* v_0.

The hard part of the proof is establishing that v_0 is nonzero; this amounts to showing that the element 1 of $U(\mathfrak{g})$ is not in I_μ. For purposes of constructing the irreducible, finite-dimensional representations of $\mathfrak{g}$, Theorem 9.13 is sufficient. Our method of proof, however, gives more information about the structure of W_μ, which we will make use of in Chapter 10.

Let $\mathfrak{n}^+$ denote the span of the root vectors $X_\alpha \in \mathfrak{g}_\alpha$ with $\alpha \in R^+$, and let $\mathfrak{n}^-$ denote the span of the root vectors $Y_\alpha \in \mathfrak{g}_{-\alpha}$ with $\alpha \in R^+$. Because $[\mathfrak{g}_\alpha, \mathfrak{g}_\beta] \subset \mathfrak{g}_{\alpha+\beta}$, both $\mathfrak{n}^+$ and $\mathfrak{n}^-$ are subalgebras of $\mathfrak{g}$.

Theorem 9.14. *If* $Y_1, \ldots, Y_k$ *form a basis for* $\mathfrak{n}^-$, *then the elements*

$$\pi_\mu(Y_1)^{n_1}\pi_\mu(Y_2)^{n_2}\cdots\pi_\mu(Y_k)^{n_k}v_0, \tag{9.21}$$

where each n_j *is a non-negative integer, form a basis for* W_μ.

The theorem, together with the PBW theorem, tells us that there is a vector space isomorphism between $U(\mathfrak{n}^-)$ and W_μ given by $\alpha \mapsto \pi_\mu(\alpha)v_0$, where π_μ is the action of $U(\mathfrak{g})$ on W_μ, given by (9.20).

Lemma 9.15. *Let* J_μ *denote the left ideal in* $U(\mathfrak{b}) \subset U(\mathfrak{g})$ *generated by elements of the form (9.18) and (9.19). Then 1 does not belong to* J_μ.

Proof. Let $\mathfrak{b}$ be the direct sum (as a vector space) of $\mathfrak{n}^+$ and $\mathfrak{h}$, which is easily seen to be a subalgebra of $\mathfrak{g}$. Let us define a one-dimensional representation σ_μ of $\mathfrak{b}$, acting on $\mathbb{C}$, by the formula

$$\sigma_\mu(X + H) = \langle \mu, H \rangle .$$

(That is to say, $\sigma_\mu(X + H)$ is the 1×1 matrix with entry $\langle \mu, H \rangle$.) To see that σ_μ is actually a representation, we note that all 1×1 matrices commute. On the other hand, the commutator of two elements Z_1 and Z_2 of $\mathfrak{b}$ will lie in $\mathfrak{n}^+$, and σ_μ is defined to be zero on $\mathfrak{n}$. Thus, $\sigma_\mu([Z_1, Z_2])$ and $[\sigma_\mu(Z_1), \sigma_\mu(Z_2)]$ are both zero.

By Proposition 9.9, the representation σ_μ of $\mathfrak{b}$ extends to a representation $\tilde{\sigma}_\mu$ of $U(\mathfrak{b})$ satisfying $\tilde{\sigma}_\mu(1) = 1$. Now, the kernel of $\tilde{\sigma}_\mu$ is easily seen to be a left ideal in $U(\mathfrak{b})$, and by construction, the kernel of $\tilde{\sigma}_\mu$ will contain all elements of the form (9.18) and (9.19). But since $\tilde{\sigma}_\mu(1) = 1$, the element 1 does not belong to the kernel of $\tilde{\sigma}_\mu$. Thus, $\ker(\tilde{\sigma}_\mu)$ is a left ideal in $U(\mathfrak{b})$ containing all elements of the form (9.18) and (9.19), which means that $\ker(\tilde{\sigma}_\mu)$ contains J_μ. Since $\ker(\tilde{\sigma}_\mu)$ does not contain 1, neither does J_μ. $\qquad\qquad\square$

Proof of Theorems 9.13 and 9.14. Note that for any $H \in \mathfrak{h}$, we have

$$(\pi_\mu(H - \langle \mu, H \rangle\, 1))v_0 = [H - \langle \mu, H \rangle\, 1] = 0,$$

because $H - \langle \mu, H \rangle\, 1$ belongs to I_μ. Thus, $\pi_\mu(H)v_0 = \langle \mu, H \rangle\, v_0$. Similarly, for any $\alpha \in R^+$, we have

$$\pi_\mu(X_\alpha)v_0 = [X_\alpha] = 0.$$

Since elements of $U(\mathfrak{g})$ are linear combinations of products of elements of $\mathfrak{g}$, any invariant subspace for the action of $\mathfrak{g}$ on W_μ is also invariant under the action of $U(\mathfrak{g})$ on W_μ. Suppose, then, that U is an invariant subspace of W_μ containing $v_0 = [1]$. Then for any $[\alpha] \in W_\mu$, we have $\pi_\mu(\alpha)v_0 = [\alpha]$, and so $U = W_\mu$. Thus, v_0 is a cyclic vector for W_μ. To prove Theorem 9.13, it remains only to show that $v_0 \neq 0$.

The Lie algebra $\mathfrak{g}$ decomposes as a vector space direct sum of $\mathfrak{n}^-$ and $\mathfrak{b}$, where $\mathfrak{n}^-$ is the span of the root spaces corresponding to roots in R^- and where $\mathfrak{b}$ is the span of $\mathfrak{h}$ and the root spaces corresponding to roots in R^+. Let us choose a basis $Y_1, \ldots, Y_k, Z_1, \ldots, Z_l$ for $\mathfrak{g}$ consisting of a basis $Y_1, \ldots, Y_k$ for $\mathfrak{n}^-$ together with a basis $Z_1, \ldots, Z_l$ for $\mathfrak{b}$. By applying the PBW theorem to this basis, we can easily see (Exercise 4) that every element α of $U(\mathfrak{g})$ can be expressed *uniquely* in the form

$$\alpha = \sum_{n_1,\ldots,n_k=0}^{\infty} Y_1^{n_1} Y_2^{n_2} \cdots Y_k^{n_k} a_{n_1,\ldots,n_k}, \qquad (9.22)$$

where each $a_{n_1,\ldots,n_k}$ belongs to $U(\mathfrak{b}) \subset U(\mathfrak{g})$. (For each $\alpha \in U(\mathfrak{g})$, only finitely many of the $a_{n_1,\ldots,n_k}$'s will be nonzero.)

Suppose now that α belongs to I_μ, which means that α is a linear combination of terms of the form $\beta(H - \langle \mu, H \rangle 1)$ and βX_α, with β in $U(\mathfrak{g})$, H in $\mathfrak{h}$, and X_α in $\mathfrak{g}_\alpha$, $\alpha \in R^+$. By writing each β as in (9.22), we see that α is a linear combination of terms of the form

$$Y_1^{n_1} Y_2^{n_2} \cdots Y_k^{n_k} b_{n_1,\ldots,n_k}(H - \langle \mu, H \rangle 1)$$

and

$$Y_1^{n_1} Y_2^{n_2} \cdots Y_k^{n_k} b_{n_1,\ldots,n_k} X_\alpha,$$

with $b_{n_1,\ldots,n_k}$ in $U(\mathfrak{b})$. Note that $b_{n_1,\ldots,n_k}(H - \langle \mu, H \rangle 1)$ and $b_{n_1,\ldots,n_k} X_\alpha$ belong to the left ideal $J_\mu \subset U(\mathfrak{b})$. Thus, if α is in I_μ, each $a_{n_1,\ldots,n_k}$ in the (unique) expansion (9.22) of α must belong to J_μ.

Now, by the uniqueness of the expansion, the only way the element α in (9.22) can equal 1 is if there is only one term, the one with $n_1 = \cdots = n_k = 0$, and if $a_{0,\ldots,0} = 1$. On the other hand, for α to be in I_μ, each $a_{n_1,\ldots,n_k}$ must belong to J_μ. Since (Lemma 9.15) 1 is not in J_μ, we see that 1 is not in I_μ, so that $v_0 = [1]$ is nonzero in $U(\mathfrak{g})/I_\mu$.

We now argue that the vectors in (9.21) are linearly independent in $U(\mathfrak{g})/I_\mu$. Suppose, then, that a linear combination of these vectors, with coefficients $c_{n_1,\ldots,n_k}$ equals zero. Then the corresponding linear combination of elements in $U(\mathfrak{g})$, namely

$$\alpha := \sum_{n_1,\ldots,n_k=0}^{\infty} Y_1^{n_1} Y_2^{n_2} \cdots Y_k^{n_k} c_{n_1,\ldots,n_k},$$

belongs to I_μ. But as shown above, for α to be in I_μ, each of the constants $c_{n_1,\ldots,n_k}$ must be in J_μ. Thus, by Lemma 9.15, each of the constants $c_{n_1,\ldots,n_k}$ is zero. $\square$

The main role of the PBW theorem in the preceding proof is to establish the uniqueness of the expansion (9.22).

9.6 Irreducible Quotient Modules

In this section, we show that every Verma module has a largest proper invariant subspace U_μ and that the quotient space $V_\mu := W_\mu/U_\mu$ is irreducible with highest weight μ. In the next section, we will show that if μ is dominant integral, this quotient space is finite dimensional.

It is easy to see (Exercise 6) that the Verma module W_μ is the direct sum of its weight spaces. It therefore makes sense to talk about the component of a vector $v \in W_\mu$ in the one-dimensional subspace spanned by v_0, which we refer to as the v_0-component of v. As in the previous section, we let $\mathfrak{n}^+$ denote the subalgebra of $\mathfrak{g}$ spanned by weight vectors $X_\alpha \in \mathfrak{g}_\alpha$, with $\alpha \in R^+$.

Definition 9.16. For any Verma module W_μ, let U_μ be the subspace of W_μ consisting of all vectors v such that the v_0-component of v is zero and such that the v_0-component of

$$\pi_\mu(X_1)\cdots\pi_\mu(X_N)v$$

is also zero for any collection of vectors $X^1,\ldots,X^N$ in $\mathfrak{n}^+$.

That is to say, a vector v belongs to U_μ if we cannot "get to" v_0 from v by applying "raising operators" $X \in \mathfrak{g}_\alpha$, $\alpha \in R^+$. Certainly the zero vector is in U_μ; for some $\mathfrak{g}$'s and μ's, it happens that $U_\mu = \{0\}$.

Proposition 9.17. *The space U_μ is an invariant subspace for the action of $\mathfrak{g}$.*

Proof. Suppose that v is in U_μ and that Z is some element of $\mathfrak{g}$. We want to show that $\pi_\mu(Z)v$ is also in U_μ. Thus, we consider

$$\pi_\mu(X^1)\cdots\pi_\mu(X^l)\pi_\mu(Z)v \tag{9.23}$$

and we must show that the v_0-component of this vector is zero. Using the reordering lemma (Lemma 6.12), we may rewrite the vector in (9.23) as a linear combination of vectors of the form

$$\pi_\mu(Y^1)\cdots\pi_\mu(Y^j)\pi_\mu(H^1)\cdots\pi_\mu(H^k)\pi_\mu(\tilde{X}^1)\cdots\pi_\mu(\tilde{X}^m)v, \tag{9.24}$$

where the Y's are in $\mathfrak{n}^-$, the H's are in $\mathfrak{h}$, and the $\tilde{X}$'s are in $\mathfrak{n}^+$. However, since v is in U_μ, the v_0-component of

$$\pi_\mu(\tilde{X}^1)\cdots\pi_\mu(\tilde{X}^m)v \tag{9.25}$$

is zero; thus, this vector is a linear combination of weight vectors with weight lower than μ. Then applying elements of $\mathfrak{h}$ and $\mathfrak{n}^-$ to the vector in (9.25) will only keep the weights the same or lower them. Thus, the v_0-component of the vector in (9.24), and hence also the v_0-component of the vector in (9.23), is zero. This shows that $\pi_\mu(Z)v$ is, again, in U_μ. □

Since U_μ is an invariant subspace of W_μ, the quotient vectors space W_μ/U_μ carries a natural action of $\mathfrak{g}$ and thus constitutes a representation of $\mathfrak{g}$.

Proposition 9.18. *The quotient space $V_\mu := W_\mu/U_\mu$ is an irreducible representation of $\mathfrak{g}$.*

Proof. A simple argument shows that the invariant subspaces of the representation W_μ/U_μ are in one-to-one correspondence with the invariant subspaces of W_μ that contain U_μ. Thus, proving that V_μ is irreducible is equivalent to showing that any invariant subspace of W_μ that contains U_μ is either U_μ or W_μ. Suppose, then, that X is an invariant subspace that contains U_μ and at least one vector v that is not in U_μ. This means that X also contains a vector $u = \pi_\mu(X_1)\cdots\pi_\mu(X_k)v$ whose v_0-component is nonzero.

We now claim that X must contain v_0 itself. To see this, we decompose u as a nonzero multiple of v_0 plus a sum of weight vectors corresponding to weights $\lambda \neq \mu$. Since $\lambda \neq \mu$, we can find H in $\mathfrak{h}$ with $\langle \lambda, H \rangle \neq \langle \mu, H \rangle$ and then we may apply to u the operator $\pi_\mu(H) - \langle \lambda, H \rangle I$. This operator will keep us in X and will "kill" the component of u that is in the weight space corresponding to the weight λ while leaving the v_0-component of u nonzero. We can then continue applying operators of this form until we have killed all the components of u in weight spaces different from μ, giving us a nonzero multiple of v_0. We conclude, then, that X contains v_0 and, therefore, all of W_μ. Thus, any invariant subspace of W_μ that properly contains U_μ must be equal to W_μ. □

Since for each $u \in U_\mu$ the v_0-component of u is zero, the vector v_0 is not in U_μ. Thus, the quotient space W_μ/U_μ is still a highest weight cyclic representation with highest weight μ and with highest weight vector being the image of v_0 in the quotient.

Example 9.19. Let α be an element of Δ and let $\mathfrak{s}^\alpha = \langle X_\alpha, Y_\alpha, H_\alpha \rangle$ be as in Theorem 7.19. If $\langle \mu, H_\alpha \rangle$ is a non-negative integer m, then the vector $v := \pi(Y_\alpha)^{m+1}v_0$ belongs to U_μ.

This result is illustrated in Figure 9.2.

Proof. By the argument in proof of Theorem 4.32, the analog of (4.15) will hold here:

$$\pi(X_\alpha)\pi(Y_\alpha)^j v_0 = j(m - (j-1))\pi(Y_\alpha)^{j-1}v_0.$$

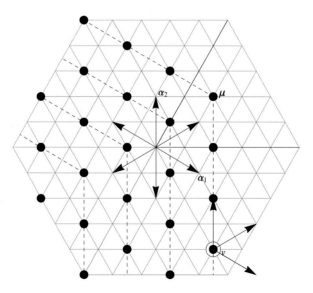

Fig. 9.2 Example 9.19 in the case $\alpha = \alpha_2$ and $m = 2$. The vector v belongs to U_μ

Thus,

$$\pi(X_\alpha)v = m(m-m)\pi(Y_\alpha)^m v_0 = 0.$$

Meanwhile, if $\beta \in R^+$ with $\beta \neq \alpha$, then for $X_\beta \in \mathfrak{g}_\beta$, if $\pi(X_\beta)v$ were nonzero, it would be a weight vector with weight

$$\lambda = \mu - (m+1)\alpha + \beta.$$

Since λ is not lower than μ, we must have $\pi(X_\beta)v = 0$. Thus, the v_0-component of v is zero and $\pi(X_\eta)v = 0$ for all $\eta \in R^+$, which implies that v is in U_μ. $\qquad\square$

9.7 Finite-Dimensional Quotient Modules

Throughout this section, we assume that μ is a dominant integral element. We will now show that, in this case, the irreducible quotient space $V_\mu := W_\mu/U_\mu$ is finite dimensional. Our strategy is to show that the set of weights for V_μ is invariant under the action of the Weyl group on $\mathfrak{h}$. Now, if μ is dominant integral, then every weight λ of W_μ—and thus also of V_μ—must be integral, since $\mu - \lambda$ is an integer combination of roots. Hence, if the weights of V_μ are invariant under W, we must have $w \cdot \lambda \preceq \mu$ for all $w \in W$. But it is not hard to show that there are only finitely many integral elements this property. We will conclude, then, that there are only finitely many weights in V_μ. Since (even in the Verma module) each weight has finite multiplicity, this will show that W_μ/U_μ is finite dimensional.

How, then, do we construct an action of the Weyl group on V_μ? If we attempt to follow the proof of Theorem 9.3, we must contend with the fact that V_μ is not yet known to be finite dimensional. Thus, we need a method of exponentiating operators on a possibly infinite-dimensional space.

Definition 9.20. A linear operator X on a vector space V is **locally nilpotent** if for each $v \in V$, there exists a positive integer k such that $X^k v = 0$.

If V is finite dimensional, then a locally nilpotent operator must actually be nilpotent, that is, there must exist a single k such that $X^k v = 0$ for all v. In the infinite-dimensional case, the value of k depends on v and there may be no single value of k that works for all v. If X is locally nilpotent, then we define e^X to be the operator satisfying

$$e^X v = \sum_{k=0}^{\infty} \frac{X^k}{k!} v, \tag{9.26}$$

where for each $v \in V$, the series on the right terminates.

Proposition 9.21. *For each $\alpha \in \Delta$, let $\mathfrak{s}^\alpha = \langle X_\alpha, Y_\alpha, H_\alpha \rangle$ be as in Theorem 7.19. If μ is dominant integral, then X_α and Y_α act in a locally nilpotent fashion on the quotient space V_μ.*

Proof. For any $X \in \mathfrak{g}$, we use $\tilde{X}$ as an abbreviation for the action of X on the quotient space V_μ. We say that a vector in V_μ is $\mathfrak{s}^\alpha$-**finite** if it is contained in a finite-dimensional, $\mathfrak{s}^\alpha$-invariant subspace. Let

$$m = \langle \mu, H_\alpha \rangle,$$

which is a non-negative integer because μ is dominant integral. Let $\tilde{v}_0$ denote the image in V_μ of the highest vector $v_0 \in W_\mu$, and consider the vectors

$$\tilde{v}_k := \tilde{Y}_\alpha^k \tilde{v}_0, \quad k = 0, 1, 2, \ldots.$$

By the calculations in Sect. 4.6, the span of the $\tilde{v}_k$'s is invariant under the action of $\mathfrak{s}^\alpha$. On the other hand, Example 9.19 shows that $\pi(Y_\alpha)^{m+1} v_0$ is in U_μ, which means that $\tilde{v}_{m+1} = \tilde{Y}_\alpha^{m+1} \tilde{v}_0$ is the zero element of V_μ. Thus, $\langle \tilde{v}_0, \ldots, \tilde{v}_m \rangle$ is a finite-dimensional, $\mathfrak{s}^\alpha$-invariant subspace (Figure 9.3). In particular, there exists a nonzero, $\mathfrak{s}^\alpha$-finite vector in V_μ.

Now let $T_\alpha \subset V_\mu$ be the space of all $\mathfrak{s}^\alpha$-finite vectors, which we have just shown to be nonzero. We now claim that T_α is invariant under the action of $\mathfrak{g}$. To see this, fix a vector v in T_α and an element X of $\mathfrak{g}$. Let S be a finite-dimensional, $\mathfrak{s}^\alpha$-invariant subspace containing v and let S' be the span of all vectors of the form

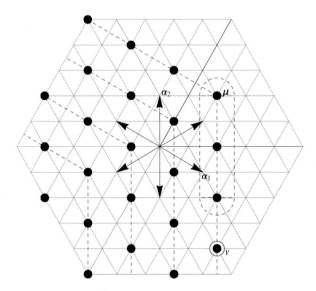

Fig. 9.3 Since the vector v is in U_μ (Figure 9.2), the circled weights span a $\mathfrak{s}^{\alpha_2}$-invariant subspace of W_μ / U_μ

$\tilde{Y}w$ with $Y \in \mathfrak{g}$ and $w \in S$. Then S' is finite dimensional, having dimension at most $(\dim \mathfrak{g})(\dim S)$. Furthermore, if $Z \in \mathfrak{s}^{\alpha}$, then for all $w \in S$, we have

$$\tilde{Z}\tilde{Y}w = \tilde{Y}\tilde{Z}w + \widetilde{[Z, Y]}w,$$

which belongs to S', because $\tilde{Z}w$ is again in S. Thus, S' is also invariant under the action of $\mathfrak{s}^{\alpha}$. We see, then, that $\tilde{X}v$ is contained in the finite-dimensional, $\mathfrak{s}^{\alpha}$-invariant subspace S'; that is, $\tilde{X}v \in T_{\alpha}$. Since V_{μ} is irreducible and T_{α} is nonzero and invariant under the action of $\mathfrak{g}$, we have $T_{\alpha} = V_{\mu}$.

We conclude that every $v \in V_{\mu}$ is contained in a finite-dimensional, $\mathfrak{s}^{\alpha}$-invariant subspace. It then follows from Point 2 of Theorem 4.34 that $(\tilde{X}_{\alpha})^{k}v = (\tilde{Y}_{\alpha})^{k}v = 0$ for some k, showing that $\tilde{X}_{\alpha}$ and $\tilde{Y}_{\alpha}$ are locally nilpotent. □

Proposition 9.22. *If μ is dominant integral, the set of weights for V_{μ} is invariant under the action of the Weyl group on $\mathfrak{h}$.*

Proof. We continue the notation from the proof of Proposition 9.21. Since (Proposition 8.24) W is generated by the reflections s_{α} with $w \in \Delta$, it suffices to show that the weights of W_{μ}/U_{μ} are invariant under each such reflection. By Proposition 9.21, $\tilde{X}_{\alpha}$ and $\tilde{Y}_{\alpha}$ are locally nilpotent, and thus it makes sense to define operators S_{α} by

$$S_{\alpha} = e^{\tilde{X}_{\alpha}}e^{-\tilde{Y}_{\alpha}}e^{\tilde{X}_{\alpha}}.$$

We may now imitate the proof of Theorem 9.3 as follows. If $H \in \mathfrak{h}$ satisfies $\langle \alpha, H \rangle = 0$, then $[H, X_{\alpha}] = [H, Y_{\alpha}] = 0$, which means that $\tilde{H}$ commutes with $\tilde{X}_{\alpha}$ and $\tilde{Y}_{\alpha}$ and, thus, with S_{α}. Meanwhile, for any $v \in V_{\mu}$, we may find a finite-dimensional, $\mathfrak{s}^{\alpha}$-invariant subspace S containing v. In the space S, we may apply Point 3 of Theorem 4.34 to show that

$$S_{\alpha}\tilde{H}_{\alpha}S_{\alpha}^{-1}v = -\tilde{H}_{\alpha}v.$$

We conclude that for all $H \in \mathfrak{h}$, we have

$$S_{\alpha}\tilde{H}S_{\alpha}^{-1} = s_{\alpha} \cdot \tilde{H}.$$

From this point on, the proof of Theorem 9.3 applies without change. □

Figure 9.4 illustrates the result of Proposition 9.22 in the case of $\mathsf{sl}(2; \mathbb{C})$ and highest weight 3. If v is a weight vector with weight -5, then $\pi(X)v = 0$, by (4.15) in the proof of Theorem 4.32. Thus, the span of the weight vectors with weights $l \leq -5$ is invariant. The quotient space has weights ranging from -3 to 3 in increments of 2 and is, thus, invariant under the action of $W = \{I, -I\}$.

We are now ready for the proof of the existence of finite-dimensional, irreducible representations.

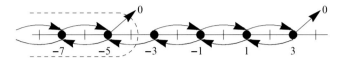

Fig. 9.4 In the Verma module for $\mathsf{sl}(2; \mathbb{C})$ with highest weight 3, the span of the weight vectors with weights $-5, -7, \ldots$, is invariant

Proof of Theorem 9.5. The quotient space $V_\mu := W_\mu / U_\mu$ is irreducible and has highest weight μ. Every weight λ of V_μ is integral and satisfies $\lambda \preceq \mu$. By Proposition 9.22, the weights are also invariant under the action of W, which means that every weight λ satisfies $w \cdot \lambda \preceq \mu$ for all $w \in W$. Thus, by Proposition 8.44, λ must be in the convex hull of the W-orbit of μ, which implies that $\|\lambda\| \leq \|\mu\|$. Since there are only finitely many integral elements λ with this property, we conclude that V_μ has only finitely many weights.

Now, V_μ has at least one weight, namely μ. Since V_μ is irreducible, it must be the direct sum of its weight spaces. Now, since the elements in (9.21) form a basis for the Verma module W_μ, the corresponding elements of V_μ certainly span V_μ. But for a given weight λ, there are only finitely many choices of the exponents $n_1, \ldots, n_k$ in (9.21) that give a weight vector with weight λ. (After all, if any of the n_j's is large, the weight of the vector in (9.21) will be much lower than μ.) Thus, each weight of V_μ has finite multiplicity. Since, also, there are only finitely many weights, V_μ is finite dimensional. $\qquad\square$

9.8 Exercises

1. Suppose $(i, U(\mathfrak{g}))$ and $(i', U'(\mathfrak{g}))$ are algebras as in Theorem 9.7. Show that there is an isomorphism $\Phi : U(\mathfrak{g}) \to U'(\mathfrak{g})$ such that $\Phi(1) = 1$ and such that

$$\Phi(i(X)) = i'(X)$$

for all $X \in \mathfrak{g}$.
   Hint: Use the defining property of $U(\mathfrak{g})$ to construct Φ.
2. Suppose that $\mathfrak{g} \subset M_n(\mathbb{C})$ is a Lie algebra of matrices (with bracket given by $XY - YX$). Prove, without appealing to the PBW theorem, that the map $i : \mathfrak{g} \to U(\mathfrak{g})$ in Theorem 9.7 is injective.
3. Suppose $\mathfrak{g}$ is a Lie algebra and $\mathfrak{h}$ is a subalgebra. Apply Theorem 9.7 to the Lie algebra $\mathfrak{h}$ with $\mathcal{A} = U(\mathfrak{g})$ and with j being the inclusion of $\mathfrak{h}$ into $\mathfrak{g} \subset U(\mathfrak{g})$. If $\phi : U(\mathfrak{h}) \to U(\mathfrak{g})$ is the associated algebra homomorphism, show that ϕ is injective.
   Hint: Use the PBW theorem.

4. Using the PBW theorem for $\mathfrak{b}$ (applied to the basis $Z_1, \ldots, Z_l$) and for $\mathfrak{g}$ (applied to the basis $Y_1, \ldots, Y_k, Z_1, \ldots, Z_l$), establish first the existence and then the uniqueness of the expansion in (9.22).

   Hint: For the uniqueness result, first prove that if α is a nonzero element of $U(\mathfrak{b})$, then $Y_1^{n_1} \cdots Y_k^{n_k} \alpha$ is a nonzero element of $U(\mathfrak{g})$, for any sequence $n_1, \ldots, n_k$ of non-negative integers. Then prove that a linear combination as in (9.22) cannot be zero unless each of the elements $a_{n_1, \ldots n_k}$ in $U(\mathfrak{b})$ is zero.

5. Let μ be any element of $\mathfrak{h}$ and let $W_\mu := U(\mathfrak{g})/I_\mu$ be the Verma module with highest weight μ. Now let σ_μ be any other highest weight cyclic representation of $\mathfrak{g}$ with highest weight μ, acting on a vector space W_μ. Show that there is a surjective intertwining map ϕ of V_μ onto W_μ.

   Note: It follows that W_μ is isomorphic to the quotient space $V_\mu/\ker(\phi)$. Thus, V_μ is maximal among highest weight cyclic representations with highest weight μ, in the sense that every other such representation is a quotient of V_μ.

   Hint: If $\tilde{\sigma}_\mu$ is the extension of σ_μ to $U(\mathfrak{g})$, as in Proposition 9.9, construct a map $\psi : U(\mathfrak{g}) \to W_\mu$ by mapping $\alpha \in U(\mathfrak{g})$ to $\tilde{\sigma}_\mu(\alpha)w_0$, where w_0 is a highest weight vector for W_μ.

6. Let $W_\mu := U(\mathfrak{g})/I_\mu$ be the Verma module with highest weight μ and highest weight vector v_0. Let X be the subspace of W_μ consisting of all those vectors that can be expressed as finite linear combinations of weight vectors. Show that X contains v_0 and is invariant under the action of $\mathfrak{g}$ on W_μ. Conclude that W_μ is the direct sum of its weight spaces.

7. Let μ be any element of $\mathfrak{h}$ and let W_μ be the associated Verma module. Suppose $\lambda \in \mathfrak{h}$ can be expressed in the form

$$\lambda = \mu - n_1\alpha_1 - \cdots - n_k\alpha_k \tag{9.27}$$

where $\alpha_1, \ldots, \alpha_k$ are the positive roots and $n_1, \ldots, n_k$ are non-negative integers. Show that the multiplicity of λ in W_μ is equal to the number of ways that $\mu - \lambda$ can be expressed as a linear combination of positive roots with non-negative integer coefficients. That is to say, the multiplicity of λ is the number of k-tuples of non-negative integers $(n_1, \ldots, n_k)$ for which (9.27) holds.

Chapter 10
Further Properties of the Representations

In this chapter we derive several important properties of the representations we constructed in the previous chapter. Throughout the chapter, $\mathfrak{g} = \mathfrak{k}_{\mathbb{C}}$ denotes a complex semisimple Lie algebra, $\mathfrak{h} = \mathfrak{t}_{\mathbb{C}}$ denotes a fixed Cartan subalgebra of $\mathfrak{g}$, and R denotes the set of roots for $\mathfrak{g}$ relative to $\mathfrak{h}$. We let W denote the Weyl group, we let Δ denote a fixed base for R, and we let R^+ and R^- denote the positive and negative roots with respect to Δ, respectively.

10.1   The Structure of the Weights

In this section, we establish the general version of Theorems 6.24 and 6.25 in the case of $\mathsf{sl}(3; \mathbb{C})$, which tells us which integral elements appear as weights of a fixed finite-dimensional irreducible representation. In Sect. 10.6, we will establish a formula for the multiplicities of the weights, as a consequence of the Weyl character formula.

Recall that the weights of a finite-dimensional representation of $\mathfrak{g}$ are integral elements (Proposition 9.2) and that the weights and their multiplicities are invariant under the action of W (Theorem 9.3). We now determine which weights occur in the representation with highest weight μ. Recall from Definition 6.23 the notion of the convex hull of a collection of vectors.

A previous version of this book was inadvertently published without the middle initial of the author's name as "Brian Hall". For this reason an erratum has been published, correcting the mistake in the previous version and showing the correct name as Brian C. Hall (see DOI http://dx.doi.org/10.1007/978-3-319-13467-3_14). The version readers currently see is the corrected version. The Publisher would like to apologize for the earlier mistake.

© Springer International Publishing Switzerland 2015   267
B.C. Hall, *Lie Groups, Lie Algebras, and Representations*, Graduate
Texts in Mathematics 222, DOI 10.1007/978-3-319-13467-3_10

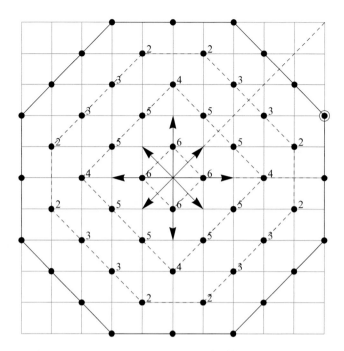

Fig. 10.1 Typical weight diagram for the Lie algebra $\mathsf{so}(5;\mathbb{C})$

Theorem 10.1. *Let (π, V_μ) be an irreducible finite-dimensional representation of $\mathfrak{g}$ with highest weight μ. An integral element λ is a weight of V_μ if and only if the following two conditions are satisfied.*

1. *λ belongs to the convex hull of the Weyl-group orbit of μ.*
2. *$\mu - \lambda$ can be expressed as an integer combination of roots.*

   Figure 10.1 shows a typical example for the Lie algebra $\mathsf{so}(5;\mathbb{C})$. In the figure, the square lattice indicates the set of integral elements, the highest weight is circled, and the black dots indicate the weights of the representation. A number next to a dot indicates the multiplicity, with an unnumbered dot representing a weight of multiplicity 1. The multiplicities can be calculated using the Kostant multiplicity formula (Sect. 10.6). Note that the multiplicities do not have the sort of simple pattern that we saw in Sect. 6.7 for the case of $\mathsf{sl}(3;\mathbb{C})$; that is, the multiplicities for $\mathsf{so}(5;\mathbb{C})$ are not constant on the "rings" in the weight diagram.

   We will use the notation $W \cdot \mu$ to denote the Weyl group orbit of an element μ, and the notation $\mathrm{Conv}(E)$ to denote the convex hull of E. The following result is the key step on the way to proving Theorem 10.1.

Proposition 10.2. *Let μ be a dominant integral element. Suppose λ is dominant, λ is lower than μ, and $\mu - \lambda$ can be expressed as an integer combination of roots. Then λ is a weight of the irreducible representation with highest weight μ.*

Lemma 10.3 ("No Holes" Lemma). *Suppose (π, V) is a finite-dimensional repre-
sentation of $\mathfrak{g}$. Suppose that λ is a weight of V and that $\langle \lambda, \alpha \rangle > 0$ for some root α.
If we define j by*

$$j = \langle \lambda, H_\alpha \rangle = 2 \frac{\langle \lambda, \alpha \rangle}{\langle \alpha, \alpha \rangle},$$

*then $\lambda - k\alpha$ is a weight of π for every integer k with $0 \le k \le j$. In particular, $\lambda - \alpha$
is a weight of V.*

Note that since λ is integral, j must be a (positive) integer.

Proof. Let $\mathfrak{s}_\alpha = \langle X_\alpha, Y_\alpha, H_\alpha \rangle$ be the copy of $\mathsf{sl}(2; \mathbb{C})$ corresponding to the weight α
(Theorem 7.19). Let U be the subspace of V spanned by weight spaces with weights
η of the form $\eta = \lambda - k\alpha$ for $k \in \mathbb{Z}$. Since X_α and Y_α shift weights by $\pm\alpha$, the
space U is invariant under $\mathfrak{s}_\alpha$. Note that since $\langle \alpha, H_\alpha \rangle = 2$, we have

$$\langle \lambda - k\alpha, H_\alpha \rangle = j - 2k.$$

That is, the weight space corresponding to weight $\lambda - k\alpha$ is precisely the eigenspace
for $\pi(H_\alpha)$ inside U corresponding to the eigenvalue $j - 2k$.

By Point 4 of Theorem 4.34, if $j > 0$ is an eigenvalue for $\pi(H_\alpha)$ inside U, then
all of the integers $j - 2k, 0 \le k \le j$, must also be eigenvalues for $\pi(H_\alpha)$ inside U.
Thus, $\lambda - k\alpha$ must be a weight of π for $0 \le k \le j$. Since j is a positive integer, j
must be at least 1, and, thus, $\lambda - \alpha$ must be a weight of π. $\qquad\qquad\square$

Figure 10.2 illustrates the results of the "no holes" lemma. For the indicated
weight λ, the orthogonal projection of λ onto α equals $(3/2)\alpha$, so that $j = 3$. Thus,
$\lambda - \alpha, \lambda - 2\alpha$, and $\lambda - 3\alpha$ must also be weights.

Proof of Proposition 10.2. Since $\mu - \lambda$ is an integer combination of roots, $\mu - \lambda$
is also an integer combination of the positive simple roots $\alpha_1, \ldots, \alpha_r$. Since, also,
$\lambda \preceq \mu$, we have

$$\mu = \lambda + \sum_{j=1}^{r} k_j \alpha_j$$

for some non-negative integers $k_1, \ldots, k_r$. Consider now the following set P of
integral elements,

$$P = \left\{ \eta = \lambda + \sum_{j=1}^{r} l_j \alpha_j \,\middle|\, 0 \le l_j \le k_j \right\}. \qquad (10.1)$$

The elements of P form a discrete parallelepiped.

We *do not* claim that every element of P is a weight of π, which is not, in
general, true (Figure 10.3). Rather, we will show that if $\eta \ne \lambda$ is a weight of π in P,
then there is another weight of V_μ in P that is "closer" to λ. Specifically, for each

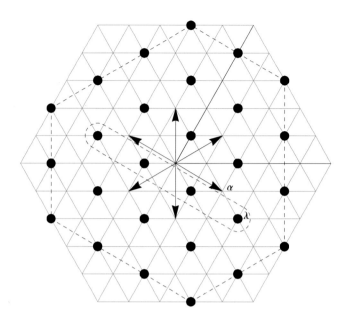

Fig. 10.2 Since λ is a weight, each of the *circled elements* must also be a weight

element η of P as in (10.1), let $L(\eta) = \sum_j l_j$. Starting from μ, we will construct a sequence of weights of V_μ in P with decreasing values of $L(\eta)$, until we reach one with $L(\eta) = 0$, which means that $\eta = \lambda$.

Suppose then that η is a weight of π in P with $L(\eta) > 0$. In that case, the second term in the formula for η is nonzero, and, thus,

$$\left\langle \sum_{j=1}^{r} l_j \alpha_j, \sum_{k=1}^{r} l_k \alpha_k \right\rangle = \sum_{k=1}^{r} l_k \left\langle \sum_{j=1}^{r} l_j \alpha_j, \alpha_k \right\rangle > 0.$$

Since each l_k is non-negative, there must be some α_k for which $l_k > 0$ and for which

$$\left\langle \sum_{j=1}^{r} l_j \alpha_j, \alpha_k \right\rangle > 0.$$

On the other hand, since λ is dominant, $\langle \lambda, \alpha_k \rangle \geq 0$, and we conclude that $\langle \eta, \alpha_k \rangle > 0$. Thus, by the "no holes" lemma, $\eta - \alpha_k$ must also be a weight of π.

Now, since l_k is positive, $l_k - 1$ is non-negative, meaning that $\eta - \alpha_k$ is still in P, where all the l_j's are unchanged except that l_k is replaced by $l_k - 1$. Thus, $L(\eta - \alpha_k) = L(\eta) - 1$. We can then repeat the process starting with $\eta - \alpha_k$ and obtain a sequence of weights of π with successively smaller values of L, until we reach $L = 0$, which corresponds to $\eta = \lambda$.   □

Figure 10.3 illustrates the proof of Proposition 10.2. Starting at μ, we look for a sequence of weights of π in P. Each weight in the sequence has positive inner

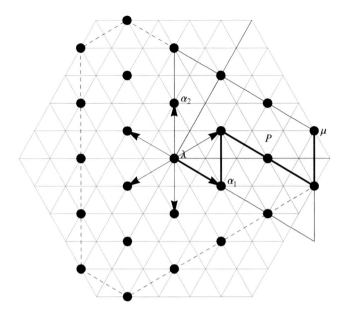

Fig. 10.3 The *thick line* indicates a path of weights in P connecting μ to λ

product either with α_1 or with α_2, allowing us to move in the direction of $-\alpha_1$ or $-\alpha_2$ to another weight of π in P, until we reach λ.

Proof of Theorem 10.1. Let $X \subset V_\mu$ denote the span of all weight vectors whose weights differ from μ by a linear combination of roots. Then X is easily seen to be invariant under the action of $\mathfrak{g}$ and X contains v_0, so $X = V_\mu$. Thus, every weight of π must satisfy Point 2 of Theorem 10.1. Furthermore, if λ is a weight of π, then $w \cdot \lambda \preceq \mu$ for all $w \in W$. Thus, by Proposition 8.44, $\lambda \in \mathrm{Conv}(W \cdot \mu)$.

Conversely, suppose that λ satisfies the two conditions of Theorem 10.1. We can choose $w \in W$ so that $\lambda' := w \cdot \lambda$ is dominant. Clearly, λ' still belongs to $\mathrm{Conv}(W \cdot \mu)$. Furthermore, since λ is integral, $w \cdot \lambda - \lambda$ is an element of the root lattice. After all, the definition of integrality implies that $s_\alpha \cdot \lambda - \lambda$ is an integer multiple of α; since the s_α's generate W, the result holds for all $w \in W$. Thus, $\mu - \lambda' = \mu - \lambda + \lambda - \lambda'$ is an element of the root lattice, which means that λ' also satisfies the two conditions of Theorem 10.1. Thus, by Proposition 10.2, λ' is a weight of π, which means that $\lambda = w^{-1} \cdot \lambda'$ is also a weight. $\qquad\square$

10.2 The Casimir Element

In this section, we construct an element of $U(\mathfrak{g})$ known as the Casimir, which belongs to the center of $U(\mathfrak{g})$. The Casimir element is important in its own right and also plays a crucial role in the proof of complete reducibility (Sect. 10.3) and of the Weyl character formula (Sect. 10.8).

Definition 10.4. Let X_j be an orthonormal basis for $\mathfrak{k}$. Then the **Casimir element** C of $U(\mathfrak{g})$ is given by

$$C = -\sum_j X_j^2.$$

Proposition 10.5. *1. The value of C is independent of the choice of orthonormal basis for $\mathfrak{k}$.*
2. The element C is in the center of $U(\mathfrak{g})$.

Proof. If $\{X_j\}$ and $\{Y_j\}$ are two different orthonormal bases for $\mathfrak{k}$, then there is an orthogonal matrix R such that

$$Y_j = \sum_k R_{kj} X_k.$$

Then

$$\sum_j Y_j^2 = \sum_{j,k,l} R_{kj} X_k R_{lj} X_l$$

$$= \sum_{k,l} \sum_j R_{kj} (R^{tr})_{jl} X_k X_l$$

$$= \sum_{k,l} \delta_{kl} X_k X_l$$

$$= \sum_k X_k^2.$$

This shows that C is independent of the choice of basis.

Meanwhile, if $\{X_j\}$ is an orthonormal basis, let c_{jkl} be the associated structure constants:

$$[X_j, X_k] = \sum_l c_{jkl} X_l.$$

Note that for a fixed j, the matrix A^j given by $(A^j)_{kl} = c_{jlk}$ is the matrix representing the operator ad_{X_j} in the chosen basis. Since the inner product on $\mathfrak{k}$ is Ad-K-invariant, ad_{X_j} is a skew operator, which means that c_{jkl} is skew symmetric in k and l for a fixed j. If we compute the commutator of some X_j with C in $U(\mathfrak{g})$, we obtain

$$[X_j, C] = \sum_k [X_j, X_k^2]$$

$$= \sum_k ([X_j, X_k] X_k + X_k [X_j, X_k])$$

$$= \sum_{k,l} c_{jkl} X_l X_k + \sum_{k,l} c_{jkl} X_k X_l. \tag{10.2}$$

In the first sum in the last line of (10.2), we may reverse the labeling of the summation variables and use the skew symmetry of c_{jkl} in k and l to obtain

$$[X_j, C] = \sum_{k,l}(-c_{jkl} + c_{jkl})X_l X_k = 0.$$

Thus, C commutes the each X_j. But since $U(\mathfrak{g})$ is generated by elements of $\mathfrak{g}$, we see that C actually commutes with every element of $U(\mathfrak{g})$.   □

Let π be a finite-dimensional, irreducible representation of $\mathfrak{g}$. By Proposition 9.9, we can extend π to a representation of $U(\mathfrak{g})$, which we also denote by π. We now show $\pi(C)$ is a constant multiple of the identity. The formula for the constant involves the element δ (Definition 8.37), equal to half the sum of the positive roots. The element δ also arises in our discussion of the Weyl character formula, the Weyl dimension formula, and the Kostant multiplicity formula.

Proposition 10.6. *Let (π, V) be a finite-dimensional irreducible representation of $\mathfrak{g}$ (extended to $U(\mathfrak{g})$) with highest weight μ. Then we have*

$$\pi(C) = -\sum_j \pi(X_j)^2 = c_\mu I,$$

where c_μ is a constant given by

$$c_\mu = \langle \mu + \delta, \mu + \delta \rangle - \langle \delta, \delta \rangle.$$

Furthermore, $c_\mu \geq 0$ with $c_\mu = 0$ only if $\mu = 0$.

Lemma 10.7. *Let $X \in \mathfrak{g}_\alpha$ be a unit vector, so that $X^* \in \mathfrak{g}_{-\alpha}$. Then under our usual identification of $\mathfrak{h}$ with $\mathfrak{h}^*$, we have*

$$[X, X^*] = \alpha.$$

Proof. According to Lemma 7.22, we have

$$\langle [X, X^*], H_\alpha \rangle = \langle \alpha, H_\alpha \rangle \langle X^*, X^* \rangle = \langle \alpha, H_\alpha \rangle, \tag{10.3}$$

since X (and thus, also, X^*) is a unit vector. On the other hand, we know that the commutator of any element of $\mathfrak{g}_\alpha$ with any element of $\mathfrak{g}_{-\alpha}$ is a multiple of H_α, which is (under our identification of $\mathfrak{h}$ with $\mathfrak{h}^*$) a multiple of α. But if $[X, X^*] = c\alpha$, (10.3) tells us that c must equal 1.   □

Proof of Proposition 10.6. Since C is in the center of the universal enveloping algebra, $\pi(C)$ commutes with each $\pi(X)$, $X \in \mathfrak{g}$. Thus, by Schur's lemma, $\pi(C)$ must act as a constant multiple c_μ of the identity operator.

To compute the constant c_μ, we choose an orthonormal basis for $\mathfrak{k}$ as follows. Take an orthonormal basis $H_1, \ldots, H_j$ for $\mathfrak{t}$. Then for each $\alpha \in R^+$, choose a unit vector X_α in $\mathfrak{g}_\alpha$, so that X_α^* is a unit vector in $\mathfrak{g}_{-\alpha}$. Then the elements

$$Y_\alpha := (X_\alpha + X_\alpha^*)/(\sqrt{2}i); \quad Z_\alpha := (X_\alpha - X_\alpha^*)/\sqrt{2},$$

satisfy $Y_\alpha^* = -Y_\alpha$ and $Z_\alpha^* = -Z_\alpha$, which shows that these elements belong to $\mathfrak{k}$. Since $\mathfrak{g}_\alpha$ is orthogonal to $\mathfrak{g}_{-\alpha}$, it is easy to see that these vectors are also unit vectors and orthogonal to each other. The set of vectors of the form H_j, $j = 1, \ldots, r$, and $Y_\alpha, \alpha \in R^+$, and $Z_\alpha, \alpha \in R^+$, form an orthonormal basis for $\mathfrak{k}$.

We compute that

$$Y_\alpha^2 = -\frac{1}{2}(X_\alpha^2 + X_\alpha X_\alpha^* + X_\alpha^* X_\alpha + (X_\alpha^*)^2)$$

$$Z_\alpha^2 = \frac{1}{2}(X_\alpha^2 - X_\alpha X_\alpha^* - X_\alpha^* X_\alpha + (X_\alpha^*)^2),$$

so that

$$-Y_\alpha^2 - Z_\alpha^2 = X_\alpha X_\alpha^* + X_\alpha^* X_\alpha$$
$$= 2X_\alpha^* X_\alpha + [X_\alpha, X_\alpha^*].$$

Thus, the Casimir element C may be computed as

$$C = \sum_{\alpha \in R^+} (2X_\alpha^* X_\alpha + [X_\alpha, X_\alpha^*]) - \sum_{j=1}^r H_j^2.$$

Suppose now that v is a highest weight vector, and compute that

$$\pi(C)v = -\sum_{j=1}^r \pi(H_j)^2 v + 2 \sum_{\alpha \in R^+} \pi(X_\alpha^*)\pi(X_\alpha)v$$
$$+ \sum_{\alpha \in R^+} \pi([X_\alpha, X_\alpha^*])v.$$

Since v is a highest weight vector, $\pi(X_\alpha)v = 0$, and since $H_j \in \mathfrak{t} \subset \mathfrak{h}$, we have $\pi(H_j)^2 v = \langle \mu, H_j \rangle^2 v$. Now, since the roots live in $i\mathfrak{t}$ and $\langle \mu, \alpha \rangle$ is real for all roots α, we see that μ also lives in $i\mathfrak{t}$. Thus, $\langle \mu, H_j \rangle$ is pure imaginary and $\langle \mu, H_j \rangle^2 = -|\langle \mu, H_j \rangle|^2$. Using Lemma 10.7, we then see that

$$\pi(C)v = \left(\sum_{j=1}^r |\langle \mu, H_j \rangle|^2 + \sum_{\alpha \in R^+} \langle \mu, \alpha \rangle \right) v, \qquad (10.4)$$

where the coefficient of v on the right-hand side of (10.4) must equal c_μ.

Now, since $\{H_j\}_{j=1}^r$ is an orthonormal basis for $\mathfrak{t}$, the first term in the coefficient of v equals $\langle \mu, \mu \rangle$. Moving the sum over α inside the inner product in the second term gives

$$c_\mu = \langle \mu, \mu \rangle + \langle \mu, 2\delta \rangle, \qquad (10.5)$$

which is the same as $\langle \mu + \delta, \mu + \delta \rangle - \langle \delta, \delta \rangle$. Finally, we note that since μ is dominant, $\langle \mu, \alpha \rangle \geq 0$ for every positive root, from which it follows that $\langle \mu, 2\delta \rangle = \sum_{\alpha \in R^+} \langle \mu, \alpha \rangle$ is non-negative. Thus, $c_\mu \geq 0$ for all μ and $c_\mu > 0$ if $\mu \neq 0$.   □

10.3   Complete Reducibility

Let $\mathfrak{g}$ be a complex semisimple Lie algebra. Then $\mathfrak{g}$ is isomorphic to the complexification of a the Lie algebra of a compact matrix Lie group K. (This is, for us, true by definition; see Definition 7.1.) Actually, it is possible to show that there exists a *simply connected* compact Lie group K with Lie algebra $\mathfrak{k}$ such that $\mathfrak{g} \cong \mathfrak{k}_\mathbb{C}$. (This claim follows from Theorems 4.11.6 and 4.11.10 of [Var].) Assuming this result, we can see that every finite-dimensional representation π of $\mathfrak{g}$ gives rise to a representation of K by restricting π to $\mathfrak{k}$ and then applying Theorem 5.6. Theorem 4.28 then tells us that every finite-dimensional representation of $\mathfrak{g}$ is completely reducible; compare Corollary 4.11.11 in [Var].

Rather than relying on the existence of a simply connected K, we now give an algebraic proof of complete reducibility. Our proof makes use of the Casimir element and, in particular, the fact that the eigenvalue of the Casimir is nonzero in each nontrivial irreducible representation.

Proposition 10.8. *If (π, V) is a one-dimensional representation of $\mathfrak{g}$, then $\pi(X) = 0$ for all $X \in \mathfrak{g}$.*

Proof. By Theorem 7.8, $\mathfrak{g}$ decomposes as a Lie algebra direct sum of simple algebras $\mathfrak{g}_j$. Since the kernel of $\pi|_{\mathfrak{g}_j}$ is a ideal, the restriction of π to $\mathfrak{g}_j$ must be either zero or injective. But since $\dim \mathfrak{g}_j \geq 2$ and $\dim(\mathrm{End}(V)) = 1$, this restriction cannot be injective, so it must be zero for each j.   □

Theorem 10.9. *Every finite-dimensional representation of a semisimple Lie algebra is completely reducible.*

We begin by considering what appears to be a very special case of the theorem.

Lemma 10.10. *Suppose (π, V) is a finite-dimensional representation of $\mathfrak{g}$ and that W is an invariant subspace of V of codimension 1. Then V decomposes as $W \oplus U$ for some invariant subspace U of V.*

Proof. We consider first the case in which W is irreducible. If W is one-dimensional, then by Proposition 10.8, the restriction of π to W is zero. Since W is one-dimensional and has codimension 1, V must be two dimensional. The space of linear operators on V that are zero on W then has dimension 2. (Pick a basis for V consisting of a nonzero element of W and another linearly independent vector; in this basis, any such operator will have first column equal to zero.) On the other hand, each of the simple summands $\mathfrak{g}_j$ in the decomposition of $\mathfrak{g}$ has dimension at least 3, since an algebra of dimension 2 cannot be simple. (An algebra of dimension 2 is either commutative or has a commutator ideal of dimension 1.) Thus, the restriction of π to $\mathfrak{g}_j$ cannot be injective and thus must be zero. We conclude, then, that π is identically zero, and we may take U to be any subspace of V complementary to W.

Assume now that W is irreducible and nontrivial. Let C be the Casimir element of $U(\mathfrak{g})$ and let $\pi(C)$ denote the action of C on V, by means of the extension of π to $U(\mathfrak{g})$. By Proposition 10.6, the restriction of $\pi(C)$ to W is a nonzero multiple c of the identity. On the other hand, since V/W is one dimensional, the action of $\mathfrak{g}$ on V/W is trivial, by Proposition 10.8. Thus, the action of $\pi(C)$ on V/W is zero, from which it follows that $\pi(C)$ must have a nonzero kernel. (If we pick a basis for V consisting of a basis for W together with one other vector, the bottom row of the matrix of $\pi(C)$ in this basis be identically zero.) Because $\pi(C)$ commutes with each $\pi(X)$, this kernel is an invariant subspace of V. Furthermore, $\ker(\pi(C)) \cap W = \{0\}$, because $\pi(C)$ acts as a nonzero scalar on W. Thus, $U := \ker(\pi(C))$ is the desired invariant complement to W.

We consider next the case in which W has a nontrivial invariant subspace W', which is, of course, also an invariant subspace of V. Then W/W' is a codimension-one invariant subspace of V/W'. Thus, by induction on the dimension of W, we may assume that W/W' has a one-dimensional invariant complement, say Y/W'. Then W' is a codimension-one invariant subspace of Y. Since $\dim W' < \dim W$, we may apply induction again to find a one-dimensional invariant complement U to W' in Y, so that $Y = W' \oplus U$. Now, $Y \cap W = W'$ and $U \cap W' = 0$, from which it follows that $U \cap W = \{0\}$. Thus, U is the desired complement to W in V. $\square$

Proof of Theorem 10.9. Let (π, V) be a finite-dimensional representation of $\mathfrak{g}$ and let W be nontrivial invariant subspace of V. We now look for an invariant complement to W, that is, an invariant subspace U such that $V = W \oplus U$. If we can always find such an U, then we may proceed by induction on the dimension to establish complete reducibility. If $A : V \to W$ is an intertwining map, the kernel of A will be an invariant subspace of V. If, in addition, the restriction of A to W is injective, then $\ker(A) \cap W = \{0\}$ and, by a dimension count, V will decompose as $W \oplus \ker(A)$. Now, the simplest way to ensure that A is injective on W is to assume that $A|_W$ is a nonzero multiple of the identity (If, for example, the restriction of A to W is the identity, we may think of A as a "projection" of V onto W.) If W is irreducible, nontrivial, and of codimension 1, we may take $A = \pi(C)$ as in the proof of Lemma 10.10.

To construct A in general, we proceed as follows. Let $\mathrm{Hom}(V, W)$ denote the space of linear maps of V to W (not necessarily intertwining maps). This space can be viewed as a representation of $\mathfrak{g}$ by means of the action

$$X \cdot A = \pi(X)A - A\pi(X) \tag{10.6}$$

for all $X \in \mathfrak{g}$ and $A \in \mathrm{Hom}(V, W)$. Let $\mathcal{V}$ denote the subspace of $\mathrm{Hom}(V, W)$ consisting of those maps A whose restriction to W is a scalar multiple c_A of the identity and let $\mathcal{W}$ denote the subspace of $\mathrm{Hom}(V, W)$ consisting of those maps whose restriction to W is zero. The map $A \mapsto c_A$ is a linear functional on $\mathcal{V}$ which is easily seen not to be identically zero. The space $\mathcal{W}$ is the kernel of this linear functional and is, thus, a codimension-one subspace of $\mathcal{V}$.

We now claim that both V and W are invariant subspaces of $\mathrm{Hom}(V, W)$. To see this, suppose $A \in \mathrm{Hom}(V, W)$ is equal to cI on W. Then for $w \in W$, we have

$$(X \cdot A)w = \pi(X)Aw - A\pi(X)w$$
$$= \pi(X)cw - c\pi(X)w$$
$$= 0,$$

because $\pi(X)w$ is again in W, showing that $X \cdot A$ is actually in W. Thus, the action of $\mathfrak{g}$ maps V into W, showing that both V and W are invariant.

Since W has codimension 1 in V, we are in the situation of Lemma 10.10. Thus, there exists an invariant complement $\mathcal{U}$ to W in V. Let A be a nonzero element of $\mathcal{U}$. Since A is not in W, the restriction of A to W is a *nonzero* scalar. Furthermore, since $\mathcal{U}$ is one dimensional, Proposition 10.8 tells us that the action of $\mathfrak{g}$ on A is zero, meaning that A commutes with each $\pi(X)$. That is to say, A is an intertwining map of V to W. Thus, A is precisely the operator we were looking for: an intertwining map of V to W whose restriction to W is a nonzero multiple of the identity. Now, A maps V into W, and actually maps *onto* W, since A acts as a nonzero scalar on W itself. Thus, $\dim(\ker(A)) = \dim V - \dim W$ and since $\ker(A) \cap W = \{0\}$, we conclude that $U := \ker(A)$ is an invariant complement to W.

We conclude, then, that every nontrivial invariant subspace W of V has an invariant complement U. By induction on the dimension, we may then assume that both W and U decompose as direct sums of irreducible invariant subspaces, in which case, V also has such a decomposition. $\qquad\square$

10.4   The Weyl Character Formula

The character formula is a major result in the structure of the irreducible representations of $\mathfrak{g}$. Its consequences include a formula for the dimension of an irreducible representation (Sect. 10.5) and a formula for the multiplicities of the weights in an irreducible representation (Sect. 10.6).

We now introduce the notion of the *character* of a finite-dimensional representation of a group.

Definition 10.11. Suppose (π, V) is a finite-dimensional representation of a complex semisimple Lie algebra $\mathfrak{g}$. The **character** of π is the function $\chi_\pi : \mathfrak{g} \to \mathbb{C}$ given by

$$\chi_\pi(X) = \mathrm{trace}(e^{\pi(X)}).$$

It turns out that the character of π encodes many interesting properties of π. We will give a formula for the character of an irreducible representation of a semisimple Lie algebra in terms of the highest weight of the representation. This section gives the statement of the character formula, Sects. 10.5 and 10.6 give consequences of it, and Sect. 10.8 gives the proof.

If the representation π of $\mathfrak{g} = \mathfrak{k}_\mathbb{C}$ comes from a representation Π of the compact group K, then for $X \in \mathfrak{k}$, we have $\chi_\pi(X) = \mathrm{trace}(\Pi(e^X))$. In Chapter 12, the group version of a character, namely, the function given by $x \mapsto \mathrm{trace}(\Pi(x))$, $x \in K$, plays a key role in the compact group approach to representation theory.

Proposition 10.12. *If (π, V) is a finite-dimensional representation of $\mathfrak{g}$, we have the following results.*

1. *The dimension of V is equal to the value of χ_π at the origin:*

$$\dim(V) = \chi_\pi(0).$$

2. *Suppose V decomposes as a direct sum of weight spaces V_λ with multiplicity $\mathrm{mult}(\lambda)$. Then for $H \in \mathfrak{h}$ we have*

$$\chi_\pi(H) = \sum_\lambda \mathrm{mult}(\lambda)e^{\langle \lambda, H \rangle}. \tag{10.7}$$

Proof. The first point holds because $\mathrm{trace}(I) = \dim V$. Meanwhile, since $\pi(H)$ acts as $\langle \lambda, H \rangle I$ in each weight space V_λ, the second point follows from the definition of χ_π. □

Example 10.13. Let π denote the irreducible representation of $\mathrm{sl}(2; \mathbb{C})$ of dimension $m + 1$ and let

$$H = \begin{pmatrix} 1 & 0 \\ 0 & -1 \end{pmatrix}.$$

Then

$$\chi_\pi(aH) = \chi_\pi\left(\begin{pmatrix} a & 0 \\ 0 & -a \end{pmatrix}\right)$$

$$= e^{ma} + e^{(m-2)a} + \cdots + e^{-ma}. \tag{10.8}$$

We may also compute χ_π as

$$\chi_\pi(aH) = \frac{\sinh((m+1)a)}{\sinh(a)} \tag{10.9}$$

whenever a not an integer multiple of $i\pi$.

Proof. The eigenvalues of $\pi_m(H)$ are $m, m-2, \ldots, -m$, from which (10.8) follows. To obtain (10.9), we note that

$$(e^a - e^{-a})\chi_\pi(aH)$$

$$= e^{(m+1)a} + e^{(m-1)a} + \cdots + e^{-(m-1)a}$$

$$- e^{(m-1)a} - \cdots - e^{-(m-1)a} - e^{-(m+1)a}$$

$$= e^{(m+1)a} - e^{-(m+1)a}, \tag{10.10}$$

so that

$$\chi_\pi(aH) = \frac{e^{(m+1)a} - e^{-(m+1)a}}{e^a - e^{-a}} = \frac{\sinh((m+1)a)}{\sinh(a)},$$

as claimed.   □

Note that in deriving (10.9) from (10.8), we multiplied the character by a cleverly chosen combination of exponentials $(e^a - e^{-a})$, leading to a large cancellation, so that only two terms remain in (10.10). The Weyl character formula asserts that we can perform a similar trick for the characters of irreducible representations of arbitrary semisimple Lie algebras.

Recall that each element w of the Weyl group W acts as an orthogonal linear transformation of $i\mathfrak{t} \subset \mathfrak{h}$. We let $\det(w)$ denote the determinant of this transformation, so that $\det(w) = \pm 1$. Recall from Definition 8.37 that δ denotes half the sum of the positive roots. We are now ready to state the main result of this chapter.

Theorem 10.14 (Weyl Character Formula). *If (π, V_μ) is an irreducible representation of $\mathfrak{g}$ with highest weight μ, then*

$$\chi_\pi(H) = \frac{\sum_{w \in W} \det(w) e^{\langle w \cdot (\mu+\delta), H \rangle}}{\sum_{w \in W} \det(w) e^{\langle w \cdot \delta, H \rangle}} \qquad (10.11)$$

for all $H \in \mathfrak{h}$ for which the denominator is nonzero.

Since we will have frequent occasion to refer to the function in the denominator in (10.11), we give it a name.

Definition 10.15. Let $q : \mathfrak{h} \to \mathbb{C}$ be the function given by

$$q(H) = \sum_{w \in W} \det(w) e^{\langle w \cdot \delta, H \rangle}.$$

The function q is called the **Weyl denominator**.

The character formula may also be written as

$$q(H)\chi_\pi(H) = \sum_{w \in W} \det(w) e^{\langle w \cdot (\mu+\delta), H \rangle}. \qquad (10.12)$$

Let us pause for a moment to reflect on what is going on in (10.12). The Weyl denominator $q(H)$ is a sum of $|W|$ exponentials with coefficients equal to ± 1. Meanwhile, the character $\chi_\pi(H)$ is a large sum of exponentials with positive integer coefficients, as in (10.7). When we multiply these two functions, we seemingly obtain an even larger sum of exponentials of the form

$$e^{\langle w \cdot \delta + \lambda, H \rangle},$$

for $w \in W$ and λ a weight of π, with integer coefficients.

The character formula, however, asserts that most of these terms are not actually present. Specifically, the only exponentials that actually appear are those of the form $e^{\langle w\cdot(\mu+\delta),H\rangle}$, which occurs with a coefficient of $\det(w)$. The point is that, in most cases, if a weight η can be written in the form $\eta = w\cdot\delta + \lambda$, with λ a weight of π, then η can be written in this form in *more than one way*. The character formula asserts that unless η is in the Weyl-group orbit of $\delta + \mu$, the coefficient of $e^{\langle\eta,H\rangle}$, after all the different contributions are taken into account, ends up being zero.

By contrast, the weight $\eta = \delta+\mu$ only occurs once, since it corresponds to taking the highest weight occurring in q (namely δ) and the highest weight occurring in χ_π (namely the highest weight μ of π). (Note that by Propositions 8.38 and 8.27, the elements of the form $w\cdot\delta$ are all distinct. Then by Proposition 8.42, $w\cdot\delta \preceq \delta$ for all w.) Furthermore, since the weights occurring in both q and χ_π are Weyl invariant, the weight $w\cdot(\delta+\mu)$ also occurs only once. The Weyl character formula, then, can be expressed as stating that if we compute the product $q\chi_\pi$, a huge cancellation occurs: Every exponential in the product ends canceling out to zero, except for those that occur only once, namely those of the form $w\cdot(\delta+\mu)$.

In the case of $\mathsf{sl}(2;\mathbb{C})$, we have already observed how this cancellation occurs, in (10.10). The Weyl denominator is equal to $e^a - e^{-a}$ in this case. Each exponential e^{la} occurring in the product $(e^a-e^{-a})\chi_\pi(aH)$ will occur once with a plus sign (from $e^a e^{(l-1)a}$) and once with a minus sign (from $e^{-a}e^{(l+1)a}$), except for the extreme cases $l = \pm(m+1)$. This cancellation occurs because the multiplicity of the weight $l-1$ equals the multiplicity of the weight $l+1$, namely 1.

Figure 10.4, meanwhile, illustrates the case of the irreducible representation of $\mathsf{sl}(3;\mathbb{C})$ with highest weight $(1,2)$. The top part of the figure indicates the six exponentials occurring in the Weyl denominator, with alternating signs. The middle part of the figure indicates the exponentials in the character of the representation with highest weight $(1,2)$. The bottom part of the figure shows the product of the Weyl denominator and the character, in which only the six exponentials indicated by black dots survive. The white dots in the bottom part of the figure indicate exponentials that occur at least once in the product, but which end up with a coefficient of zero.

The cancellation inherent in the Weyl character formula reflects a very special structure to the multiplicities of the various weights that occur. For an integral element λ, each product of the form $e^{\langle w\cdot\delta,H\rangle}e^{\langle\eta,H\rangle}$, where $\eta = \lambda - w\cdot\delta$, makes a contribution of $\det(w)\mathrm{mult}(\eta)$ to the coefficient of $e^{\langle\lambda,H\rangle}$. Thus, if λ is *not* in the Weyl orbit of $\mu+\delta$, the Weyl character formula implies that

$$\sum_{w\in W}\det(w)\mathrm{mult}(\lambda - w\cdot\delta) = 0, \quad \lambda \notin W\cdot(\mu+\delta). \tag{10.13}$$

(In (10.13), some of elements of the form $\lambda - w\cdot\delta$ may not actually be weights of π, in which case, the multiplicity should be considered to be zero.) In the case of $\mathsf{sl}(3;\mathbb{C})$, the weights of the form $\lambda - w\cdot\delta$ form a small hexagon around the weight λ. Figure 10.5 illustrates how the alternating sum of multiplicities around one such hexagon equals zero.

Fig. 10.4 The product of the
Weyl denominator (*top*) and
the character of a
representation (*middle*)
produces an alternating sum
of exponentials (*bottom*). The
white dots indicate
exponentials that occur at
least once in the product but
that end up with a coefficient
of zero

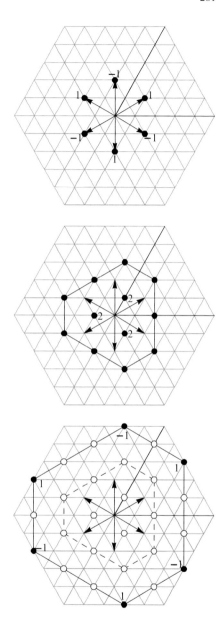

We will see in Sect. 10.6 that the character formula leads to a formula for the
multiplicities of all the weights occurring in a particular irreducible representation.

Before concluding this section, we establish a technical result that we will use in
the remainder of the chapter.

Proposition 10.16. *The exponential functions* $H \mapsto e^{\langle \lambda, H \rangle}$, *with* $\lambda \in \mathfrak{h}$, *are
linearly independent in* $C^{\infty}(\mathfrak{h})$.

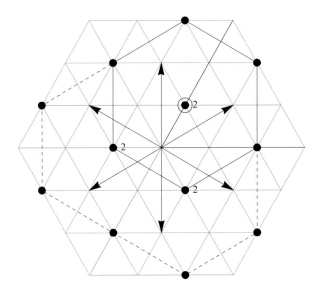

Fig. 10.5 We compute the alternating sum of multiplicities around the hexagon enclosing the circled weight, beginning in the fundamental Weyl chamber and proceeding counterclockwise. The result is $1 - 1 + 1 - 2 + 2 - 1 = 0$

The proposition means that if a function $f \in C^\infty(\mathfrak{h})$ can be expressed as a finite linear combination of exponentials, it has a *unique* such expression.

Proof. We need to show that if the function $f : \mathfrak{h} \to \mathbb{C}$ given by

$$f(H) = c_1 e^{\langle \lambda_1, H \rangle} + \cdots + c_n e^{\langle \lambda_n, H \rangle}$$

is identically zero, where $\lambda_1, \ldots, \lambda_n$ are distinct elements of $\mathfrak{h}$, then $c_1 = \cdots = c_n = 0$. If $n = 1$, we evaluate at $H = 0$ and conclude that c_1 must be zero. If $n > 1$, we choose, for each $k = 2, \ldots, n$, some $H_k \in \mathfrak{h}$ such that $\langle \lambda_1, H_k \rangle \neq \langle \lambda_k, H_k \rangle$. Since f is identically zero, so is the function

$$g := (D_{H_2} - \langle \lambda_2, H_2 \rangle) \cdots (D_{H_n} - \langle \lambda_n, H_n \rangle) f$$

where D_X denotes the directional derivative in the direction of X:

$$(D_X f)(H) = \frac{d}{dt} f(H + tX) \bigg|_{t=0}. \tag{10.14}$$

Direct calculation then shows that

$$g(H) = \sum_{j=1}^{n} c_j \left(\prod_{k=2}^{n} (\langle \lambda_j, H_k \rangle - \langle \lambda_k, H_k \rangle) \right) e^{\langle \lambda_j, H \rangle}$$

$$= c_1 \left(\prod_{k=2}^{n} (\langle \lambda_1, H_k \rangle - \langle \lambda_k, H_k \rangle) \right) e^{\langle \lambda_1, H \rangle}. \tag{10.15}$$

By evaluating at $H = 0$ and noting that, by construction, the product in the second line of (10.15) is nonzero, we conclude that $c_1 = 0$. An entirely similar argument then shows that each $c_j = 0$ as well.   □

Corollary 10.17. *Suppose* $f \in C^{\infty}(\mathfrak{h})$ *can be expressed as a finite linear combination of exponentials* $e^{\langle \lambda, H \rangle}$, *with* λ *integral,*

$$f(H) = \sum_{\lambda} c_{\lambda} e^{\langle \lambda, H \rangle}.$$

If f *satisfies* $f(w \cdot H) = \det(w) f(H)$, *then*

$$c_{w \cdot \lambda} = \det(w) c_{\lambda}$$

for each λ *occurring in the expansion of* f.

Proof. On the one hand,

$$f(w \cdot H) = \sum_{\lambda} c_{\lambda} e^{\langle w^{-1} \cdot \lambda, H \rangle} = \sum_{\eta} c_{w \cdot \eta} e^{\langle \eta, H \rangle}. \tag{10.16}$$

On the other hand,

$$f(w \cdot H) = \det(w) f(H) = \sum_{\eta} \det(w) c_{\eta} e^{\langle \eta, H \rangle}. \tag{10.17}$$

By the linear independence of the exponentials, the only way the expansions in (10.16) and (10.17) can agree is if $c_{w \cdot \eta} = \det(w) c_{\eta}$ for all η.   □

10.5   The Weyl Dimension Formula

Before coming to the proof of the Weyl character formula, we derive two important consequences of it, the Weyl dimension formula (described in this section) and the Kostant multiplicity formula (described in the next section).

   The dimension of a representation is equal to the value of the character at the identity (Proposition 10.12). In the Weyl character formula, however, both the numerator and the denominator are equal to zero when $H = 0$. In the case of $\mathsf{sl}(2; \mathbb{C})$ case, for example, the character formula reads

$$\chi_\pi(aH) = \frac{\sinh((m+1)a)}{\sinh a}.$$

The limit of this expression as θ tends to zero may be computed by l'Hospital's rule to be $m + 1$, which is, of course, the dimension of V_m.

In the general case, we will expand both numerator and denominator of the character formula in a power series. We will see that in both numerator and denominator, the first nonzero term has degree k, where

$$k = \text{the number of positive roots.}$$

To evaluate the limit of this expression at the origin, we will develop a version of l'Hospital's rule. The limit is then computed as the ratio of a certain k-fold derivative of the numerator and the corresponding k-fold derivative of the denominator, evaluated at the origin. The result of this analysis is expressed in the following theorem.

Theorem 10.18. *If (π_μ, V_μ) is the irreducible representation of $\mathfrak{g}$ with highest weight μ, then the dimension of V_μ may be computed as*

$$\dim(V_\mu) = \frac{\prod_{\alpha \in R^+} \langle \alpha, \mu + \delta \rangle}{\prod_{\alpha \in R^+} \langle \alpha, \delta \rangle}.$$

Note that both $\mu + \delta$ and δ are strictly dominant elements, so that all the factors in both the numerator and the denominator are nonzero.

A function P on $\mathfrak{h}$ is called a **polynomial** if for every basis of $\mathfrak{h}$, the function P is a polynomial in the coordinates $z_1, \ldots, z_r$ associated to that basis. That is to say, P should be expressible as a finite linear combination of terms of the form $z_1^{n_1} z_2^{n_2} \cdots z_r^{n_r}$, where $n_1, \ldots, n_r$ are non-negative integers. It is easy to see that if P is a polynomial in any one basis, then it is also a polynomial in every other basis as well. A polynomial P is said to be homogeneous of degree l if $P(cH) = c^l P(H)$ for all constants c and all $H \in \mathfrak{h}$.

Definition 10.19. Let $P : \mathfrak{h} \to \mathbb{C}$ be the function given by

$$P(H) = \prod_{\alpha \in R^+} \langle \alpha, H \rangle.$$

Note that P is a product of k linear functions and is, thus, a homogeneous polynomial of degree k. The dimension formula may be restated in terms of P as

$$\dim(V_\mu) = \frac{P(\mu + \delta)}{P(\delta)}.$$

A key property of the polynomial P is its behavior under the action of the Weyl group.

Definition 10.20. A function $f : \mathfrak{h} \to \mathbb{C}$ is said to be **Weyl alternating** if

$$f(w \cdot H) = \det(w) f(H)$$

for all $w \in W$ and $H \in \mathfrak{h}$.

It is easy to see, for example, that the Weyl denominator q is Weyl alternating (Exercise 3).

Proposition 10.21. *1. The function P is Weyl alternating.*
2. If $f : \mathfrak{h} \to \mathbb{C}$ is a Weyl-alternating polynomial, there is a polynomial $g : \mathfrak{h} \to \mathbb{C}$ such that

$$f(H) = P(H)g(H).$$

In particular, if f is homogeneous of degree $l < k$, then f must be identically zero, and if f is homogeneous of degree k, then f must be a constant multiple of P.

Proof. For any $w \in W$, consider the collection of roots of the form $w^{-1} \cdot \alpha$ for $\alpha \in R^+$. Since R^+ contains exactly one element out of each pair $\pm \alpha$ of roots, the same is true of the collection of $w^{-1} \cdot \alpha$'s, with w fixed and α varying over R^+. Thus,

$$
\begin{aligned}
P(w \cdot H) &= \prod_{\alpha \in R^+} \langle \alpha, w \cdot H \rangle \\
&= \prod_{\alpha \in R^+} \langle w^{-1} \cdot \alpha, H \rangle \\
&= (-1)^j \prod_{\alpha \in R^+} \langle \alpha, H \rangle,
\end{aligned}
$$

where j is the number of negative roots in the collection $\{w^{-1} \cdot \alpha\}_{\alpha \in R^+}$.
Suppose first that $w = w^{-1} = s_\alpha$, where α is a positive simple root. According to Proposition 8.30, s_α permutes the positive roots different from α, whereas $s_\alpha \cdot \alpha = -\alpha$. Thus, $j = 1$ in this case, and so

$$P(w \cdot H) = -P(H) = \det(w) P(H),$$

since the determinant of a reflection is -1. By Proposition 8.24, every element w of W is a product of reflections associated to positive simple roots, and so

$$
\begin{aligned}
P(w \cdot H) &= P(s_{\alpha_{j_1}} \cdots s_{\alpha_{j_N}} \cdot H) \\
&= (-1)^N P(H) \\
&= \det(s_{\alpha_{j_1}} \cdots s_{\alpha_{j_N}}) P(H),
\end{aligned}
$$

showing that P is alternating.

Suppose now that f is any Weyl-alternating polynomial. Then for any positive root α, if $\langle \alpha, H \rangle = 0$, we will have

$$f(H) = f(s_\alpha \cdot H) = -f(H),$$

since $\det(s_\alpha) = -1$. Thus, f must vanish on the hyperplane orthogonal to α, which we denote as V_α. It is then not hard to show (Exercise 4) that f is divisible in the space of polynomials by the linear function $\langle \alpha, H \rangle$, that is, $f(H) = \langle \alpha, H \rangle f'(H)$ for some polynomial f'. Now, if β is any positive root different from α, the polynomial f' must vanish at least on the portion of V_β not contained in V_α. But since β is not a multiple of α, V_β is distinct from V_α, so that $V_\beta \cap V_\alpha$ is a subspace of dimension $r - 2$. Thus, $V_\beta - (V_\alpha \cap V_\beta)$ is dense in V_β. Since f' is continuous, it must actually vanish on all of V_β.

It follows that f' is divisible in the space of polynomials by $\langle \beta, H \rangle$, so that

$$f(H) = \langle \alpha, H \rangle \langle \beta, H \rangle f''(H)$$

for some polynomial f''. Proceeding on in the same way, we see that f contains a factor of $\langle \alpha, H \rangle$ for each positive root α, meaning that

$$f(H) = \left(\prod_{\alpha \in R^+} \langle \alpha, H \rangle \right) g(H)$$

$$= P(H)g(H)$$

for some polynomial g, as claimed. $\square$

Recall the notion of directional derivative D_X, defined in (10.14).

Lemma 10.22. *Let $\mathcal{A}$ denote the differential operator*

$$\mathcal{A} = \prod_{\alpha \in R^+} D_\alpha.$$

For any $\lambda \in \mathfrak{h}$, let $f_\lambda : \mathfrak{h} \to \mathbb{C}$ be the function given by

$$f_\lambda(H) = \sum_{w \in W} \det(w) e^{\langle w \cdot \lambda, H \rangle}.$$

Then f_λ is Weyl alternating and is given by a convergent power series of the form

$$f_\lambda(H) = c_\lambda P(H) + \text{terms of degree at least } k + 1 \tag{10.18}$$

for some constant c_λ. Furthermore, $(\mathcal{A}P)(0) \neq 0$ and the constant c_λ may be computed as

$$c_\lambda = \frac{(\mathcal{A}f_\lambda)(0)}{(\mathcal{A}P)(0)},$$

where

$$(\mathcal{A}f_\lambda)(0) = |W| P(\lambda).$$

Proof. The proof that f_λ is Weyl alternating is elementary. Since f_λ is a sum of exponentials, it is a real-analytic function, meaning that it can be expanded in a convergent power series in the coordinates $x_1, \ldots, x_r$ associated to any basis for $\mathfrak{h}$. In the power-series expansion of f_λ, we collect together all the terms that are homogeneous of degree l. Thus, f_λ is expressible as the sum of homogeneous polynomials $q_{\lambda,l}$ of degree l. Since f_λ is Weyl-alternating, it is not hard to show (Exercise 5) that each of the polynomials $q_{\lambda,l}$ is also Weyl-alternating. Thus, by Proposition 10.21, all the polynomials $q_{\lambda,l}$ with $l < k$ must zero, and the polynomial $q_{\lambda,k}$ must be a constant multiple of $P(H)$. This establishes the claimed form of the series for f_λ.

On the one hand, applying $\mathcal{A}$ to a homogeneous term of degree $l > k$ gives a homogeneous term of degree $l - k > 0$, which will evaluate to zero at the origin. Thus,

$$(\mathcal{A}f_\lambda)(0) = c_\lambda(\mathcal{A}P)(0).$$

On the other hand, by directly differentiating the exponentials in the definition of f_λ, we get

$$(\mathcal{A}f_\lambda)(0) = \sum_{w \in W} \det(w) \prod_{\alpha \in R^+} \langle w \cdot \lambda, \alpha \rangle$$

$$= \sum_{w \in W} \det(w) P(w \cdot \lambda)$$

$$= |W| P(\lambda).$$

This shows that

$$c_\lambda(\mathcal{A}P)(0) = |W| P(\lambda). \tag{10.19}$$

Now, if λ is strictly dominant, each factor in the formula for $P(\lambda)$ is nonzero, so that $P(\lambda) \neq 0$. Applying (10.19) in such a case shows that $(\mathcal{A}P)(0) \neq 0$. We can thus solve (10.19) for c_λ to obtain the claimed formula.   □

Proof of the Weyl Dimension Formula. As we have noted already, $P(\mu + \delta)$ and $P(\delta)$ are nonzero, so that $c_{\mu+\delta}$ and c_δ are also nonzero. For any $H \in \mathfrak{h}$, we have

$$\chi_\pi(tH) = \frac{f_{\mu+\delta}(tH)}{f_\delta(tH)}$$

$$= \frac{c_{\mu+\delta}t^k P(H) + O(t^{k+1})}{c_\delta t^k P(H) + O(t^{k+1})}$$

$$= \frac{c_{\mu+\delta} P(H) + O(t)}{c_\delta P(H) + O(t)} \qquad (10.20)$$

for any t for which the numerator of the last expression is nonzero.

Now, we know from the definition of a character, that χ_π is a continuous function. To determine the value of χ_π at the identity, we choose H to be in the open fundamental Weyl chamber, so that $P(H) \neq 0$, in which case the denominator in (10.20) is nonzero for all sufficiently small nonzero t. Thus,

$$\dim(V_\mu) = \lim_{t \to 0} \chi_\pi(tH)$$

$$= \frac{c_{\mu+\delta}}{c_\delta}$$

$$= \frac{P(\mu + \delta)}{P(\delta)},$$

where we have used the formula for c_λ in Lemma 10.22. Recalling the definition of P gives the dimension formula as stated in Theorem 10.18.   □

Example 10.23. If μ_1 and μ_2 denote the two fundamental weights for $\mathsf{sl}(3; \mathbb{C})$ then the dimension of the representation with highest weight $m_1\mu_1 + m_2\mu_2$ is given by

$$\frac{1}{2}(m_1 + 1)(m_2 + 1)(m_1 + m_2 + 2).$$

See Exercises 7 and 8 for the analogous formulas for B_2 and G_2.

Proof. Note that scaling the inner product on $\mathfrak{h}$ by a constant does not affect the right-hand side of the dimension formula, since the inner product occurs an equal number of times in the numerator and denominator. Let us then normalize the inner product so that all roots α satisfy $\langle \alpha, \alpha \rangle = 2$. With this normalization, $H_\alpha = \alpha$ and we have

$$m_1 = \langle \mu, H_1 \rangle = \langle \alpha_1, \mu \rangle$$
$$m_2 = \langle \mu, H_2 \rangle = \langle \alpha_2, \mu \rangle.$$

Letting $\alpha_3 = \alpha_1 + \alpha_2$, we have $\delta = \frac{1}{2}(\alpha_1 + \alpha_2 + \alpha_3) = \alpha_3$. We then note that $\langle \alpha_1, \delta \rangle = 1$, $\langle \alpha_2, \delta \rangle = 1$, and $\langle \alpha_3, \delta \rangle = 2$. Thus, the numerator in the dimension formula is

$$(\langle \alpha_1, \mu \rangle + \langle \alpha_1, \delta \rangle) \, (\langle \alpha_2, \mu \rangle + \langle \alpha_2, \delta \rangle) \, (\langle \alpha_3, \mu \rangle + \langle \alpha_3, \delta \rangle)$$
$$= (m_1 + 1)(m_2 + 1)(m_1 + m_2 + 2)$$

and the denominator is $(1)(1)(2)$. □

10.6 The Kostant Multiplicity Formula

We will obtain Kostant's multiplicity formula from the Weyl character formula by developing a method for dividing by the Weyl denominator q. We now illustrate this method in the case of $\mathsf{sl}(2; \mathbb{C})$, where the character formula takes the form

$$\chi_m(aH) = \frac{e^{(m+1)a} - e^{-(m+1)a}}{e^a - e^{-a}}. \tag{10.21}$$

One way to divide the numerator of (10.21) by the denominator is to use the geometric series $1/(1 - x) = 1 + x + x^2 + \cdots$. Applying this formally with $x = e^{-2a}$ gives the following result:

$$\frac{1}{e^a - e^{-a}} = \frac{1}{e^a(1 - e^{-2a})} = e^{-a}(1 + e^{-2a} + e^{-4a} + \cdots). \tag{10.22}$$

If the real part of a is negative, the series will not converge in the ordinary sense. Nevertheless, if we treat the right-hand side (10.22) as simply a formal series, then (10.22) holds in the sense that if we multiply the right-hand side by $e^a - e^{-a}$, we get 1. Using (10.22), we have

$$\frac{e^{(m+1)a} - e^{-(m+1)a}}{e^a - e^{-a}}$$
$$= e^{-a}e^{(m+1)a}(1 + e^{-2a} + \cdots) - e^{-a}e^{-(m+1)a}(1 + e^{-2a} + \cdots)$$
$$= (e^{ma} + e^{(m-2)a} + \cdots) - (e^{-(m+2)a} + e^{-(m+4)a} + \cdots)$$
$$= e^{ma} + e^{(m-2)a} + \cdots + e^{-ma}.$$

This last expression is, indeed, the character of V_m. From this formula, we can read off that each weight of V_m has multiplicity 1 (Proposition 10.12).

We now develop a similar method for dividing by the Weyl denominator in the general case.

Definition 10.24. A **formal exponential series** is a formal series of the form

$$\sum_{\lambda} c_{\lambda} e^{\langle \lambda, H \rangle},$$

where each λ is an integral element and where the c_{λ}'s are complex numbers. A **finite exponential series** is a series of the same form where all but finitely many of the c_{λ}'s are equal to zero.

Since we make no restrictions on the coefficients c_{λ}, a formal exponential series may not converge. Thus, the series should properly be thought of not as a function of H but simply as a list of coefficients c_{λ}. If f is a formal exponential series with coefficients c_{λ} and g is a *finite* exponential series with coefficients d_{λ}, the product of f and g is a well-defined formal exponential series with coefficients e_{λ} given by

$$e_{\lambda} = \sum_{\lambda'} c_{\lambda - \lambda'} d_{\lambda'}. \tag{10.23}$$

Note that only finitely many of the terms in (10.23) are nonzero, since g is a finite exponential series. (The product of two formal exponential series is, in general, not defined, because the sum defining the coefficients in the product might be divergent.)

Definition 10.25. If λ is an integral element, we let $p(\lambda)$ denote the number of ways (possibly zero) that λ can be expressed as a non-negative integer combination of positive roots. The function p is known as the **Kostant partition function**.

More explicitly, if the positive roots are $\alpha_1, \ldots, \alpha_k$, then $p(\lambda)$ is the number of k-tuples of non-negative integers $(n_1, \ldots, n_k)$ such that $n_1 \alpha_1 + \cdots + n_k \alpha_k = \lambda$.

Example 10.26. If α_1 and α_2 are the two positive simple roots for $\mathsf{sl}(3; \mathbb{C})$, then for any two non-negative integers m_1 and m_2, we have

$$p(m_1 \alpha_1 + m_2 \alpha_2) = 1 + \min(m_1, m_2).$$

If λ is not a non-negative integer combination of α_1 and α_2, then $p(\lambda) = 0$.

The result of Example 10.26 is shown graphically in Figure 10.6.

Proof. We have the two positive simple roots α_1 and α_2, together with one other positive root $\alpha_3 = \alpha_1 + \alpha_2$. Thus, if λ can be expressed as a non-negative integer combination of $\alpha_1 + \alpha_2 + \alpha_3$, we can rewrite α_3 as $\alpha_1 + \alpha_2$ to express λ as $\lambda = m_1 \alpha_1 + m_2 \alpha_2$, with $m_1, m_2 \geq 0$. Every expression for λ is then of the form

$$\lambda = (m_1 - k)\alpha_1 + (m_2 - k)\alpha_2 + k\alpha_3,$$

for $0 \leq k \leq \min(m_1, m_2)$. $\square$

We are now ready to explain how to invert the Weyl denominator.

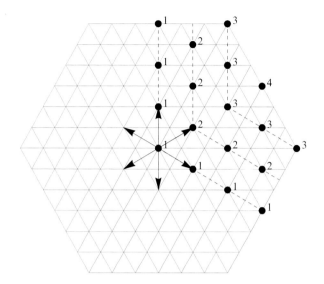

Fig. 10.6 The Kostant partition function p for A_2. Unlabeled points λ have $p(\lambda) = 0$

Proposition 10.27 (Formal Reciprocal of the Weyl Denominator). *At the level of formal exponential series, we have*

$$\frac{1}{q(H)} = \sum_{\eta \geq 0} p(\eta) e^{-\langle \eta + \delta, H \rangle}. \tag{10.24}$$

Here the sum is nominally over all integral elements η with $\eta \succeq 0$, but $p(\eta) = 0$ unless η is an integer combination of roots.

The proposition means, more precisely, that the product of $q(H)$ and the formal exponential series on the right-hand side of (10.24) is equal to 1 (i.e., to e^0). To prove Proposition 10.24, we first rewrite q as a product.

Lemma 10.28. *The Weyl denominator may be computed as*

$$q(H) = \prod_{\alpha \in R^+} (e^{\langle \alpha, H \rangle / 2} - e^{-\langle \alpha, H \rangle / 2}). \tag{10.25}$$

See Exercise 9 for the explicit form of this identity in the case $\mathfrak{g} = \mathsf{sl}(n + 1; \mathbb{C})$.

Proof. Let $\tilde{q}$ denote the product on the right-hand side of (10.25). If we expand out the product in the definition of $\tilde{q}$, there will be a term equal to

$$\prod_{\alpha \in R^+} e^{\langle \alpha, H \rangle / 2} = e^{\langle \delta, H \rangle},$$

where δ is half the sum of the positive roots. Note that even though $\alpha/2$ is not necessarily integral, the element δ is integral (Proposition 8.38). We now claim that every other exponential in the expansion will be of the form $\pm e^{\langle \lambda, H \rangle}$ with λ integral and strictly lower than δ. To see this, note that every time we take $e^{-\langle \alpha, H \rangle/2}$ instead of $e^{\langle \alpha, H \rangle/2}$, we lower the exponent by α. Thus, any λ appearing will be of the form

$$\lambda = \delta - \sum_{\alpha \in E} \alpha,$$

for some subset E of R^+. Such a λ is integral and lower than δ.

Meanwhile, by precisely the same argument as in the proof of Point 1 of Proposition 10.21, the function $\tilde{q}$ is alternating with respect to the action of W. Thus, if we write

$$\tilde{q}(H) = \sum_{\lambda} a_\lambda e^{\langle \lambda, H \rangle}, \tag{10.26}$$

then by Corollary 10.17, the coefficients a_λ must satisfy

$$a_{w \cdot \lambda} = \det(w) a_\lambda. \tag{10.27}$$

Since the exponential $e^{\langle \delta, H \rangle}$ occurs in the expansion (10.26) with a coefficient of 1, $e^{\langle w \cdot \delta, H \rangle}$ must occur with a coefficient of $\det(w^{-1}) = \det(w)$. Thus, it remains only to show that no other exponentials can occur in (10.26). To see this, note that if any exponential $e^{\langle \lambda, H \rangle}$ occurs for which λ is not in the W-orbit of δ, then by (10.27), another exponential $e^{\langle \lambda', H \rangle}$ must appear with λ' dominant but strictly lower than δ. Since δ is the minimal strictly dominant integral element, λ' cannot be strictly dominant (see Proposition 8.43) and thus must be orthogonal to one of the positive simple roots α_j. Thus, $s_{\alpha_j} \cdot \lambda' = \lambda'$, where $\det(s_{\alpha_j}) = -1$. Applying (10.27) to this case shows that the coefficient of $e^{\langle \lambda', H \rangle}$ must be zero. $\qquad\square$

Figure 10.7 illustrates the proof of Lemma 10.28 for the root system G_2. The white dots indicate weights of exponentials that do not, in fact, occur in the expansion of $\tilde{q}$. Each of these white dots lies on the line orthogonal to some root, which means that the corresponding exponential cannot occur in the expansion of a Weyl-alternating function.

Proof of Proposition 10.27. As in the $\mathsf{sl}(2; \mathbb{C})$ example considered at the beginning of this section, we have

$$\frac{1}{e^{\langle \alpha, H \rangle/2} - e^{-\langle \alpha, H \rangle/2}} = e^{-\langle \alpha, H \rangle/2}(1 + e^{-\langle \alpha, H \rangle} + e^{-2\langle \alpha, H \rangle} + \cdots),$$

at the level of formal exponential series. Taking a product over $\alpha \in R^+$ gives

$$\frac{1}{q(H)} = e^{-\langle \delta, H \rangle} \prod_{\alpha \in R^+} (1 + e^{-\langle \alpha, H \rangle} + e^{-2\langle \alpha, H \rangle} + \cdots).$$

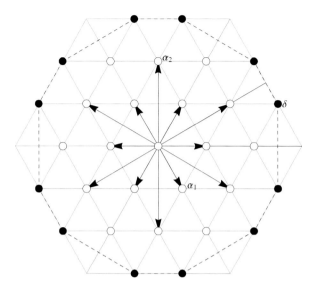

Fig. 10.7 The *black dots* indicate the W-orbit of δ for G_2. The *white dots* indicate weights of exponentials that *do not* occur in the expansion of $\tilde{q}$

In the product, a term of the form $e^{-\langle \eta, H \rangle}$ will occur precisely as many times there are ways to express η as a non-negative integer combination of the α's, namely, $p(\eta)$ times. $\qquad\square$

Theorem 10.29 (Kostant's Multiplicty Formula). *Suppose μ is a dominant integral element and V_μ is the finite-dimensional irreducible representation with highest weight μ. Then if λ is a weight of V_μ, the multiplicity of λ is given by*

$$\mathrm{mult}(\lambda) = \sum_{w \in W} \det(w) \, p(w \cdot (\mu + \delta) - (\lambda + \delta)).$$

Proof. By the Weyl character formula and Proposition 10.27, we have

$$\chi_\pi(H) = \left(\sum_{\eta \geq 0} p(\eta) e^{-\langle \eta + \delta, H \rangle} \right) \left(\sum_{w \in W} \det(w) e^{\langle w \cdot (\mu + \delta), H \rangle} \right).$$

For a fixed weight λ, the coefficient of $e^{\langle \lambda, H \rangle}$ in the character χ_π is just the multiplicity of λ in V_μ (Proposition 10.12). This coefficient is the sum of the quantity

$$p(\eta) \det(w) \qquad\qquad (10.28)$$

over all pairs (η, w) for which

$$-\eta - \delta + w \cdot (\mu + \delta) = \lambda,$$

or

$$\eta = w \cdot (\mu + \delta) - (\lambda + \delta). \tag{10.29}$$

Substituting (10.29) into (10.28) and summing over W gives Kostant's formula. □

If μ is well in the interior of the fundamental Weyl chamber and λ is very close to μ, then for all nontrivial elements of W, $w \cdot (\mu + \delta)$ will fail to be higher than $\lambda + \delta$. In those cases, there is only one nonzero term in the formula and the multiplicity of λ will be simply $p(\mu - \lambda)$. (By Exercise 7 in Chapter 9, $p(\mu - \lambda)$ is the multiplicity of λ in the Verma module W_μ.) In general, it suffices to compute the multiplicities of *dominant* weights λ. For any dominant λ, there will be many elements w of W for which $w \cdot (\mu + \delta)$ will fail to be higher than $\lambda + \delta$. Nevertheless, in high-rank examples, the order of the Weyl group is very large and the number of nonzero terms in Kostant's formula can be large, even if it is not as large as the order of W.

Figure 10.8 carries out the multiplicity calculation for $\mathsf{sl}(3; \mathbb{C})$ in the irreducible representation with highest weight $(2, 9)$. (These multiplicities were presented without proof in Figure 6.5.) The calculation is done only for the weights in the fundamental Weyl chamber; all other multiplicities are determined by Weyl invariance. The term involving $w \in W$ makes a nonzero contribution to the multiplicity of λ only if $\lambda + \delta$ is lower than $w \cdot (\mu + \delta)$, or equivalently if λ is lower than $w \cdot (\mu + \delta) - \delta$. For most weights λ in the fundamental chamber, only the $w = 1$ term makes a nonzero contribution. In those cases, the multiplicity of λ is simply $p(\mu - \lambda)$.

Now, by Example 10.26 and Figure 10.6, $p(\mu - \lambda)$ increases by 1 each time λ moves from one "ring" of weights to the ring immediately inside. For the weights indicated by white dots, however, there are two nonzero terms, the second being the one in which w is the reflection about the vertical root α_1. Since the determinant of the reflection is -1, the second term enters with a minus sign. On the medium-sized triangle, the first term is 4 and the second term is -1, while on the small triangle, the two terms are 5 and -2. Thus, in the end, all of the weights in all three of the triangles end up with a multiplicity of 3.

It is not hard to see that the pattern in Figure 10.8 holds in general for representations of $\mathsf{sl}(3; \mathbb{C})$: As we move inward from one "ring" of weights to the next, the multiplicities increase by 1 at each step, until the rings become triangles, at which point the multiplicities become constant. (There is an increase in multiplicity from the last hexagon to the first triangle, but not from the first triangle to any of the subsequent triangles.) In particular, if the highest weight is on the edge of the fundamental Weyl chamber, all of the rings will be triangles and, thus, all of the weights have multiplicity 1. A small piece of this pattern of multiplicities was determined in a more elementary way in Exercises 11 and 12 in Chapter 6.

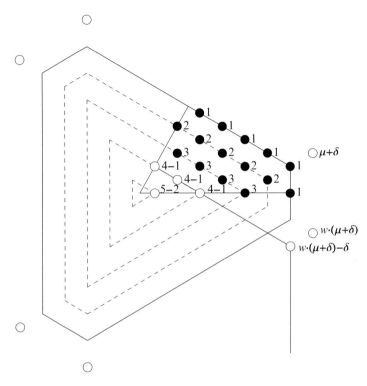

Fig. 10.8 Multiplicity calculation for the representation with highest weight $(2, 9)$

For other low-rank Lie algebras, the pattern of multiplicities is more complicated, but can, in principle, be computed explicitly. In particular, the Kostant partition function can computed explicitly for B_2 and G_2, at which point one only needs to work out which terms in the multiplicity formula contribute, for weights in the fundamental chamber. (See, for example, [Tar] or [Cap] and compare [CT], which gives an elegant graphical method of computing the multiplicities in the $B_2 \cong C_2$ case.) For Lie algebras of rank higher than two, [BBCV] gives efficient algorithms for computing the partition function for the classical Lie algebras, either numerically or symbolically. If the order of the Weyl group is not too large, one can then implement Kostant's formula to read off the multiplicities.

For higher-rank cases, the order of the Weyl group can be very large, in which case it is not feasible to use Kostant's formula, even if the partition function is known. Freudenthal's formula (Sect. 22.3 in [Hum]) gives an alternative method of computing the multiplicities, which is preferable in these cases, because it does not involve a sum over the Weyl group.

10.7   The Character Formula for Verma Modules

Before coming to the proof of the Weyl character formula, we consider a "warm up case," that of the character of a Verma module. In the character formula for Verma modules, there is an even larger cancellation than in the Weyl character formula. The product of the Weyl denominator and the character would appear to be an *infinite* sum of exponentials, and yet all but one of these exponentials cancels out! The proof of this character formula follows a very natural course, consisting of writing down explicitly the multiplicities for the various weight spaces and then using the formula for the multiplicities to establish the desired cancellation. The character formula for Verma modules is a key ingredient in the proof of the character formula for finite-dimensional representations.

Since a Verma module (π, V_μ) is infinite dimensional, the operator $e^{\pi(X)}$ may not have a well-defined trace. On the other hand, Point 2 of Proposition 10.12 gives an expression for the character of a finite-dimensional representation, evaluated at a point in $\mathfrak{h}$, in terms of the weights for the representation. This observation allows us to define the character of any representation that decomposes as a direct sum of weight spaces, as a formal exponential series on $\mathfrak{h}$. (Recall Definition 10.24.)

Definition 10.30. Let V be any representation (possibly infinite dimensional) that decomposes as a direct sum of integral weight spaces of finite multiplicity. We define the **formal character** of V by the formula

$$Q_V(H) = \sum_\lambda \text{mult}(\lambda) e^{\langle \lambda, H \rangle}, \quad H \in \mathfrak{h},$$

where the Q_V is interpreted as a formal exponential series.

Proposition 10.31 (Character Formula for Verma Modules). *For any integral element λ, the formal character of the Verma module is given by*

$$Q_{W_\lambda}(H) = \sum_{\eta \geq 0} p(\eta) e^{\langle \lambda - \eta, H \rangle}, \tag{10.30}$$

where p is the Kostant partition function in Definition 10.25. This formal character may also be expressed as

$$Q_{W_\lambda}(H) = e^{\langle \lambda + \delta, H \rangle} \left(\frac{1}{q(H)} \right) \tag{10.31}$$

where $1/q(H)$ is the formal reciprocal of the Weyl denominator given in Proposition 10.27.

The formula (10.31) is similar to the Weyl character formula, except that in the numerator we have only a single exponential rather than an alternating sum of exponentials. Of course, (10.31) implies that

$$q(H) Q_{W_\lambda}(H) = e^{\langle \lambda + \delta, H \rangle}. \tag{10.32}$$

This result means that when we multiply the Weyl denominator (an alternating finite sum of exponentials) and the formal character Q_{W_λ} (an *infinite* sum of exponentials), the result is just a *single* exponential, with all the other terms cancelling out. As usual, it is easy to see this cancellation explicitly in the case of $\mathsf{sl}(n;\mathbb{C})$. There, if m is any integer (possibly negative) and H is the usual diagonal basis element, we have

$$Q_{W_m}(aH) = e^{ma} + e^{(m-2)a} + e^{(m-4)a} + \cdots .$$

Since $q(aH) = e^a + e^{-a}$, we obtain

$$q(aH)Q_{W_m}(aH) = e^{(m+1)a} + e^{(m-1)a} + e^{(m-3)a} + \cdots$$
$$- e^{(m-1)a} - e^{(m-3)a} - \cdots$$
$$= e^{(m+1)a}.$$

Proof. Enumerate the positive roots as $\alpha_1,\ldots,\alpha_k$ and let $\{Y_j\}$ be nonzero elements of $\mathfrak{g}_{-\alpha_j}$, $j = 1,\ldots,k$. By Theorem 9.14, the elements of the form

$$\pi(Y_1)^{n_1} \cdots \pi(Y_k)^{n_k} v_0 \tag{10.33}$$

form a basis for the Verma module. An element of the form (10.33) is a weight vector with weight

$$\xi = \lambda - n_1\alpha_1 - \cdots - n_k\alpha_k.$$

The number of times the weight ξ will occur is the number of ways that $\lambda - \xi$ can be written as a non-negative integer combinations of the positive roots. Thus, we obtain the first expression for Q_{W_λ}. The second expression the follows easily from the first expression and the formula (Proposition 10.27) for the formal inverse of the Weyl denominator. [Compare (10.30) to (10.24).]   □

10.8 Proof of the Character Formula

Our strategy in proving the character formula is as follows. We will show first that the (formal) character of a Verma module W_μ can be expressed as a finite linear combination of characters of irreducible highest weight cyclic representations V_η. (The V_η's may be infinite dimensional.) By using the action of the Casimir element, we will see that only the only η's appearing in this decomposition are those satisfying $|\eta + \rho|^2 = |\mu + \rho|^2$. We will then invert this relationship and express the character of an irreducible representation V_μ as a linear combination of characters of Verma modules W_η, with η again satisfying $|\eta + \rho|^2 = |\mu + \rho|^2$. We will then specialize to the case in which μ is dominant integral, where V_μ is finite dimensional and its character χ^μ is a finite sum of exponentials. By the character formula for Verma modules from Sect. 10.7, we obtain the following conclusion: The product $q(H)\chi^\mu(H)$ is a finite linear combination of exponentials $e^{i\langle\lambda,H\rangle}$, where each $\lambda = \eta + \rho$ satisfies $|\lambda|^2 = |\mu + \rho|^2$. From this point, it is a short step to show that only λ's of the form $\lambda = w \cdot (\mu + \rho)$ occur and then to prove the character formula.

We know from general principles that the product $q(H)\chi^\mu(H)$ is a finite sum of exponentials. We need to show that the *only* exponential that actually occur in this product (with a nonzero coefficient) are those of the form $e^{\langle w\cdot(\mu+\delta),H\rangle}$, and that such exponentials occur with a coefficient of $\det(w)$. We begin with a simple observation that limits which exponentials could possibly occur in the product.

Proposition 10.32. *If an exponential $e^{\langle\lambda,H\rangle}$ occurs with nonzero coefficient in the product $q(H)\chi^\mu(H)$, then λ must be in the convex hull of the W-orbit of $\mu + \delta$ and λ must differ from $\mu + \delta$ by an element of the root lattice.*

In the last image in Figure 10.4, the white and black circles indicate weights consistent with Proposition 10.32. Only a small fraction of these weights (the black circles) actually occur in $q(H)\chi^\mu(H)$.

Proof. If λ is an integral element, then for each root α,

$$s_\alpha \cdot \lambda = \lambda - 2\frac{\langle\lambda,\alpha\rangle}{\langle\alpha,\alpha\rangle}\alpha$$

will differ from λ by an integer multiple of α. Since roots are integral, we conclude that $s_\alpha\cdot\lambda$ is, again, integral. It follows that for any $w \in W$, the element $w\cdot\lambda$ is integral and differs from λ by an integer combination of roots. Thus, by Proposition 8.38, $w \cdot \delta$ is integral and differs from δ by an integer combination of roots. Similarly, each weight of π is integral and differs from μ by an integer combination of roots (Theorem 10.1). Thus, for each exponential $e^{\langle\lambda,H\rangle}$ occurring in the product $q(H)\chi^\mu(H)$, the element λ will be integral and will differ from $\delta + \mu$ by an integer combination of roots.

Meanwhile, since each exponential in q is in the Weyl-orbit of δ and each exponential in χ^μ is in the convex hull of the Weyl-orbit of μ (Theorem 10.1), each exponential in the product will be in the convex hull of the Weyl-orbit of $\mu + \delta$, as claimed. $\qquad\square$

Our next result is the key to the proof of the character formula. In fact, we will see that it, in conjunction with Proposition 10.32, limits the exponentials that can occur in $q(H)\chi^\mu(H)$ to only those whose weights are in the Weyl orbit of $\mu + \delta$.

Proposition 10.33. *If an exponential $e^{\langle\lambda,H\rangle}$ occurs with nonzero coefficient in the product $q(H)\chi^\mu(h)$, then λ must satisfy*

$$\langle\lambda,\lambda\rangle = \langle\mu + \delta, \mu + \delta\rangle. \tag{10.34}$$

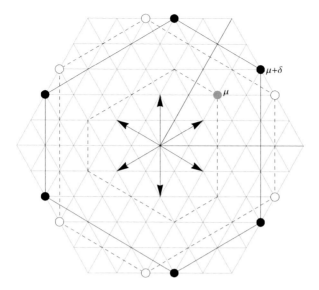

Fig. 10.9 The weights indicated by *white dots* satisfy the condition in Proposition 10.33 but not the condition in Proposition 10.32. The weights indicated by *black dots* satisfy both conditions

In the last image in Figure 10.4, for example, the exponentials λ represented by the white dots do not satisfy (10.34) and, thus, cannot occur in the product $q(H)\chi^{\mu}(h)$. Much of the rest of this section will be occupied with the proof of Proposition 10.33. Before turning to this task, however, let us see how Propositions 10.32 and 10.33 together imply the character formula. The claim is that the only weights satisfying the conditions in both of these propositions are the ones in the W-orbit of $\mu + \delta$. See Figure 10.9.

Proof of Character Formula Assuming Proposition 10.33. Suppose $v_1, \ldots, v_m$ are distinct elements of a real inner product space, all of which have the same norm, S. The convex hull of these elements is the space of vectors of the form

$$\sum_{j=1}^{m} a_j v_j$$

with $0 \leq a_j \leq 1$ and $\sum_j a_j = 1$. It is then not hard to show by induction on m that the only way such a convex combination can have norm S is if one of the a_j's is equal to 1 and all the others are zero. Applying this with the v_j's equal to $w \cdot (\mu + \delta), w \in W$, shows that the only exponentials $e^{\langle \lambda, H \rangle}$ satisfying both Propositions 10.32 and 10.33 are those of the form $\lambda = w \cdot (\mu + \delta)$. Thus, the only exponentials appearing in the product $q\chi^{\mu}$ are those of the form $ce^{\langle w \cdot (\mu + \delta), H \rangle}, w \in W$.

Now, the exponential $e^{\langle \mu + \delta, H \rangle}$ occurs in the product exactly once, since it corresponds to taking the highest weight from q (i.e., δ) and the highest weight from χ^{μ} (i.e., μ) and the coefficient of $e^{\langle \mu+\delta, H\rangle}$ is 1. (Note that since δ is strictly dominant, $w\cdot\delta$ is strictly lower than δ for all nontrivial $w \in W$, by Proposition 8.42.) Since the character is Weyl invariant and the Weyl denominator is Weyl alternating, their product is Weyl alternating. Thus, by Corollary 10.17, the coefficients in the expansion of $\chi^{\mu}q$ are alternating; that is, the coefficient of $e^{\langle w\cdot(\mu+\delta), H\rangle}$ must equal $\det(w)$. $\square$

We now begin the process of proving Proposition 10.33. A crucial role in the proof is played by the Casimir element, which we have already introduced in Sect. 10.2. (See Definition 10.4.) We now extend the result of Proposition 10.6 to highest-weight cyclic representations that may not be irreducible or finite dimensional.

Proposition 10.34. *Let (π, V) be a highest-weight cyclic representation of $\mathfrak{g}$ (possibly infinite dimensional) with highest weight λ and let $\tilde{\pi}$ be the extension of π to $U(\mathfrak{g})$. Then $\tilde{\pi}(C) = c_{\lambda}I$, where*

$$c_{\lambda} = \langle \lambda + \delta, \lambda + \delta\rangle - \langle \delta, \delta\rangle .$$

Proposition 10.34 applies, in particular, to the Verma module W_{λ} and to the irreducible representation V_{λ}. A key consequence of the proposition is that if two highest weight cyclic representations, with highest weights λ_1 and λ_2, have the same eigenvalue of the Casimir, then $\langle \lambda_1 + \delta, \lambda_1 + \delta\rangle$ must coincide with $\langle \lambda_2 + \delta, \lambda_2 + \delta\rangle$.

Proof. The same argument as in the proof of Proposition 10.6 shows if v_0 is the highest weight vector for V, then $\tilde{\pi}(C)v_0 = c_{\lambda}v_0$. Now let U be the space of all $v \in V$ for which $\tilde{\pi}(C)v = c_{\lambda}v$. Since C is in the center of $U(\mathfrak{g})$ (Proposition 10.5), we see that if $v \in U$, then

$$\tilde{\pi}(C)(\pi(X)v) = \pi(X)\tilde{\pi}(C)v = c_{\lambda}\pi(X)v,$$

showing that $\pi(X)v$ is again in U. Thus, U is an invariant subspace of V containing v_0, which means that $U = V$, that is, that $\tilde{\pi}(C)v = c_{\lambda}v$ for all $v \in V$. $\square$

Definition 10.35. Let Λ denote the set of integral elements. If μ is a dominant integral element, define a set $S_{\mu} \subset \Lambda$ as follows:

$$S_{\mu} = \{\eta \in \Lambda \,|\, \langle \eta + \delta, \eta + \delta\rangle = \langle \mu + \delta, \mu + \delta\rangle\} .$$

Note that S_{μ} is the intersection of the integral lattice Λ with sphere of radius $\|\mu + \delta\|$ centered at $-\delta$; see Figure 10.10. Since there are only finitely many elements of Λ in any bounded region, the set S_{μ} is finite, for any fixed μ. Our strategy for the proof of Proposition 10.33 is as discussed at the beginning of this section. We will decompose the formal character of each Verma module W_{η} with $\eta \in S_{\mu}$ as a finite sum of formal characters of irreducible representations V_{γ}, with each γ also belonging to S_{μ}. This expansion turns out to be of an "upper

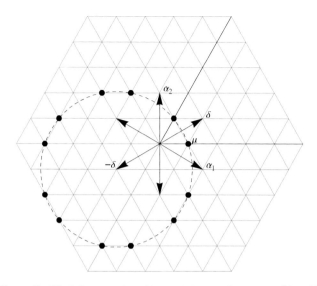

Fig. 10.10 The set S_μ (*black dots*) consists of integral elements λ for which $\|\lambda + \delta\| = \|\mu + \delta\|$

triangular with ones on the diagonal" form, allowing us to invert the expansion to express the formal character of irreducible representations in terms of formal characters of Verma modules. In particular, the character of the finite-dimensional representation V_μ will be expressed as a linear combination of formal characters of Verma modules W_η with $\eta \in S_\mu$. When we multiply both sides of this formula by the Weyl denominator and use the character formula for the Verma module (Proposition 10.31), we obtain the claimed form for $q\chi^\mu$.

Of course, a key point in the argument is to verify that whenever the character of an irreducible representation V_γ appears in the expansion of the character of $W_\eta, \eta \in S_\mu$, the highest weight γ is also in S_μ. This claim holds because any subrepresentation V_γ occurring in the decomposition of W_η must have the same eigenvalue of the Casimir as W_η. In light of Proposition 10.34, this means that $\langle \gamma + \delta, \gamma + \delta \rangle$ must equal $\langle \eta + \delta, \eta + \delta \rangle$, which is assumed to be equal to $\langle \mu + \delta, \mu + \delta \rangle$.

Proposition 10.36. *For each η in S_μ, the formal character of the Verma module W_η can be expressed as a linear combination of formal characters of irreducible representations V_γ with γ in S_μ and $\gamma \preceq \eta$:*

$$Q_{W_\eta} = \sum_{\substack{\gamma \in S_\mu \\ \gamma \preceq \eta}} a_\gamma^\eta Q_{V_\gamma} \qquad (10.35)$$

Furthermore, the coefficient a_η^η of Q_{V_η} in this decomposition is equal to 1.

See Figure 10.11.

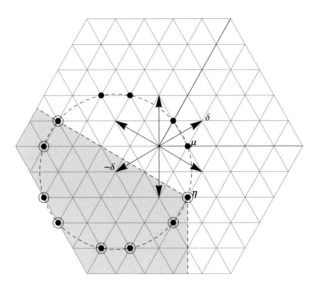

Fig. 10.11 For $\eta \in S_\mu$, the character of W_η is a linear combination of characters of irreducible representations V_γ with highest weights $\gamma \preceq \eta$ in S_μ

Lemma 10.37. *Let (π, V) be a representation of $\mathfrak{g}$, possibly infinite dimensional, that decomposes as direct sum of weight spaces of finite multiplicity, and let U be a nonzero invariant subspace of V. Then both U and the quotient representation V/U decompose as a direct sum of weight spaces. Furthermore, the multiplicity of any weight in V is the sum of its multiplicity in U and its multiplicity in V/U.*

Proof. Let u be an element of U. By assumption, we can decompose u as $u = v_1 + \cdots + v_j$, where the v_k's belong to weight spaces in V corresponding to distinct weights $\lambda_1, \ldots, \lambda_j$. We wish to show that each v_k actually belongs to U. If $j = 1$, there is nothing to prove. If $j > 1$, then $\lambda_j \neq \lambda_1$, which means that there is some $H \in \mathfrak{h}$ for which $\langle \lambda_j, H \rangle \neq \langle \lambda_1, H \rangle$. Then apply to u the operator $\pi(H) - \langle \lambda_1, H \rangle I$

$$(\pi(H) - \langle \lambda_1, H \rangle I)u = \sum_{k=1}^{j}(\langle \lambda_k, H \rangle - \langle \lambda_1, H \rangle)v_k. \tag{10.36}$$

Since the coefficient of v_1 is zero, the vector in (10.36) is the sum of fewer than j weight vectors. Thus, by induction on j, we can assume that each term on the right-hand side of (10.36) belongs to U. In particular, a nonzero multiple of v_j belongs to U, which means v_j itself belongs to U. Now, if v_j is in U, then $u - v_j = v_1 + \cdots + v_{j-1}$ is also in U. Thus, using induction again, we see that each of $v_1, \ldots, v_{j-1}$ belongs to U.

We conclude that the sum of the weight spaces in U is all of U. Since weight vectors with distinct weights are linearly independent (Proposition A.17), the sum must be direct. We turn, then, to the quotient space V/U. It is evident that the images of the weight spaces in V are weight spaces in V/U with the same weight. Thus, the sum of the weight spaces in V/U is all of V/U and, again, the sum must be direct.

Finally, consider a fixed weight λ occurring in V, and let V_λ be the associated weight space. Let q_λ be the restriction to V_λ of the quotient map $q : V \to V/U$. The kernel of q_λ consists precisely of the weight vectors with weight λ in U. Thus, the dimension of the image of q_λ, which is the weight space in V/U with weight λ, is equal to $\dim V_\lambda - \dim(V_\lambda \cap U)$. The claim about multiplicities in V, U, and V/U follows. $\qquad\square$

Proof of Proposition 10.36. We actually prove a stronger result, that the formal character of any highest-weight cyclic representation U_η with $\eta \in S_\mu$ can be decomposed as in (10.35). As the proof of Proposition 6.11, any such U_η decomposes as a direct sum of weight spaces with weights lower than η and with the multiplicity of the weight η being 1. For any such U_η, let

$$M = \sum_{\gamma \in S_\mu} \mathrm{mult}(\gamma).$$

Our proof will be by induction on M.

We first argue that if $M = 1$, then U_η must be irreducible. If not, U_η would have a nontrivial invariant subspace X, and this subspace would, by Lemma 10.37, decompose as a direct sum of weight spaces, all of which are lower than η. Thus, X would have to contain a weight vector w that is annihilated by each raising operator $\pi(X_\alpha), \alpha \in R^+$. Thus, X would contain a highest weight cyclic subspace X' with some highest weight γ. By Proposition 10.34, the Casimir would act as the scalar $\langle \gamma + \delta, \gamma + \delta \rangle - \langle \delta, \delta \rangle$ in X'. On the other hand, since X' is contained in U_η, the Casimir has to act as $\langle \eta + \delta, \eta + \delta \rangle - \langle \delta, \delta \rangle$ in X', which, since $\eta \in S_\mu$, is equal to $\langle \mu + \delta, \mu + \delta \rangle - \langle \delta, \delta \rangle$. Thus, γ must belong to S_μ. Meanwhile, γ cannot equal η or else X' would be all of U_η. Thus, both η and γ would have to have positive multiplicities and M would have to be at least 2.

Thus, when $M = 1$, the representation U_η is irreducible, in which case, Proposition 10.36 holds trivially. Assume now that the proposition holds for highest weight cyclic representations with $M \le M_0$, and consider a representation U_η with $M = M_0 + 1$. If U_η is irreducible, there is nothing to prove. If not, then as we argued in the previous paragraph, U_η must contain a nontrivial invariant subspace X' that is highest weight cyclic with some highest weight γ that belongs to S_μ and is strictly lower than η. We can then form the quotient vector space U_η/X', which will still be highest weight cyclic with highest weight η. By Lemma 10.37, the multiplicity of ξ in U_η is the sum of the multiplicities of ξ in X' and in U_η/X'. Thus,

$$Q_{U_\eta} = Q_{X'} + Q_{U_\eta/X'}.$$

Now, both X' and U_η/X' contain at least one weight in S_μ with nonzero multiplicity, namely γ for X' and η for U_η/X'. Thus, both of these spaces must have $M \leq M_0$ and we may assume, by induction, that their formal characters decompose as a sum of characters of irreducible representations with highest weights in S_μ. These highest weights will be lower than μ, in fact lower than γ in the case of X'. Thus, the character of V_η will not occur in the expansion of $Q_{X'}$. But since U_η/X' still has highest weight η, we may assume by induction that the character of V_η will occur exactly once in the expansion of $Q_{U_\eta/X'}$, and, thus, exactly once in the expansion of Q_{U_η}.   □

Proposition 10.38. *If we enumerate the elements of S_μ in nondecreasing order as $S_\mu = \{\eta_1, \ldots, \eta_l\}$, then the matrix $A_{jk} := a_{\eta_k}^{\eta_j}$ is upper triangular with ones on the diagonal. Thus, A is invertible, and the inverse of A is also upper triangular with ones on the diagonal. It follows that we can invert the decomposition in (10.35) to a decomposition of the form*

$$Q_{V_\eta} = \sum_{\gamma \in S_\mu} b_\gamma^\eta Q_{W_\gamma}, \tag{10.37}$$

where $b_\eta^\eta = 1$.

It is easy to check (using, say, the formula for the inverse of a matrix in terms of cofactors) that the inverse of an upper triangular matrix with ones on the diagonal is again upper triangular with ones on the diagonal.

Proof. For any finite partially ordered set, it is possible (Exercise 6) to enumerate the elements in nondecreasing order. In our case, this means that we can enumerate the elements of S_μ as $\eta_1, \ldots, \eta_l$ in such a way that if $\eta_j \preceq \eta_k$ then $j \leq k$. If we expand $Q_{W_{\eta_k}}$ in terms of $Q_{V_{\lambda_j}}$ as in Proposition 10.38, the only nonzero coefficients are those with $\eta_j \preceq \eta_k$, which means that $j \leq k$. Thus, the matrix is upper triangular. Since, also, the coefficient of $Q_{V_{\eta_k}}$ in the expansion of $Q_{W_{\eta_k}}$ is 1, the expansion has ones on the diagonal.   □

Proof of Proposition 10.33. We apply (10.37) with $\eta = \mu$, so that Q_{V_μ} is the character of $\chi^\mu(H)$ of the finite-dimensional, irreducible representation with highest weight μ. We then multiply both sides of (10.37) by the Weyl denominator q. Using the character formula for Verma modules [Proposition 10.31 and Eq. (10.32)], we obtain

$$q(H)\chi^\mu(H) = \sum_{\gamma \in S_\mu} b_\gamma^\mu e^{\langle \gamma + \delta, H \rangle}. \tag{10.38}$$

Since each γ belongs to S_μ, each weight $\lambda := \gamma + \delta$ occurring on the right-hand side of (10.38) satisfies

$$\langle \lambda, \lambda \rangle = \langle \gamma + \delta, \gamma + \delta \rangle = \langle \mu + \delta, \mu + \delta \rangle.$$

Thus, we have expressed $q(H)\chi^{\mu}(H)$ as a linear combination of exponentials with weights λ satisfying Proposition 10.33. Since any such decomposition is unique by Proposition 10.16, it must be the one obtained by multiplying together the exponentials in q and χ^{μ}.   ☐

10.9   Exercises

1. Suppose $\mathfrak{g}$ is a complex Lie algebra that is reductive (Definition 7.1) but not semisimple. Show that there exists a finite-dimensional representation of $\mathfrak{g}$ that is not completely reducible. (Compare Theorem 10.9 in the semisimple case.)

2. Suppose $\mathfrak{g}$ is a complex semisimple Lie algebra with the property that every finite-dimensional representation of $\mathfrak{g}$ is completely reducible. Show that $\mathfrak{g}$ decomposes as a direct sum of simple algebras.
   Hint: Consider the adjoint representation of $\mathfrak{g}$.

3. Using Proposition 10.16, show that the Weyl denominator is a Weyl-alternating function.

4. Suppose that $f : \mathfrak{h} \rightarrow \mathbb{C}$ is a polynomial and that $f(H) = 0$ whenever $\langle \alpha, H \rangle = 0$. Show that f is divisible in the space of polynomial functions by $\langle \alpha, H \rangle$.
   Hint: Choose coordinates $z_1, \ldots, z_r$ on $\mathfrak{h}$ for which $\langle \alpha, H \rangle = z_1$.

5. Suppose f is an analytic function on $\mathfrak{h}$, meaning that f can be expressed in coordinates in a globally convergent power series. Collect together all the terms in this power series that are homogeneous of degree k, so that f is the sum of homogeneous polynomials p_k of degree k. Show that if f is alternating with respect to the action of W, so is each of the polynomials p_k.
   Hint: Show that the composition of a homogeneous polynomial with a linear transformation is again a homogeneous polynomial.

6. Show that any finite partially ordered set E can be enumerated in nondecreasing order.
   Hint: Every finite partially ordered set has a minimal element, that is, an element $x \in E$ such that no $y \neq x$ in E is smaller than x.

7. Let $\{\alpha_1, \alpha_2\}$ be a base for the B_2 root system, with α_1 being the shorter root, as in the left-hand side of Figure 8.7. Let μ_1 and μ_2 be the associated fundamental weights (Definition 8.36), so that every dominant integral element is uniquely expressible as

$$\mu = m_1\mu_1 + m_2\mu_2,$$

with m_1 and m_2 being non-negative integers. Show that the dimension of the irreducible representation with highest weight μ is

$$\frac{1}{6}(m_1 + 1)(m_2 + 1)(m_1 + m_2 + 2)(m_1 + 2m_2 + 3).$$

Hint: Imitate the calculations in Example 10.23.

8. Let the notation be as in Exercise 7, but with B_2 replaced by G_2. (See Figures 8.6 and 8.11.) Show that the dimension of the irreducible representation with highest weight μ is

$$\frac{1}{5!}(m_1 + 1)(m_2 + 1)(m_1 + m_2 + 2)$$

$$\times (m_1 + 2m_2 + 3)(m_1 + 3m_2 + 4)(2m_1 + 3m_2 + 5).$$

Show that the smallest nontrivial irreducible representation of G_2 has dimension 7.

9. According to Lemma 10.28, we have the identity

$$\sum_{w \in W} \det(w) e^{i\langle w \cdot \delta, H\rangle} = \prod_{\alpha \in R^+} (e^{\langle \alpha, H\rangle/2} - e^{-\langle \alpha, H\rangle/2}). \tag{10.39}$$

Work out the explicit form of this identity for the case of the Lie algebra $\mathsf{sl}(n + 1; \mathbb{C})$, using the Cartan subalgebra $\mathfrak{h}$ described in Sect. 7.7.1 and the system of positive roots described in Sect. 8.10.1. If a typical element of $\mathfrak{h}$ has the form $(a_0, \ldots, a_n)$, introduce the variables $z_j = e^{a_j}$. Show that after multiplying both sides by $(z_0 \cdots z_n)^{n/2}$, the identity (10.39) takes the form of a Vandermonde determinant:

$$\det \begin{pmatrix} z_0^n & z_0^{n-1} & \cdots & 1 \\ z_1^n & z_1^{n-1} & \cdots & 1 \\ \vdots & \vdots & & \vdots \\ z_n^n & z_n^{n-1} & \cdots & 1 \end{pmatrix} = \prod_{j<k}(z_j - z_k).$$

10. Let π be an irreducible, finite-dimensional representation of $\mathfrak{g}$ with highest weight μ, and let π^* be the dual representation to π.

   (a) Show that the weights of π^* are the negative of the weights of π.
   (b) Let w_0 be the unique element of W that maps the fundamental Weyl chamber C to $-C$. Show that the highest weight μ^* of π^* may be computed as

$$\mu^* = -w_0 \cdot \mu.$$

   (Compare Exercises 2 and 3 in Chapter 6.)
   (c) Show that if $-I$ is an element of the Weyl group, then every representation of $\mathfrak{g}$ is isomorphic to its dual.

Part III
Compact Lie Groups

Chapter 11
Compact Lie Groups and Maximal Tori

In this chapter and Chapter 12 we develop the representation theory of a connected, compact matrix Lie group K. The main result is a "theorem of the highest weight," which is very similar to our main results for semisimple Lie algebras. If we let $\mathfrak{k}$ be the Lie algebra of K and we let $\mathfrak{g}$ be the complexification of $\mathfrak{k}$, then $\mathfrak{g}$ is reductive, which means (Proposition 7.6) that $\mathfrak{g}$ is the direct sum of a semisimple algebra and a commutative algebra. We can, therefore, draw on our structure results for semisimple Lie algebras to introduce the notions of roots, weights, and the Weyl group. We will, however, give a completely different proof of the theorem of the highest weight. In particular, our proof of the hard part of the theorem, the existence of a irreducible representation for each weight of the appropriate sort, will be based on decomposing the space of functions on K under the left and right action of K. This argument is independent of the Lie-algebraic construction using Verma modules.

In the present chapter, we develop the structures needed to formulate a theorem of the highest weight for K, and we develop some key tools that will aid is in the proof of the theorem. The representations themselves will appear in the next chapter. The key results of this chapter are the torus theorem and the Weyl integral formula. Although parts of the chapter assume familiarity with the theory of manifolds and differential forms, the reader who is not familiar with that theory can still follow the statements of the key results. Furthermore, the torus theorem can easily be proved by hand in the case of $\mathsf{SU}(n)$. The reader who is willing to take the results of this chapter on faith can proceed on to Chapter 12, where they are applied to prove the compact-group versions of the Weyl character formula and the theorem of the

A previous version of this book was inadvertently published without the middle initial of the author's name as "Brian Hall". For this reason an erratum has been published, correcting the mistake in the previous version and showing the correct name as Brian C. Hall (see DOI http://dx.doi.org/10.1007/978-3-319-13467-3_14). The version readers currently see is the corrected version. The Publisher would like to apologize for the earlier mistake.

© Springer International Publishing Switzerland 2015
B.C. Hall, *Lie Groups, Lie Algebras, and Representations*, Graduate Texts in Mathematics 222, DOI 10.1007/978-3-319-13467-3_11

highest weight. Finally, in Chapter 13, we will take a close look at the fundamental group of K. We will prove, among other things, that when K is simply connected, the notion of "dominant integral element" for K coincides with the analogous notion for the Lie algebra $\mathfrak{g}$.

Throughout the chapter, we assume that K is a connected, compact matrix Lie group with Lie algebra $\mathfrak{k}$. We allow $\mathfrak{k}$, and thus also $\mathfrak{g} := \mathfrak{k}_{\mathbb{C}}$, to have a nontrivial center, which means that $\mathfrak{g}$ is reductive but not necessarily semisimple. As in Proposition 7.4, we fix on $\mathfrak{g}$ an inner product that is real on $\mathfrak{k}$ and invariant under the adjoint action of K.

11.1 Tori

In this section, we consider tori, that is, groups isomorphic to a product of copies of S^1. In the rest of the chapter, we will be interested in tori that arise as subgroups of connected compact groups.

Definition 11.1. A matrix Lie group T is a **torus** if T is isomorphic to the direct product of k copies of the group $S^1 \cong \mathsf{U}(1)$, for some k.

Consider, for example, the group T of diagonal, unitary $n \times n$ matrices with determinant 1. Every element of T can be expressed uniquely as

$$\mathrm{diag}(u_1, \ldots, u_{n-1}, (u_1 \cdots u_{n-1})^{-1})$$

for some complex numbers $u_1, \ldots, u_{n-1}$ with absolute value 1. Thus, T is isomorphic to $n-1$ copies of S^1.

Theorem 11.2. *Every compact, connected, commutative matrix Lie group is a torus.*

Recall that a subset E of a topological space is *discrete* if every element e of E has a neighborhood containing no points of E other than e.

Lemma 11.3. *Let V be a finite-dimensional inner product space over $\mathbb{R}$, viewed as a group under vector addition, and let Γ be a discrete subgroup of V. Then there exist linearly independent vectors $v_1, \ldots, v_k$ in V such that Γ is precisely the set of vectors of the form*

$$m_1 v_1 + \cdots + m_k v_k,$$

with each $m_j \in \mathbb{Z}$.

Proof. Since Γ is discrete, there is some $\varepsilon > 0$ such that the only point γ in Γ with $\|\gamma\| < \varepsilon$ is $\gamma = 0$. For any $\gamma, \gamma' \in \Gamma$, if $\|\gamma' - \gamma\| < \varepsilon$, then since $\gamma' - \gamma$ is also in Γ, we must have $\gamma' = \gamma$. It then follows easily that there can be only finitely many points in Γ in any bounded region of V. If $\Gamma = \{0\}$, the result holds with $k = 0$. Otherwise, we can find some nonzero $\gamma_0 \in \Gamma$ with such that $\|\gamma_0\|$ is minimal among

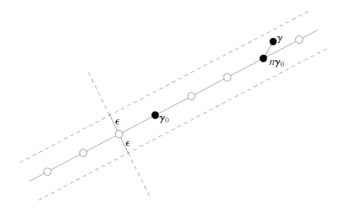

Fig. 11.1 If γ is very close to the line through γ_0, then $\|\gamma - n\gamma_0\|$ is smaller than $\|\gamma_0\|$ for some n

nonzero elements of Γ. Let W denote the orthogonal complement of the span of γ_0, let P denote the orthogonal projection of V onto W, and let Γ' denote the image of Γ under P. Since P is linear, Γ' will be a subgroup of W. We now claim that Γ' is discrete in W.

Suppose, toward a contradiction that Γ' is not discrete. Then for every $\varepsilon > 0$, there must exist $\delta \neq \delta' \in \Gamma'$ with $\|\delta' - \delta\| < \varepsilon$. Thus, $\delta' - \delta$ is a nonzero element of Γ' with norm less than ε. There then exists some $\gamma \in \Gamma$ with $P(\gamma) = \delta' - \delta$. Since $\|\delta' - \delta\| < \varepsilon$, the distance from γ to the span of γ_0 is less than ε. Now let β be the orthogonal projection of γ onto the span of γ_0, which is the point closest to γ in line through γ_0. Then β lies between $m\gamma_0$ and $(m + 1)\gamma_0$ for some integer m. By taking $n = m$ or $n = m + 1$, we can assume that the distance between $n\gamma_0$ and β is at most half the length of γ_0. Meanwhile, the distance from β to γ is at most ε. Thus, the distance from $n\gamma_0$ to γ is at most $\varepsilon + \|\gamma_0\|/2$, which is less than $\|\gamma_0\|$, if ε is small enough. But then $\gamma - n\gamma_0$ is a nonzero element of Γ with norm less than the norm of γ_0, contradicting the minimality of γ_0. (See Figure 11.1.)

Now that Γ' is known to be discrete, we can apply induction on the dimension of V with the base case corresponding to dimension zero. Thus, there exist linearly independent vectors $u_1, \ldots, u_{k-1}$ in Γ' such that Γ' is precisely the set of integer linear combinations of $u_1, \ldots, u_{k-1}$. Let us then choose vectors $v_1, \ldots, v_{k-1}$ in Γ such that $P(v_j) = u_j$. Since $P(v_1), \ldots, P(v_{k-1})$ are linearly independent in W, it is easy to see that $v_1, \ldots, v_{k-1}$ and γ_0 are linearly independent in V. For any $\gamma \in \Gamma$, the element $P(\gamma)$ is of the form $m_1 u_1 + \cdots + m_{k-1} u_{k-1}$. Thus, γ must be equal to $m_1 v_1 + \cdots + m_{k-1} v_{k-1} + \sigma$, where $\sigma \in \Gamma$ satisfies $P(\sigma) = 0$, meaning that σ is a multiple of γ_0. But then σ must be an *integer* multiple of γ_0, or else, by an argument similar to the one in the previous paragraph, there would be an element of Γ in the span of γ_0 with norm less than $\|\gamma_0\|$. We conclude, then that

$$\gamma = m_1 v_1 + \cdots + m_{k-1} v_{k-1} + m_k \gamma_0,$$

establishing the desired form of γ.   □

Proof of Theorem 11.2. Let T be compact, connected, and compact, and let t be
the Lie algebra of T. Then t is also commutative (Proposition 3.22), in which
case Corollary 3.47 tells us that the exponential map exp : t $\to$ T is surjective.
Furthermore, the exponential map for T is a Lie group homomorphism and its
kernel Γ must be discrete, since the exponential map is injective in a neighborhood
of the origin. Thus, by Lemma 11.3, Γ is the set of integer linear combinations
of independent vectors $v_1, \ldots, v_k$. If dim t $= n$, then exp descends to a bijective
homomorphism of t$/\Gamma \cong (S^1)^k \times \mathbb{R}^{n-k}$ with T. Now, the Lie algebra map associated
to this homomorphism is invertible (since the Lie algebra of t is t), which means that
the inverse map is also continuous. Thus, T is homeomorphic to $(S^1)^k \times \mathbb{R}^{n-k}$. Since
T is compact, this can only happen if $k = n$, in which case, T is the torus $(S^1)^n$. $\square$

At various points in the developments in later chapters, it will be useful to
consider elements t in a torus T for which the subgroup of S generated by s is
dense in s as Proposition 11.4 in The following result guarantees the existence of
such elements.

Proposition 11.4. *Let* $T = (S^1)^k$ *and let* $t = (e^{2\pi\theta_1}, \ldots, e^{2\pi\theta_k})$ *be an element of*
T. *Then t generates a dense subgroup of T if and only if the numbers*

$$1, \theta_1, \ldots, \theta_k$$

are linearly independent over the field $\mathbb{Q}$ *of rational numbers.*

The $k = 1$ case of this result is Exercise 9 in Chapter 1. In particular, if x is
any transcendental real number and we define $\theta_j = x^j, j = 1, \ldots, k$, then t will
generate a dense subgroup of T. See Figure 11.2 for an example of an element that
generates a dense subgroup of $S^1 \times S^1$.

Lemma 11.5. *If T is a torus and t is an element of T, then the subgroup generated
by t is* not *dense in T if and only if there exists a nonconstant homomorphism* Φ :
$T \to S^1$ *such that* $\Phi(t) = 1$.

Proof. Suppose first that there exists a nonconstant homomorphism $\Phi : T \to S^1$
with $T(t) = 1$. Then ker(Φ) is a closed subgroup of T that contains t and thus the
group generated by t. But since Φ is nonconstant, ker$(\Phi) \neq T$, which means that
the closure of the group generated by t is not all of T.

In the other direction, let S be the closure of the group generated by t and suppose
that S is not all of T. We will proceed by describing the preimage of S under the
exponential map, using an extension of Lemma 11.3. Thus, let Λ be the set of all
$H \in$ t such that $e^{2\pi H} \in S$. Since S is a closed subgroup of T, the set Λ will be a
closed subgroup of the additive group t. Now let Λ_0 be the identity component of
Λ, which must be a subspace of t. (Indeed, by Corollaries 3.47 and 3.52, Λ_0 must
be equal to the Lie algebra of Λ.) Since S is not all of T, the subspace Λ_0 cannot be
all of t.

The entire group Λ now decomposes as $\Lambda_0 \times \Lambda_1$, where

$$\Lambda_1 := \Lambda \cap (\Lambda_0)^\perp$$

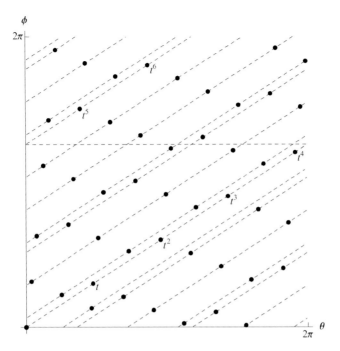

Fig. 11.2 A portion of the dense subgroup generated by t in $S^1 \times S^1$

is a closed subgroup of $(\Lambda_0)^\perp$. Furthermore, the identity component of Λ_1 must be trivial, which means that the Lie algebra of Λ_1 must be $\{0\}$. Thus, by Theorem 3.42, Λ_1 is discrete. Let us now define a homomorphism $\phi : \mathfrak{t} \to S^1$ by setting

$$\phi(H) = e^{2\pi i \xi(H)}$$

for some linear functional ξ on $\mathfrak{t}$. By Lemma 11.3, Λ_1 is the integer span of linearly independent vectors $v_1, \ldots, v_l$. Since $\Lambda_0 \neq \mathfrak{t}$, we can arrange things so that ξ is zero on Λ_0, the values of ξ on $v_1, \ldots, v_l$ are integers, but ξ is not identically zero.

Then $\ker(\phi)$ will contain Λ, and, in particular, the kernel of the map $H \mapsto e^{2\pi H}$. Thus, there is a nonconstant, continuous homomorphism $\Phi : T \to S^1$ satisfying

$$\Phi(e^{2\pi H}) = \phi(H)$$

for all $H \in \mathfrak{t}$. If we choose H so that $e^{2\pi H} = t \in S$, then $H \in \Lambda$, which means that

$$\Phi(t) = \Phi(e^{2\pi H}) = 1,$$

but Φ is not constant. $\qquad\qquad\qquad\square$

Proof of Proposition 11.4. In light of Lemma 11.5, we may reformulate the proposition as follows: The numbers $1, \theta_1, \ldots, \theta_k$ are linearly dependent over $\mathbb{Q}$ if and only if there exists a nonconstant homomorphism $\Phi : T \to S^1$ with $(e^{2\pi i \theta_1}, \ldots, e^{2\pi i \theta_k}) \in \ker(\Phi)$. Suppose first that there is a dependence relation among $1, \theta_1, \ldots, \theta_k$ over $\mathbb{Q}$. Then after clearing the denominators from this relation, we find that there exist integers $m_1, \ldots, m_k$, not all zero, such that

$$m_1 \theta_1 + \cdots + m_k \theta_k \in \mathbb{Z}.$$

Thus, we may define a nonconstant $\Phi : T \to S^1$ by

$$\Phi(u_1, \ldots, u_k) = u_1^{m_1} \cdots u_k^{m_k} \tag{11.1}$$

and the kernel of Φ will contain $(e^{2\pi i \theta_1}, \ldots, e^{2\pi i \theta_k})$.

In the other direction, Exercise 2 tells us that *every* continuous homomorphism $\Phi : T \to S^1$ is of the form (11.1) for some set of integers $m_1, \ldots, m_k$. Furthermore, if Φ is nonconstant, these integers cannot all be zero. Thus, if

$$1 = \Phi(e^{2\pi i \theta_1}, \ldots, e^{2\pi i \theta_k}) = e^{2\pi i (m_1 \theta_1 + \cdots + m_k \theta_k)},$$

we must have $m_1 \theta_1 + \cdots + m_k \theta_k = n$ for some integer n, which implies that $1, \theta_1, \ldots, \theta_k$ are linearly dependent over $\mathbb{Q}$. $\qquad\qquad\square$

11.2   Maximal Tori and the Weyl Group

In this section, we introduce a the concept of a maximal torus, which plays the same role in the compact group approach to representation theory as the Cartan subalgebra plays in the Lie algebra approach.

Definition 11.6. A subgroup T of K is a **torus** if T is isomorphic to $(S^1)^k$ for some k. A subgroup T of K is a **maximal torus** if it is a torus and is not properly contained in any other torus in K.

If $K = \mathsf{SU}(n)$, we may consider

$$T = \left\{ \begin{pmatrix} e^{i\theta_1} & & & \\ & \ddots & & \\ & & e^{i\theta_{n-1}} & \\ & & & e^{-i(\theta_1 + \cdots + \theta_{n-1})} \end{pmatrix} \middle| \; \theta_j \in \mathbb{R} \right\},$$

which is a torus of dimension $n - 1$. If T is contained in another torus $S \subset \mathsf{SU}(n)$, then every element s of S would commute with every element t of T. If we choose

t to have distinct eigenvalues, then by Proposition A.2, s would have to be diagonal in the standard basis, meaning that $s \in T$. Thus, T is actually a maximal torus.

Proposition 11.7. *If T is a maximal torus, the Lie algebra t of T is a maximal commutative subalgebra of $\mathfrak{k}$. Conversely, if t is a maximal commutative subalgebra of $\mathfrak{k}$, the connected Lie subgroup T of $\mathfrak{k}$ with Lie algebra t is a maximal torus.*

Proof. If T is a maximal torus, it is commutative, which means that its Lie algebra t is also commutative (Proposition 3.22). Suppose t is contained in a commutative subalgebra $\mathfrak{s}$. Then it is also contained in a maximal commutative subalgebra $\mathfrak{s}'$ containing $\mathfrak{s}$. The connected Lie subgroup S' with Lie algebra $\mathfrak{s}'$ must be commutative (since S' is generated by exponentials of elements of $\mathfrak{s}'$) and closed (Proposition 5.24) and hence compact. Thus, by Theorem 11.2, S' is a torus. Since T is a maximal torus, we must have $S' = T$ and thus $\mathfrak{s}' = \mathfrak{s} = t$, showing that t is maximal commutative.

In the other direction, if t is maximal commutative, the connected Lie subgroup T with Lie algebra t is closed (Proposition 5.24), hence compact. But T is also commutative and connected, hence a torus, by Theorem 11.2. If T is contained in a torus S, then t is contained in the commutative Lie algebra $\mathfrak{s}$ of S. Since t is maximal commutative, we have $\mathfrak{s} = t$ and since S is connected, $S = T$, showing that T is a maximal torus. $\qquad\square$

Definition 11.8. If T is a maximal torus in K, then the **normalizer** of T, denoted $N(T)$, is the group of elements $x \in K$ such that $xTx^{-1} = T$. The quotient group

$$W := N(T)/T$$

is the **Weyl group** of T.

Note that T is, almost by definition, a normal subgroup of $N(T)$. If w is an element of W represented by $x \in N(T)$, then w acts on T by the formula

$$w \cdot t = xtx^{-1}, \quad t \in T.$$

If $x \in N(T)$, the conjugation action of x maps T onto T. It follows that Ad_x maps the Lie algebra t of T into itself. We define an action of W on t by

$$w \cdot H = \mathrm{Ad}_x(H), \quad H \in t. \tag{11.2}$$

Since our inner product is invariant under the adjoint action of K, the action of W on t is by orthogonal linear transformations.

We will see in Sect. 11.7 that the centralizer of T—that is, the group of those $x \in K$ such that $xtx^{-1} = t$ for all $t \in T$—is equal to T. It follows that W acts effectively on T, meaning that if $w \cdot t = t$ for all $t \in T$, then w is the identity element of W. It then follows from Corollary 3.49 (with $G = H = T$) that W also acts effectively on t. Thus, W may be identified with the group of orthogonal linear transformations of t of the form $H \mapsto w \cdot H$. We will also show in Sect. 11.7

that this group of linear transformations coincides with the group generated by the reflections through the hyperplanes orthogonal to the roots. Thus, the Weyl group as defined in Definition 11.8 is naturally isomorphic to the Weyl group associated to the Lie algebra $\mathfrak{g} := \mathfrak{k}_{\mathbb{C}}$ in Sect. 7.4. Exercise 3, meanwhile, asks the reader to verify directly in the case of $\mathsf{SU}(n)$ that the centralizer of T is T and that $N(T)/T$ is the permutation group on n entries, thus agreeing with the Weyl group for $\mathsf{sl}(n; \mathbb{C})$ computed in Sect. 7.7.1.

The following "torus theorem" is a key result that underlies many of the developments in this chapter and the next two chapters.

Theorem 11.9 (Torus Theorem). *If K is a connected, compact matrix Lie group, the following results hold.*

1. *If S and T are maximal tori in K, there exists an element x of K such that $T = xSx^{-1}$.*
2. *Every element of K is contained in some maximal torus.*

The torus theorem has many important consequences; we will mention just two of these now.

Corollary 11.10. *If K is a connected, compact matrix Lie group, the exponential map for K is surjective.*

Proof. For any $x \in K$, choose a maximal torus T containing x. Since the exponential map for $T \cong (S^1)^k$ is surjective, x can be expressed as the exponential of an element of the Lie algebra of T. $\square$

Corollary 11.11. *Let K be a connected, compact matrix Lie group and let x an arbitrary element of K. Then x belongs to the center of K if and only if x belongs to every maximal torus in K.*

Proof. Assume first that x belongs to the center $Z(K)$ of K, and let T be any maximal torus in K. By the torus theorem, x is contained in a maximal torus S, and this torus is conjugate to T. Thus, there is some $y \in K$ such that $S = yTy^{-1}$. Since $x \in S$, we have $x = yty^{-1}$ for some $t \in T$, and thus $t = y^{-1}xy$. But we are assuming that x is central, and so, actually, $t = x$, showing that x belongs to T.

In the other direction, assume x belongs to every maximal torus in K. Then for any $y \in K$, we can find some torus T containing y, and this torus will also contain x. Since T is commutative, we conclude that x and y commute, showing that x is in $Z(K)$. $\square$

The torus theorem follows from the following result.

Lemma 11.12. *Let T be a fixed maximal torus in K. Then every $y \in K$ can be written in the form*

$$y = xtx^{-1}$$

for some $x \in K$ and $t \in T$.

If $K = \mathsf{SU}(n)$ and T is the diagonal subgroup of K, Lemma 11.12 follows easily from the fact that every unitary matrix has an orthonormal basis of eigenvectors. The proof of the general case of Lemma 11.12 requires substantial preparation and is given in Sect. 11.5.

Proof of torus theorem assuming Lemma 11.12. Since each $y \in K$ can be written as $y = xtx^{-1}$, we see that y belongs to the maximal torus xTx^{-1}.

Next we show that every maximal torus S in K is conjugate to T, from which it follows any two maximal tori S_1 and S_2 are conjugate to each other. Suppose s is an element of S such that the subgroup of S generated by s is dense in S, as in Proposition 11.4. Then we can choose some $x \in K$ and $t \in T$ such that $s = xtx^{-1}$ and $t = x^{-1}sx$. Thus, $x^{-1}s^k x = t^k \in T$ for all integers k. Since the set of elements of the form s^k is dense in S, we must have $x^{-1}Sx \subset T$. But since S is maximal, we actually have $x^{-1}Sx = T$.   □

11.3 Mapping Degrees

We now introduce a method that we will use in proving the torus theorem in Sect. 11.5. The current section requires greater familiarity with manifolds than elsewhere in the book. In addition to differential forms (Appendix B), we make use of the exterior derivative ("d"), the pullback of a form by a smooth map, and Stoke's theorem. See Chapters 14 and 16 in [Lee] for more information. For our purposes, the main result of this section is Corollary 11.17, which gives a condition guaranteeing that a map between two manifolds of the same dimension is surjective. The reader who is not familiar with manifold theory should still be able to get an idea of what is going on from the example in Figures 11.3 and 11.4.

If V is a finite-dimensional vector space over $\mathbb{R}$, we may define an equivalence relation on ordered bases of V by declaring two ordered bases $(v_1, \ldots, v_n)$ and $(v'_1, \ldots, v'_n)$ to be equivalent if the unique linear transformation L mapping v_j to v'_j has positive determinant. The set of ordered bases for V then consists of exactly two equivalence classes. An **orientation** for V is a choice of one of these two equivalence classes. Once an orientation of V has been fixed, an ordered basis for V is said to be **oriented** if it belongs to the chosen equivalence class. An **orientation** on a smooth manifold M is then a continuous choice of orientation on each tangent space to M. A smooth manifold together with a choice of orientation is called an **oriented manifold**.

We consider manifolds that are closed—that is, compact, connected, and without boundary—and oriented. We will be interested in smooth maps between two closed, oriented manifolds of the same dimension.

Definition 11.13. Let X and Y be closed, oriented manifolds of dimension $n \geq 1$ and let $f : X \to Y$ be a smooth map. A point $y \in Y$ is a **regular value** of f if for all $x \in X$ such that $f(x) = y$, the differential $f_*(x)$ of f at x is invertible.

A point $y \in Y$ is a **singular value** of f if there exists some $x \in X$ such that $f(x) = y$ and the differential $f_*(x)$ of f at x is not invertible.

It is important to note that if y is not in the range of f, then y is a regular value. After all, if there is no x with $f(x) = y$, then it is vacuously true that for every x with $f(x) = y$, the differential $f_*(x)$ is invertible.

Proposition 11.14. *Let X, Y, and f be as in Definition 11.13. If y is a regular value of f, then y has only finitely many preimages under f.*

Proof. If $f^{-1}(\{y\})$ were infinite, the set would have to have an accumulation point x_0, by the assumed compactness of X. Then by continuity, we would have $f(x_0) = y$. Since y is a regular value of f, then $f_*(x_0)$ would be invertible. But then the inverse function theorem would say that f is injective in a neighborhood of x_0, which is impossible since every neighborhood of x_0 contains infinitely many points x with $f(x) = y$. $\square$

Saying that X and Y are oriented means that we have chosen a consistent orientation on each tangent space to X and to Y. If $f : X \to Y$ is smooth and the differential $f_*(x)$ of f at x is invertible, then $f_*(x)$ is either an orientation preserving or an orientation reversing map of $T_x(X)$ to $T_{f(x)}(Y)$. Since f' is assumed to be continuous, if f_* is invertible and orientation preserving at x, it is invertible and orientation preserving in a neighborhood of x, and similarly if f_* is invertible and orientation reversing at x.

Definition 11.15. If y is a regular value of f, let the **signed number of preimages** of y denote the number of preimages, where $x \in f^{-1}(\{y\})$ counted with a plus sign if $f_*(x)$ is orientation preserving and with a minus sign if $f_*(x)$ is orientation reversing.

The main result of the section is the following.

Theorem 11.16. *Let X and Y be closed, oriented manifolds of dimension $n \geq 1$ and let $f : X \to Y$ be a smooth map. Then there exists an integer k such that for every regular value y of f, the signed number of preimages of y is equal to k.*

If there *are*, in fact, any regular values of f, the integer k is unique and is called the **mapping degree** of f. (Actually, Sard's theorem guarantees that every such f has a nonempty set of regular values, but we do not need to know this, since in our application of Theorem 11.16 in Sect. 11.5, we will find regular values of the relevant map "by hand.") See the "Degree Theory" section in Chapter 17 of [Lee] for more information.

Corollary 11.17. *Let X, Y, and f be as in Theorem 11.16. If there exists a regular value y of f for which the signed number of preimages is nonzero, then f must map onto Y.*

Proof. If there existed some y' that is not in the range of f, then y' would be a regular value and the (signed) number of preimages of y' would be zero. This would contradict Theorem 11.16. $\square$

Fig. 11.3 The graph of a
map f from S^1 to S^1. The
signed number of preimages
of each regular value is 1

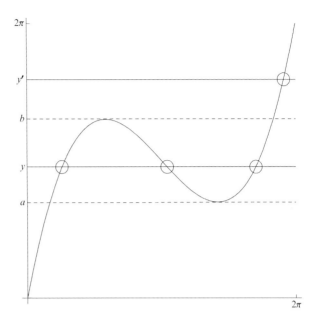

Figure 11.3 illustrates Theorem 11.16. The figure shows the graph of a map f
from S^1 (which we think of as $[0, 2\pi]$ with ends identified) to itself. The singular
values are the points marked a and b on the y-axis; all other values are regular. For
two different values y and y', we compute the signed number of preimages. The
point y has three preimages, but f' is negative at one of these, so that the signed
number of preimages is $1 - 1 + 1 = 1$. Meanwhile, the point y' has one preimage,
at which f' is positive so that the signed number of preimages is 1. The mapping
degree of f is 1 and, consistent with Corollary 11.17, f is surjective. Meanwhile,
Figure 11.4 shows the same map in a more geometric way, as a map between two
manifolds X and Y, each of which is diffeomorphic to S^1.

We now turn to the proof of Theorem 11.16; see also Theorem 17.35 in
[Lee]. Using the inverse function theorem, it is not hard to show that the signed
number of preimages *and* the unsigned number of preimages are both constant in a
neighborhood of any regular value y. This result, however, does not really help us,
because the set of regular values may be disconnected. (See Figure 11.3.) Indeed,
Figure 11.3 shows that the *unsigned* number of preimages may not be constant; we
need a creative method to show that the *signed* number of preimages is constant on
the set of regular values.

Our tool for proving this result is that of differential forms. If $f : X \to Y$ is
an orientation-preserving diffeomorphism, then for any n-form α on Y, the integral
of $f^*(\alpha)$ over X will equal the integral of α over Y. If, on the other hand, f is an
orientation-reversing diffeomorphism, the integral of $f^*(\alpha)$ will be the negative of
the integral α. Suppose now that $f : X \to Y$ is a smooth map, not necessarily a
diffeomorphism. Suppose that y is a regular value of f and that $x_1, \ldots, x_N$ are the

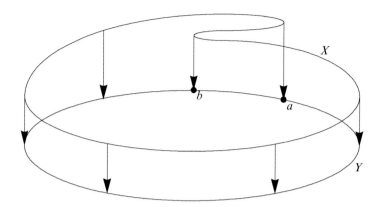

Fig. 11.4 The map indicated by the *arrows* is the same one as in Figure 11.3, but shown more geometrically

elements of $f^{-1}(\{y\})$. Then we can find a neighborhood V of y such that $f^{-1}(V)$ is a disjoint union of neighborhoods $U_1, \ldots, U_N$ of $x_1, \ldots, x_N$ and such that f maps each U_j diffeomorphically onto V. Furthermore, by shrinking V if necessary, we can assume that for each j, the differential $f_*(x)$ is either orientation preserving at every point of U_j or orientation reversing at every point of U_j. Then for any n-form α supported in V, we see that

$$\int_X f^*(\alpha) = k \int_Y \alpha, \tag{11.3}$$

where k is the signed number of preimages of y.

If α is chosen so that $\int_Y \alpha \neq 0$, then (11.3) becomes

$$k = \frac{\int_X f^*(\alpha)}{\int_Y \alpha}. \tag{11.4}$$

The right-hand side of (11.4) gives us an analytic method for determining the signed number of preimages. Our goal is to use (11.4) to show that k is constant on the set of regular values. To this end, we will establish a key result: The value of the right-hand side of (11.4) is unchanged if α is "deformed" by pulling it back by a family of diffeomorphisms of Y.

Suppose, then, that y and y' are regular values of f with the signed number of preimages being k and k', respectively. We will construct a family $\alpha_t, 0 \leq t \leq 1$, of n-forms on Y such that α_0 is supported near y and α_1 is supported near y'. For all t, the expression

$$\frac{\int_X f^*(\alpha_t)}{\int_Y \alpha_t} \tag{11.5}$$

makes sense, even if the support of α_t contains singular values of f. Furthermore, the values of both integrals in (11.5) are—as we will show—independent of t. Thus, we will conclude that

$$k = \frac{\int_X f^*(\alpha_0)}{\int_Y \alpha_0} = \frac{\int_X f^*(\alpha_1)}{\int_Y \alpha_1} = k',$$

as claimed.

Lemma 11.18. *Suppose Ψ_t is a continuous, piecewise-smooth family of orientation-preserving diffeomorphisms of Y, with Ψ_0 being the identity map. For any n-form α on Y, let $\alpha_t = \Psi_t^*(\alpha)$. Then for all t, we have*

$$\int_Y \alpha_t = \int_Y \alpha \qquad (11.6)$$

and

$$\int_X f^*(\alpha_t) = \int_X f^*(\alpha). \qquad (11.7)$$

Proof. Saying that Ψ_t is piecewise smooth means that it we can divide $[0, T]$ into finitely many subintervals on each of which $\Psi_t(x)$ is smooth in x and t. Since Ψ_t is continuous, it suffices to prove the result on each of the subintervals, that is, in the case where $\Psi_t(x)$ is smooth, which we now assume. The result (11.6) holds because Ψ_t is an orientation preserving diffeomorphism.

To establish (11.7), we show that the left-hand side of (11.7) is independent of t. Note that

$$f^*(\alpha_t) = f^*(\Psi_t^*(\alpha)) = (\Psi_t \circ f)^*(\alpha).$$

Thus, if we define $g : X \times [0, T] \to Y$ by

$$g(x, t) = \Psi_t(f(x)),$$

we have

$$\int_X f^*(\alpha_T) - \int_X f^*(\alpha_0) = \int_{\partial(X \times [0,T])} g^*(\alpha).$$

Using Stoke's theorem and a standard result relating pullbacks and exterior derivatives, we obtain

$$\int_X f^*(\alpha_T) - \int_X f(\alpha_0) = \int_{X \times [0,T]} d(g^*(\alpha))$$

$$= \int_{X \times [0,T]} g^*(d\alpha).$$

But since α is a top-degree form on Y, we must have $d\alpha = 0$, showing that $\int_X f^*(\alpha_T) = \int_X f(\alpha_0)$. $\qquad\qquad\square$

Proof of Theorem 11.16. To complete the proof of Theorem 11.16, it remains only to address the existence of a continuous, piecewise smooth, orientation-preserving family Ψ_t of diffeomorphisms of Y such that Ψ_0 is the identity and such that $\Psi_1(y') = y$. (Thus, if α is supported near y, then $\Psi_1^*(\alpha)$ will be supported near y'.) We actually only require this in the case that Y is a compact Lie group, in which case, the diffeomorphisms can easily be constructed using the group structure on Y. Nevertheless, we will outline an argument for the general result. Let U be a neighborhood of y' that is a rectangle in some local coordinate system around y'. Then it is not hard to construct a family of diffeomorphisms of Y that are the identity on $Y \setminus U$ and that map y' to any desired point of U. (See Exercise 7.)

If $y \in U$, we are done. If not, we consider the set E of points $z \in Y$ such that y' can be moved to z by a family of diffeomorphisms of the desired sort. If $z \in E$, we have, by assumption, a family moving y' to z. We can then use the argument in the preceding paragraph to move z to any point z' in a neighborhood of z. Thus, E is open and contains y'. We now claim that E is closed. If z is a limit point of E, then in any neighborhood V of z, there is an element z' of E. Thus, by the argument in the preceding paragraph, we can move z' to z by a family of diffeomorphisms. Since E is both open and closed and nonempty (because it contains y), E must be all of Y. $\qquad\qquad\square$

Proposition 11.19. *Let X, Y, and f be as in Theorem 11.16, and suppose f has mapping degree k. Then for every n-form α on Y, we have*

$$\int_X f^*(\alpha) = k \int_Y \alpha. \tag{11.8}$$

In Figure 11.4, for example, the map f indicated by the arrows has mapping degree 1. Thus, for every form α on Y, the integral of $f^*(\alpha)$ over X is the same as the integral of α over Y, even though f is not a diffeomorphism. When pulling back the part of α between a and b, we get three separate integrals on X, but one of these integrals occurs with a minus sign, because f_* is orientation reversing on the middle of the three intervals over $[a, b]$.

Proof. We have already noted in (11.3) that if y is a regular value of f, there is a neighborhood U of y such that (11.8) holds for all α supported in U. By the deformation argument in the proof of Theorem 11.16, the same result holds for *any* y in Y. Since Y is compact, we can then cover Y by a finite number of open sets U_j

such that (11.8) holds whenever α is supported in U_j. Using a partition of unity, we can express any form α as a sum of forms α_j such that α_j is supported on U_j, and the general result follows. $\square$

11.4   Quotient Manifolds

Before coming the proof of the torus theorem, we require one more preparatory concept, that of the quotient of a matrix Lie group by a closed subgroup. Throughout this section, we assume that G is a matrix Lie group with Lie algebra $\mathfrak{g}$ and that H is a *closed* subgroup of G with Lie algebra $\mathfrak{h}$. Even if H is not normal, we can still consider the quotient G/H as a set (the set of left cosets gH of H). We now show that G/H has the structure of a smooth manifold. We will let $[g]$ denote the coset gH of H in G and we will let $Q : G \to G/H$ denote the quotient map. Recall that a topological structure on a set E is **Hausdorff** if for every pair of distinct points $x, y \in E$, there exist disjoint open sets U and V in E with $x \in U$ and $y \in V$.

Lemma 11.20. *Define a topology on G/H by decreeing that a set U in G/H is open if and only if $Q^{-1}(U)$ is open in G. Then G/H is Hausdorff with respect to this topology. Furthermore, G/H has a countable dense subset.*

   If we did not assume that H is closed, the Hausdorff condition for G/H would, in general, fail.

Proof. We begin by noting that the quotient map $Q : G \to G/H$ is open; that is, $Q(U)$ is open in G/H whenever U is open in G. To establish this, we must show that $Q^{-1}(Q(U))$ is open in G. But

$$Q^{-1}(Q(U)) = \bigcup_{h \in H} Uh,$$

where $Uh = \{uh \mid u \in U\}$. But since right-multiplication by h is a homeomorphism of G with itself, each Uh is open, so that $Q^{-1}(Q(U))$ is also open. Next, let $\rho : G \times G \to G$ be the continuous map given by $\rho(x, y) = x^{-1}y$. Note that $[x] = [y]$ if and only if $\rho(x, y) \in H$. Suppose now that $[x]$ and $[y]$ are distinct points in G/H, so that $\rho(x, y) \notin H$.

   Then since H is closed, $G \setminus H$ is open, so that $\rho^{-1}(G \setminus H)$ is open in $G \times G$. Thus, we can find an open rectangle $U \times V$ in $\rho^{-1}(G \setminus H)$, where U and V are open sets in G containing x and y, respectively. For all $u \in U$ and $v \in V$, we have $\rho(u, v) \notin H$, meaning that $[u] \neq [v]$. It follows that $Q(U)$ and $Q(V)$ are disjoint open sets in G/H containing $[x]$ and $[y]$, respectively.

   Finally, since G inherits its topology from the separable metric space $M_n(\mathbb{C}) \cong \mathbb{R}^{2n^2}$, it follows that G has a countable dense subset E. Furthermore, the quotient map $Q : G \to G/H$ is surjective and (by the definition of the topology on G/H) continuous. Thus, $Q(E)$ will be dense in G/H. $\square$

Fig. 11.5 The *gray* region
indicates the set of points of
the form gh with $g \in \exp(U)$
and $h \in H$. This set is
diffeomorphic to $\exp(U) \times H$

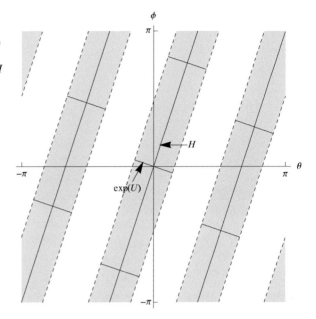

Lemma 11.21 (Slice Lemma). *Let G be a matrix Lie group with Lie algebra $\mathfrak{g}$,
let H be a closed subgroup of G, and let $\mathfrak{h}$ be the Lie algebra of H. Decompose
$\mathfrak{g}$ as a vector space as $\mathfrak{g} = \mathfrak{h} \oplus \mathfrak{f}$ for some subspace $\mathfrak{f}$ of $\mathfrak{g}$ and define a map
$\Lambda : \mathfrak{f} \times H \to G$ by*

$$\Lambda(X, h) = e^X h.$$

*Then there is a neighborhood U of 0 in $\mathfrak{f}$ such that Λ is injective on $U \times H$ and Λ
maps $U \times H$ diffeomorphically onto an open subset of G. In particular, if X_1 and
X_2 are distinct elements of U, then e^{X_1} and e^{X_2} belong to distinct cosets of G/H.*

The term "slice lemma" refers to the fact that the map sending $X \in U$ to e^X
slices across the different cosets of H. Figure 11.5 illustrates the slice lemma in the
case in which $G = S^1 \times S^1$ and H is the subgroup consisting of points of the form
$(e^{i\theta}, e^{3i\theta})$.

Proof. We identify the tangent spaces at the identity to both $\mathfrak{f} \times H$ and G with $\mathfrak{f} \oplus \mathfrak{h}$.
If $X(t)$ is a smooth curve in $\mathfrak{f}$ passing through 0 at $t = 0$, we have

$$\frac{d}{dt}\Lambda(X(t), I)\Big|_{t=0} = \frac{d}{dt}e^{X(t)}\Big|_{t=0} = X'(0).$$

Meanwhile, if $h(t)$ is a smooth curve in H passing through I at $t = 0$, we have

$$\frac{d}{dt}\Lambda(0, h(t))\Big|_{t=0} = \frac{d}{dt}h(t)\Big|_{t=0} = h'(0).$$

From this calculation, and the linearity of the differential, we can see that the differential Λ_* of Λ at $(0, I)$ is the identity map of $\mathfrak{f} \oplus \mathfrak{h}$ to itself. Thus, by continuity, $\Lambda_*(X, e)$ is invertible for X in some neighborhood U of 0 in $\mathfrak{f}$.

Meanwhile, the map Λ commutes with the right action of H:

$$\Lambda(X, hh_0) = \Lambda(X, h)h_0.$$

From this, it is easy to see that if $\Lambda_*(X, I)$ is invertible, then $\Lambda_*(X, h)$ is invertible for all h. We conclude, then, that $\Lambda_*(X, h)$ is invertible for all $(X, h) \in U \times H$. By the inverse function theorem, then, Λ maps a small neighborhood of each point in $U \times H$ injectively onto an open set in G. In particular, the image of $U \times H$ under Λ is an open subset of G.

We must now show that by shrinking U as necessary, we can ensure that Λ is globally injective on $U \times H$. By the inverse function theorem, there are neighborhoods U' of 0 in $\mathfrak{f}$ and V of I in H such that Λ maps $U' \times V$ injectively into G. If we choose a small subneighborhood U'' of 0 in $\mathfrak{f}$ and X and X' are in U'', then $e^{-X'}e^X$ will be close to the identity in G. Indeed, if U'' is small enough, then whenever $e^{-X'}e^X$ happens to be in H, the element $e^{-X'}e^X$ will actually be in V.

We now claim that Λ is injective on $U'' \times H$. To see this, suppose $e^X h = e^{X'}h'$ with $X, X' \in U''$ and $h, h' \in H$. Then

$$h'h^{-1} = e^{-X'}e^X \in V,$$

by our choice of U''. But then

$$e^X = e^{X'}(h'h^{-1}),$$

and since $h'h^{-1} \in V$, we must have $X = X'$ and $I = h'h^{-1}$, by the injectivity of Λ on $U' \times V$. Thus, actually, $X = X'$ and $h = h'$, establishing the desired injectivity. □

It is instructive to contemplate the role in the preceding proof played by the assumption that H is closed. It is evident from Figure 1.1 that the slice lemma can fail if H is not closed. (For example, even for very small nonzero $X \in \mathfrak{f}$, the element e^X can be in H.) On the other hand, even if H is not closed, Theorem 5.23 says that there is a new topology on H and an atlas of coordinate neighborhoods making H into a smooth manifold, in such a way that the inclusion of H into G is smooth. If we use this structure on H, the map Λ in Lemma 11.21 is smooth and much of the proof proceeds in the same way as when H is closed. The proof of global injectivity of Λ, however, breaks down because the new topology on H does not agree with the topology that H inherits from G. Thus, even if $e^{-X'}e^X$ belongs to H and is very close to the identity in G, this element may not be close to the identity in the new topology on H. (See Figure 5.4.) Thus, we cannot conclude that $e^{-X'}e^X$ is in V, and the proof of injectivity fails.

Theorem 11.22. *If G is a matrix Lie group and H a closed subgroup of G, then G/H can be given the structure of a smooth manifold with*

$$\dim(G/H) = \dim G - \dim H$$

in such a say that (1) the quotient map $Q : G \to G/H$ is smooth, (2) the differential of Q at the identity maps $T_I(G)$ onto $T_{[I]}(G/H)$ with kernel $\mathfrak{h}$, and (3) the left action of G on G/H is smooth.

Proof. Let $U \subset \mathfrak{f}$ be as in Lemma 11.21. Then for each $g \in G$, let Λ_g be the map from $U \times H$ into G given by

$$\Lambda_g(X, h) = ge^X h.$$

Combining Lemma 11.21 with a translation by g, we see that Λ_g is a diffeomorphism of $U \times H$ onto its image, and that $\Lambda_g(X, h)$ and $\Lambda_g(X', h')$ are in distinct cosets of H provided that $X \neq X'$. Let W_g be the (open) image of $U \times H$ under Λ_g and let $V_g = Q(W_g)$, that is,

$$V_g = \left\{ [ge^X] \in G/H \,\middle|\, X \in U \right\}.$$

Then $Q^{-1}(V_g) = W_g$, showing that V_g is open in G/H. By the above properties of Λ_g, the map $X \mapsto [ge^X]$ is an injective map of U onto V_g.

We now propose to use the maps $X \mapsto [ge^X]$ as local coordinates on G/H, where $U \subset \mathfrak{f}$ may be identified with $\mathbb{R}^k$, with $k = \dim \mathfrak{f} = \dim \mathfrak{g} - \dim \mathfrak{h}$. By the way the topology on G/H is defined, the map Q is continuous and a function f on G/H is continuous if and only if $f \circ Q$ is continuous on G. Thus, the map $X \mapsto Q(ge^X) = [ge^X]$ is continuous. Furthermore, if we compose the inverse map $[ge^X] \mapsto X$ with Q, we obtain the map $ge^X h \mapsto X$. This map is continuous because it consists of the inverse of the diffeomorphism $(X, h) \mapsto ge^X h$, combined with the map $(X, h) \mapsto X$.

Thus, G/H is locally homeomorphic to $\mathbb{R}^k$. Since, also, G/H is Hausdorff and has a countable dense subset (Lemma 11.20), we see that G/H is a topological manifold.

Now, the coordinate patches $[ge^X]$ clearly cover G/H. If two such patches overlap, the change of coordinates map is the map $X \mapsto X'$, where $[ge^X] = [g'e^{X'}]$. This map can be computed by mapping X to $ge^X \in G$ and then applying $(\Lambda_{g'})^{-1}$ to write ge^X as $ge^X = g'e^{X'}h'$. Since $(\Lambda_{g'})^{-1}$ is smooth, we see that the change of coordinates map is smooth. Thus, we may give a smooth structure to G/H using these coordinates.

It is now a straightforward matter to check the remaining claims about the smooth structure on G/H. To see, for example, that Q is smooth, pick some g in G and write points near g as $ge^X h$, with $X \in U$ and $h \in H$. Then $Q(ge^X h) = [ge^X]$. Thus, Q can be written locally as the inverse of Λ_g followed by the map $(X, h) \mapsto [ge^X]$, which is smooth in our local identification of G/H with U. The remaining claims are left as an exercise to the reader (Exercise 8). □

Proposition 11.23. *Suppose there exists an inner product on $\mathfrak{g}$ that is invariant under the adjoint action of H, and let V denote the orthogonal complement of $\mathfrak{h}$ with respect to this inner product. Then we may identify the tangent space at each point $[g]$ of G/H with V by writing $v \in T_{[g]}(G/H)$ as*

$$v = \frac{d}{dt}[ge^{tX}]\Big|_{t=0}, \quad X \in V.$$

This identification of $T_{[g]}(G/H)$ with V is unique up to the adjoint action of H on V.

Note that since the adjoint action of H on $\mathfrak{g}$ is orthogonal and preserves $\mathfrak{h}$, this action also preserves V, by Proposition A.10.

Proof. Since $\mathfrak{g} = \mathfrak{h} \oplus V$, Point 2 of Theorem 11.22 tells us that every tangent vector v to G/H at the identity coset can be expressed uniquely as

$$v = \frac{d}{dt}[e^{tX}]\Big|_{t=0}, \quad X \in V.$$

For any $[g] \in G/H$, we identify $T_{[g]}(G/H)$ with $T_{[I]}(G/H) \cong V$ by using the left action of g. Thus, each $v \in T_{[g]}(G/H)$ can be written uniquely as

$$v = \frac{d}{dt}[ge^{tX}]\Big|_{t=0}, \quad X \in V.$$

If we use a different element $gh, h \in H$, of the same coset, we get a different identification of $T_{[g]}(G/H)$ with V, as follows:

$$[ghe^{tX}] = [ghe^{tX}h^{-1}] = [ge^{tX'}], \tag{11.9}$$

where $X' = \mathrm{Ad}_h(X)$. Differentiating (11.9) shows that the two identifications differ by the adjoint action of H. $\qquad\square$

A **volume form** on a manifold M of dimension n is a nowhere-vanishing n-form on M. As we have already discussed in the proof of Theorem 4.28, each matrix Lie group G has a volume form that is invariant under the right action of G on itself. The same argument shows that G has a volume form invariant under the left action of G on itself. (For some groups G, it is possible to find a *single* volume form that is invariant under *both* the left and right action of G on itself, but this is not always the case.) We now address the existence of an invariant volume form on a quotient manifold.

Proposition 11.24. *If G is a matrix Lie group and H is a connected compact subgroup of G, there exists a volume form on G/H that is invariant under the left action of G. This form is unique up to multiplication by a constant.*

In the case $H = \{I\}$, we conclude that there is a left-invariant volume form on G itself and that this form is unique up to a constant. (In this chapter, it is convenient to use a *left*-invariant volume form on G, rather than a right-invariant form as in Sect. 4.4.)

Proof. Since H is compact, there exists an inner product on $\mathfrak{g}$ that is invariant under the adjoint action of H. Let V denote the orthogonal complement of $\mathfrak{h}$ in $\mathfrak{g}$, so that V is invariant under the adjoint action of H. Since H is connected, the restriction to V of Ad_h will actually be in $\mathsf{SO}(V)$ for all $h \in H$.

Now pick an orientation on V and let α be the standard volume form on V, that is, the unique one for which $\alpha(e_1, \ldots, e_N) = 1$ whenever $(e_1, \ldots, e_N)$ is an oriented orthonormal basis for V. Since the action of Ad_x on V has determinant 1, we have

$$\alpha(\mathrm{Ad}_x(v_1), \ldots, \mathrm{Ad}_x(v_N)) = \alpha(v_1, \ldots, v_N)$$

for all $v_1, \ldots, v_N \in V$.

Now, the tangent space to G/H at the identity coset is identified with V. We define a form on G/H as follows. At the identity coset, we take it to be α. At any other coset $[g]$, we use the action of $g \in G$ to transport α from $[I]$ to $[g]$. If $[g] = [g']$, then $g' = xh$ for $h \in H$. The action of h on $T_{[I]}(G/H) \cong V$ is the adjoint action, which preserves α. Thus, the resulting form at $[g]$ is independent of the choice of g.

Finally, to address the uniqueness, note that *any* two top degree forms on G/H must agree up to a constant at the identity coset $[I]$. But the since the left action of G on G/H is transitive, the value of the form at $[I]$ uniquely determines the form everywhere. Thus, any two left-invariant forms on G/H must be equal up to an overall constant. $\square$

11.5   Proof of the Torus Theorem

Having made the required preparations in Sects. 11.3 and 11.4, we are now ready to complete the proof of Theorem 11.9. It remains only to prove Lemma 11.12; to that end, we define a key map.

Definition 11.25. Let T be a fixed maximal torus in K. Let

$$\Phi : T \times (K/T) \to K$$

be defined by

$$\Phi(t, [x]) = xtx^{-1}, \qquad (11.10)$$

where $[x]$ denotes the coset xT in K/T.

Note that if $s \in T$, then since T is commutative, we have

$$(xs)t(xs)^{-1} = xsts^{-1}x^{-1} = xtx^{-1},$$

showing that Φ is well defined as a map of $T \times (K/T)$ into K. Lemma 11.12 amounts to saying that Φ is surjective. Since $T \times (K/T)$ has the same dimension as K, we may apply Theorem 11.16 and Corollary 11.17. Thus, if there is even one regular value of Φ for which the signed number of preimages is nonzero, Φ must be surjective. Our strategy will be to find a certain class of points $y \in K$ for which we can (1) determine all of the preimages of y under Φ, and (2) verify that Φ_* is invertible and orientation preserving at each of the preimages.

Lemma 11.26. *Let $t \in T$ be such that the subgroup generated by t is dense in T (Proposition 11.4). Then $\Phi^{-1}(\{t\})$ of t consists precisely of elements of the form $(x^{-1}tx, [x])$ with $[x]$ belonging to $W = N(T)/T$. In particular, if $xsx^{-1} = t$ for some $x \in K$ and $s \in T$, then s must be of the form $s = w^{-1} \cdot t$ for some $w \in W$.*

Note that if x and y in $N(T)$ represent distinct elements of $W = N(T)/T$, then $[x]$ and $[y]$ are distinct elements of K/T. Thus, the lemma tells us that there is a one-to-one correspondence between $\Phi^{-1}(\{t\})$ and W.

Proof. If $x \in N(T)$, then $x^{-1}tx \in T$ and we can see that $\Phi(x^{-1}tx, [x]) = t$. In the other direction, if $xsx^{-1} = t$, then

$$x^{-1}t^m x = s^m \in T$$

for all integers m, so that $x^{-1}Tx \subset T$ by our assumption on t. Since $x^{-1}Tx$ is again a maximal torus, we must actually have $x^{-1}Tx = T$ and, thus, $T = xTx^{-1}$, showing that $x \in N(T)$. Furthermore, since $xsx^{-1} = t$, we have $s = x^{-1}tx = w^{-1} \cdot t$, where $w = [x]$. □

We now compute the differential of Φ. Using Proposition 11.23 with (G, H) equal to $(T, \{I\}), (K, T)$, and $(K, \{I\})$, we identify the tangent space at each point in $T \times (K/T)$ with $\mathfrak{t} \oplus \mathfrak{f} \cong \mathfrak{k}$ and the tangent space at each point in K with $\mathfrak{k}$. Since we are trying to determine the *signed* number of preimages of Φ, we must choose orientations on $T \times (K/T)$ and on K. To this end, we choose orientations on the vector spaces $\mathfrak{t}$ and $\mathfrak{f}$ and use the obvious associated orientation on $\mathfrak{k} \cong \mathfrak{t} \oplus \mathfrak{f}$. We then define orientations on $T \times (K/T)$ and K using the above identifications of the tangent spaces with $\mathfrak{t} \oplus \mathfrak{f} \cong \mathfrak{k}$. The identification of the tangent spaces to K/T with $\mathfrak{f}$ is unique up to the adjoint action of T (Proposition 11.23). Since T is connected, this action will have positive determinant, showing that the orientation on $T \times (K/T)$ is well defined. Recall that the (orthogonal) adjoint action of T on $\mathfrak{k}$ preserves $\mathfrak{t}$ and thus, also, $\mathfrak{f} := \mathfrak{t}^{\perp}$.

Proposition 11.27. *Let $(t, [x])$ be a fixed point in $T \times (K/T)$. If we identify the tangent spaces to $T \times (K/T)$ and to K with $\mathfrak{t} \oplus \mathfrak{f} \cong \mathfrak{k}$, then the differential of Φ at $(t, [x])$ is represented by the following operator:*

$$\Phi_* = (\mathrm{Ad}_x) \begin{pmatrix} I & 0 \\ 0 & \mathrm{Ad}'_{t^{-1}} - I \end{pmatrix}, \tag{11.11}$$

where $\mathrm{Ad}'_{t^{-1}}$ denotes the restriction of $\mathrm{Ad}_{t^{-1}}$ to $\mathfrak{f}$.

Proof. For $H \in \mathfrak{t}$, we compute that

$$\frac{d}{d\tau}\Phi(te^{\tau H}, [x])\Big|_{\tau=0} = \frac{d}{d\tau}xte^{\tau H}x^{-1}\Big|_{\tau=0}$$
$$= xtHx^{-1}$$
$$= (xtx^{-1})(\mathrm{Ad}_x(H)).$$

Since we identify the tangent space to K at xtx^{-1} with $\mathfrak{k}$ using the left action of xtx^{-1}, we see that $\Phi_*((H, 0)) = \mathrm{Ad}_x(H)$.

Meanwhile, if $X \in \mathfrak{f}$, we compute that

$$\frac{d}{d\tau}\Phi(t, [xe^{\tau X}])\Big|_{\tau=0} = \frac{d}{d\tau}xe^{\tau X}te^{-\tau X}x^{-1}\Big|_{\tau=0}$$
$$= xXtx^{-1} - xtXx^{-1}$$
$$= xtx^{-1}(xt^{-1}Xtx^{-1} - xXx^{-1})$$
$$= (xtx^{-1})[\mathrm{Ad}_x(\mathrm{Ad}_{t^{-1}}(X) - X)],$$

so that $\Phi_*((0, X)) = \mathrm{Ad}_x(\mathrm{Ad}_{t^{-1}}(X) - X)$. These two calculations, together with the linearity of the differential, establish the claimed form of Φ_*. □

We now wish to determine when $\Phi_*(t, [x])$ is invertible. Since Ad_x is invertible, the question becomes whether $\mathrm{Ad}'_{t^{-1}} - I$ is invertible. When $\Phi_*(t, [x])$ is invertible, we would like to know whether this linear map is orientation preserving or orientation reversing. In light of the way our orientations on $T \times (K/T)$ and K have been chosen, the orientation behavior of Φ_* will be determined by the sign of the determinant of Φ_* as a linear map of $\mathfrak{k}$ to itself. Now, since K is connected and our inner product on $\mathfrak{k}$ is Ad_K-invariant, we see that $\mathrm{Ad}_x \in SO(\mathfrak{k})$ for every x. Thus, $\det(\mathrm{Ad}_x) = 1$, which means that we only need to calculate the determinant of the second factor on the right-hand side of (11.11).

Lemma 11.28. *For $t \in T$, let $\mathrm{Ad}'_{t^{-1}}$ denote the restriction of $\mathrm{Ad}_{t^{-1}}$ to $\mathfrak{f}$.*

1. *If t generates a dense subgroup of T, then $\mathrm{Ad}'_{t^{-1}} - I$ is an invertible linear transformation of $\mathfrak{f}$.*
2. *For all $w \in W$ and $t \in T$, we have*

$$\det(\mathrm{Ad}'_{w\cdot t^{-1}} - I) = \det(\mathrm{Ad}'_{t^{-1}} - I).$$

Proof. The operator $\mathrm{Ad}'_{t^{-1}} - I$ is invertible provided that the restriction of $\mathrm{Ad}_{t^{-1}}$ to $\mathfrak{f}$ does not have an eigenvalue of 1. Suppose, then, that $\mathrm{Ad}_{t^{-1}}(X) = X$ for some $X \in \mathfrak{f}$. Then for every integer m, we will have $\mathrm{Ad}_{t^m}(X) = X$. If t generates a dense subgroup of T, then by taking limits, we conclude that $\mathrm{Ad}_s(X) = X$ for all $s \in T$. But then for all $H \in \mathfrak{t}$, we have

$$[H, X] = \frac{d}{d\tau} \mathrm{Ad}_{e^{\tau H}}(X) \Big|_{\tau=0} = 0.$$

Since $\mathfrak{t}$ is maximal commutative (Proposition 11.7), we conclude that $X \in \mathfrak{f} \cap \mathfrak{t} = \{0\}$. Thus, there is no nonzero $X \in \mathfrak{f}$ for which $\mathrm{Ad}_{t^{-1}}(X) = X$.

For the second point, if $w \in W$ is represented by $x \in N(T)$, we have

$$\mathrm{Ad}'_{w \cdot t^{-1}} - I = \mathrm{Ad}'_{xt^{-1}x^{-1}} - I$$
$$= \mathrm{Ad}_x (\mathrm{Ad}'_{t^{-1}} - I) \mathrm{Ad}_{x^{-1}}.$$

Thus, $\mathrm{Ad}'_{w \cdot t^{-1}} - I$ and $\mathrm{Ad}'_{t^{-1}} - I$ are similar and have the same determinant. □

We are now ready for the proof of Lemma 11.12, which will complete the proof of the torus theorem.

Proof of Lemma 11.12. By Proposition 11.4, we can choose $t \in T$ so that the subgroup generated by t is dense in T. Then by Lemma 11.26, the preimages of t are in one-to-one correspondence with elements of W. Furthermore, by Proposition 11.27 and Point 1 of Lemma 11.28, Φ_* is nondegenerate at each preimage of t. Finally, by Point 2 of Lemma 11.28, Φ_* has the same orientation behavior at each point of $\Phi^{-1}(\{t\})$. Thus, t is a regular value of Φ and the signed number of preimages of t under Φ is either $|W|$ or $-|W|$. It then follows from Corollary 11.17 that Φ is surjective, which is the content of Lemma 11.12. □

Corollary 11.29. *The Weyl group W is finite and the orientations on $T \times (K/T)$ and K can be chosen so that the mapping degree of Φ is $|W|$, the order of the Weyl group.*

Proof. If t generates a dense subgroup of T, then by Lemma 11.26, $\Phi^{-1}(\{t\})$ is in one-to-one correspondence with W. Furthermore, Point 1 of Lemma 11.28 then tells us that such a t is a regular value of Φ. Thus, by Proposition 11.14, $\Phi^{-1}(\{t\})$ is finite, and W is thus also finite. Meanwhile, we already noted in the proof of Lemma 11.12 that Φ has mapping degree equal to $\pm|W|$. By reversing the orientation on K as necessary, we can ensure that the mapping degree is $|W|$. □

11.6   The Weyl Integral Formula

In this section, we apply Proposition 11.19 to the map Φ to obtain an integration formula for functions f on K. Of particular importance will be the special case in which f satisfies $f(yxy^{-1}) = f(x)$ for all $x, y \in K$. This special case of the Weyl integral formula will be a main ingredient in our analytic proof of the Weyl character formula in Sect. 12.4.

Recall that we have decomposed $\mathfrak{k}$ as $\mathfrak{t} \oplus \mathfrak{f}$, where $\mathfrak{f}$ is the orthogonal complement of $\mathfrak{t}$, and that the adjoint action of T on $\mathfrak{k}$ preserves both $\mathfrak{t}$ and $\mathfrak{f}$. Define a function $\rho : T \to \mathbb{R}$ by

$$\rho(t) = \det(\mathrm{Ad}'_{t^{-1}} - I), \tag{11.12}$$

where $\mathrm{Ad}'_{t^{-1}}$ is the restriction of $\mathrm{Ad}_{t^{-1}}$ to $\mathfrak{f}$. Using Proposition 11.24, we can construct volume forms on K, T, and K/T that are invariant under the left action of K, T, and K, respectively. Since each of these manifolds is compact, the total volume is finite, and we can normalize this volume to equal 1.

Theorem 11.30 (Weyl Integral Formula). *For all continuous functions f on K, we have*

$$\int_K f(x)\, dx = \frac{1}{|W|} \int_T \rho(t) \int_{K/T} f(yty^{-1})\, d[y]\, dt, \tag{11.13}$$

where dx, dt, and $d[y]$ are the normalized, left-invariant volume forms on K, T, and K/T, respectively and $|W|$ is the order of the Weyl group.

In Sect. 12.4, we will compute ρ explicitly and relate it to the Weyl denominator.

Proof of Theorem 11.30, up to a constant. Since Φ has mapping degree $|W|$, Theorem 11.16 tells us that

$$|W| \int_K f(x)\, dx = \int_{T \times (K/T)} \Phi^*(f(x)\, dx)$$

$$= \int_{T \times (K/T)} (f \circ \Phi)\, \Phi^*(dx), \tag{11.14}$$

for any smooth function f. Since, by the Stone–Weierstrass theorem (Theorem 7.33 in [Rud1]), every continuous function on K can be uniformly approximated by smooth functions, (11.14) continues to hold when f is continuous. Thus, to establish (11.13), it suffices to show that $\Phi^*(dx) = \rho(t)\, d[y] \wedge dt$.

Pick orthonormal bases $H_1, \ldots, H_r$ for $\mathfrak{t}$ and $X_1, \ldots, X_N$ for $\mathfrak{f}$. Then by the proof of Proposition 11.24, we can find invariant volume forms α_1 on T, α_2 on K/T, and β on K such that at each point, we have

$$\alpha_1(H_1, \ldots, H_r) = \alpha_2(X_1, \ldots, X_N) = \beta(H_1, \ldots, H_r, X_1, \ldots, X_N) = 1,$$

so that

$$(\alpha_1 \wedge \alpha_2)(H_1, \ldots, H_r, X_1, \ldots, X_N) = 1.$$

By the uniqueness in Proposition 11.24, α_1, α_2, and β will coincide, *up to multiplication by a constant*, with the normalized volume forms $dt, d[y]$, and dx, respectively.

Now, at each point, the matrix of Φ_*, with respect to the chosen bases for $T(T \times (K/T))$ and for $T(K)$, is given by the matrix in (11.11). Thus, using the definition of the pulled-back form $\Phi^*(\beta)$, we have

$$\Phi^*(\beta)(H_1, \ldots, H_r, X_1, \ldots, X_N)$$
$$= \beta(\Phi_*(H_1), \ldots, \Phi_*(H_r), \Phi_*(X_1), \ldots, \Phi_*(X_N))$$
$$= \det(\Phi^*)\beta(H_1, \ldots, H_r, X_1, \ldots, X_N)$$
$$= \rho(t)(\alpha_1 \wedge \alpha_2)(H_1, \ldots, H_r, X_1, \ldots, X_N),$$

where in the third line, we use (B.1) in Appendix B. Since dx coincides with β up to a constant and $d[y] \wedge dt$ coincides with $\alpha_1 \wedge \alpha_2$ up to a constant, (11.14) then becomes

$$|W| \int_K f(x)\, dx = C \int_{T \times (K/T)} \rho(t) f(yty^{-1})\, d[y]\, dt.$$

It remains only to show that $C = 1$. We postpone the proof of this claim until Sect. 12.4.   □

It is possible to verify that $C = 1$ directly, using Lemma 11.21. According to that result, if U is a small open set in K/T, then $q^{-1}(U)$ is diffeomorphic to $U \times T$. It is not hard to check that under the diffeomorphism between $q^{-1}(U)$ and $U \times T$, the volume form β decomposes as the product of α_2 and α_2. Thus, for any (nice enough) $E \subset U$, the volume of $E \times T = q^{-1}(E)$ is the product of the volume of E and the volume of T. From this, it is not hard to show that for any (nice enough) set $E \subset K/T$, the volume of $q^{-1}(E)$ equals the volume of E (with respect to α_2) times the volume of T (with respect to α_1). In particular, the volume of $q^{-1}(K/T) = K$ is the product of the volume of K/T and the volume of T. Thus, if we choose our inner products on $\mathfrak{t}$ and on $\mathfrak{f}$ so that the volume forms α_1 and α_2 are normalized, the volume form β will also be normalized. In that case, the above computation holds on the nose, without any undetermined constant. Since we will offer a different proof of the normalization constant in Sect. 12.4, we omit the details of this argument.

We now consider an important special case of Theorem 11.30.

Definition 11.31. A function $f : K \to \mathbb{C}$ is called a **class function** if $f(yxy^{-1}) = f(x)$ for all $x, y \in K$.

That is to say, a function is a class function if it is constant on each conjugacy class.

Corollary 11.32. *If f is a continuous class function on K, then*

$$\int_K f(x)\, dx = \frac{1}{|W|} \int_T \rho(t) f(t)\, dt. \tag{11.15}$$

Proof. If f is a class function, then $f(yty^{-1}) = f(t)$ for all $y \in K$ and $t \in T$. Since the volume form on K/T is normalized, we have

$$\int_K f(yty^{-1})\, d[y] = f(t),$$

in which case, the Weyl integral formula reduces to (11.15). □

Example 11.33. Suppose $K = \mathsf{SU}(2)$ and T is the diagonal subgroup. Then Corollary 11.32 takes the form

$$\int_{\mathsf{SU}(2)} f(x)\, dx = \frac{1}{2} \int_{-\pi}^{\pi} f(\mathrm{diag}(e^{i\theta}, e^{-i\theta}))\, 4\sin^2(\theta)\, \frac{d\theta}{2\pi}, \tag{11.16}$$

where $|W| = 2$ and the normalized volume measure on T is $d\theta/(2\pi)$.

Note that if $f \equiv 1$, both sides of (11.16) integrate to 1. See also Exercise 9 in Chapter 12 for an explicit version of the Weyl integral formula for $\mathsf{U}(n)$.

Proof. If we use the Hilbert–Schmidt inner product on $\mathfrak{su}(2)$, the orthogonal complement of $\mathfrak{t}$ in $\mathfrak{su}(2)$ is the space of matrices X of the form

$$X = \begin{pmatrix} 0 & x+iy \\ -x+iy & 0 \end{pmatrix},$$

with $x, y \in \mathbb{R}$. Direct computation then shows that if $t = \mathrm{diag}(e^{i\theta}, e^{-i\theta})$, then

$$\mathrm{Ad}_{t^{-1}}(X) = \begin{pmatrix} 0 & e^{-2i\theta}(x+iy) \\ e^{2i\theta}(-x+iy) & 0 \end{pmatrix}.$$

Thus, $\mathrm{Ad}_{t^{-1}}$ acts as a rotation by angle -2θ in $\mathbb{C} = \mathbb{R}^2$. It follows that

$$\det(\mathrm{Ad}_{t^{-1}} - I) = \det \begin{pmatrix} \cos(2\theta)-1 & \sin(2\theta) \\ -\sin(2\theta) & \cos(2\theta)-1 \end{pmatrix}.$$

This determinant simplifies by elementary trigonometric identities to $4\sin^2\theta$. Finally, since $W = \{I, -I\}$, we have $|W| = 2$. □

Note that the matrix $\mathrm{diag}(e^{i\theta}, e^{-i\theta})$ is conjugate in $\mathsf{SU}(2)$ to the matrix $\mathrm{diag}(e^{-i\theta}, e^{i\theta})$:

$$\begin{pmatrix} e^{-i\theta} & 0 \\ 0 & e^{i\theta} \end{pmatrix} = \begin{pmatrix} 0 & 1 \\ -1 & 0 \end{pmatrix} \begin{pmatrix} e^{i\theta} & 0 \\ 0 & e^{-i\theta} \end{pmatrix} \begin{pmatrix} 0 & -1 \\ 1 & 0 \end{pmatrix}.$$

Thus, if f is a class function on $\mathsf{SU}(2)$, the value of f at $\mathrm{diag}(e^{-i\theta}, e^{i\theta})$ is the same its value at $\mathrm{diag}(e^{i\theta}, e^{-i\theta})$. We may therefore rewrite the right-hand side of (11.16) as

$$\int_0^\pi f(\mathrm{diag}(e^{i\theta}, e^{-i\theta}))\, 4\sin^2(\theta)\, \frac{d\theta}{2\pi}. \tag{11.17}$$

Meanwhile, recall from Exercise 5 in Chapter 1 that $\mathsf{SU}(2)$ can be identified with the unit sphere $S^3 \subset \mathbb{C}^2$. By Exercise 9, a function f on $\mathsf{SU}(2)$ is a class function if and only if the associated function on S^3 depends only on the polar angle. Furthermore, for $0 \le \theta \le \pi$, the polar angle associated to $\mathrm{diag}(e^{i\theta}, e^{-i\theta})$ is simply θ. With this perspective, (11.17) is simply the formula for integration in spherical coordinates on S^3, in the special case in which the function depends only on the polar angle. (Apply the $m = 4$ case of Eq. (21.15) in [Has] to a function that depends only on the polar angle.)

11.7 Roots and the Structure of the Weyl Group

In the context of compact Lie groups, it is convenient and customary to redefine the notion of "root" by a factor of i so that the roots will now live in $\mathfrak{t}$ rather than in $i\mathfrak{t}$.

Definition 11.34. An element α of $\mathfrak{t}$ is **real root** of $\mathfrak{g}$ with respect to $\mathfrak{t}$ if $\alpha \ne 0$ and there exists a nonzero element X of $\mathfrak{g}$ such that

$$[H, X] = i\,\langle \alpha, H \rangle\, X$$

for all $H \in \mathfrak{t}$. For each real root α, we consider also the associated **real coroot** H_α given by

$$H_\alpha = 2\frac{\alpha}{\langle \alpha, \alpha \rangle}.$$

When working with the group K and its Lie algebra $\mathfrak{k}$, the use of real roots (and, later, real weights for representations) is convenient because it makes the factors of i explicit, rather than hiding them in the fact that the roots live in $i\mathfrak{t}$. If, for example, we wish to compute the complex conjugate of the expression $e^{i\langle \alpha, H \rangle}$, where α is a real root and H is in $\mathfrak{t}$, the explicit factor of i makes it obvious that the conjugate is $e^{-i\langle \alpha, H \rangle}$.

If our compact group K is simply connected, then $\mathfrak{g} := \mathfrak{k}_\mathbb{C}$ is semisimple, by Proposition 7.7. In general, $\mathfrak{k}$ decomposes as $\mathfrak{k} = \mathfrak{k}_1 \oplus \mathfrak{z}$ where $\mathfrak{z}$ is the center of $\mathfrak{k}$ and where $\mathfrak{g}_1 := (\mathfrak{k}_1)_\mathbb{C}$ is semisimple. (See the proof of Proposition 7.6.) Furthermore, as in the proof of Theorem 7.35, every maximal commutative subalgebra $\mathfrak{t}$ of $\mathfrak{k}$ will be of the form $\mathfrak{t}_1 \oplus \mathfrak{z}$, where $\mathfrak{t}_1$ is a maximal commutative subalgebra of $\mathfrak{k}_1$. All the results from Chapter 7 then apply—with slight modifications to account for the use of real roots—*except* that the roots may not span $\mathfrak{t}$. Nevertheless, the roots form a root system in their span, namely the space $\mathfrak{t}_1$. Throughout the section, we will let $R \subset \mathfrak{t}$ denote the set of real roots, Δ denote a fixed base for R, and R^+ denote the associated set of positive (real) roots.

Now that we have introduced the (real) roots for K, it makes sense to compare the Weyl group in the compact group sense (the group $N(T)/T$) to the Weyl group in the Lie algebra sense (the group generated by reflections about the hyperplanes perpendicular to the roots). As it turns out, these two groups are isomorphic. It is not hard to show that for each reflection there is an associated element of the Weyl group. The harder part of the proof is to show that these elements generate all of $N(T)/T$. This last claim is proved by making a clever use of the torus theorem.

Proposition 11.35. *For each $\alpha \in R$, there is an element x in $N(T)$ such that*

$$\mathrm{Ad}_x(H_\alpha) = -H_\alpha$$

and such that

$$\mathrm{Ad}_x(H) = H$$

for all $H \in \mathfrak{t}$ for which $\langle \alpha, H \rangle = 0$. Thus, the adjoint action of x on $\mathfrak{t}$ is the reflection s_α about the hyperplane orthogonal to α.

Proof. Choose X_α and Y_α as in Theorem 7.19, with $Y_\alpha = X_\alpha^*$. Then $(X_\alpha - Y_\alpha)^* = -(X_\alpha - Y_\alpha)$, from which it follows that $X_\alpha - Y_\alpha \in \mathfrak{k}$. Let us define $x \in K$ by

$$x = \exp\left[\frac{\pi}{2}(X_\alpha - Y_\alpha)\right]$$

(where π is the number $3.14\cdots$, not a representation). Then by the relationship between Ad and ad (Proposition 3.35), we have

$$\mathrm{Ad}_x(H) = \exp\left[\frac{\pi}{2}(\mathrm{ad}_{X_\alpha} - \mathrm{ad}_{Y_\alpha})\right](H) \qquad (11.18)$$

for all $H \in \mathfrak{t}$. If $\langle \alpha, H \rangle = 0$, then $(\mathrm{ad}_{X_\alpha} - \mathrm{ad}_{Y_\alpha})(H) = 0$, so that $\mathrm{Ad}_x(H) = H$.

Consider, then, the case $H = H_\alpha$. In that case, the entire calculation on the right-hand side of (11.18) taking place in the subalgebra $\mathfrak{s}^\alpha = \langle X_\alpha, Y_\alpha, H_\alpha \rangle$ of $\mathfrak{g}$. In $\mathfrak{s}^\alpha$, the elements $X_\alpha - Y_\alpha$, $iX_\alpha + iY_\alpha + H_\alpha$, and $iX_\alpha + iY_\alpha - H_\alpha$ are eigenvectors for $\mathrm{ad}_{X_\alpha} - \mathrm{ad}_{Y_\alpha}$ with eigenvalues $0, 2i$, and $-2i$, respectively. Since H_α is half the difference of the last two vectors, we have

$$\exp\left[\frac{\pi}{2}(\mathrm{ad}_{X_\alpha} - \mathrm{ad}_{Y_\alpha})\right](H_\alpha)$$

$$= e^{i\pi}(iX_\alpha + iY_\alpha + H_\alpha)/2 - e^{-i\pi}(iX_\alpha + iY_\alpha - H_\alpha)/2$$

$$= -H_\alpha. \tag{11.19}$$

Thus, Ad_x maps H_α to $-H_\alpha$ and is the identity on the orthogonal complement of α. Note that $\mathrm{Ad}_x(H)$ belongs to $\mathfrak{t}$ for all H in $\mathfrak{t}$. It follows that xtx^{-1} belongs to T for all t in T, showing that x is in $N(T)$. □

See Exercise 10 for an alternative approach to verifying (11.19). We now proceed to show that the Weyl group is generated by the reflections s_α, $\alpha \in R$. We let $Z(T)$ denote the **centralizer** of T, that is

$$Z(T) = \{x \in K \mid xt = tx, \text{ for all } t \in T\}.$$

Theorem 11.36. *If T is a maximal torus in K, the following results hold.*

1. *$Z(T) = T$.*
2. *The Weyl group acts effectively on $\mathfrak{t}$ and this action is generated by the reflections s_α, $\alpha \in R$, in Proposition 11.35.*

Since $Z(T) = T$, it follows that T is a maximal commutative subgroup of T (i.e., there is no commutative subgroup of K properly containing T). Nevertheless, there may exist maximal commutative subgroups of K that are not maximal tori; see Exercise 5.

It is not hard to verify Theorem 11.36 directly in the case of $\mathsf{SU}(n)$; see Exercise 3. The following lemma is the key technical result in the proof of Theorem 11.36 in general.

Lemma 11.37. *Suppose S is a connected, commutative subgroup of K. If x belongs to $Z(S)$, then there is a maximal torus S' containing both S and x.*

Proof. Let x be in $Z(S)$, let B be the subgroup of K generated by S and x, and let $\bar{B}$ be the closure of B. We are going to show that there is an element y of $\bar{B}$ such that the group generated by y is dense in $\bar{B}$. The torus theorem will then tell us that there is a maximal torus S' containing y and, thus, both S and x.

Since $\bar{B}$ is compact and commutative, Theorem 11.2 implies that the identity component $\bar{B}_0$ of $\bar{B}$ is a torus. Since $\bar{B}$ is compact, it has only finitely many components (Exercise 15 in Chapter 3), which means that the quotient group $\bar{B}/\bar{B}_0$ is finite.

Now, every element y of $\bar{B}$ is the limit of sequence of the form $x^{n_k} s_k$ for some integers n_k and elements $s_k \in S$. Thus, for some large k, the element $x^{n_k} s_k$ will be in the same component of $\bar{B}$ as y. But since S is connected, x^{n_k} must also be in the same component of $\bar{B}$ as y. It follows that $[y]$ and $[x^{n_k}]$ represent the same element of the quotient group $\bar{B}/\bar{B}_0$. We conclude, then, that $\bar{B}/\bar{B}_0$ is a cyclic group generated by $[x]$. Since also $\bar{B}/\bar{B}_0$ is finite, it must be isomorphic to $\mathbb{Z}/m$ for some positive integer m.

It follows that x^m belongs to the torus $\bar{B}_0$. Choose some $t \in \bar{B}_0$ such that the subgroup generated by t is dense in $\bar{B}_0$ (Proposition 11.4), and choose $g \in \bar{B}_0$ so that $g^m = x^{-m}t$. (Since the exponential map for the torus $\bar{B}_0$ is surjective, $x^{-m}t \in \bar{B}_0$ has an mth root in $\bar{B}_0$.) Now set $y = gx$, so that y is in the same component of $\bar{B}$ as x. Since $\bar{B}$ is commutative, we have

$$y^m = g^m x^m = t,$$

which means that the set of elements of the form $y^{nm} = t^n$ is dense in $\bar{B}_0$. Now, since $\bar{B}/\bar{B}_0$ is cyclic with generator $[x]$, each component of $\bar{B}$ is of the form $x^k \bar{B}_0$ for some k. Furthermore, the set of elements of the form $y^{nm+k} = t^n x^k g^k$, with k and m fixed and n varying, is dense in $\bar{B}_0$.

We see, then, that the group generated by y contains a dense subset of each component of $\bar{B}$ and is, thus, dense in $\bar{B}$. By the torus theorem, there is a maximal torus S' that contains y. It follows that S' must contain $\bar{B}$ and, in particular, both S and x.   □

Figure 11.6 illustrates the proof of Lemma 11.37 in the case where $\bar{B}/\bar{B}_0$ is cyclic of order 5. We choose y in the same component of $\bar{B}$ as x in such a way that $t := y^5$ generates a dense subgroup of $\bar{B}_0$. Then y generates a dense subgroup of the whole group $\bar{B}$.

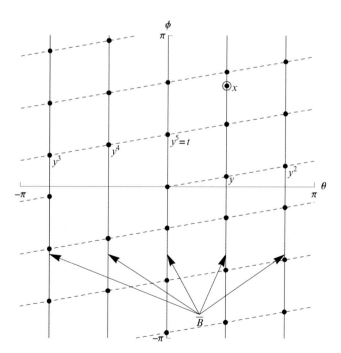

Fig. 11.6 The element y generates a dense subgroup of $\bar{B}$

Proof of Point 1 of Theorem 11.36. If we apply Lemma 11.37 with $S = T$, we see that any element x of $Z(T)$ must be contained in a maximal torus S' that contains T. But since T is also maximal, we must have $S' = T$, so that $x \in T$.   □

Lemma 11.38. *Suppose C is the fundamental Weyl chamber with respect to Δ and that w is an element of W that maps C to C. Then $w = 1$.*

For any $w \in W$, the action of W constitutes a symmetry of the root system R, that is, an orthogonal linear transformation that maps R to itself. In some cases, such as the root system B_2, there is no nontrivial symmetry of R that maps C to C. In the case of A_2, on the other hand, there *is* a nontrivial symmetry of R that maps C to C, namely the unique linear transformation that interchanges α_1 and α_2. (This map is just the reflection about the line through the root $\alpha_3 = \alpha_1 + \alpha_2$.) The lemma asserts that although this map is a symmetry of W, it is *not* given by the action of a Weyl group element. In the A_2 case, of course, we have an explicit description of the Weyl group, and we can easily check that there is no $w \in W$ with $w \cdot \alpha_1 = \alpha_2$ and $w \cdot \alpha_2 = \alpha_1$. (See Figure 11.7. where the Weyl group is the symmetry group of the indicated triangle.) Nevertheless, we need an argument that works in the general case.

The idea of the proof is as follows. We want to show that if $x \in N(T)$ and $\mathrm{Ad}_x(C) \subset C$, then $x \in T$. The idea is to show that x must commute with some nonzero $H \in \mathfrak{t}$, and that this H can be chosen to be "nice." *If H could be chosen so that the group $\exp(tH)$ were dense in T, then x would have to commute with*

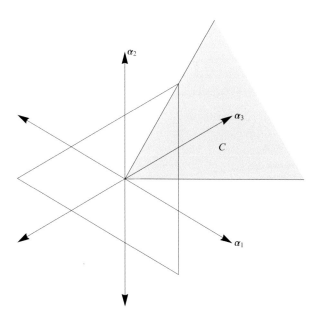

Fig. 11.7 The reflection about the line through α_3 is a symmetry of R that maps C to itself, but that is not an element of W

every element of T, so that x would belong to $Z(T) = T$. Although we cannot, in general, choose H to be as nice as that, we can choose H to be in the interior of C, which means that $\langle \alpha, H \rangle \neq 0$ for all $\alpha \in R$. This turns out to be sufficient to show that $x \in T$.

Proof. Let $x \in K$ be a representative of the element $w \in W$. Take any H_0 in the interior of C and average H_0 over the action of the (finite) subgroup of W generated by w. The resulting vector H is still in the interior of C (which is convex) and is now fixed by w, meaning that $\mathrm{Ad}_x(H) = H$. Thus, x commutes with every element of the one-parameter subgroup $S := \{\exp(tH) | t \in \mathbb{R}\}$. By Lemma 11.37, there is a maximal torus S' containing x and S. Suppose, toward a contradiction, that x is not in T. Then S' cannot equal T, and, since S' is maximal, S' cannot be contained in T. Thus, there is some $X \in \mathfrak{k}$ that is in the Lie algebra $\mathfrak{s}'$ of S' but not in $\mathfrak{t}$, and this X commutes with $\exp(tH), t \in \mathbb{R}$, and hence with H.

On the other hand, suppose we decompose $X \in \mathfrak{k} \subset \mathfrak{g}$ as a sum of an element of $\mathfrak{h}$ and elements of the various root spaces $\mathfrak{g}_\alpha$. Now, $[H, X] = 0$ and $\langle \alpha, H \rangle$ is nonzero for all α, since H is in the interior of C. Thus, the component of X in each $\mathfrak{g}_\alpha$ must be zero, meaning that $X \in \mathfrak{h} \cap \mathfrak{k} = \mathfrak{t}$, which is a contradiction. Thus, actually, x must be in T, which means that w is the identity element of W. $\square$

Proof of Point 2 of Theorem 11.36. We let $W' \subset W$ be the group generated by reflections. By Proposition 8.23, W' acts transitively on the set of Weyl chambers. Thus, for any $w \in W$, we can find $w' \in W'$ mapping $W(C)$ back to C, so that $w'w$ maps C to C. Then by Lemma 11.38, we have that $w'w = 1$, which means that $w = (w')^{-1}$ belongs to W'. Thus, every element of W actually belongs to W'. $\square$

Theorem 11.39. *If two elements s and t of T are conjugate in K, then there exists an element w of W such that $w \cdot s = t$.*

We may restate the theorem equivalently as follows: If $t = xsx^{-1}$ for some $x \in K$, then t can *also* be expressed as $t = ysy^{-1}$ for some $y \in N(T)$.

Proof. Suppose s and t are in T and $t = xsx^{-1}$ for some $x \in K$. Let $Z(t)$ be the centralizer of t. Since, xux^{-1} commutes with $t = xsx^{-1}$ for all $u \in T$, we see that $xTx^{-1} \subset Z(t)$. Thus, both T and xTx^{-1} are tori in $Z(t)$. Actually, since T and xTx^{-1} are connected, they must be contained in the identity component $Z(t)_0$ of $Z(t)$. Furthermore, since T and xTx^{-1} are *maximal* tori in K, they must be maximal tori in $Z(t)_0$. Thus, we may apply the torus theorem to $Z(t)_0$ to conclude that there is some $z \in Z(t)_0$ such that

$$zxTx^{-1}z^{-1} = T. \tag{11.20}$$

Now, (11.20) says that zx is in the normalizer $N(T)$ of T. Furthermore, since $z \in Z(t)_0$ commutes with t, we have

$$(zx)s(zx)^{-1} = z(xsx^{-1})z^{-1} = ztz^{-1} = t.$$

Thus, $y := zx$ is the desired element of $N(T)$ such that $t = ysy^{-1}$. $\square$

Corollary 11.40. *If f is a continuous Weyl-invariant function on T, then f extends uniquely to a continuous class function on K.*

Proof. By the torus theorem, each conjugacy class in K intersects T in at least one point. By Theorem 11.39, each conjugacy class intersects T in a single orbit of W. Thus, if f is a W-invariant function on T, we can unambiguously (and uniquely) extend f to a class function F on K by making F constant on each conjugacy class.

It remains to show that the extended function F is continuous on K. Suppose, then that $\langle x_n \rangle$ is a sequence in K converging to some x. We can write each x_n as $x_n = y_n t_n y_n^{-1}$, with $y_n \in K$ and $t_n \in T$. Since both K and T are compact, we can—after passing to a subsequence—assume that y_n converges to some $y \in K$ and t_n converges to some $t \in T$. It follows that

$$x = \lim_{n\to\infty} x_n = \lim_{n\to\infty} y_n t_n y_n^{-1} = y t y^{-1}.$$

Now, by our construction of F, we have $F(x_n) = f(t_n)$ and $F(x) = f(t)$. Thus, since f is assumed to be continuous on T, we see that

$$F(x_n) = f(t_n) \to f(t) = F(x),$$

showing that F is continuous on K. $\qquad\square$

11.8 Exercises

1. Let Γ denote the set of all vectors in $\mathbb{R}^2$ that can be expressed in the form

$$a(1,1) + b(3,1) + c(2,-4),$$

   for $a,b,c \in \mathbb{Z}$. Then Γ is a subgroup of $\mathbb{R}^2$ and Γ is discrete, since it is contained in $\mathbb{Z}^2$. Find linearly independent vectors v_1 and v_2 in Γ such that Γ consists precisely of the set of integer linear combinations of v_1 and v_2. (Compare Lemma 11.3.)
2. (a) Show that every continuous homomorphism from S^1 to S^1 is of the form $u \mapsto u^m$ for some $m \in \mathbb{Z}$.
      Hint: Use Theorem 3.28.
   (b) Show that every continuous homomorphism from $(S^1)^k \to S^1$ is of the form

$$(u_1, \ldots, u_k) \mapsto u_1^{m_1} \cdots u_k^{m_k}$$

      for some integers $m_1, \ldots, m_k$.
3. Consider the group $\mathsf{SU}(n)$, with maximal torus T being the intersection of $\mathsf{SU}(n)$ with the space of diagonal matrices. Prove directly (without appealing

to Theorem 11.36) that $Z(T) = T$ and that the $N(T)/T$ is isomorphic to the Weyl group of the root system A_{n-1}. (Compare Sect. 7.7.1.)

Hint: Imitate the calculations in Sect. 6.6.

4. Give an example of closed, oriented manifolds M and N of the same dimension and a smooth map $f : M \to N$ such that f has mapping degree zero but f is, nevertheless, surjective.

5. Let $K = SO(n)$, where $n \geq 3$. Let H be the (commutative) subgroup of K consisting of the diagonal matrices in $SO(n)$. (Of course, the diagonal entries have to be ± 1 and the number of diagonal entries equal to -1 must be even.) Show that H is a maximal commutative subgroup of K and that H is not contained in a maximal torus.

Hint: Use Proposition A.2.

Note: This example shows that in Lemma 11.37, the assumption that S be connected cannot be omitted. (Otherwise, we could take $S = H$ and $x = I$ and we would conclude that there is a maximal torus S' containing H.)

6. Suppose $K = SU(n)$ and H is any commutative subgroup of K. Show that H is conjugate to a subgroup of the diagonal subgroup of K and thus that H is contained in a maximal torus. This result should be contrasted with the result of Exercise 5.

7. (a) For any interval $(a, b) \subset \mathbb{R}$ and any $x, y \in (a, b)$, show that there exists a smooth family of diffeomorphisms $f_t : (a, b) \to (a, b), 0 \leq t \leq 1$, with the following properties. First, $f_0(z) = z$ for all z. Second, there is some $\varepsilon > 0$ such that $f_t(z) = z$ for all $z \in (a, a + \varepsilon)$ and for all $z \in (b - \varepsilon, b)$. Third, $f_1(x) = y$.

   Hint: Take

$$f_t(z) = z + t \int_a^z g(u)\, du$$

   for some carefully chosen function g.

   (b) If $R \subset \mathbb{R}^n$ is an open rectangle and x and y belong to R, show that there is a smooth family of diffeomorphisms $\Psi_t : R \to R$ such that (1) Ψ_0 is the identity map, (2) each Ψ_t is the identity in a neighborhood of ∂R, and (3) $\Psi_1(x) = y$.

8. (a) Show that the left action of G on G/H is smooth with respect to the collection of coordinate patches on G/H described in the proof of Theorem 11.22.

   (b) Show that the kernel of the differential of the quotient map $Q : G \to G/H$ at the identity is precisely $\mathfrak{h}$.

9. According to Exercise 5 in Chapter 1, each element of $SU(2)$ can be written uniquely as

$$U = \begin{pmatrix} \alpha & -\bar{\beta} \\ \beta & \bar{\alpha} \end{pmatrix},$$

where $(\alpha, \beta) \in \mathbb{C}^2$ belongs to the unit sphere S^3. For each $U \in \mathsf{SU}(2)$, let v_U denote the corresponding unit vector (α, β). Now, the angle θ between $(\alpha, \beta) \in S^3$ and the "north pole" $(1, 0)$ satisfies

$$\cos \theta = \mathrm{Re}(\langle (\alpha, \beta), (1, 0) \rangle) = \mathrm{Re}(\alpha),$$

and this relation uniquely determines θ, if we take $0 \le \theta \le \pi$. In spherical coordinates on S^3, the angle θ is the **polar angle**.

(a) Suppose $U \in \mathsf{SU}(2)$ has eigenvalues $e^{i\theta}$ and $e^{-i\theta}$. Show that

$$\mathrm{Re}(\langle v_U, (1, 0) \rangle) = \cos \theta.$$

   Hint: Use the trace.
(b) Conclude that U_1 and U_2 are conjugate in $\mathsf{SU}(2)$ if and only if v_{U_1} and v_{U_2} have the same polar angle.

10. In this exercise, we give an alternative verification of the identity (11.19). Since the left-hand side of (11.19) is expressed in purely Lie-algebraic terms, we may do the calculation in any Lie algebra isomorphic to $\langle X_\alpha, Y_\alpha, H_\alpha \rangle$, for example, in $\mathsf{sl}(2; \mathbb{C})$ itself. That is to say, it suffices to prove the formula with $X_\alpha = X$, $Y_\alpha = Y$, and $H_\alpha = H$, where X, Y, and H are the usual basis elements for $\mathsf{sl}(2; \mathbb{C})$.
   Show that

$$\exp\left[\frac{\pi}{2}(\mathrm{ad}_X - \mathrm{ad}_Y)\right](H) = e^{\frac{\pi}{2}(X-Y)} H e^{-\frac{\pi}{2}(X-Y)} = -H,$$

as claimed.

Chapter 12
The Compact Group Approach to Representation Theory

In this chapter, we follow Hermann Weyl's original approach to establishing the Weyl character formula and the theorem of the highest weight. Throughout the chapter, we assume K is a connected, compact matrix Lie group, with Lie algebra $\mathfrak{k}$. We fix on $\mathfrak{k}$ an inner product that is invariant under the adjoint action of K, constructed as in the proof of Theorem 4.28. Throughout the chapter, we let T denote a fixed maximal torus in K and we let $\mathfrak{t}$ denote the Lie algebra of T. We let R denote the set of *real* roots for $\mathfrak{k}$ relative to $\mathfrak{t}$ (Definition 11.34), we let Δ be a fixed base for R, and we let R^+ denote the positive real roots relative to Δ. We also let $W := N(T)/T$ denote the Weyl group for K relative to T. In light of Theorem 11.36, W is isomorphic to the subgroup of $\mathsf{O}(\mathfrak{t})$ generated by the reflections about the hyperplanes orthogonal to the roots.

12.1 Representations

All representations of K considered in this chapter are assumed to be *finite dimensional* and defined on a vector space over $\mathbb{C}$, unless otherwise stated. Although we are studying in this chapter the representations of the compact group K, it is convenient to describe the weights of such a representation in terms of the associated representation of $\mathfrak{g} = \mathfrak{k}_{\mathbb{C}}$. Since we are using real roots for $\mathfrak{g}$, we will also use real weights for representations of K.

A previous version of this book was inadvertently published without the middle initial of the author's name as "Brian Hall". For this reason an erratum has been published, correcting the mistake in the previous version and showing the correct name as Brian C. Hall (see DOI http://dx.doi.org/10.1007/978-3-319-13467-3_14). The version readers currently see is the corrected version. The Publisher would like to apologize for the earlier mistake.

© Springer International Publishing Switzerland 2015
B.C. Hall, *Lie Groups, Lie Algebras, and Representations*, Graduate
Texts in Mathematics 222, DOI 10.1007/978-3-319-13467-3_12

Definition 12.1. Let (Π, V) be a finite-dimensional representation of K and π the associated representation of $\mathfrak{g}$. An element λ of $\mathfrak{t}$ is called a **real weight** of V if there exists a nonzero element v of V such that

$$\pi(H)v = i \langle \lambda, H \rangle v \tag{12.1}$$

for all $H \in \mathfrak{t}$. The **weight space** with weight λ is the set of all $v \in V$ satisfying (12.1) and the **multiplicity** of λ is the dimension of the corresponding weight space.

We now consider some elementary properties of the weights of a representation.

Proposition 12.2. *If (Π, V) is a finite-dimensional representation of K, the real weights for Π and their multiplicities are invariant under the action of the Weyl group.*

Proof. Following the proof of Theorem 6.22, we can show that if $w \in W$ is represented by $x \in N(T)$, then $\Pi(x)$ will map the weight space with weight λ isomorphically onto the weight space with weight $w \cdot \lambda$. $\qquad\square$

The weights of representation of K satisfy an integrality condition that does not, in general, coincide with the notion of integrality in Definition 8.34.

Definition 12.3. Let Γ be the subset of $\mathfrak{t}$ given by

$$\Gamma = \left\{ H \in \mathfrak{t} \,\middle|\, e^{2\pi H} = I \right\}.$$

We refer to Γ as the **kernel of the exponential map** for $\mathfrak{t}$.

The set Γ should, more precisely, be referred to as the kernel of the exponential map scaled by a factor of 2π. Note that if $w \in W$ is represented by $x \in N(T)$, then for all $H \in \Gamma$, we have

$$e^{2\pi w \cdot H} = e^{2\pi x H x^{-1}} = x e^{2\pi H} x^{-1} = I.$$

Thus, Γ is invariant under the action of W on $\mathfrak{t}$.

Definition 12.4. An element of λ of $\mathfrak{t}$ is an **analytically integral element** if

$$\langle \lambda, H \rangle \in \mathbb{Z}$$

for all H in Γ. An element λ of $\mathfrak{t}$ is an **algebraically integral element** if

$$\langle \lambda, H_\alpha \rangle = 2 \frac{\langle \lambda, \alpha \rangle}{\langle \alpha, \alpha \rangle} \in \mathbb{Z}$$

for each real root α. An element λ of $\mathfrak{t}$ is **dominant** if

$$\langle \lambda, \alpha \rangle \geq 0$$

for all $\alpha \in \Delta$. Finally, if $\Delta = \{\alpha_1, \ldots, \alpha_r\}$ and μ and λ are two elements of $\mathfrak{t}$, we say that μ is **higher** than λ if

$$\mu - \lambda = c_1\alpha_1 + \cdots + c_r\alpha_r$$

with $c_j \geq 0$. We denote this relation by $\mu \succeq \lambda$.

The notion of an algebraically integral element is essentially the one we used in Chapters 8 and 9 in the context of semisimple Lie algebras. Specifically, if $\mathfrak{g} := \mathfrak{k}_\mathbb{C}$ is semisimple, then $\lambda \in \mathfrak{t}$ is algebraically integral if and only if $i\lambda$ is integral in the sense of Definition 8.34. We will see in Sect. 12.2 that every algebraically integral element is analytically integral, but not vice versa. In Chapter 13, we will show that when K is simply connected, the two notions of integrality coincide.

Proposition 12.5. *Let (Σ, V) be a representation of K and let σ be the associated representation of $\mathfrak{k}$. If $\lambda \in \mathfrak{t}$ is a real weight of σ, then λ is an analytically integral element.*

Proof. If v is a weight vector with weight λ and H is an element of Γ, then on the one hand,

$$\Sigma(e^{2\pi H})v = Iv = v,$$

while on the other hand,

$$\Sigma(e^{2\pi H})v = e^{2\pi\sigma(H)}v = e^{2\pi i \langle \lambda, H \rangle}v.$$

Thus, $e^{2\pi i \langle \lambda, H \rangle} = 1$, which implies that $\langle \lambda, H \rangle \in \mathbb{Z}$. $\square$

We are now ready to state the theorem of the highest weight for (finite-dimensional) representations of a connected compact group.

Theorem 12.6 (Theorem of the Highest Weight). *If K is a connected, compact matrix Lie group and T is a maximal torus in K, the following results hold.*

1. *Every irreducible representation of K has a highest weight.*
2. *Two irreducible representations of K with the same highest weight are isomorphic.*
3. *The highest weight of each irreducible representation of K is dominant and analytically integral.*
4. *If μ is a dominant, analytically integral element, there exists an irreducible representation of K with highest weight μ.*

Let us suppose now that $\mathfrak{g} := \mathfrak{k}_\mathbb{C}$ is semisimple. Even in this case, the set of analytically integral elements may not coincide with the set of algebraically integral elements, as we will see in several examples in Sect. 12.2. Thus, the theorem of the highest weight for $\mathfrak{g}$ (Theorems 9.4 and 9.5) will, in general, have a different set of possible highest weights than the theorem of the highest weight for K. This discrepancy arises because a representation of $\mathfrak{g}$ may not come from a representation of K, unless K is simply connected. In the simply connected case, there is no such discrepancy; according to Corollary 13.20, when K is simply connected, every algebraically integral element is analytically integral.

We will develop the tools for proving Theorem 12.6 in Sects. 12.2–12.4, with the proof itself coming in Sect. 12.5. The hard part of the theorem is Point 4; this will be established by appealing to a completeness result for characters.

12.2   Analytically Integral Elements

In this section, we establish some elementary properties of the set of analytically integral elements (Definition 12.4) and consider several examples. One additional key result that we will establish in Chapter 13 is Corollary 13.20, which says that when K is simply connected, the set of analytically integral elements coincides with the set of algebraically integral elements.

Proposition 12.7.

1. *The set of analytically integral elements is invariant under the action of the Weyl group.*
2. *Every analytically integral element is algebraically integral.*
3. *Every real root is analytically integral.*

We begin with an important lemma.

Lemma 12.8. *If $\alpha \in \mathfrak{t}$ is a real root and $H_\alpha = 2\alpha/\langle \alpha, \alpha \rangle$ is the associated real coroot, we have*

$$e^{2\pi H_\alpha} = I$$

in K. That is to say, H_α belongs to the kernel Γ of the exponential map.

Proof. By Corollary 7.20, there is a Lie algebra homomorphism $\phi : \mathsf{su}(2) \to \mathfrak{k}$ such that the element $iH = \operatorname{diag}(i, -i)$ in $\mathsf{su}(2)$ maps to the real coroot H_α. (In the notation of the corollary, $H_\alpha = 2E_1^\alpha$.) Since $\mathsf{SU}(2)$ is simply connected, there is (Theorem 5.6) a homomorphism $\Phi : \mathsf{SU}(2) \to K$ for which the associated Lie algebra homomorphism is ϕ. Now, the element $iH \in \mathsf{su}(2)$ satisfies $e^{2\pi iH} = I$, and thus

$$I = \Phi(e^{2\pi iH}) = e^{2\pi \phi(iH)} = e^{2\pi H_\alpha},$$

as claimed.   □

Proof of Proposition 12.7. For Point 1, we have already shown (following Definition 12.3) that Γ is invariant under the action of W. Thus, if λ is analytically integral and w is the Weyl group element represented by x, we have $\langle w \cdot \lambda, H \rangle = \langle \lambda, w^{-1} \cdot H \rangle \in \mathbb{Z}$, since $w^{-1} \cdot H$ is in Γ. For Point 2, note that for each α, we have $H_\alpha \in \Gamma$ by Lemma 12.8. Thus, if λ is analytically integral, $\langle \lambda, H_\alpha \rangle \in \mathbb{Z}$, showing that λ is algebraically integral. For Point 3, we note that the real roots are real weights of the adjoint representation, which is a representation of the *group* K, and not just its Lie algebra. Thus, by Proposition 12.5, the real roots are analytically integral.   □

Proposition 12.9. *If λ is an analytically integral element, there is a well-defined function $f_\lambda : T \to \mathbb{C}$ such that*

$$f_\lambda(e^H) = e^{i\langle \lambda, H \rangle} \tag{12.2}$$

for all $H \in \mathfrak{t}$. Conversely, for any $\lambda \in \mathfrak{t}$, if there is a well-defined function on T given by the right-hand side of (12.2), then λ must be analytically integral.

Proof. Replacing H by $2\pi H$, we can equivalently write (12.2) as

$$f_\lambda(e^{2\pi H}) = e^{2\pi i\langle \lambda, H \rangle}, \quad H \in \mathfrak{t}. \tag{12.3}$$

Now, since T is connected and commutative, every $t \in T$ can be written as $t = e^{2\pi H}$ for some $H \in \mathfrak{t}$. Furthermore, $e^{2\pi(H+H')} = e^{2\pi H}$ if and only if $e^{2\pi H'} = I$, that is, if and only if $H' \in \Gamma$. Thus, the right-hand side of (12.3) defines a function on T if and only if $e^{2\pi i\langle \lambda, H+H' \rangle} = e^{2\pi i\langle \lambda, H \rangle}$ for all $H' \in \Gamma$. This happens if and only if $e^{2\pi i\langle \lambda, H' \rangle} = 1$ for all $H' \in \Gamma$, that is, if and only if $\langle \lambda, H' \rangle \in \mathbb{Z}$ for all $H' \in \Gamma$. $\qquad\square$

Proposition 12.10. *The exponentials f_λ in (12.2), as λ ranges over the set of analytically integral elements, are orthonormal with respect to the normalized volume form dt on T:*

$$\int_T \overline{f_\lambda(t)} f_{\lambda'}(t) \, dt = \delta_{\lambda, \lambda'}. \tag{12.4}$$

Proof. Let us identify T with $(S^1)^k$ for some k, so that $\mathfrak{t}$ is identified with $\mathbb{R}^k$ and the scaled exponential map is given by

$$(\theta_1, \dots, \theta_n) \mapsto (e^{2\pi i \theta_1}, \dots, e^{2\pi i \theta_k}).$$

The kernel Γ of the exponential is the integer lattice inside $\mathbb{R}^k$. The lattice of analytically integral elements (points having integer inner product with each element of Γ) is then also the integer lattice. Thus, the exponentials in the proposition are the functions of the form

$$f_\lambda(e^{i\theta_1}, \dots, e^{i\theta_k}) = e^{i\lambda_1\theta_1} \cdots e^{i\lambda_k\theta_k},$$

with $\lambda = (\lambda_1, \dots, \lambda_k) \in \mathbb{Z}^k$.

Meanwhile, if we use the coordinates $\theta_1, \dots, \theta_k$ on T, then any k-form on T can be represented as a density ρ times $d\theta_1 \wedge \cdots \wedge d\theta_n$. Since the volume form dt on T is translation invariant, the density ρ must be constant. Thus, the normalized integral in (12.4) becomes

$$\int_T \overline{f_\lambda(t)} f_{\lambda'}(t) \, dt$$

$$= (2\pi)^{-k} \int_0^{2\pi} \cdots \int_0^{2\pi} \overline{e^{i\lambda_1\theta_1} \cdots e^{i\lambda_k\theta_k}} \, e^{i\lambda_1'\theta_1} \cdots e^{i\lambda_k'\theta_k} \, d\theta_1 \cdots d\theta_k$$

Fig. 12.1 Dominant, analytically integral elements (*black dots*) and dominant, algebraically integral elements (*black* and *white dots*) for SO(3)

$$= (2\pi)^{-k} \left(\int_0^{2\pi} e^{i(\lambda_1' - \lambda_1)\theta_1} d\theta_1 \right) \cdots \left(\int_0^{2\pi} e^{i(\lambda_k' - \lambda_k)\theta_k} d\theta_k \right).$$

The claimed orthonormality then follows from direct computation. $\square$

We now calculate the algebraically integral and analytically integral elements in several examples, with an eye toward clarifying the distinction between the two notions. When K is simply connected, Corollary 13.20 shows that the set of analytically integral and algebraically integral elements coincide. Thus, in the simply connected case, the calculations in Sects. 6.7 and 8.7 provide examples of the set of analytically integral elements. We consider now three groups that are not simply connected.

Example 12.11. Consider the group SO(3) and let $\mathfrak{t}$ be the maximal commutative subalgebra spanned by the element F_3 in Example 3.27. Let the unique positive root α be chosen so that $\langle \alpha, F_3 \rangle = 1$. Then $\mu \in \mathfrak{t}$ is dominant and algebraically integral if and only if $\mu = m\alpha/2$, where m is a non-negative integer, and $\mu \in \mathfrak{t}$ is dominant and analytically integral if and only if $\mu = m\alpha$, where m is a non-negative integer.

See Figure 12.1. Note that in this case, δ (half the sum of the positive roots) is equal to $\alpha/2$, which is not an analytically integral element.

Proof. Following Sect. 7.7, but adjusting for our current convention of using real roots (Definition 11.34), we identify $\mathfrak{t}$ with $\mathbb{R}$ by mapping aF_3 to a. The roots are then the numbers ± 1, where we take 1 as our positive root α, so that $\langle \alpha, F_3 \rangle = 1$. Then $\mu \in \mathfrak{t} \cong \mathbb{R}$ is dominant if and only if $\mu \geq 0$. Furthermore, μ is algebraically integral if and only if

$$2 \langle \mu, \alpha \rangle = 2(\mu)(1) \in \mathbb{Z},$$

that is, if and only if 2μ is an integer. Thus, the dominant, algebraically integral elements are the numbers of the form $m/2 = m\alpha/2$.

Now, $e^{2\pi aF_3} = I$ if and only if a is an integer. Thus, μ is analytically integral if and only if $(\mu)(a) \in \mathbb{Z}$ for all $a \in \mathbb{Z}$, that is, if and only if μ is an integer. Thus, the dominant, analytically integral elements are the numbers of the form $m = m\alpha$. $\square$

We consider next the group U(2), with $\mathfrak{t}$ consisting of diagonal matrices with pure imaginary diagonal entries. We identify $\mathfrak{t}$ with $\mathbb{R}^2$ by mapping diag(ia, ib) to (a, b). The roots are then the elements of the form $(1, 1)$ and $(-1, -1)$, and we select $\alpha := (1, 1)$ as our positive root. We now decompose every $H \in \Gamma$ as a linear combination of the vectors

$$\alpha = (1, 1); \quad \beta = (1, -1). \tag{12.5}$$

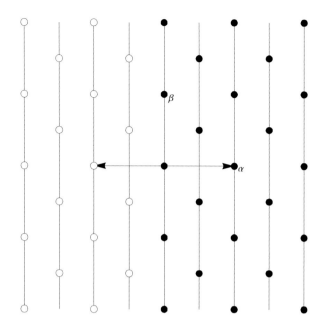

Fig. 12.2 The dominant, analytically integral elements (*black dots*) and nondominant, analytically integral elements (*white dots*) for U(2). The *vertical lines* indicate the algebraically integral elements

Example 12.12. Let t be the diagonal subalgebra of u(2), and write every element $\lambda \in$ t as

$$\lambda = c\alpha + d\beta,$$

with α and β as in (12.5). Then λ is analytically integral if and only if *either* c and d are both integers *or* c and d are both of the form integer plus one-half. Furthermore, λ is dominant if and only if $c \geq 0$. On the other hand, λ is algebraically integral if and only if c is either an integer or an integer plus one-half.

In Figure 12.2, the black dots are the dominant, analytically integral elements and the white dots are the nondominant, analytically integral elements. All the points in the vertical lines are algebraically integral.

Proof. If $H = \mathrm{diag}(ia, ib)$, then $e^{2\pi iH} = I$ if and only if a and b are both integers. Thus, when we identify t with $\mathbb{R}^2$, the kernel Γ of the exponential corresponds to the integer lattice $\mathbb{Z}^2$. The lattice of analytically integral elements is then also identified with $\mathbb{Z}^2$. Now, it is straightforward to check that $c\alpha + d\beta$ is in $\mathbb{Z}^2$ if and only if either c and d are both integers or c and d are both integers plus one-half, accounting for the claimed form of the analytically integral elements. Since β is orthogonal to α, an element $\lambda = c\alpha + d\beta$ has non-negative inner product with α if and only if $c \geq 0$, accounting for the claimed condition for λ to be dominant. Finally, λ is algebraically integral if and only if $2 \langle \lambda, \alpha \rangle / \langle \alpha, \alpha \rangle$ is an integer, which happens if c is either an integer or an integer plus one-half, with no restriction on d.   □

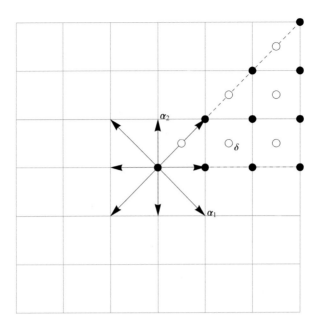

Fig. 12.3 Dominant, analytically integral elements (*black dots*) and dominant, algebraically integral elements (*black* and *white dots*) for SO(5)

Example 12.13. The dominant, analytically integral elements for SO(5) are as shown in Figure 12.3.

The figure shows the dominant, analytically integral elements (black dots). The white dots are dominant, algebraically integral elements that are not analytically integral. The background square lattice is the set of all analytically integral elements. Note that in Figure 12.3, the B_2 root system is rotated by $\pi/4$ compared to Figure 8.11. If we rotate Figure 12.3 clockwise by $\pi/4$ and then reflect across the *x*-axis, the set of dominant *algebraically* integral elements in Figure 12.3 (black and white dots) will match the set of "dominant integral" elements in Figure 8.11. Note that δ (half the sum of the positive roots) is not analytically integral.

Proof. Elements of t are of the form

$$
H = \begin{pmatrix} 0 & a & & & \\ -a & 0 & & & \\ & & 0 & b & \\ & & -b & 0 & \\ & & & & 0 \end{pmatrix},
\tag{12.6}
$$

with $a, b \in \mathbb{R}$. Following Sect. 7.7, but adjusting for our current convention of using real roots, we identify t with $\mathbb{R}^2$ by means of the map $H \mapsto (a, b)$. The roots are then the elements $(\pm 1, 0), (0, \pm 1)$, and $(\pm 1, \pm 1)$. As a base, we take $\alpha_1 = (1, -1)$ and $\alpha_2 = (0, 1)$. Furthermore, (x, y) is algebraically integral if

$$2\frac{\langle(1, -1), (x, y)\rangle}{2} = x - y \in \mathbb{Z}$$

and

$$2\frac{\langle(0, 1), (x, y)\rangle}{1} = 2y \in \mathbb{Z}.$$

These conditions hold if and only if *either* x and y are both integers *or* x and y are both of the form "integer plus one-half."

Now, if H is as in (12.6), $e^{2\pi H} = I$ if and only if a and b are integers. Thus, under our identification of $\mathfrak{t}$ with $\mathbb{R}^2$, the kernel of the exponential map is the set of elements of the form (a, b), with $a, b \in \mathbb{Z}$. Thus, (x, y) is analytically integral if

$$\langle(a, b), (x, y)\rangle = ax + by \in \mathbb{Z}$$

for all $a, b \in \mathbb{Z}$. This condition holds if and only if x and y are both integers.

Finally, (x, y) is dominant if it has non-negative inner product with each of α_1 and α_2, which happens if (x, y) is in the 45-degree sector indicated by dashed lines in Figure 12.3. □

12.3 Orthonormality and Completeness for Characters

In this section, we show that the characters of irreducible representations form an orthonormal set and that these characters are complete in the space of continuous class functions on K.

Definition 12.14. Suppose (Π, V) is representation of K. Then the **character** of Π is the function $\chi_\Pi : K \to \mathbb{C}$ given by

$$\chi_\Pi(x) = \text{trace}(\Pi(x)).$$

Note that we now consider the character as a function on the group K, rather than on the Lie algebra $\mathfrak{g} = \mathfrak{k}_\mathbb{C}$, as in Chapter 10. If π is the associated representation of $\mathfrak{g}$, then the character χ_π of π (Definition 10.11) is related to the character χ_Π of Π by

$$\chi_\Pi(e^H) = \chi_\pi(H), \quad H \in \mathfrak{k}.$$

Note that each character is a class function on K:

$$\chi_\Pi(yxy^{-1}) = \text{trace}(\Pi(y)\Pi(x)\Pi(y)^{-1}) = \text{trace}(\Pi(x)).$$

The following theorem says that the characters of irreducible representations form an orthonormal set in the space of class functions.

Theorem 12.15. *If (Π, V) and (Σ, W) are irreducible representations of K, then*

$$\int_K \overline{\text{trace}(\Pi(x))}\,\text{trace}(\Sigma(x))\,dx = \begin{cases} 1 & \text{if } V \cong W \\ 0 & \text{if } V \not\cong W \end{cases},$$

where dx is the normalized left-invariant volume form on K.

If (Π, V) is a representation of K, let V^K denote the space given by

$$V^K = \{v \in V \,|\, \Pi(x)v = v \text{ for all } x \in K\}.$$

Lemma 12.16. *Suppose (Π, V) is a finite-dimensional representation of K, and let P be the operator on V given by*

$$P = \int_K \Pi(x)\,dx.$$

Then P is a projection onto V^K. That is to say, P maps V into V^K and $Pv = v$ for all $v \in V^K$.

Clearly, V^K is an invariant subspace for Π. If we pick an inner product on V for which Π is unitary, then $(V^K)^\perp$ is also invariant under each $\Pi(x)$ and thus under P. But since P maps into V^K, the map P must be zero on $(V^K)^\perp$; thus, P is actually the *orthogonal* projection onto V^K.

Proof. For any $y \in K$ and $v \in V$, we have

$$\Pi(y)Pv = \Pi(y)\left(\int_K \Pi(x)\,dx\right)v$$
$$= \left(\int_K \Pi(yx)\,dx\right)v$$
$$= Pv,$$

by the left-invariance of the form dx. This shows that Pv belongs to V^K. Meanwhile, if $v \in V^K$, then

$$Pv = \int_K \Pi(x)v\,dx$$
$$= \left(\int_K dx\right)v$$
$$= v,$$

by the normalization of the volume form dx.   □

Note that if V is irreducible and nontrivial, then $V^K = \{0\}$. In this case, the proposition says that $\int_K \Pi(x) \, dx = 0$.

Lemma 12.17. *For $A : V \to V$ and $B : W \to W$, we have*

$$\text{trace}(A)\text{trace}(B) = \text{trace}(A \otimes B),$$

where $A \otimes B : V \otimes W \to V \otimes W$ is as in Proposition 4.16.

Proof. If $\{v_j\}$ and $\{w_l\}$ are bases for V and W, respectively, then $\{v_j \otimes w_l\}$ is a basis for $V \otimes W$. If A_{jk} and B_{lm} are the matrices of A and B with respect to $\{v_j\}$ and $\{w_l\}$, respectively, then the matrix of $A \otimes B$ with respect to $\{v_j \otimes w_l\}$ is easily seen to be

$$(A \otimes B)_{(j,l)(k,m)} = A_{jk} B_{lm}.$$

Thus,

$$\text{trace}(A \otimes B) = \sum_{j,l} A_{jj} B_{ll} = \text{trace}(A)\text{trace}(B),$$

as claimed.   □

Proof of Theorem 12.15. We know that there exists an inner product on V for which each $\Pi(x)$ is unitary. Thus,

$$\overline{\text{trace}(\Pi(x))} = \text{trace}(\Pi(x)^*) = \text{trace}(\Pi(x^{-1})). \qquad (12.7)$$

Recall from Sect. 4.3.3 that for any $A : V \to V$, we have the transpose operator $A^{tr} : V^* \to V^*$. Since the matrix of A^{tr} with respect to the dual of any basis $\{v_j\}$ of V is the transpose of the matrix of A with respect to $\{v_j\}$, we see that $\text{trace}(A^{tr}) = \text{trace}(A)$. Thus,

$$\overline{\text{trace}(\Pi(x))} = \text{trace}(\Pi(x^{-1})^{tr}) = \text{trace}(\Pi^{tr}(x)),$$

where Π^{tr} is the dual representation to Π. Thus, the complex conjugate of the character of Π is the character of the dual representation Π^{tr} of Π.
    Using Lemma 12.17, we then obtain

$$\overline{\text{trace}(\Pi(x))}\text{trace}(\Sigma(x)) = \text{trace}(\Pi^{tr}(x) \otimes \Sigma(x))$$
$$= \text{trace}((\Pi^{tr} \otimes \Sigma)(x)).$$

By Lemma 12.16, this becomes

$$\int_K \overline{\text{trace}(\Pi(x))}\text{trace}(\Sigma(x))\, dx = \int_K \text{trace}((\Pi^{tr} \otimes \Sigma)(x))\, dx$$

$$= \text{trace}\left(\int_K (\Pi^{tr} \otimes \Sigma)(x)\, dx\right)$$

$$= \text{trace}(P)$$

$$= \dim((V^* \otimes W)^K) \tag{12.8}$$

where P is a projection of $V^* \otimes W$ onto $(V^* \otimes W)^K$.

Now, for any two finite-dimensional vector spaces V and W, there is a natural isomorphism between $V^* \otimes W$ and $\text{End}(V, W)$, the space of linear maps from V to W. This isomorphism is actually an intertwining map of representations, where $x \in K$ acts on $A \in \text{End}(V, W)$ by

$$x \cdot A = \Sigma(x) A \Pi(x)^{-1}.$$

Finally, under this isomorphism, $(V^* \otimes W)^K$ maps to the space of intertwining maps of V to W. (See Exercises 3 and 4 for the proofs of the preceding claims.) By Schur's lemma, the space of intertwining maps has dimension 1 if $V \cong W$ and dimension 0 otherwise. Thus, (12.8) reduces to the claimed result.   □

Our next result says that the characters form a *complete* orthonormal set in the space of class functions on K.

Theorem 12.18. *Suppose f is a continuous class function on K and that for every finite-dimensional, irreducible representation Π of K, the function f is orthogonal to the character of Π:*

$$\int_K \overline{f(x)}\text{trace}(\Pi(x))\, dx = 0.$$

Then f is identically zero.

If $K = S^1$, the irreducible representations are one dimensional and of the form $\Pi(e^{i\theta}) = e^{im\theta} I$, $m \in \mathbb{Z}$, so that $\chi_\Pi(e^{i\theta}) = e^{im\theta}$. In this case, the completeness result for characters reduces to a standard result about Fourier series.

The proof given in this section assumes in an essential way that K is a compact *matrix* Lie group. (Of course, we work throughout the book with matrix Lie groups, but most of the proofs we give extend with minor modifications to arbitrary Lie groups.) In Appendix D, we sketch a proof of Theorem 12.18 that does not rely on the assumption that K is a matrix group. That proof, however, requires a bit more functional analysis than the proof given in this section.

We now consider a class of functions called matrix entries, that include as a special case the characters of representations. We will prove a completeness result for matrix entries and then specialize this result to class functions in order to obtain completeness for characters.

Definition 12.19. If (Π, V) is a representation of K and $\{v_j\}$ is a basis for V, the functions $f : K \to \mathbb{C}$ of the form

$$f(x) = (\Pi(x))_{jk} \tag{12.9}$$

are called **matrix entries for** Π. Here $(\Pi(x))_{jk}$ denotes the (j, k) entry of the matrix of $\Pi(x)$ in the basis $\{v_j\}$.

In a slight abuse of notation, we will also call f a matrix entry for Π if f is expressible as a *linear combination* of the functions in (12.9):

$$f(x) = \sum_{j,k} c_{jk}(\Pi(x))_{jk}. \tag{12.10}$$

We may write functions of the form (12.10) in a basis-independent way as

$$f(x) = \text{trace}(\Pi(x)A),$$

where A is the operator on V whose matrix in the basis $\{v_j\}$ is $A_{jk} = c_{kj}$. (The sum over k computes the product of $\Pi(x)$ with A and the sum over j computes the trace.) Note that if $A = I$ then f is the character of Π.

We will actually prove an orthogonality and completeness result for matrix entries, known as the **Peter–Weyl theorem**, which will imply the desired completeness result for characters. According to Exercise 5, matrix entries for nonisomorphic irreducible representations are orthogonal. We will show a completeness result for matrix entries as well: If f is continuous function on K and f is orthogonal to every matrix entry, then f is identically zero. The Peter–Weyl theorem is an important result in the study of functions on a compact group, quite apart from its role in the proof of the theorem of the highest weight.

Lemma 12.20. *If* (Π, V) *is an irreducible representation of* K, *then for each operator* A *on* V, *we have*

$$\int_K \Pi(y)A\Pi(y)^{-1}\,dy = cI \tag{12.11}$$

for some constant c.

Proof. Let B denote the operator on the left-hand side of (12.11). By the left-invariance of the integral, we have

$$\Pi(x)B\Pi(x)^{-1} = \int_K \Pi(xy)A\Pi(xy)^{-1}\,dy$$

$$= \int_K \Pi(y)A\Pi(y)^{-1}\,dy$$

$$= B.$$

Thus, B commutes with each $\Pi(x)$, which, by Schur's lemma, implies that B is a multiple of the identity. $\square$

We are now ready for the proof of our completeness result for characters.

Proof of Theorem 12.18. Let $\mathcal{A}$ denote the space of continuous functions on K that can be expressed as a linear combination of matrix entries, for some finite collection of representations of K. We claim that $\mathcal{A}$ satisfies the hypotheses of the complex version of the Stone–Weierstrass theorem, namely, that is an algebra, that it vanishes nowhere, that it is closed under complex conjugation, and that it separates points. (See Theorem 7.33 in [Rud1].) First, using Lemma 12.17, we see that the product of two matrix entries is a matrix entry for the tensor product representation, which decomposes as a direct sum of irreducibles. Thus, the product of two matrix entries is expressible as a linear combination of matrix entries, showing that $\mathcal{A}$ is an algebra. Second, the matrix entries of the trivial representation are nonzero constants, showing that $\mathcal{A}$ is nowhere vanishing. Third, by a simple extension of (12.7), we can show that the complex conjugate of a matrix entry is a matrix entry for the dual representation. Last, since K is a matrix Lie group, it has, by definition, a faithful finite-dimensional representation. The matrix entries of any such representation separate points in K.

Thus, the complex version of the Stone–Weierstrass theorem applies to $\mathcal{A}$, meaning that if f is continuous, we can find $g \in \mathcal{A}$ such that g is everywhere within ε of f. If f is a class function, then for all $x, y \in K$, we have

$$\left| f(x) - g(yxy^{-1}) \right| = \left| f(yxy^{-1}) - g(yxy^{-1}) \right| < \varepsilon.$$

Thus, if h is given by

$$h(x) = \int_K g(y^{-1}xy) \, dy,$$

then h will also be everywhere within ε of f.

Now, by assumption, g can be represented as

$$g(x) = \sum_j \operatorname{trace}(\Pi_j(x) A_j)$$

for some family of representations (Π_j, V_j) of K and some operators $A_j \in \operatorname{End}(V_j)$. We can then easily compute that

$$h(x) = \sum_j \operatorname{trace}(\Pi_j(x) B_j),$$

where

$$B_j = \int_K \Pi(y) A_j \Pi(y)^{-1} \, dy.$$

But by Lemma 12.20, each B_j is a multiple of the identity, which means that h is a linear combination of characters.

We conclude, then, that every continuous class function f can be uniformly approximated by a sequence of functions h_n, where each h_n is a linear combinations of characters. If f is orthogonal to every character, f is orthogonal to each h_n. Then, by letting n tend to infinity, we find that f is orthogonal to itself, meaning that $\int_K |f(x)|^2 \, dx = 0$. Since f is continuous, this can only happen if f is identically zero.   □

We used the assumption that K is a matrix Lie group to show that the algebra $\mathcal{A}$ separates points in K, which allowed us to prove (using the Stone–Weierstrass theorem) that $\mathcal{A}$ is dense in the space of continuous functions on K. In Appendix D, we sketch a proof of Theorem 12.18 that does not assume ahead of time that K is a matrix group.

12.4   The Analytic Proof of the Weyl Character Formula

In order to simplify certain parts of the analysis, we make the following temporary assumption concerning δ (half the sum of the real, positive roots).

Assumption 12.21 *In this section and in Sect. 12.5, we assume that the element δ is analytically integral.*

As we will show in Chapter 13 (Corollary 13.20), δ is analytically integral whenever K is simply connected. In Sect. 12.6, we describe the modifications needed to the arguments when δ is not analytically integral. We will give, in this section, an an analytic proof of the Weyl character formula, as an alternative to the algebraic argument in Sect. 10.8. The proof is based on a more-explicit version of the Weyl integral formula, obtained by computing the weight function $\rho(t)$ occurring in Theorem 11.30. We will see that the function ρ is the square of another function, which turns out to be our old friend, the Weyl denominator. This computation will provide a crucial link between the Weyl integral formula and the Weyl character formula.

If δ denotes half the sum of the positive *real* roots, then the Weyl denominator function (Definition 10.15) takes the form

$$q(H) = \sum_{w \in W} \det(w) e^{i \langle w \cdot \delta, H \rangle}, \quad H \in \mathfrak{t}. \tag{12.12}$$

By Assumption 12.21 and Proposition 12.9, each exponential $e^{i \langle w \cdot \delta, H \rangle}$, $w \in W$, defines a function on T. Thus, there is a unique function $Q : T \to \mathbb{C}$ satisfying

$$Q(e^H) = q(H), \quad H \in \mathfrak{t}. \tag{12.13}$$

We now state Weyl's formula for the character (Definition 12.14) χ_Π of Π.

Theorem 12.22 (Weyl Character Formula). *Suppose (Π, V) is an irreducible representation of K and that μ is a maximal weight for Π. Then μ is dominant and analytically integral, and μ is actually the highest weight of Π. Furthermore, the character of Π is given by the formula*

$$\chi_\Pi(e^H) = \frac{\sum_{w \in W} \det(w) e^{i \langle w \cdot (\mu + \delta), H \rangle}}{q(H)}, \tag{12.14}$$

at all points $e^H \in T$ for which $q(H) \neq 0$. *In particular, every irreducible representation has a highest weight that is dominant and analytically integral.*

The character formula amounts to saying that $Q(t) \chi_\Pi(t)$ is an alternating sum of exponentials. The right-hand side of (12.14) is the same expression as in Theorem 10.14, adjusted for our current convention of using real roots and real weights.

Example 12.23. Let $K = \mathsf{SU}(2)$, let T be the diagonal subgroup, and let Π_m be the irreducible representation for which the largest eigenvalue of $\pi_m(H)$ is m. Then the character formula for Π_m takes the form

$$\chi_{\Pi_m}\left(\begin{pmatrix} e^{i\theta} & 0 \\ 0 & e^{-i\theta} \end{pmatrix}\right) = \frac{\sin((m+1)\theta)}{\sin\theta}.$$

This formula may also be obtained from Example 10.13 with $a = i\theta$.

Proof. If $H = \mathrm{diag}(1, -1)$, then we may choose the unique positive root α to satisfy $\langle \alpha, H \rangle = 2$. The highest weight μ of the representation then satisfies $\langle \mu, H \rangle = m$. Note that $\mathrm{diag}(e^{i\theta}, e^{-i\theta}) = e^{i\theta H}$, that δ satisfies $\langle \delta, H \rangle = \langle \alpha/2, H \rangle = 1$, and that $W = \{1, -1\}$. Thus, the character formula reads

$$\chi_{\Pi_m}(\mathrm{diag}(e^{i\theta}, e^{-i\theta})) = \frac{e^{i(m+1)\theta} - e^{-i(m+1)\theta}}{e^{i\theta} - e^{-i\theta}},$$

which simplifies to the claimed expression.   □

The hard part of the proof of the character formula is showing that in the product $q(H) \chi_\Pi(e^H)$ of the Weyl denominator and a character, no *other* exponentials occur besides the ones of the form $e^{i \langle w \cdot (\mu + \delta), H \rangle}$, $w \in W$. (Compare Section 10.4.) Now, Theorem 12.15 tells us that the norm of χ_Π is 1:

$$\int_K |\chi_\Pi(x)|^2 \, dx = 1. \tag{12.15}$$

In the analytic approach, the unwanted exponentials will be ruled out by applying the Weyl integral formula to show that if any other exponentials did occur, the integral on the left-hand side of (12.15) would be greater than 1. To make this argument work, we must work out the Weyl integral formula in a more explicit form.

Proposition 12.24. *If δ is analytically integral, the Weyl integral formula takes the form*

$$\int_K f(x) \, dx = \frac{1}{|W|} \int_T |Q(t)|^2 \int_{K/T} f(yty^{-1}) \, d[y] \, dt, \tag{12.16}$$

where dx, d[y], and dt are the normalized volume forms on K, K/T, and T, respectively, and where Q is as in (12.13).

In the proof of this proposition, we also verify the correctness of the normalization constant in the Weyl integral formula (Theorem 11.30), a point that was not addressed in Sect. 11.6. (We will address the normalization issue when δ is not integral in Sect. 12.6.) As we have already indicated above, knowing the correct normalization constant is an essential part of our analytic proof of the Weyl integral formula.

Proof. We may extend $\mathrm{Ad}_{t^{-1}}$ to a complex-linear operator on $\mathfrak{g} = \mathfrak{k}_{\mathbb{C}}$, where $\mathfrak{g}$ is the direct sum of $\mathfrak{h} := \mathfrak{t}_{\mathbb{C}}$ and $\mathfrak{f}_{\mathbb{C}}$, where $\mathfrak{f}_{\mathbb{C}}$ is the orthogonal complement of $\mathfrak{h}$ in $\mathfrak{g}$. Meanwhile, $\mathfrak{g}$ also decomposes as the direct sum of $\mathfrak{h}$ and the root spaces $\mathfrak{g}_\alpha$. We now claim that each $\mathfrak{g}_\alpha$ is orthogonal to $\mathfrak{h}$. To see this, choose $H \in \mathfrak{t}$ for which $\langle \alpha, H \rangle \neq 0$. Then each $H' \in \mathfrak{h}$ is an eigenvector for ad_H with eigenvalue 0, whereas $X \in \mathfrak{g}_\alpha$ is an eigenvector for ad_H with a nonzero eigenvalue. Since ad_H is skew self-adjoint, H' and X must be orthogonal. Thus, $\mathfrak{f}_{\mathbb{C}}$ is actually the direct sum of the $\mathfrak{g}_\alpha$'s.

Now, if α is a real root and X belongs to $\mathfrak{g}_\alpha$, we have

$$\mathrm{Ad}_{e^{-H}}(X) = e^{-\mathrm{ad}_H}(X) = e^{-i\langle \alpha, H \rangle} X$$

for $X \in \mathfrak{g}_\alpha$. Thus, letting $\mathrm{Ad}'_{e^{-H}}$ denote the restriction of $\mathrm{Ad}_{e^{-H}}$ to $\mathfrak{f}_{\mathbb{C}}$, we have

$$\det(\mathrm{Ad}'_{e^{-H}} - I) = \prod_{\alpha \in R}(e^{-i\langle \alpha, H \rangle} - 1)$$

$$= \prod_{\alpha \in R^+}(e^{-i\langle \alpha, H \rangle} - 1)(e^{i\langle \alpha, H \rangle} - 1).$$

Since $(e^{-i\theta} - 1)(e^{i\theta} - 1) = \left| e^{i\theta/2} - e^{-i\theta/2} \right|^2$, we have

$$\det(\mathrm{Ad}_{e^{-H}} - I) = \left| \prod_{\alpha \in R^+}(e^{i\langle \alpha, H \rangle/2} - e^{-i\langle \alpha, H \rangle/2}) \right|^2$$

$$= |q(H)|^2.$$

Thus, the Weyl integral formula (Theorem 11.30) takes the claimed form. Here we have used Lemma 10.28, adapted to our current convention of using real roots.

It remains only to verify the correct normalization constant in the Weyl integral formula. To see that (12.16) is properly normalized, it suffices to consider $f \equiv 1$, in which case (12.16) says that

$$1 = \frac{1}{|W|} \int_T |Q(t)|^2 \, dt, \tag{12.17}$$

where dt is the normalized volume form on T. Now, by Proposition 8.38, δ belongs to the open fundamental Weyl chamber, which means (Proposition 8.27) that the exponentials $e^{i\langle w\cdot\delta, H\rangle}$, $w \in W$, are all distinct. Proposition 12.10 then tells us that these exponentials are orthonormal, so that the integral of $|Q(t)|^2$ over T equals $|W|$, verifying (12.17). □

We are now ready for our analytic proof of the Weyl character formula.

Proof of Theorem 12.22. Since Π is finite dimensional, it has only finitely many weights. Thus, there is a maximal weight μ (i.e., one such that no other weight of Π is higher than μ), with multiplicity mult$(\mu) \geq 1$. We then claim that in the product $q(H)\chi_\Pi(e^H)$, the exponential $e^{i\langle\mu+\delta, H\rangle}$ occurs only once, by taking the term $e^{i\langle\delta, H\rangle}$ from q and the term mult$(\mu)e^{i\langle\mu, H\rangle}$ from χ_Π. To see this, suppose we take $e^{i\langle w\cdot\delta, H\rangle}$ from q and mult$(\lambda)e^{i\langle\lambda, H\rangle}$ from χ_Π, and that

$$\lambda + w\cdot\delta = \mu + \delta,$$

so that

$$\lambda = \mu + (\delta - w\cdot\delta).$$

But by Proposition 8.42, $\delta \succeq w\cdot\delta$ and, thus, $\lambda \succeq \mu$. Since μ is maximal, we conclude that $\lambda = \mu$, in which case we must also have $w = I$.

We conclude, then, that $e^{i\langle\mu+\delta, H\rangle}$ occurs with multiplicity exactly mult(μ) in the product. Now, $\chi_\Pi(e^H)$ is a Weyl-invariant function, while $q(H)$ is Weyl-alternating. The product of these two functions is then Weyl alternating. Thus, by Corollary 10.17, each of the exponentials $e^{i\langle w\cdot(\mu+\delta), H\rangle}$ occurs with multiplicity det(w)mult(μ).

We now claim that mult$(\mu) = 1$ and that the *only* exponentials in the product $q\chi_\Pi$ are those of the form det$(w)e^{i\langle w\cdot(\mu+\delta), H\rangle}$. To see this, recall from Theorem 12.15 that

$$\int_K |\chi_\Pi(x)|^2 \, dx = 1.$$

Thus, by Proposition 12.24 (as applied to a class function), we have

$$\int_T |Q(t)\chi_\Pi(t)|^2 \, dt = |W|, \tag{12.18}$$

where $|W|$ is the order of the Weyl group.

Now, the product $Q\chi_\Pi$ is a sum of exponentials and those exponentials are orthonormal on T (Proposition 12.10). Since at least $|W|$ exponentials *do* occur in the product, namely those of the form det$(w)e^{i\langle w\cdot(\mu+\delta), H\rangle}$, these exponentials make a contribution of mult$(\mu)|W|$ to the integral in (12.18). By orthonormality, any remaining exponentials would make a positive contribution to the integral. Thus, the only way (12.18) can hold is if mult$(\mu) = 1$ and there are no exponentials in $Q\chi_\Pi$ other than those of the form det$(w)e^{i\langle w\cdot(\mu+\delta), H\rangle}$. Thus, after dividing by $Q(t)$, we obtain the Weyl character formula.

Since μ is a weight of Π, it must be analytically integral. It remains to show that μ is dominant and that μ is the highest weight of Π. If μ were not dominant, there would be some $w \in W$ for which $w \cdot \mu$ is dominant, in which case (by Proposition 8.42), $\mu = w^{-1} \cdot (w \cdot \mu)$ would be lower than $w \cdot \mu$, contradicting the maximality of μ. Meanwhile, if μ were not the highest weight of Π, we could choose a maximal element μ' among the weights of Π that are not lower than μ, and this μ' would be another maximal weight for Π. As we have just shown, μ' would also have to be dominant, in which case (Proposition 8.29), the Weyl-orbit of μ' would be disjoint from the Weyl-orbit of μ. But then by the reasoning above, both the exponentials $e^{i\langle w \cdot (\mu'+\delta), H\rangle}$ and the exponentials $e^{i\langle w \cdot (\mu+\delta), H\rangle}$ would occur with nonzero multiplicity in the product $Q\chi_\Pi$, which would force the integral in (12.18) to be larger than $|W|$. □

In Chapter 10, we used the Casimir element and the character formula for Verma modules to show that any exponential $e^{i\langle \lambda, H\rangle}$ that appears in the product $q(H)\chi_\pi(H)$ must satisfy $|\lambda| = |\mu + \delta|$. This was the key step in proving that only exponentials of the form $\lambda = w \cdot (\mu + \delta)$ appear in $q\chi^\mu$. In this chapter, we have instead used the Weyl integral formula to show that if any unwanted exponentials appeared, the integral of $|\chi_\Pi|^2$ over K would be greater than 1, which would contradict Theorem 12.15.

Proposition 12.25. *Two irreducible representations of K with the same highest weight are isomorphic.*

Proof. If Π and Σ both have highest weight μ, then the Weyl character formula shows that the characters χ_Π and χ_Σ must be equal. But if Π and Σ were not isomorphic, Theorem 12.15 would imply that χ_Π and $\chi_\Sigma = \chi_\Pi$ are orthogonal. Thus, χ_Π would have to be identically zero, which would contradict the normalization result in Theorem 12.15. □

12.5 Constructing the Representations

In our proof of Theorem 12.6, we showed as part of the Weyl character formula that every irreducible representation of K has a highest weight, and that this highest weight is dominant and analytically integral. We also showed in Proposition 12.25 that two irreducible representations of K with the same highest weight are isomorphic. Thus, in proving Theorem 12.6, it remains only to prove that every dominant, analytically integral element arises as the highest weight of a representation. We continue to assume that δ (half the sum of the real, positive roots) is an analytically integral element. This assumption is lifted in Sect. 12.6.

Suppose now that μ is a dominant, analytically integral element. We do not yet know that μ is the highest weight of an irreducible representation of K. Nevertheless, as we now demonstrate, there is a well-defined function ϕ_μ on K whose restriction to T is given by the right-hand side of the Weyl character formula.

Lemma 12.26. *For each dominant, analytically integral element μ, there is a unique continuous function $\phi_\mu : T \to \mathbb{C}$ satisfying*

$$\phi_\mu(e^H) = \frac{\sum_{w \in W} \det(w) e^{i \langle w \cdot (\mu + \delta), H \rangle}}{q(H)}, \qquad H \in \mathfrak{t}, \tag{12.19}$$

whenever $q(H) \neq 0$.

In preparation for the proof of this lemma, recall from Lemma 10.28 (adjusted for using real roots) that the Weyl denominator can be written as

$$q(H) := \sum_{w \in W} \det(w) e^{i \langle w \cdot \delta, H \rangle} = (2i)^k \prod_{\alpha \in R^+} \sin(\langle \alpha, H \rangle / 2)$$

where k is the number of positive roots. For each $\alpha \in R$ and each integer n, define a hyperplane (not necessarily through the origin) by

$$V_{\alpha,n} = \{ H \in \mathfrak{t} \,|\, \langle \alpha, H \rangle = 2\pi n \}. \tag{12.20}$$

Then the zero-set of the Weyl denominator q is the union of all the $V_{\alpha,n}$'s, as α ranges over the set of positive real roots and n ranges over the integers. Furthermore, q has a "simple zero" on each of these hyperplanes. Symmetry properties of the numerator on the right-hand side of (12.19) will then force this function to be zero each $V_{\alpha,n}$, so that the zeros in the numerator cancel the zeros in the denominator.

Proof. By Assumption 12.21, the element δ is analytically integral, which means that each of $w \cdot (\mu + \delta)$ and $w \cdot \delta, w \in W$, is also analytically integral. Thus, by Proposition 12.9, all of the exponentials involved are well-defined functions on T. It follows that the right-hand side of (12.19) is a well-defined function T, outside of the set where the denominator is zero. We now argue that the right-hand side of (12.19) extends uniquely to a continuous function on $\mathfrak{t}$.

Let $\psi_\mu(H), H \in \mathfrak{t}$, denote the numerator on the right-hand side of (12.19):

$$\psi_\mu(H) = \sum_{w \in W} \det(w) e^{i \langle w \cdot (\mu + \delta), H \rangle}.$$

We claim that ψ_μ vanishes on each of the hyperplanes $V_{\alpha,n}$ in (12.20). To verify this claim, note that each weight $\lambda = w \cdot (\mu + \delta)$ occurring in ψ_μ is analytically integral, and thus also algebraically integral, by Proposition 12.7. Thus, $\langle \lambda, H_\alpha \rangle$ is an integer for each $\alpha \in R$. It follows that each exponential in ψ_μ—and, thus, ψ_μ itself—is invariant under translation by $2\pi H_\alpha$:

$$\psi_\mu(H + 2\pi H_\alpha) = \psi_\mu(H)$$

for all $H \in \mathfrak{t}$ and all $\alpha \in R$.

Meanwhile, the coefficients of $\det(w)$ in the formula for ψ_μ guarantee that ψ_μ is alternating with respect to the action of W. Finally, if $H \in V_{\alpha,n}$, we have

$$s_\alpha \cdot H = H - 2\frac{\langle \alpha, H \rangle}{\langle \alpha, \alpha \rangle}\alpha = H - 2\pi n H_\alpha.$$

Putting these observations together, we have, for $H \in V_{\alpha,n}$,

$$\psi_\mu(H) = -\psi_\mu(s_\alpha \cdot H) = -\psi_\mu(H - 2\pi n H_\alpha) = -\psi_\mu(H),$$

which can happen only if $\psi_\mu(H) = 0$. (See Figure 12.4, where the gray lines are the hyperplanes $V_{\alpha,n}$ with $|n| \le 2$. Compare also Exercise 6.)

Now, for each α and n, let $L_{\alpha,n} : \mathfrak{t} \to \mathbb{R}$ be given by

$$L_{\alpha,n}(H) = \langle \alpha, H \rangle - 2\pi n,$$

so that the zero-set of $L_{\alpha,n}$ is precisely $V_{\alpha,n}$. Then the function

$$\frac{L_{\alpha,n}(H)}{\sin(\langle \alpha, H \rangle /2)} \qquad (12.21)$$

extends to a continuous function on a neighborhood of $V_{\alpha,n}$ in $\mathfrak{t}$. On the other hand, we have shown that ψ_μ vanishes on $V_{\alpha,n}$ for each α and n. Furthermore, since ψ_μ is given by a globally convergent power series, it is not hard to prove that ψ_μ can be expressed as

$$\psi_\mu(H) = f(H)L_{\alpha,n}(H)$$

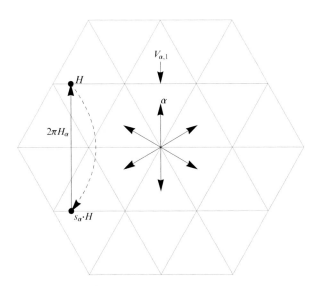

Fig. 12.4 The function ψ_μ changes sign under the reflection s_α and is unchanged under translation by $2\pi H_\alpha$, forcing ψ_μ to be zero on each $V_{\alpha,n}$

for some smooth function f, where f is also given by a globally convergent power series and where f still vanishes on each hyperplane $V_{\beta,m}$ with $(\beta, m) \neq (\alpha, n)$. (See Exercise 7.)

Now let H be an arbitrary point in t. Since the hyperplanes $V_{\alpha,n}$ are disjoint for α fixed and n varying, H is contained in only finitely many of these hyperplanes, say $V_{\alpha_1,n_1}, \ldots, V_{\alpha_m,n_m}$. Since ψ_μ vanishes on each of these hyperplanes, we can show inductively that ψ_μ can be expressed as

$$\psi_\mu = L_{\alpha_1,n_1} \cdots L_{\alpha_m,n_m} g \tag{12.22}$$

for some smooth function g. Since the function in (12.21) is nonsingular in a neighborhood of $V_{\alpha,n}$, we conclude that the factors of L_{α_j,n_j} in (12.22) cancel all the zeros in $q(H)$, showing that $\phi_\mu(e^H) = \psi_\mu(H)/q(H)$ extends to a continuous function in a neighborhood of H.

Finally, in local exponential coordinates on T, each point H is contained in at most finitely many of the hyperplanes on which q vanishes. Thus, H is a limit of points on which $q \neq 0$, showing the uniqueness of the continuous extension.   □

Lemma 12.27. *Let ϕ_μ be as in Lemma 12.26 and let $\Phi_\mu : K \to \mathbb{C}$ be the unique continuous class function on K such that $\Phi_\mu\big|_T = \phi_\mu$ (Corollary 11.40). Then as μ ranges over the set of dominant, analytically integral elements, the functions Φ_μ form an orthonormal set:*

$$\int_K \overline{\Phi_\mu(x)} \Phi_{\mu'}(x) \, dx = \delta_{\mu,\mu'}.$$

Note that we do not know, at the moment, that each Φ_μ is actually the character of a representation of K. Thus, we cannot appeal to Theorem 12.15.

Proof. Since the denominator in the definition of ϕ_μ is the Weyl denominator, Corollary 11.32 to the Weyl integral formula tells us that

$$\int_K \overline{\Phi_\mu(x)} \Phi_{\mu'}(x) \, dx$$

$$= \frac{1}{|W|} \int_T |Q(t)|^2 \, \overline{\phi_\mu(t)} \phi_{\mu'}(t) \, dt$$

$$= \frac{1}{|W|} \sum_{w \in W} \sum_{w' \in W} \int_T \overline{\left(\det(w) e^{i \langle w \cdot (\mu+\delta), H \rangle} \right)} \left(\det(w') e^{i \langle w' \cdot (\mu'+\delta), H \rangle} \right) dH.$$

$$\tag{12.23}$$

Now, since $\mu + \delta$ and $\mu' + \delta$ are strictly dominant, W acts freely on these elements. If $\mu \neq \mu'$, the W-orbit of $\mu + \delta$ will be disjoint from the W-orbit of $\mu' + \delta$. Thus, the exponentials occurring in ϕ_μ will be disjoint from those in $\phi_{\mu'}$, which means, by Proposition 12.10, that (12.23) is zero. If, on the other hand, $\mu = \mu'$, we have the norm-squared of $|W|$ distinct, orthonormal exponentials, so that the right-hand side of (12.23) reduces to 1.   □

We are now ready to complete the proof of Theorem 12.6, in the case in which the element δ is analytically integral. It remains only to prove that for each dominant, analytically integral element μ, there is an irreducible representation of K with highest weight μ.

Proof of Theorem 12.6. Here is what we know so far about the characters of irreducible representations. First, by Theorem 12.22, the character of each irreducible representation Π must be equal to Φ_μ, where μ is the highest weight of Π. Second, by Lemma 12.26, *all* the Φ_μ's, with μ dominant and analytically integral—whether or not they are characters of a representation—form an orthonormal set. Last, by Theorem 12.18, the characters of the irreducible representations form a *complete* orthonormal set.

Let μ_0 be a fixed dominant, analytically integral element and suppose, toward a contradiction, that there did not exist an irreducible representation with highest weight μ_0. Then the function Φ_{μ_0} would be a continuous class function on K and Φ_{μ_0} would be orthogonal to the character of *every* irreducible representation. (After all, every irreducible character is of the form Φ_μ, where μ is the highest weight of the representation, and we are assuming that μ_0 is not the highest weight of any representation.) But Theorem 12.18 says that a continuous class function that is orthogonal to every irreducible character must be identically zero. Thus, Φ_{μ_0} would have to be the zero function, which is impossible, since $\left|\Phi_{\mu_0}\right|^2$ integrates to 1 over K.   □

We may put the argument in a different way as follows. Theorem 12.18 says that the characters of irreducible representations form a *complete* orthonormal set inside the space of continuous class functions on K. The Weyl character formula, meanwhile, says that the characters form a subset of the set of Φ_μ's. Finally, the Weyl integral formula says that the collection of all Φ_μ's, with μ dominant and analytically integral, are orthonormal. If there were some such μ for which Φ_μ was not a character, then the set of characters would be a proper subset of an orthonormal set, in which case, the characters could not be complete.

Now, the preceding proof of the theorem of the highest weight is not very constructive, in contrast to the Lie-algebraic proof, in which we gave a direct construction of each finite-dimensional representation as a quotient of a Verma module. If one looks carefully at the proof of Theorem 12.18, however, one sees a hint of a more direct description of the representations from the compact group perspective. Specifically, let $C(K)$ denote the space of continuous functions on K, and define a left and a right action of K on $C(X)$ as follows. For each $x \in K$, define L_x and R_x as operators on $C(X)$ as by

$$(L_x f)(y) = f(x^{-1}y)$$
$$(R_x f)(y) = f(yx).$$

One may easily check that $L_{xy} = L_x L_y$ and $R_{xy} = R_x R_y$. Thus, both $L.$ and $R.$ define representations of K acting on the infinite-dimensional space $C(K)$.

We now show that each irreducible representation of K occurs as a finite-dimensional subspace of $C(K)$ that is invariant under the right action of K. For each representation irreducible (Π, V) of K, we can fix some nonzero vector $v_0 \in V$, which might, for example, be a highest weight vector. Then for each $v \in V$, we can consider the function $f_v : K \to \mathbb{C}$ given by

$$f_v(x) = \langle v_0, \Pi(x)v \rangle. \qquad (12.24)$$

(The function f_v is a special sort of *matrix entry* for Π, in the sense of the proof of Theorem 12.18.) It is not hard to check that the map $v \mapsto f_v$ is injective; see Exercise 8.

Let W denote the space of all functions of the form f_v, with $v \in V$. We can compute that

$$(R_x f_v)(y) = \langle v_0, \Pi(yx)v \rangle = \langle v_0, \Pi(y)(\Pi(x)v) \rangle,$$

which means that

$$R_x f_v = f_{\Pi(x)v}.$$

Thus, W is a finite-dimensional invariant subspace for the right action of K on $C(K)$. Indeed, the map $v \mapsto f_v$ is a bijective intertwining map between V and W.

Conclusion 12.28. *For each irreducible representation (Π, V) of K, fix a nonzero vector $v_0 \in V$ and let W denote the subspace of $C(X)$ consisting of functions of the form (12.24), with $v \in V$. Then W is invariant under the right action of K and is isomorphic, as a representation of K, to V.*

One can pursue this line of analysis further by choosing v_0 to be a highest weight vector and then attempting to describe the space W—without referring to the representation Π—by means of its behavior under certain differential operators on K. See Sect. 4.12 of [DK] for more information.

12.6   The Case in Which δ is Not Analytically Integral

For a general connected compact group K, the element δ may not be analytically integral. (See Sect. 12.2.) If δ is not analytically integral, many of the functions we have been working with will not be well-defined functions on T. Specifically, exponentials of the form $e^{i \langle w \cdot \delta, H \rangle}$ and $e^{i \langle w \cdot (\mu + \delta), H \rangle}$, with $w \in W$ and μ a dominant, analytically integral element, no longer define functions on T. Fortunately, all of our calculations involve *products* of such exponentials, and these products turn out to be well-defined functions on T. Thus, all the arguments in the previous three sections will go through with minor modifications.

In all cases, the quantity $2\delta = \sum_{\alpha \in R^+} \alpha$ is analytically integral, by Point 3 of Proposition 12.7. Thus, for each $H \in \mathfrak{t}$ for which $e^{2\pi H} = I$, the quantity $\langle \delta, H \rangle$ must either be an integer or a half integer. If δ is not analytically integral, there must exist some $H \in \mathfrak{t}$ with $e^{2\pi H} = I$ for which $\langle \delta, H \rangle$ is a half integer (but not an integer). With this observation in mind, we make the following definition.

Definition 12.29. If δ is not analytically integral, we say that $\lambda \in \mathfrak{t}$ is **half integral** if $\lambda - \delta$ is analytically integral.

That is to say, the half integral elements are those of the form $\lambda = \delta + \lambda'$, with λ' being analytically integral.

Proposition 12.30. *If λ and η are half integral, then $\lambda + \eta$ is analytically integral. If λ is half integral, then $-\lambda$ is also half integral and $w \cdot \lambda$ is half integral for all $w \in W$.*

Proof. If $\lambda = \delta + \lambda'$ and $\eta = \delta + \eta'$ are half integral, then $\lambda + \eta = 2\delta + \lambda' + \eta'$ is analytically integral. If $\lambda = \delta + \lambda'$ is half integral, so is

$$-\lambda = -\delta - \lambda' = \delta - 2\delta - \lambda'.$$

For each $w \in W$, the set $w \cdot R^+$ will consist of a certain subset S of the positive roots, together with the negatives of the roots in $R^+ \setminus S$. Thus, $w \cdot \delta$ will consist of half the sum of the elements of S minus half the sum of the elements of $R^+ \setminus S$. It follows that

$$\delta - w \cdot \delta = \sum_{\alpha \in R^+ \setminus S} \alpha,$$

showing that $w \cdot \delta$ is again half integral. (Recall that each root is analytically integral, by Proposition 12.7.) More generally, if $\lambda = \delta + \lambda'$ is half integral, so is $w \cdot \lambda = w \cdot \delta + w \cdot \lambda'$. $\square$

Note that exponentials of the form $e^{i\langle \lambda, H \rangle}$, with λ being half integral, do not descend to functions on T. Our next result says that, nevertheless, the *product* of two such exponentials (possibly conjugated) does descend to T. Furthermore, such exponentials are still "orthonormal on T," as in Proposition 12.10 in the integral case.

Proposition 12.31. *If λ and η are half integral, there is a well-defined function f on T such that*

$$f(e^H) = \overline{e^{i\langle \lambda, H \rangle}} e^{i\langle \eta, H \rangle}$$

and

$$\int_T \overline{e^{i\langle\lambda,H\rangle}} e^{i\langle\eta,H\rangle}\, dH = \delta_{\lambda,\eta}.$$

Proof. If λ and η are half integral, then $-\lambda$ is half integral, so that $\eta - \lambda$ is analytically integral. Thus, by Proposition 12.9, there is a well-defined function $f : T \to \mathbb{C}$ satisfying

$$f(e^H) = \overline{e^{i\langle\lambda,H\rangle}} e^{i\langle\eta,H\rangle} = e^{i\langle\eta-\lambda,H\rangle}.$$

Furthermore, if $\lambda = \delta + \lambda'$ and $\eta = \delta + \eta'$ are half integral, we have

$$\int_T \overline{e^{i\langle\lambda,H\rangle}} e^{i\langle\eta,H\rangle}\, dH = \int_T e^{-i\langle\delta+\lambda',H\rangle} e^{i\langle\delta+\eta',H\rangle}\, dH$$

$$= \int_T e^{-i\langle\lambda',H\rangle} e^{i\langle\eta',H\rangle}\, dH$$

$$= \delta_{\lambda',\delta'},$$

by Proposition 12.10. Since $\lambda = \eta$ if and only if $\lambda' = \delta'$, we have the desired "orthonormality" result. $\qquad\qquad\square$

We now discuss how the results of Sects. 11.6, 12.4, and 12.5 should be modified when δ is not analytically integral. In the case of the Weyl integral formula (Theorem 11.30), the function $q(H)$, $H \in \mathfrak{t}$, does not descend to T when δ is not integral. That is to say, there is no function $Q(t)$ on T such that $Q(e^H) = q(H)$. Nevertheless, the function $|q(H)|^2$ *does* descend to T, since $|q(H)|^2$ is a sum of products of half integral exponentials. The Weyl integral formula, with the same proof, then holds even if δ is not analytically integral, provided that the expression $|Q(t)|^2$ is interpreted as the function $e^H \mapsto |q(H)|^2$. We may then consider the case $f \equiv 1$ and use Proposition 12.31 to verify the correctness of the normalization in the Weyl integral formula.

In the case of the Weyl character formula, we claim that the right-hand side of (12.14) descends to a function on T. To see this, note that we can pull a factor of $e^{i\langle\delta,H\rangle}$ out of each exponential in the numerator and each exponential in the denominator. After canceling these factors, we are left with exponentials in both the numerator and denominator that descend to T. Meanwhile, in the proof of the character formula, although the function $Q(t)\chi_\Pi(t)$ is not well defined on T, the function $|Q(t)\chi_\Pi(t)|^2$ is well defined. The Weyl integral formula (interpreted as in the previous paragraph) tells us that the integral of $|Q(t)\chi_\Pi(t)|^2$ over T is equal to $|W|$, as in the case where δ is analytically integral. If we then apply the orthonormality result in Proposition 12.31, we see that, just as in the integral case, the only exponentials present in the product $q(H)\chi_\Pi(e^H)$ are those in the numerator of the character formula.

Finally, we consider the proof that every dominant, analytically integral element is the highest weight of a representation. If δ is not analytically integral, then neither the numerator nor the denominator on the right-hand side of (12.19) descends to function on T. Nevertheless, the ratio of these functions does descend to T, by the argument in the preceding paragraph. The argument that ϕ_μ extends to a *continuous* function on T then goes through without change. (This argument requires only that each weight $\lambda = w \cdot (\mu + \delta)$ in ψ_μ be *algebraically* integral, which holds even if δ is not analytically integral, by Proposition 8.38.) Thus, we may apply the half-integral version of the Weyl integral formula to show that the functions Φ_μ on K are orthonormal, as μ ranges over the set of dominant, analytically integral elements. The rest of the argument then proceeds without change.

12.7 Exercises

1. Let $K = \mathsf{SU}(2)$ and let $\mathfrak{t}$ be the diagonal subalgebra of $\mathsf{su}(2)$. Prove directly that every algebraically integral element is analytically integral.
   Note: Since $\mathsf{SU}(2)$ is simply connected, this claim also follows from the general result in Corollary 13.20.

2. This exercise asks you to use the theory of Fourier series to give a direct proof of the completeness result for characters (Theorem 12.18), in the case $K = \mathsf{SU}(2)$. To this end, suppose f is a continuous class function on $\mathsf{SU}(2)$ that f is orthogonal to the character of every representation.

   (a) Using the explicit form of the Weyl integral formula for $\mathsf{SU}(2)$ (Example 11.33) and the explicit form of the characters for $\mathsf{SU}(2)$ (Example 12.23), show that

   $$\int_{-\pi}^{\pi} \overline{f(\mathrm{diag}(e^{i\theta}, e^{-i\theta}))}(\sin\theta)\sin((m+1)\theta)\,d\theta = 0$$

   for every non-negative integer m.

   (b) Show that the function $\theta \mapsto f(\mathrm{diag}(e^{i\theta}, e^{-i\theta}))(\sin\theta)$ is an odd function of θ.

   (c) Using standard results from the theory of Fourier series, conclude that f must be identically zero.

3. Suppose (Π, V) and (Σ, W) are representations of a group G, and let $\mathrm{Hom}(V, W)$ denote the space of all linear maps from V to W. Let G act on $\mathrm{Hom}(V, W)$ by

   $$g \cdot A = \Sigma(g)A\Pi(g)^{-1}, \qquad\qquad (12.25)$$

   for all $g \in G$ and $A \in \mathrm{Hom}(W, V)$. Show that A is an intertwining map of V to W if and only if $g \cdot A = A$ for all $g \in G$.

4. If V and W are finite-dimensional vector spaces, let $\Phi : V^* \otimes W \to \mathrm{Hom}(V, W)$ be the unique linear map such that for all $\xi \in V^*$ and $w \in W$, we have

$$\Phi(\xi \otimes w)(v) = \xi(v)w, \qquad v \in V.$$

(a) Show that Φ is an isomorphism.
(b) Let (Π, V) and (Σ, W) be representations of a group G, let G act on V^* as in Sect. 4.3.3, and let G act on $\mathrm{Hom}(V, W)$ as in (12.25). Show that the map $\Phi : V^* \otimes W \to \mathrm{Hom}(V, W)$ in Part (a) is an intertwining map.

5. Suppose $f(x) := \mathrm{trace}(\Pi(x)A)$ and $g(x) := \mathrm{trace}(\Sigma(x)B)$ are matrix entries for nonisomorphic, irreducible representations Π and Σ of K (Definition 12.19). Show that f and g are orthogonal:

$$\int_K \overline{f(x)}g(x)\, dx = 0.$$

Hint: Imitate the proof of Theorem 12.15.

6. Let $\{V_{\alpha,n}\}$ denote the collection of hyperplanes in (12.20). If $H_\alpha = 2\alpha/\langle \alpha, \alpha \rangle$ is the real coroot associated to a real root α, show that

$$V_{\alpha,n} + \pi m H_\alpha = V_{\alpha,n+m}.$$

7. Let V be a hyperplane in $\mathbb{R}^n$, not necessarily through the origin, and let $L : \mathbb{R}^n \to \mathbb{R}$ be an affine function whose zero-set is precisely V. Suppose $g : \mathbb{R}^n \to \mathbb{C}$ is given by a globally convergent power series in n variables and that g vanishes on V.

(a) Show that g can be expressed as $g = Lh$ for some function h, where h is also given by a globally convergent power series.
Hint: Choose a coordinate system $y_1, \ldots, y_n$ on $\mathbb{R}^n$ with origin in L such that $L(y) = y_1$.
(b) Suppose g also vanishes on some hyperplane V' distinct from V. Show that the function h in Part (a) vanishes on V'.

8. Let (Π, V) be an irreducible representation of K and let v_0 be a nonzero element of V. For each $v \in V$, let f_v be the function given in (12.24). Show that if f_v is the zero function, then $v = 0$.

9. Let $K = \mathsf{U}(n)$ and let T be the diagonal subgroup of K. Show that the density $\rho(\cdot)$ in the Weyl integral formula (Theorem 11.30) can be computed explicitly as

$$\rho(\mathrm{diag}(e^{i\theta_1}, \ldots, e^{i\theta_n})) = \prod_{1 \le j \le k \le n} \left| e^{i\theta_j} - e^{i\theta_k} \right|^2.$$

Hint: Use Proposition 12.24, as interpreted in Sect. 12.6 in the case where δ is not necessarily analytically integral.

Chapter 13
Fundamental Groups of Compact Lie Groups

13.1 The Fundamental Group

In this section, we briefly review the notion of the fundamental group of a topological space. For a more detailed treatment, the reader should consult any standard book on algebraic topology, such as [Hat, Chapter 1]. Let X be any path-connected Hausdorff topological space and let x_0 be a fixed point in X (the "basepoint"). We consider **loops in X based at** x_0 (i.e., continuous maps $l : [0, 1] \to X$ with the property that $l(0) = l(1) = x_0$). The choice of the basepoint makes no substantive difference to the constructions that follow. From now on, "based loop" will mean "loop based at x_0." Ultimately, we are interested in the case that X is a matrix Lie group.

If l_1 and l_2 are two based loops, then we define the **concatenation** of l_1 and l_2 to be the loop $l_1 \cdot l_2$ given by

$$l_1 \cdot l_2(t) = \begin{cases} l_1(2t), & 0 \leq t \leq \frac{1}{2} \\ l_2(2t - 1), & \frac{1}{2} \leq t \leq 1; \end{cases}$$

that is, $l_1 \cdot l_2$ traverses l_1 as t goes from 0 to $1/2$ and then traverses l_2 as t goes from $1/2$ to 1.

Two based loops l_1 and l_2 are said to be **homotopic** if one can be "continuously deformed" into the other. More precisely, this means that there exists a continuous map $A : [0, 1] \times [0, 1] \to X$ such that $A(0, t) = l_1(t)$ and $A(1, t) = l_2(t)$ for all

A previous version of this book was inadvertently published without the middle initial of the author's name as "Brian Hall". For this reason an erratum has been published, correcting the mistake in the previous version and showing the correct name as Brian C. Hall (see DOI http://dx.doi.org/10.1007/978-3-319-13467-3_14). The version readers currently see is the corrected version. The Publisher would like to apologize for the earlier mistake.

© Springer International Publishing Switzerland 2015
B.C. Hall, *Lie Groups, Lie Algebras, and Representations*, Graduate
Texts in Mathematics 222, DOI 10.1007/978-3-319-13467-3_13

$t \in [0, 1]$ and such that $A(s, 0) = A(s, 1) = x_0$ for all $s \in [0, 1]$. One should think of $A(s, t)$ as a family of loops parameterized by s. In some cases, we may use the notation $l_s(t)$ in place of $A(s, t)$ to emphasize this point of view.

A based loop is said to be **null homotopic** if it is homotopic to the constant loop (i.e., the loop l^0 for which $l^0(t) = x_0$ for all $t \in [0, 1]$). If all loops in X based at x_0 are null homotopic, then X is said to be **simply connected**. Since we are assuming X is path connected, it is not hard to show that if all loops at one basepoint are null homotopic, the same it true for every other basepoint. Furthermore, if a loop based at x_0 can be shrunk to a point *without* fixing the basepoint (i.e., requiring only that $A(s, 0) = A(s, 1)$), then it can also be shrunk to a point with basepoint fixed (i.e., requiring $A(s, 0) = A(s, 1) = x_0$).

The notion of homotopy is an equivalence relation on loops based at x_0. The **homotopy class** of a loop l is then the set of all loops that are homotopic to l, and each loop belongs to one and only one homotopy class. The concatenation operation "respects homotopy," meaning that if l_1 is homotopic to l_2 and m_1 is homotopic to m_2, then $l_1 \cdot m_1$ is homotopic to $l_2 \cdot m_2$. As a result, it makes sense to define the concatenation operation on equivalence classes.

The operation of concatenation makes the set of homotopy classes of loops based at x_0 into a group, called the **fundamental group** of X and denoted $\pi_1(X)$. To verify associativity, we note that although $(l_1 \cdot l_2) \cdot l_3$ is *not* the same as $l_1 \cdot (l_2 \cdot l_3)$, the second of these two loops is a reparameterization of the first, from which it is not hard to see that the loops are homotopic. Meanwhile, the identity in $\pi_1(X)$ is the constant loop l^0. This is not an identity at the level of loops but is at the level of homotopy classes; that is, $l \cdot l^0$ and $l^0 \cdot l$ are not equal to l, but they are both homotopic to l, since both are reparameterizations of l. Finally, for inverses, the inverse to a homotopy class $[l]$ is the homotopy class $[l']$ where $l'(t) = l(1 - t)$. (It is not hard to see that both $l \cdot l'$ and $l' \cdot l$ are null homotopic.) A topological space X is simply connected precisely if its fundamental group is the trivial group.

Some standard examples of fundamental groups are as follows: $\mathbb{R}^n$ is simply connected for all n, S^n is simply connected for $n \geq 2$, and the fundamental group of S^1 is isomorphic to $\mathbb{Z}$.

Definition 13.1. If X and Y are Hausdorff topological space, a continuous map $\pi : Y \rightarrow X$ is a **covering map** if (1) π maps Y onto X and (2) for each $x \in X$, there is a neighborhood V of x such that $\pi^{-1}(V)$ is a disjoint union of open sets U_α, where the restriction of π to each U_α is a homeomorphism of U_α onto V. A **cover** of X is a pair (Y, π), where $\pi : Y \rightarrow X$ is a covering map. If (Y, π) is a cover of X and Y is simply connected, then (Y, π) is a **universal cover** of X.

If (Y, π) is a cover of X and $f : Z \rightarrow X$ is a continuous map, then a map $\tilde{f} : Z \rightarrow Y$ is a **lift** of f if $\tilde{f}$ is continuous and $\pi \circ \tilde{f} = f$.

It is known that every connected manifold (indeed, every reasonably nice connected topological space) has a universal cover, and that this universal cover is unique up to a "canonical" homeomorphism, that is, one that intertwines the covering maps. (See, for example, pp. 63–66 in [Hat].) Thus, we may speak about "the" universal cover of any connected manifold. A key property of a covering maps

$\pi : Y \rightarrow X$ is that lifts of reasonable maps into X always exist, as described in the next two results. (See Proposition 1.30 in [Hat].)

Proposition 13.2 (Path Lifting Property). *Suppose (Y, π) is a cover of X and that $p : [0, 1] \rightarrow X$ is a (continuous) path with $p(0) = x$. Then for each $y \in \pi^{-1}(\{x\})$, there is a unique lift $\tilde{p}$ of p for which $\tilde{p}(0) = y$.*

Proposition 13.3 (Homotopy Lifting Property). *Suppose that l is a loop in X, that a path p in Y is a lift of l, and that l_s is a homotopy of l in X with basepoint fixed. Then there is a unique lift of l_s to a homotopy p_s of p in Y with endpoints fixed.*

If we can find a universal cover (Y, π) of a space X, the cover gives a simple criterion for determining when a loop in X is null homotopic.

Corollary 13.4. *Suppose that (Y, π) is a universal cover of X, that l is a loop in X, and that p is a lift of l to Y. Then l is null homotopic in X if and only if p is a loop in Y, that is, if and only if $p(1) = p(0)$.*

Proof. If the lift p of l is a loop, then since Y is simply connected, there is a homotopy p_s of p to a point with basepoint fixed. Then $l_s := \pi \circ p_s$ is a homotopy of l to a point in X. In the other direction, if there is a homotopy l_s of l to a point in X, then by Proposition 13.3, we can lift this to a homotopy p_s with endpoints fixed. Now, if l_1 is the constant loop at x_0, then p_1, which is a lift of l_1, must live entirely in $\pi^{-1}(\{x_0\})$, which is a discrete set. Thus, actually, p_1 must be constant, and, in particular, has equal endpoints. But since p_s is a homotopy with endpoints fixed, the endpoints of each p_s must be equal. Thus, $p = p_0$ must be a loop in Y. $\qquad\square$

13.2 Fundamental Groups of Compact Classical Groups

In this section, we discuss a method of computing, inductively, the fundamental groups of the classical compact groups. The same results can also be obtained by using the results of Sects. 13.4–13.7; see Exercises 1–4. In all cases, we will find that $\pi_1(K)$ is commutative. This is not a coincidence; a general argument shows that fundamental group of any Lie group is commutative (Exercise 7). In the case of a compact matrix Lie group K, the commutativity of $\pi_1(K)$ also follows from Corollary 13.18.

For any nice topological space, one can define **higher homotopy groups** $\pi_k(X), k = 1, 2, 3, \ldots$. The group $\pi_k(X)$ is the set of homotopy classes of maps of S^k into X, where the notion of homotopy for maps of S^k into X is analogous to that for maps of S^1 into X. Although one can define a group structure on $\pi_k(X)$, this structure is not relevant to us. All that is relevant is what it means for $\pi_k(X)$ to be trivial, which is that every continuous map of the k-sphere S^k into X can be shrunk continuously to a point. We will make use of the following standard topological result (e.g., Corollary 4.9 in [Hat]).

Proposition 13.5. *For a d-sphere S^d, $\pi_k(S^d)$ is trivial if $k < d$.*

This result is plausible because for $k < d$, the image of a "typical" continuous map of S^k into S^d will not be all of S^d. However, if the image of the map omits even one point in S^d, then we can remove that point and what is left of the sphere can be contracted continuously to a point.

We now introduce the topological concept underlying all the calculations in this section, that of a fiber bundle.

Definition 13.6. Suppose that B and F are Hausdorff topological spaces. A **fiber bundle** with base B and fiber F is a Hausdorff topological space X together with a continuous map $p : X \to B$, called the **projection map**, having the following properties. First, for each b in B, the preimage $p^{-1}(b)$ of b in X is homeomorphic to F. Second, for every b in B, there is a neighborhood U of b such that $p^{-1}(U)$ is homeomorphic to $U \times F$ in such a way that the projection map is simply projection onto the first factor.

In any fiber bundle, the sets of the form $p^{-1}(b)$ are called the **fibers**. The second condition in the definition may be stated more pedantically as follows. For each $b \in B$, there should exist a neighborhood U of B and a homeomorphism Φ of $p^{-1}(U)$ with $U \times F$ having the property that $p(x) = p_1(\Phi(x))$, where $p_1 : U \times F \to U$ is the map $p_1(u, f) = u$.

The simplest sort of fiber bundle is the product space $X = B \times F$, with the projection map being simply the projection onto the first factor. Such a fiber bundle is called **trivial**. The second condition in the definition of a fiber bundle is called **local triviality** and it says that any fiber bundle must look locally like a trivial bundle. In general, X need not be globally homeomorphic to $B \times F$.

If X were a trivial fiber bundle, then the fundamental group of X would be simply the product of the fundamental group of the base B and the fundamental group of the fiber F. In particular, if X were a trivial fiber bundle and $\pi_1(B)$ were trivial, then $\pi_1(X)$ would be isomorphic to $\pi_1(F)$. The following result says that if $\pi_1(B)$ and $\pi_2(B)$ are trivial, then the same conclusion holds, even if X is nontrivial.

Theorem 13.7. *Suppose that X is a fiber bundle with base B and fiber F. If $\pi_1(B)$ and $\pi_2(B)$ are trivial, then $\pi_1(X)$ is isomorphic to $\pi_1(F)$.*

Proof. According to a standard topological result (e.g., Theorem 4.41 and Proposition 4.48 in [Hat]), there is a long exact sequence of homotopy groups for a fiber bundle. The portion of this sequence relevant to us is the following:

$$\pi_2(B) \underset{f}{\to} \pi_1(F) \underset{g}{\to} \pi_1(X) \underset{h}{\to} \pi_1(B). \tag{13.1}$$

Saying that the sequence is exact means that each map is a homomorphism and the image of each map is equal to the kernel of the following map. Since we are assuming $\pi_2(B)$ is trivial, the image of f is trivial, which means the kernel of g is also trivial. Since $\pi_1(B)$ is also trivial, the kernel of h must be $\pi_1(X)$, which means that the image of g is $\pi_1(X)$. Thus, g is an isomorphism of $\pi_1(F)$ with $\pi_1(X)$. □

Proposition 13.8. *Suppose G is a matrix Lie group and H is a closed subgroup of G. Then G has the structure of a fiber bundle with base G/H and fiber H, where the projection map $p : G \to G/H$ is given by $p(x) = [x]$, with $[x]$ denoting the coset $xH \in G/H$.*

Proof. For any coset $[x]$ in G/H, the preimage of $[x]$ under p is the set $xH \subset G$, which is clearly homeomorphic to H. Meanwhile, the required local triviality property of the bundle follows from Lemma 11.21 and Theorem 11.22. (If we take a open set U in G/H as in the proof of Theorem 11.22, Lemma 11.21 tells us that the preimage of U under p is homeomorphic to $U \times H$ in such a way that the projection p is just projection onto the first factor.) □

Proposition 13.9. *Consider the map $p : \mathsf{SO}(n) \to S^{n-1}$ given by*

$$p(R) = Re_n, \tag{13.2}$$

where $e_n = (0, \dots, 0, 1)$. Then $(\mathsf{SO}(n), p)$ is a fiber bundle with base S^{n-1} and fiber $\mathsf{SO}(n-1)$.

Proof. We think of $\mathsf{SO}(n-1)$ as the (closed) subgroup of $\mathsf{SO}(n)$ consisting of block diagonal matrices of the form

$$R = \begin{pmatrix} R' & 0 \\ 0 & 1 \end{pmatrix}$$

with $R' \in \mathsf{SO}(n-1)$. By Proposition 13.8, $\mathsf{SO}(n)$ is a fiber bundle with base $\mathsf{SO}(n)/\mathsf{SO}(n-1)$ and fiber $\mathsf{SO}(n-1)$. Now, it is easy to see that $\mathsf{SO}(n)$ acts transitively on the sphere S^{n-1}. Thus, the map p in (13.2) maps $\mathsf{SO}(n)$ onto S^{n-1}. Since $Re_n = e_n$ if and only $R \in \mathsf{SO}(n-1)$, we see that p descends to a (continuous) bijection of $\mathsf{SO}(n)/\mathsf{SO}(n-1)$ onto S^{n-1}. Since both $\mathsf{SO}(n)/\mathsf{SO}(n-1)$ and S^{n-1} are compact, this map is actually a homeomorphism (Theorem 4.17 in [Rud1]). Thus, $\mathsf{SO}(n)$ is a fiber bundle of the claimed sort. □

Proposition 13.10. *For all $n \geq 3$, the fundamental group of $\mathsf{SO}(n)$ is isomorphic to $\mathbb{Z}/2$. Meanwhile, $\pi_1(\mathsf{SO}(2)) \cong \mathbb{Z}$.*

Proof. Suppose that n is at least 4, so that $n - 1$ is at least 3. Then, by Proposition 13.5, $\pi_1(S^{n-1})$ and $\pi_2(S^{n-1})$ are trivial and, so, Theorem 13.7 and Proposition 13.9 tell us that $\pi_1(\mathsf{SO}(n))$ is isomorphic to $\pi_1(\mathsf{SO}(n-1))$. Thus, $\pi_1(\mathsf{SO}(n))$ is isomorphic to $\pi_1(\mathsf{SO}(3))$ for all $n \geq 4$. It remains to show that $\pi_1(\mathsf{SO}(3)) \cong \mathbb{Z}/2$. This can be done by noting that $\mathsf{SO}(3)$ is homeomorphic to $\mathbb{RP}^3$, as in Proposition 1.17, or by observing that the map Φ in Proposition 1.19 is a two-to-one covering map from $\mathsf{SU}(2) \sim S^3$ onto $\mathsf{SO}(3)$.

Finally, we observe that $\mathsf{SO}(2)$ is homeomorphic to the unit circle S^1, so that $\pi_1(\mathsf{SO}(2)) \cong \mathbb{Z}$ (Theorem 1.7 in [Hat]). □

If one looks into the proof of the long exact sequence of homotopy groups for a fiber bundle, one finds that the map g in (13.1) is induced by the inclusion of F into X. Thus, if l is a homotopically nontrivial loop in $SO(n)$, then after we include $SO(n)$ into $SO(n+1)$, the loop l is still homotopically nontrivial.

Meanwhile, we may take $SU(2)$ as the universal cover of $SO(n)$, with covering map being the homomorphism Φ in Proposition 1.19 (compare Exercise 8). Now, if we take l to be the loop in $SO(3)$ consisting of rotations by angle θ in the (x_2, x_3)-plane, $0 \leq \theta \leq 2\pi$, the computations in (1.15) and (1.16) show that the lift of l to $SU(2)$ is not a loop. (Rather, the lift will start at I and end at $-I$.) Thus, by Corollary 13.4, l is homotopically nontrivial in $SO(3)$. But this loop l is conjugate in $SO(3)$ to the loop of rotations in the (x_1, x_2)-plane, so that loop is also homotopically nontrivial. Thus, by the discussion in the previous paragraph, we may say that, for any $n \geq 3$, the one nontrivial homotopy class in $SO(n)$ is represented by the loop

$$l(\theta) := \begin{pmatrix} \cos\theta & -\sin\theta & & & \\ \sin\theta & \cos\theta & & & \\ & & 1 & & \\ & & & \ddots & \\ & & & & 1 \end{pmatrix}, \quad 0 \leq \theta \leq 2\pi.$$

(Compare Exercise 6.)

Proposition 13.11. *The group $SU(n)$ is simply connected for all $n \geq 2$. For all $n \geq 1$, we have that $\pi_1(U(n)) \cong \mathbb{Z}$.*

Proof. For all $n \geq 3$, the group $SU(n)$ acts transitively on the sphere S^{2n-1}. By a small modification of the proof of Proposition 13.9, $SU(n)$ is a fiber bundle with base S^{2n-1} and fiber $SU(n-1)$. Since $2n-1 > 3$ for all $n \geq 2$, Theorem 13.7 and Proposition 13.5 tell us that $\pi_1(SU(n)) \cong \pi_1(SU(n-1))$. Since $\pi_1(SU(2)) \cong \pi_1(S^3)$ is trivial, we conclude that $\pi_1(SU(n))$ is trivial for all $n \geq 2$.

The analysis of the case of $U(n)$ is similar. The fiber bundle argument shows that $\pi_1(U(n)) \cong \pi_1(U(n-1))$ for all $n \geq 2$. Since $U(1)$ is just the unit circle S^1, we have that $\pi_1(U(1)) \cong \mathbb{Z}$ (Theorem 1.7 in [Hat]). Thus, $\pi_1(U(1)) \cong \mathbb{Z}$ for all $n \geq 1$. $\square$

Proposition 13.12. *For all $n \geq 1$, the compact symplectic group $Sp(n)$ is simply connected.*

Proof. Since $Sp(n)$ is contained in $U(2n)$, it acts on the unit sphere $S^{4n-1} \subset \mathbb{C}^{2n}$. If this action is transitive, then we can imitate the arguments from the cases of $SO(n)$ and $SU(n)$. Since $4n-1 \geq 3$, we see that $\pi_1(S^{4n-1})$ and $\pi_2(S^{4n-1})$ are trivial for all $n \geq 1$. We conclude, then, that $\pi_1(Sp(n)) \cong \pi_1(Sp(n-1))$. Since $Sp(1) = SU(2)$, which is simply connected, we see that $Sp(n)$ is simply connected for all n.

It remains, then, to show that $Sp(n)$ acts transitively on S^{4n-1}. For this, it suffices to show that for all unit vectors $u \in \mathbb{C}^{2n}$, there is some $U \in Sp(n)$ with $Ue_1 = u$.

To see this, let $u_1 = u$ and $v_1 = Ju$, where J is the map in Sect. 1.2.8. Then v_1 is orthogonal to u_1 (check) and we may consider the space

$$W := (\text{span}\{u_1, v_1\})^{\perp}.$$

By the calculations in Sect. 1.2.8, W will be invariant under J. Thus, we can choose an arbitrary unit vector u_2 in W and let $v_2 = Ju_2$. Proceeding on in this way, we eventually obtain an orthonormal family

$$u_1, \ldots, u_n, v_1, \ldots, v_n \tag{13.3}$$

in $\mathbb{C}^{2n}$ where $Ju_j = v_j$. It is then straightforward to check that the matrix U whose columns are the vectors in (13.3) belongs to $\mathsf{Sp}(n)$, and that $Ue_1 = u_1 = u$. $\quad\Box$

13.3 Fundamental Groups of Noncompact Classical Groups

Using the polar decomposition (Sect. 2.5), we can reduce the computation of the fundamental group of certain noncompact groups to the computation of the fundamental group of one of the compact groups in Sect. 13.2. Theorem 2.17, for example, tells us that $\mathsf{GL}(n; \mathbb{C})$ is homeomorphic to $\mathsf{U}(n) \times V$ for a certain vector space V (the space of $n \times n$ self-adjoint matrices). Since V is simply connected, we conclude that $\pi_1(\mathsf{GL}(n; \mathbb{C})) \cong \mathbb{Z}$. Using Proposition 2.19, we can similarly show that $\pi_1(\mathsf{SL}(n; \mathbb{C})) \cong \pi_1(\mathsf{SU}(n))$ and that $\pi_1(\mathsf{SL}(n; \mathbb{R})) \cong \pi_1(\mathsf{SO}(n))$.

Conclusion 13.13. *For all $n \geq 1$, we have*

$$\pi_1(\mathsf{GL}(n; \mathbb{C})) \cong \pi_1(\mathsf{U}(n)) \cong \mathbb{Z}.$$

For all $n \geq 2$, the group $\mathsf{SL}(n; \mathbb{C})$ is simply connected. For $n \geq 3$, we have

$$\pi_1(\mathsf{SL}(n; \mathbb{R})) \cong \pi_1(\mathsf{SO}(n)) \cong \mathbb{Z}/2,$$

whereas

$$\pi_1(\mathsf{SL}(2; \mathbb{R})) \cong \pi_1(\mathsf{SO}(2)) \cong \mathbb{Z}.$$

13.4 The Fundamental Groups of K and T

In this section and the subsequent ones, we develop a different approach to computing the fundamental group of a compact group K, based on the torus theorem. In this section, we state the main results; the proofs will be developed in Sects. 13.5, 13.6, and 13.7. One important consequence of these results will be Corollary 13.20, which says that if K is simply connected, then every algebraically integral element is analytically integral. This claim allows us (in the simply connected case) to match up our theorem of the highest weight for the compact group K with the theorem of the highest weight for the Lie algebra $\mathfrak{g} := \mathfrak{k}_{\mathbb{C}}$. That is to say, when K is simply connected, the set of possible highest weights for the representations of K (namely the dominant, analytically integral elements) coincides with the set of possible highest weights for the representations of $\mathfrak{g}$ (namely the dominant, algebraically integral elements).

In these sections, we will follow the convention of composing the exponential map for T with a factor of 2π, writing an element t of T as

$$t = e^{2\pi H}, \quad H \in \mathfrak{t}.$$

(There are factors of 2π that have to go somewhere in the theory, and this seems to be the most convenient spot to put them.) Recall (Definition 12.3) that $\Gamma \subset \mathfrak{t}$ denotes the *kernel of the (scaled) exponential map*:

$$\Gamma = \left\{ H \in \mathfrak{t} \,\middle|\, e^{2\pi H} = I \right\}$$

and that Γ is invariant under the action of W.

We begin with a simple result describing the fundamental group of T.

Proposition 13.14. *Every loop in T is homotopic in T to a unique loop of the form*

$$\tau \mapsto e^{2\pi \tau \gamma}, \quad 0 \le \tau \le 1,$$

with $\gamma \in \Gamma$. Furthermore, $\pi_1(T)$ is isomorphic to Γ.

Proof. The main issue is to prove that the (scaled) exponential map $H \mapsto e^{2\pi H}$ is a covering map. Since the kernel Γ of the exponential map is discrete, there is some $\varepsilon > 0$ such that every nonzero element γ of Γ has norm at least ε. Let $B_{\varepsilon/2}(\gamma)$ be the ball of radius $\varepsilon/2$ around a point $\gamma \in \Gamma$. Now, for any $t \in T$, write t as $e^{2\pi H}$ for some $H \in \mathfrak{t}$. Then let $V \subset T$ denote the set of element of the form $e^{2\pi H'}$, where $H' \in B_{\varepsilon/2}(H)$, so that V is a neighborhood of t. The preimage of V under the exponential is the union of the balls

$$B_{\varepsilon/2}(H + \gamma), \quad \gamma \in \Gamma.$$

By the way ε was chosen, these balls are disjoint, and each ball maps homeomorphically onto V. Since we can do this for each $t \in T$, we see that the exponential is a covering map.

Since, also, $\mathfrak{t}$ is simply connected, $\mathfrak{t}$ is the universal cover of T. Now, every loop l in T based at the identity has a unique lift to a path $\tilde{l}$ in $\mathfrak{t}$ starting at 0 and ending at some point γ in Γ. The theory of covering spaces tells us that two loops l_1 and l_2 in T (based at I) are homotopic if and only if $\tilde{l}_1(1) = \tilde{l}_2(1)$. Meanwhile, if $\tilde{l}(1) = \gamma$, then (since t is simply connected) $\tilde{l}$ is homotopic with endpoints fixed to the straight-line path $\tau \mapsto \tau\gamma$, showing that l itself is homotopic to $\tau \mapsto e^{2\pi\tau\gamma}$, as claimed. Finally, if we compose two loops of the form $\tau \mapsto e^{2\pi\tau\gamma_1}$ and $\tau \mapsto e^{2\pi\tau\gamma_2}$, the lift of the composite loop will be the composition of the lifts, where the lift of the second part of the composite loop must be taken to start at γ_1. Thus, the lift of the composite loop will go from 0 to γ_1 and then (shifting the lift of the second loop by γ_1) from γ_1 to $\gamma_1 + \gamma_2$. But any path from 0 to $\gamma_1 + \gamma_2$ in t is homotopic with endpoints fixed to the straight-line path $\tau \mapsto e^{2\pi\tau(\gamma_1+\gamma_2)}$. Thus, if we identify elements of $\pi_1(T)$ with elements of Γ, the composition operator corresponds to addition in Γ.   □

We now state the first main result of this section; the proof is given in Sect. 13.7.

Theorem 13.15. *Every loop in K is homotopic to a loop in T.*

The theorem does *not* mean that $\pi_1(K)$ is isomorphic to $\pi_1(T)$, since a loop in T may be null homotopic in K even if it is not null homotopic in T. Indeed, $\pi_1(T)$ (which is isomorphic to Γ) is often very different from $\pi_1(K)$ (which is, for example, trivial when $K = \mathsf{SU}(n)$). Nevertheless, the theorem gives us a useful way to study $\pi_1(K)$, because we *understand* $\pi_1(T)$. Taking Theorem 13.15 and Proposition 13.14 together, we see that every loop in K is homotopic to a loop of the form $\tau \mapsto e^{2\pi\tau\gamma}$, with $\gamma \in \Gamma$. Thus, we are faced with a very concrete problem to calculate $\pi_1(K)$: We must only determine, for each $\gamma \in \Gamma$, whether the loop $t \mapsto e^{2\pi t\gamma}$ is null homotopic *in the compact group* K. (By Proposition 13.14, such a loop is null homotopic *in the torus* T only if $\gamma = 0$.)

The condition for $\tau \mapsto e^{2\pi\tau\gamma}$ to be null homotopic in K turns out to be related to the notion of coroots. Recall that if α is a real root for t, then (Lemma 12.8), the associated real coroot $H_\alpha = 2\alpha/\langle\alpha,\alpha\rangle$ belongs to Γ. Thus, any integer linear combination of coroots also belongs to Γ.

Definition 13.16. The **coroot lattice**, denoted I, is the set of all integer linear combinations of real coroots $H_\alpha, \alpha \in R$.

We now state the second main result of this section; the proof is also in Sect. 13.7.

Theorem 13.17. *For each $\gamma \in \Gamma$, the loop $\tau \mapsto e^{2\pi\tau\gamma}$ is null homotopic in K if and only if γ belongs to the coroot lattice I.*

If we combine this result with Theorem 13.15, we obtain the following description of the fundamental group of K.

Corollary 13.18. *The fundamental group of K is isomorphic to the quotient group Γ/I, where the quotient is of commutative groups.*

Proof. By Theorem 13.15, every loop in K is homotopic to a loop of the form $\tau \mapsto e^{2\pi\tau\gamma}$, with $\gamma \in \Gamma$. Under this correspondence, composition of loops corresponds to addition in Γ. By Theorem 13.17, two loops of the form $\tau \mapsto e^{2\pi\tau\gamma_1}$ and $\tau \mapsto e^{2\pi\tau\gamma_2}$ are homotopic if and only if $\gamma_1 - \gamma_2$ belongs to I. Thus, $\pi_1(K)$ may be identified with the set of cosets of I in Γ.   □

We now consider three examples. The first two are familiar friends, the groups $SO(5)$ and $SU(3)$. The third is the *projective unitary* group $PSU(3)$. In general, $PSU(n)$ is the quotient of $SU(n)$ by the subgroup consisting of whichever multiples of the identity have determinant 1 in $U(n)$. In the case $n = 3$, a matrix of the form $e^{i\theta}I$ has determinant 1 if and only if $e^{3i\theta} = 1$. Thus,

$$PSU(3) = SU(3)/\{I, e^{2\pi i/3}I, e^{4\pi i/3}I\}. \tag{13.4}$$

We can represent $PSU(3)$ as a matrix group by using the adjoint representation. It is easy to check that the center of $SU(3)$ is the group being divided by on the right-hand side of (13.4); thus, the image $SU(3)$ under the adjoint representation will be a subgroup of $GL(sl(3; \mathbb{C}))$ isomorphic to $PSU(3)$. Since the center of the Lie algebra $su(3)$ is trivial, the Lie algebra version of the adjoint representation is faithful; thus, we may identify the Lie algebra of $PSU(3)$ with the Lie algebra of $SU(3)$.

Example 13.19. If $K = SO(5)$, then $\Gamma/I \cong \mathbb{Z}/2$. If $K = SU(3)$, then Γ/I is trivial. Finally, if $K = PSU(3)$, then $\Gamma/I \cong \mathbb{Z}/3$.

The verification of the claims in Example 13.19 is left to the reader (Exercise 5). Corollary 13.18 together with Example 5 give another way of computing the fundamental groups of $SO(5)$ ($\mathbb{Z}/2$) and $SU(3)$ (trivial), in addition to Proposition 13.10.

Figures 13.1 and 13.2 show Γ and I in the case of the groups $SO(5)$ and $PSU(3)$. The black dots indicate points in I, whereas white dots indicate points in Γ that are

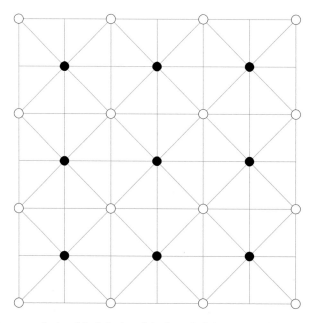

Fig. 13.1 The coroot lattice (*black dots*) and the kernel of the exponential mapping (*black and white dots*) for the group $SO(5)$

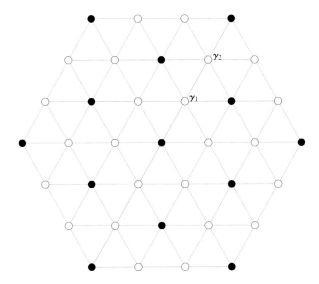

Fig. 13.2 The coroot lattice (*black dots*) and the kernel of the exponential mapping (*black and white dots*) for the group $\mathsf{PSU}(3)$

not in I. In the case of $\mathsf{SO}(5)$, it is easy to see that any two white dots differ by an element of the coroot lattice, showing that there is exactly one nontrivial element of Γ/I. In the case of $\mathsf{PSU}(3)$, note that the elements γ_1 and $\gamma_2 = 2\gamma_1$ are not in I, but $3\gamma_1$ is in I, showing that $[\gamma_1]$ is an element of order 3 in Γ/I. The reader may verify that every element of Γ is either in I, differs from γ_1 by an element of I, or differs from γ_2 by an element of I. The situation for $\mathsf{SU}(3)$ is similar to that for $\mathsf{PSU}(3)$; the coroot lattice I does not change, but Γ is now equal to I, so that Γ/I is trivial. For now, the reader may regard the lines in the figures as merely decorative; these lines will turn out to make up the "Stiefel diagram" for the relevant group. (See Sect. 13.6.)

Corollary 13.20. *If K is simply connected, then every algebraically integral element is analytically integral.*

Proof. In light of Theorem 13.17, K is simply connected if and only if $I = \Gamma$. If $\lambda \in \mathfrak{t}$ is algebraically integral, then $\langle \lambda, H_\alpha \rangle \in \mathbb{Z}$ for all α, where $H_\alpha = 2\alpha/\langle \alpha, \alpha \rangle$ is the real coroot associated to α. It follows that $\langle \lambda, \gamma \rangle \in \mathbb{Z}$ for every element of I, the set of integer linear combinations of coroots. Thus, if K is simply connected, $\langle \lambda, \gamma \rangle \in \mathbb{Z}$ for every element of $I = \Gamma$, which means that λ is analytically integral. $\square$

We may offer a completely different proof of Corollary 13.20, as follows. We first observe that if K is simply connected, then by Proposition 7.7, the complex Lie algebra $\mathfrak{g} := \mathfrak{k}_{\mathbb{C}}$ is semisimple. Then let λ be an algebraically integral element. Since the sets of analytically integral and algebraically integral elements are both invariant under the action of W, it is harmless to assume that λ is dominant. Thus, by the results of Chapter 9, there is a finite-dimensional irreducible representation (π, V_λ) of $\mathfrak{g} := \mathfrak{k}_{\mathbb{C}}$ with highest weight λ. Since K is simply connected, Theorem 5.6 then tells us that there is an associated representation Π of K acting on V_λ. Thus, λ is a weight for a representation of the group K, which implies (Proposition 12.5) that λ is analytically integral.

Although it is mathematically correct, the preceding argument may be considered as "cheating," since it depends on the whole machinery of Verma modules (to construct the representations of $\mathfrak{g}$) and on the Baker–Campbell–Hausdorff formula (to prove Theorem 5.6). In the subsequent sections, we will use techniques similar to those in the proof of the torus theorem to prove Theorem 13.17 and thus to give a more direct (but not easy!) proof of Corollary 13.20.

Corollary 13.21. *If K is simply connected, the element δ (half the sum of the real, positive roots) is analytically integral.*

Proof. According to Proposition 8.38 (translated into the language of real roots), the element δ is algebraically integral. But by Corollary 13.20, if K is simply connected, every algebraically integral element is analytically integral. $\square$

Since it is easy to do so, we will immediately prove one direction of Theorem 13.17, namely that if γ is in the coroot lattice I, then $\tau \mapsto e^{2\pi\tau\gamma}$ is homotopically trivial in K.

Proof of Theorem 13.17, one direction. We assume at first that $\gamma = H_\alpha$, a single real coroot. Then by Corollary 7.20, there is a homomorphism ϕ of $\mathsf{su}(2)$ into $\mathfrak{k}$ such that ϕ maps the element $iH = \operatorname{diag}(i, -i)$ in $\mathsf{su}(2)$ to the real coroot H_α. Since $\mathsf{SU}(2)$ is simply connected, there exists a homomorphism $\Phi : \mathsf{SU}(2) \to K$ such that $\Phi(e^X) = e^{\phi(X)}$ for all $X \in \mathsf{su}(2)$.

Consider, then, the loop l in $\mathsf{SU}(2)$ given by

$$l(\tau) = e^{2\pi\tau iH} = \begin{pmatrix} e^{2\pi i\tau} & 0 \\ 0 & e^{-2\pi i\tau} \end{pmatrix}, \quad 0 \leq \tau \leq 1.$$

Observe that $\Phi(l(\tau)) = e^{2\pi\tau H_\alpha}$, so that the image of l under Φ is the relevant loop in K. Since $\mathsf{SU}(2)$ is simply connected, there is a family l_s of loops connecting l to a constant loop in $\mathsf{SU}(2)$. Thus, the loops

$$\tau \mapsto \Phi(l_s(\tau))$$

constitute a homotopy of our original loop to a point in K.

Now, if γ is an integer linear combination of coroots, then by Proposition 13.14, the loop $\tau \mapsto e^{2\pi\tau\gamma}$ is homotopic (in T and thus in K) to a composition of loops of the form $\tau \mapsto e^{2\pi\tau H_\alpha}$, for various coroots α. Since each of those loops is null homotopic in K, so is the loop $\tau \mapsto e^{2\pi\tau\gamma}$. $\square$

13.5 Regular Elements

We will study the topology of K by writing every element x of K as

$$x = yty^{-1},$$

with $y \in K$ and $t \in T$. It is then convenient to write the variable t in the preceding expression in exponential coordinates. Thus, we may consider the map $\Psi : \mathfrak{t} \times (K/T) \to K$ given by

$$\Psi(H, [y]) = ye^{2\pi H} y^{-1}. \tag{13.5}$$

(We continue to scale the exponential map for T by a factor of 2π.) The torus theorem tells us that Ψ maps onto K. On the other hand, the behavior of Ψ is easiest to understand when the differential Ψ_* is nonsingular.

We now introduce a notion of "regular"elements in K; we will see (Proposition 13.24) that $\Psi_*(H, [y])$ is invertible if and only if $x := \Psi(H, [y])$ is regular. It will turn out that the fundamental group of K is the same as the fundamental group of the set of regular elements in K.

Definition 13.22. If $x \in K$ is contained in a unique maximal torus, x is **regular**; if x is contained in two distinct maximal tori, x is **singular**. The set of regular elements in K is denoted K_{reg} and the set of singular elements is denoted K_{sing}.

If, for example, $t \in T$ generates a dense subgroup of T, then the only maximal torus containing yty^{-1} is yTy^{-1}. Thus, for such a t, the element yty^{-1} is regular. As we will see, however, yty^{-1} can be regular even if t does not generate a dense subgroup of T.

Proposition 13.23. *Suppose $x \in K$ has the form $x = yty^{-1}$ with $t \in T$ and $y \in K$. If there exists some $X \in \mathfrak{t}$ with $X \notin \mathfrak{t}$ such that*

$$\text{Ad}_t(X) = X,$$

then x is singular. If no such X exists, x is regular.

Proof. The condition of being regular is clearly invariant under conjugation; that is, x is regular if and only if t is regular. Suppose now that there is some $X \notin \mathfrak{t}$ with $\text{Ad}_t(X) = X$. Then t commutes with $e^{\tau X}$ for all $\tau \in \mathbb{R}$. Applying Lemma 11.37 with $S = \{e^{\tau X}\}_{\tau \in \mathbb{R}}$, there is a maximal torus S' containing both t and $\{e^{\tau X}\}_{\tau \in \mathbb{R}}$. But then the Lie algebra $\mathfrak{s}'$ of S' must contain X, which is not in $\mathfrak{t}$, which means that $S' \neq T$. Thus, t (and therefore, also x) is singular.

In the other direction, if x (and therefore, also, t) is singular, then there is a maximal torus $S' \neq T$ containing t. But we cannot have $S' \subset T$, or else S' would not be maximal. Thus, there must be some X in the Lie algebra $\mathfrak{s}'$ of S' that is not in $\mathfrak{t}$. But since $t \in S'$ and S' is commutative, we have $\text{Ad}_t(X) = X$. $\qquad\qquad \square$

Proposition 13.24. *An element* $x = y e^{2\pi H} y^{-1}$ *is singular if and only if there is some root* α *for which*

$$\langle \alpha, H \rangle \in \mathbb{Z}.$$

It follows that $x = y e^{2\pi H} y^{-1}$ *is singular if and only if* Ψ_* *is singular at the point* $(H, [y])$.

Note that for each fixed $\alpha \in R$ and $n \in \mathbb{Z}$, the set of H in t for which $\langle \alpha, H \rangle = n$ is a hyperplane (not necessarily through the origin).

Proof. By Proposition 13.23, x is singular if and only if $\mathrm{Ad}_{e^{2\pi H}}(X) = X$ for some $X \notin$ t. Now, the dimension of the eigenspace of $\mathrm{Ad}_{e^{2\pi H}}$ with eigenvalue $1 \in \mathbb{R}$ is the same whether we work over $\mathbb{R}$ or over $\mathbb{C}$. The eigenvalues of $\mathrm{Ad}_{e^{2\pi H}}$ over $\mathbb{C}$ are 1 (on $\mathfrak{h}$) together with the numbers of the form

$$e^{2\pi i \langle \alpha, H \rangle}, \quad \alpha \in R,$$

(from the root space $\mathfrak{g}_\alpha$). Thus, the dimension of the 1-eigenspace is greater than $\dim \mathfrak{h}$ if and only if $e^{2\pi i \langle \alpha, H \rangle} = 1$ for some α, which holds if and only if $\langle \alpha, H \rangle \in \mathbb{Z}$.

Meanwhile, the map Ψ is just the map Φ in Definition 11.25, composed with the exponential map for T. Since T is commutative, the differential of the exponential map for T is the identity at each point. Thus, using Proposition 11.27, we see that $\Psi_*(H, [y])$ is singular if and only if the restriction of $\mathrm{Ad}_{e^{-2\pi H}}$ to $\mathfrak{t}^\perp$ (or, equivalently, to $(\mathfrak{t}^\perp)_\mathbb{C}$) has an eigenvalue of 1. Since the eigenvalues of $\mathrm{Ad}_{e^{-2\pi H}}$ on $(\mathfrak{t}^\perp)_\mathbb{C}$ are the numbers of the form $e^{-2\pi i \langle \alpha, H \rangle}$, we see that $\Psi_*(H, [y])$ is singular if and only if $\langle \alpha, H \rangle \in \mathbb{Z}$ for some α. □

Definition 13.25. An element H of t is **regular** if for all $\alpha \in R$, the quantity $\langle \alpha, H \rangle$ is *not* an integer. Otherwise, H is **singular**. The set of regular elements in t is denoted $\mathfrak{t}_{\mathrm{reg}}$.

A key issue in the proof of Theorem 13.17 is to understand the extent to which the map Ψ in (13.5) fails to be injective. There are two obvious sources of noninjectivity for Ψ. The first is the kernel of the exponential map; clearly if $\gamma \in \Gamma$, then

$$\Psi(H + \gamma, [x]) = \Psi(H, [x]).$$

Meanwhile, if $w \in W$ and $z \in N(T)$ represents w, then

$$\Psi(w \cdot H, [xz^{-1}]) = xz^{-1} e^{2\pi (w \cdot H)} z x^{-1}$$
$$= x e^{2\pi H} x^{-1}$$
$$= \Psi(H, [x]),$$

since z^{-1} represents w^{-1}. We now demonstrate that if we restrict Ψ to $\mathfrak{t}_{\mathrm{reg}} \times (K/T)$, these two sources account for all of the noninjectivity of Ψ.

Proposition 13.26. *Suppose $(H, [x])$ and $(H', [x'])$ belong to $\mathfrak{t}_{reg} \times (K/T)$. Then $\Psi(H, [x]) = \Psi(H', [x'])$ if and only if there exist some $w = [z]$ in W and some $\gamma \in \Gamma$ such that*

$$H' = w \cdot H + \gamma$$

$$[x'] = [xz^{-1}]. \tag{13.6}$$

Here $[z]$ denotes the coset containing $z \in N(T)$ in $W = N(T)/T$. Furthermore, if the elements in (13.6) satisfy $H' = H$ and $[x'] = [x]$, then $\gamma = 0$ and w is the identity element of W.

Note that if $z \in N(T)$ and $t \in T$, then

$$xtz^{-1} = xz^{-1}(ztz^{-1}),$$

where $ztz^{-1} \in T$. Thus, $[xtz^{-1}]$ and $[xz^{-1}]$ are equal in K/T. That is to say, for a fixed $z \in N(T)$, the map $[x] \mapsto [xz^{-1}]$ is a well-defined map of K/T to itself. A similar argument shows that this action depends only on the coset of z in $N(T)/T$.

Proof. If H' and x' are as in (13.6), then by the calculations preceding the statement of the proposition, we will have $\Psi(H', [x']) = \Psi(H, [x])$. In the other direction, if $\Psi(H, [x]) = \Psi(H', [x'])$, then

$$xe^{2\pi H}x^{-1} = x'e^{2\pi H'}(x')^{-1},$$

which means that

$$e^{2\pi H} = z^{-1}e^{2\pi H'}z, \tag{13.7}$$

where $z = (x')^{-1}x$.

Now, the relation (13.7) implies that $e^{2\pi H}$ belongs to the torus $z^{-1}Tz$. Since $H \in \mathfrak{t}_{reg}$, it follows from Proposition 13.24 that $z^{-1}Tz = T$, that is, that $z \in N(T)$. Then, if $w = [z]$, we have

$$e^{2\pi H} = e^{2\pi w^{-1} \cdot H'}.$$

From this, we obtain $e^{2\pi(w^{-1} \cdot H' - H)} = I$, which means that $w^{-1} \cdot H' - H$ belongs to Γ. Since Γ is invariant under the action of W, the element $\gamma := H' - w \cdot H$ also belongs to Γ, and we find that $H' = w \cdot H + \gamma$ and $x' = xz^{-1}$, as claimed.

Finally, if $H' = H$ and $[x'] = [x]$, then x' and x belong to the same coset in K/T, which means that z^{-1} must be in T. Thus, w is the identity element in $W = N(T)/T$. But once $w = e$, we see that $H' = H$ only if $\gamma = 0$. □

We now come to a key result that is essential to the proofs of Theorems 13.15 and 13.17.

Theorem 13.27. *The fundamental groups of K and K_{reg} are isomorphic. Specifically, every loop in K is homotopic to a loop in K_{reg} and a loop in K_{reg} is null homotopic in K only if it is null homotopic in K_{reg}.*

To prove this result, we will first show that the singular set in K is "small," meaning that it has codimension at least 3. (In the case $K = \mathsf{SU}(2)$, for example, the singular set is just $\{I, -I\}$, so that K_{sing} has dimension 0, whereas K has dimension 3.) We will then argue as follows. Let n be the dimension of K and suppose E and F are subsets of K of dimension k and l, respectively. If $k + l < n$, then "generically" E and F will not intersect. If E and F do intersect, then (we will show) it is possible to perturb F slightly so as to be disjoint from E. We first apply this result with $E = K_{\mathrm{sing}}$ and F being a loop in K. Then E has dimension at most $n - 3$ while F has dimension 1, so the loop F is homotopic to a loop that does not intersect K_{sing}. We next apply this result with $E = K_{\mathrm{sing}}$ and F being a homotopy of a loop $l \subset K_{\mathrm{reg}}$. Then E still has dimension at most $n - 3$ while F has dimension 2 (the image of a square), so the homotopy can be deformed to a homotopy that does not intersect K_{sing}. In the remainder of this section, we will flesh out the above argument.

Lemma 13.28. *There exist finitely many smooth compact manifolds $M_1, \ldots, M_N$ together with smooth maps $f_j : M_j \to K$ such that (1) each M_j has dimension at most $\dim K - 3$, and (2) each element of K_{sing} is in the image of f_j for some j.*

Proof. Since each root α is analytically integral (Proposition 12.7), there exists a map $f_\alpha : T \to S^1$ such that

$$f_\alpha(e^{2\pi H}) = e^{2\pi i \langle \alpha, H \rangle} \tag{13.8}$$

for all $H \in \mathfrak{t}$. Clearly, f_α is actually a homomorphism of T into S^1. Let T_α be the kernel of f_α, so that T_α is a closed subgroup of T. The Lie algebra of T_α is the set of $H \in \mathfrak{t}$ with $\langle \alpha, H \rangle = 0$; thus, T_α has dimension one less than the dimension of T. (Note that T_α may not be connected.) For each $H \in \mathfrak{t}$, we see from (13.8) that $e^{2\pi H}$ belongs to T_α if and only if $\langle \alpha, H \rangle$ is an integer. Thus, by the torus theorem and Proposition 13.24, each singular element in K is conjugate to an element of T_α, for some α.

Now fix a root α and let $C(T_\alpha)$ denote the centralizer of T_α, that is, the set of $x \in K$ such that x commutes with every element of T_α. Then $C(T_\alpha)$ is a closed subgroup of K, and the Lie algebra of $C(T_\alpha)$ consists of those $X \in \mathfrak{k}$ such that $\mathrm{Ad}_t(X) = X$ for all $t \in T_\alpha$. Suppose now that X_α belongs to the root space $\mathfrak{g}_\alpha$ and $t = e^{2\pi H}$ belongs to T_α. Then

$$\mathrm{Ad}_{e^{2\pi H}}(X_\alpha) = e^{2\pi \,\mathrm{ad}_H}(X_\alpha) = e^{2\pi i \langle \alpha, H \rangle} X_\alpha = X_\alpha,$$

since $f(e^{2\pi H}) = e^{2\pi i \langle \alpha, H \rangle} = 1$, by assumption, and similarly with X_α replaced by $Y_\alpha := X_\alpha^*$. Thus, the Lie algebra of $C(T_\alpha)$ will contain $\mathfrak{t}$ and at least two additional elements, $X_\alpha - Y_\alpha$ and $i(X_\alpha + Y_\alpha)$, that are independent of each other and of $\mathfrak{t}$. (Compare Corollary 7.20.) We conclude that the dimension of $C(T_\alpha)$ is at least $\dim T_\alpha + 2$.

Now, the map $g_\alpha : T_\alpha \times K \to K$ given by

$$g_\alpha(t, x) = xtx^{-1}$$

descends to a map (still called g_α) of $T_\alpha \times (K/C(T_\alpha))$ into K. Furthermore, we compute that

$$\dim(T_\alpha \times (K/C(T_\alpha))) = \dim T_\alpha + \dim K - \dim C(T_\alpha)$$
$$\leq \dim T - 1 + \dim K - \dim T - 2$$
$$= \dim K - 3.$$

Since, as we have said, every singular element is conjugate to an element of some T_α, we have proved the lemma, with the M's being the manifolds $T_\alpha \times (K/C(T_\alpha))$ and the f's being the maps g_α. $\square$

Lemma 13.29. *Let M and N be compact manifolds and let D be a compact manifold with boundary, with*

$$\dim M + \dim D < \dim N.$$

Let $f : M \to N$ and $g : D \to N$ be smooth maps. Suppose E is a closed subset of D such that $g(E)$ is disjoint from $f(M)$. Then g is homotopic to a map g' such that $g' = g$ on E and such that $g'(D)$ is disjoint from $f(M)$.

Since $g(E)$ is already disjoint from $f(M)$, it is plausible that we can deform g without changing its values on E to make the image disjoint from $f(M)$. Our proof will make use of the following result: If X and Y are smooth manifolds with $\dim X < \dim Y$ and $f : X \to Y$ is a smooth map, then the image of X under f is a set of measure zero in Y. (Note that we do not assume f_* is injective.) This result is a consequence of Sard's theorem; see, for example, Corollary 6.11 in [Lee]. We will use this result to show that g can be moved locally off of $f(M)$; a finite number of these local moves will then produce the desired map g'.

Proof.   Step 1: The local move. For $x \in D \setminus E$, let us choose a neighborhood U of $g(x)$ diffeomorphic to $\mathbb{R}^n$, where $n = \dim N$. We may then define a map

$$h : f^{-1}(U) \times g^{-1}(U) \to U$$

by

$$h(m, x) = f(m) - g(x),$$

where the difference is computed in $\mathbb{R}^n \cong U$. By our assumption on the dimensions, the image of h is a set of measure zero in U. Thus, in every neighborhood of 0, we can find some p that is not in the image of h.

Suppose that W is any neighborhood of x in D such that $\bar{W}$ is contained in $g^{-1}(U)$ and $\bar{W}$ is disjoint from E. Then we can choose a smooth function χ on D such that χ equals 1 on $\bar{W}$ but such that χ equals 0 both on E and on the complement of $g^{-1}(U)$. Let us define a family of maps $g_s : D \to N$ by setting

$$g_s(x) = g(x) + s\chi(x)p, \quad 0 \le s \le 1$$

for $x \in g^{-1}(U)$ and setting $g_s(x) = g(x)$ for $x \notin g^{-1}(U)$.
When $s = 1$, and $x \in \bar{W}$, we have

$$f(m) - g_1(x) = f(m) - g(x) - p = h(m, x) - p.$$

Since h never takes the value p, we see that $f(m)$ does not equal $g_1(x)$; thus, $g_1(\bar{W}) \subset U$ is disjoint from $f(M)$. We conclude that g_1 is a map homotopic to g such that $g_1 = g$ on E and $g_1(\bar{W})$ is disjoint from $f(M)$. Furthermore, since p can be chosen to be as small as we want, we can make g_1 uniformly as close to g as we like.

Step 2: The global argument. Choose a neighborhood V of E in D such that $g(\bar{V})$ is disjoint from $f(M)$, and let K be the complement of V in D. For each x in K, we can find a neighborhood $U \subset N$ of $g(x)$ such that U is diffeomorphic to $\mathbb{R}^n$. Now choose a neighborhood W of x so that $\bar{W}$ is contained in $g^{-1}(U)$ but $\bar{W}$ is disjoint from E. Since K is compact, there is some finite collection $x_1, \ldots, x_N$ of points in K such that the associated open sets $W_1, \ldots, W_N$ cover K. Thus, each $\bar{W}_j$ is disjoint from E and is contained in a set of the form $g^{-1}(U_j)$, with $U_j \subset N$ diffeomorphic to $\mathbb{R}^n$.

   By the argument in Step 1, we can find a map g_1 homotopic to g such that $g_1 = g$ on E and $g_1(\bar{W}_1)$ is disjoint from $f(M)$, but such that g is as close as we like to g. Since $g(\bar{V})$ and $f(M)$ are compact, the distance between $g(\bar{V})$ and $f(M)$ (with respect to an arbitrarily chosen Riemannian metric on N) achieves a positive minimum. Therefore, if we take g_1 close enough to g, then $g_1(\bar{V})$ will still be disjoint from $f(M)$. We can similarly ensure that g_1 still maps the compact set $\bar{W}_j$ into U_j for $j = 2, \ldots, N$.
We may now perform a similar perturbation of g_1 to a map g_2 such that $g_2 = g_1 = g$ on E but such that $g_2(\bar{W}_2)$ is disjoint from $f(M)$. By making this perturbation small enough, we can ensure that g_2 has the same properties as g_1: First, $g_2(\bar{V})$ is still disjoint from $f(M)$, second, $g_2(\bar{W}_1)$ is still disjoint from $f(M)$, and third, $g_2(\bar{W}_j)$ is still contained in U_j for $j = 3, \ldots, N$. Proceeding on in this fashion, we eventually obtain a map g_N homotopic to g such that $g_N = g$ on E and such that $g_N(\bar{V})$ and each $g_N(\bar{W}_j)$ are disjoint from $f(M)$. Then $g' := g_N$ is the desired map. $\square$

Proof of Theorem 13.27. In Lemma 13.28, it is harmless to assume that each M_j has dimension $\dim K - 3$, since if any M_j has a lower dimension, we can take a product of M_j with, say, a sphere dimension of an appropriate dimension, and then make f_j independent of the sphere variable. Once this is done, we can let M be the disjoint union of the M_j's and let f be the map equal to f_j on M_j. That is to say, the singular set is actually in the image under a smooth map f of single manifold M of dimension dim $K - 3$.

If l is any loop in K, it is not hard to prove that l is homotopic to smooth loop. (See, for example, Theorem 6.26 in [Lee].) In fact, this claim can be proved by an argument similar to the proof of Lemma 13.29; locally, any continuous map from $\mathbb{R}^n$ to $\mathbb{R}^m$ can be approximated by smooth functions (even polynomials), and one can then patch together these approximations, being careful at each stage not to disrupt the smoothness achieved at the previous stage. We may thus assume that l is smooth and apply Lemma 13.29 with $D = S^1, N = K$, and $E = \emptyset$. Since dim S^1 + dim M < dim K, the lemma says that we can deform l until it does not intersect $f(M) \supset K_{\text{sing}}$.

Next, suppose l is a loop in K_{reg}, which we can assume to be smooth. If l is null homotopic in K_{reg}, it is certainly null homotopic in K. In the other direction, suppose l is null homotopic in K. We think of the homotopy of l to a point as a map g of a 2-disk D into K, where the boundary of D corresponds to the original loop l and the center of D correspond to the point. After deforming g slightly—by the same argument as in the previous paragraph—we may assume that g is smooth. We now apply Lemma 13.29 with D equal to the 2-disk and E equal to the boundary of D. The lemma tells us that we can deform g to a map g' that agrees with g on the boundary of D but such that $g'(D)$ is disjoint from $f(M) \supset K_{\text{sing}}$. Thus, g' is a homotopy of l to a point in K_{reg}.   □

13.6   The Stiefel Diagram

In this section, we look more closely at the structure of the hyperplanes in t where $\langle \alpha, H \rangle = n$, which appear in Proposition 13.24. The main result of the section is Theorem 13.35, which constructs a cover of the regular set K_{reg}.

Definition 13.30. For each $n \in \mathbb{Z}$ and $\alpha \in R$, consider the hyperplane (not necessarily through the origin) given by

$$L_{\alpha,n} = \{H \in \mathfrak{t} \,|\, \langle \alpha, H \rangle = n\}.$$

The union of all such hyperplanes is called the **Stiefel diagram**. A connected component of the complement in t of the Stiefel diagram is called an **alcove**.

In light of Proposition 13.24 and Definition 13.25, the complement of the Stiefel diagram is just the regular set $\mathfrak{t}_{\text{reg}}$ in t. Figures 13.3, 13.4, and 13.5 show the Stiefel diagrams for A_2, B_2, and G_2, respectively. In each figure, the roots are indicated with arrows, with the long roots being normalized to have length $\sqrt{2}$. The black dots in the figure indicate the coroots and one alcove is shaded. Note that since

Fig. 13.3 The Stiefel diagram for A_2, with the roots normalized to have length $\sqrt{2}$. The *black dots* indicate the coroots

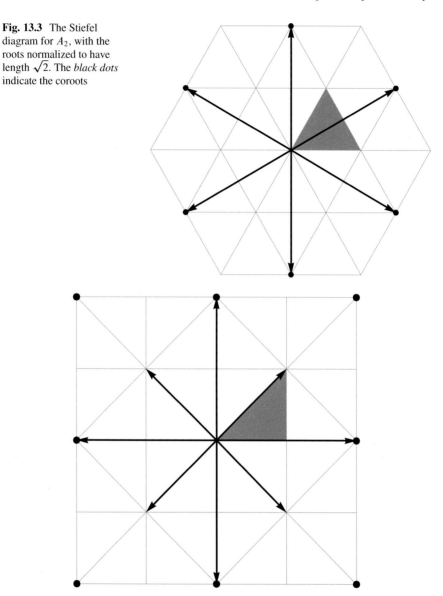

Fig. 13.4 The Stiefel diagram for B_2, with the long roots normalized to have length $\sqrt{2}$. The *black dots* indicate the coroots

$\langle \alpha, H_\alpha \rangle = 2$, if we start at the origin and travel in the α direction, the coroot H_α will be located on the *second* hyperplane orthogonal to α.

Let us check the correctness of, say, the diagram for G_2. Suppose α is a long root, which we normalize so that $\langle \alpha, \alpha \rangle = 2$. Then $\alpha/2$ belongs to the line $L_{\alpha,1}$, so that $L_{\alpha,1}$ is the (unique) line orthogonal to α passing through $\alpha/2$. Similarly, $L_{\alpha,n}$ is the

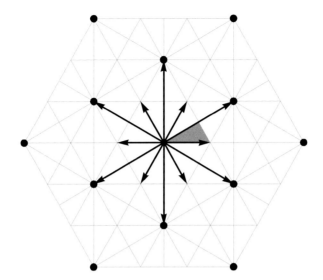

Fig. 13.5 The Stiefel diagram for G_2, with the long roots normalized to have length $\sqrt{2}$. The *black dots* indicate the coroots

line orthogonal to α passing through $n\alpha/2$. Suppose, on the other hand, that α is a short root, which we must then normalize so that $\langle \alpha, \alpha \rangle = 2/3$. Then $3\alpha/2$ belongs to $L_{\alpha,1}$, so that $L_{\alpha,1}$ is the line orthogonal to α passing through $3\alpha/2$ and $L_{\alpha,n}$ is the line orthogonal to α passing through $3n\alpha/2$. Meanwhile, for a long root α, we have $H_\alpha = 2\alpha/2 = \alpha$, whereas for a short root, we have $H_\alpha = 2\alpha/(2/3) = 3\alpha$. These results agree with what we see in Figure 13.5.

It will follow from Proposition 13.34 that every alcove is isometric to every other alcove. Furthermore, if $\mathfrak{g} := \mathfrak{k}_{\mathbb{C}}$ is simple (i.e., if the root system R spans $\mathfrak{t}$ and is irreducible), each alcove is a "simplex," that is, a bounded region in a k-dimensional space defined by $k + 1$ linear inequalities. Thus, for rank-two simple algebras, each alcove is a triangle, as can be seen from Figures 13.3, 13.4, and 13.5. In general, the structure of the alcoves for a simple algebra is described by an *extended Dynkin diagram*; see Exercises 11, 12, and 13.

Proposition 13.31. *The Stiefel diagram is invariant under the action of W and under translations by elements of Γ.*

Proof. Since W permutes the roots, invariance of the Stiefel diagram is evident. Meanwhile, if γ is in the kernel Γ of the exponential map, the adjoint action of $e^{2\pi\gamma}$ on $\mathfrak{g}$ is trivial. Thus, for $X \in \mathfrak{g}_\alpha$, we have

$$X = \mathrm{Ad}_{e^{2\pi\gamma}}(X) = e^{2\pi i \langle \alpha, \gamma \rangle} X,$$

which means that $\langle \alpha, \gamma \rangle$ is an integer, for all $\alpha \in R$. From this we can easily see that if H is in the Stiefel diagram, so $H + \gamma$. $\square$

Recall that an **affine transformation** of a vector space is a transformation that can be expressed as a combination of a linear transformation and a translation.

Proposition 13.32. *Let* $\Gamma \rtimes W$ *denote the set of affine transformations of* $\mathfrak{t}$ *that can be expressed in the form*

$$H \mapsto w \cdot H + \gamma$$

for some $\gamma \in \Gamma$ *and some* $w \in W$. *Then* $\Gamma \rtimes W$ *forms a group under composition, with the group law given by*

$$(\gamma, w) \cdot (\gamma', w') = (\gamma + w \cdot \gamma', ww').$$

The group $\Gamma \rtimes W$ is a *semidirect product* of Γ and W, with Γ being the normal factor.

Proof. We merely compute that

$$w \cdot (w' \cdot H + \gamma') + \gamma = (ww') \cdot H + \gamma + w \cdot \gamma',$$

so that the composition of the affine transformations associated to (γ, w) and (γ', w') is the affine transformation associated to $(\gamma + w \cdot \gamma', ww')$.   □

We now introduce the analog of the Weyl group for the Stiefel diagram. Even if a hyperplane $V \subset \mathfrak{t}$ does not pass through the origin, we can still speak of the reflection s about V, which is the unique affine transformation of $\mathfrak{t}$ such that

$$s(H + H') = H - H'$$

whenever H is in V and H' is orthogonal to V.

Definition 13.33. The **extended Weyl group** for K relative to $\mathfrak{t}$ is the group of affine transformations of $\mathfrak{t}$ generated by reflections about all the hyperplanes in the Stiefel diagram of $\mathfrak{t}$.

We now establish some key properties of the extended Weyl group. Recall from Definition 13.16 the notion of the coroot lattice I.

Proposition 13.34. *1. The extended Weyl group equals* $I \rtimes W$, *the semidirect product of the ordinary Weyl group* W *and the coroot lattice* I.
2. The extended Weyl group acts freely and transitively on the set of alcoves.

Note that since $I \subset \Gamma$ is invariant under the action of W, the extended Weyl group $I \rtimes W$ is a subgroup of the group $\Gamma \rtimes W$ in Proposition 13.32. That is to say, the group law in $I \rtimes W$ is the same as in Proposition 13.32, although $I \rtimes W$ will, in general, be a proper subgroup of $\Gamma \rtimes W$.

Proof. For Point 1, let $V_\alpha, \alpha \in R$, be the hyperplane through the origin orthogonal to α. Since $\langle \alpha, H_\alpha \rangle = 2$, we see that the hyperplane $L_{\alpha,n}$ in Definition 13.30 is the translate of V_α by $nH_\alpha/2$. We can then easily verify that the reflection $s_{\alpha,n}$ about $L_{\alpha,n}$ is given by

$$s_{\alpha,n}(H) = s_\alpha(H) + nH_\alpha, \tag{13.9}$$

where s_α is the reflection about V_α. Thus, $s_{\alpha,n}$ is a combination of an element s_α of W and a translation by an element of I. It follows that the extended Weyl group is contained in $I \rtimes W$. In the other direction, the extended Weyl group certainly contains W, since the Stiefel diagram contains the hyperplanes through the origin orthogonal to the roots. Furthermore, from (13.9), we see that the composition of $s_{\alpha,1}$ and $s_{\alpha,0}$ is translation by H_α. Thus, the extended Weyl group contains all of $I \rtimes W$.

For Point 2, we first argue that the orbits of $I \rtimes W$ in $\mathfrak{t}$ do not have accumulation points. For any $H \in \mathfrak{t}$, the orbit of H is the set of all vectors of the form $w \cdot H + \gamma$, with $w \in W$ and $\gamma \in I$. Since W is finite and I contains only finitely many points in each bounded region, the orbit of H also contains only finitely many points in each bounded region. Once this observation has been made, we may now repeat, almost word for word, the proof that the ordinary Weyl group acts freely and transitively on the set of open Weyl chambers. (Replace the hyperplanes orthogonal to the roots with the hyperplanes $L_{\alpha,n}$ and replace the open Weyl chambers with the alcoves.) The only point where a change is necessary is in the proof of the transitivity of the action (Proposition 8.23). To generalize that argument, we need to know that for each H and H' in $\mathfrak{t}$, the orbit $(I \rtimes W) \cdot H'$ contains a point at minimal distance from H. Although the extended Weyl group is infinite, since each orbit contains only finitely many points in each bounded region, the result still holds. The reader is invited to work through the proofs of Propositions 8.23 and 8.27 with the ordinary Weyl group replaced by the extended Weyl group, and verify that no other changes are needed.   □

Theorem 13.35. *The map* $\Psi : \mathfrak{t}_{\mathrm{reg}} \times (K/T) \to K_{\mathrm{reg}}$ *is a covering map. Furthermore, if* $A \subset \mathfrak{t}$ *is any one alcove, then* Ψ *maps* $A \times (K/T)$ *onto* K_{reg} *and*

$$\Psi : A \times (K/T) \to K_{\mathrm{reg}}$$

is also a covering map.

Recall the definition of Ψ in (13.5). We will see in the next section that K/T is simply connected. Since each alcove A is a convex set, A is contractible and therefore simply connected. We will thus conclude that $A \times (K/T)$ is the *universal cover* of K_{reg}.

Proof. Recall the map $\Phi : T \times (K/T) \to K$ given by $\Phi(t,[x]) = xtx^{-1}$, and let $T_{\mathrm{reg}} = K_{\mathrm{reg}} \cap T$. Since the exponential map for T is a local diffeomorphism, it follows from Proposition 13.24 that Φ is a local diffeomorphism on $T_{\mathrm{reg}} \times (K/T)$. Furthermore, by a trivial extension of Proposition 13.26, we have $\Phi(t,[x]) = \Phi(t',[x'])$ if and only if there is some $z \in N(T)$ such that $t' = ztz^{-1}$ and $x' = xz^{-1}$. Finally, if $W = N(T)/T$ acts on $T_{\mathrm{reg}} \times (K/T)$ by

$$w \cdot (t, [x]) = (ztz^{-1}, [xz^{-1}])$$

for each $w = [z]$ in W, then this action is free on K/T and thus free on $T_{\text{reg}} \times (K/T)$.

Fix some y in K_{reg} and pick some $(t, [x])$ in $T_{\text{reg}} \times (K/T)$ for which $\Phi(t, [x]) = y$. Then since W is finite and acts freely, we can easily find a neighborhood U of $(t, [x])$ such that $w \cdot U$ is disjoint from U for all $w \neq e$ in W. It follows that the sets $w \cdot U$ are pairwise disjoint and thus that Φ is injective on each such set. Now, since Φ is a local diffeomorphism, $\Phi(U)$ will be open in K_{reg}. The preimage of V is then the disjoint union of the sets $w \cdot U$ and the local diffeomorphism Φ will map each $w \cdot U$ homeomorphically onto V. We conclude that Φ is a covering map of $T_{\text{reg}} \times (K/T)$ onto K_{reg}.

Meanwhile, as shown in the proof of Proposition 13.14, the exponential map for T is a covering map. Thus, the map

$$(H, [x]) \mapsto (e^{2\pi H}, [x])$$

is a covering map from $\mathfrak{t}_{\text{reg}} \times (K/T)$ onto $T_{\text{reg}} \times (K/T)$. Since the composition of covering maps is a covering map, we conclude that $\Psi : \mathfrak{t}_{\text{reg}} \times (K/T)$ onto K_{reg} is a covering map, as claimed.

Finally, Proposition 13.34 shows that each point in $\mathfrak{t}_{\text{reg}}$ can be moved into A by the action of $I \rtimes W \subset \Gamma \rtimes W$. Thus, Ψ actually maps $A \times (K/T)$ onto K_{reg}. For each $y \in K_{\text{reg}}$, choose a neighborhood V of y so that $\Psi^{-1}(V)$ is a disjoint union of open sets U_α mapping homeomorphically onto V. By shrinking V if necessary, we can assume V is connected, in which case the U_α's will also be connected. Thus, each U_α is either entirely in $A \times (K/T)$ or disjoint from $A \times (K/T)$. Thus, if we restrict Ψ to $A \times (K/T)$, the preimage of V will now consist of some subset of the U_α's, each of which still maps homeomorphically onto V, showing that the restriction of Ψ to $A \times (K/T)$ is still a covering map.   □

13.7   Proofs of the Main Theorems

We now have all the necessary tools to attack the proofs of our main results. Here is an outline of our strategy in this section. We will first show that the quotient K/T is always simply connected. On the one hand, the simple connectivity of K/T leads to a proof of Theorem 13.15, that every loop in K is homotopic to a loop in T. On the other hand, the simple connectivity of K/T means that the set $A \times (K/T)$ in Theorem 13.35 is actually the *universal* cover of K_{reg}. Thus, to determine $\pi_1(K) \cong \pi_1(K_{\text{reg}})$, we merely need to determine how close to being injective the covering map $\Psi : A \times (K/T) \to K_{\text{reg}}$ is.

Now, according to Proposition 13.26, the failure of injectivity for the full map $\Psi : \mathfrak{t}_{\text{reg}} \times (K/T)$ is due to the action of the group $\Gamma \rtimes W$. If we restrict Ψ to

$A \times (K/T)$, then by Proposition 13.34, we have eliminated the failure of injectivity due to the subgroup $I \rtimes W \subset \Gamma \rtimes W$. Thus, the (possible) failure of injectivity of Ψ on $A \times (K/T)$ will be measured by the extent to which I fails to be all of Γ.

Proposition 13.36. *The quotient manifold K/T is simply connected.*

Proof. Let $[x(\tau)]$ be any loop in K/T. Let H be a regular element in T and consider the loop l in K_{reg} given by

$$l(\tau) = x(\tau)e^{2\pi H}x(\tau)^{-1}.$$

Now let $t(s)$ be any path in T connecting $e^{2\pi H}$ to I, and consider the loops

$$l_s(\tau) := x(\tau)t(s)x(\tau)^{-1}.$$

Clearly, l_s is a homotopy of l in K to the constant loop at I.

Thus, for any loop $[x(\tau)]$ in K/T, the corresponding loop $l(\tau)$ in K_{reg} is null homotopic in K. We now argue that $[x(\tau)]$ itself is null homotopic in K/T. As a first step, we use Theorem 13.27 to deform the homotopy l_s to a homotopy l'_s shrinking l to a point in K_{reg}. Now, the map Ψ is a covering map and the loop $(H, [x(\tau)])$ is a lift of l to $A \times (K/T)$. Thus, as a second step, we can lift the homotopy l'_s to $A \times (K/T)$ to a homotopy l''_s shrinking $(H, [x(\tau)])$ to a point in $A \times (K/T)$. (See Proposition 13.3.) Finally, as a third step, we can project the homotopy l''_s from $A \times (K/T)$ onto K/T to obtain a homotopy shrinking $[x(\tau)]$ to a point in K/T. □

Proposition 13.37. *Every loop in K is homotopic to a loop in T.*

Proof. By Proposition 13.8, the group K is a fiber bundle with base K/T and fiber T. Suppose now that l is a loop in K and that $l'(\tau) := [l(\tau)]$ is the corresponding loop in K/T. Since K/T is simply connected, there is a homotopy l'_s shrinking l' to a point in K/T. Furthermore, since K/T is connected, it is harmless to assume that l'_s shrinks to the point $[I]$ in K/T. Now, fiber bundles are known to have the *homotopy lifting property* (Proposition 4.48 in [Hat]), which in our case means that there is a homotopy l_s in K such that $l_0 = l$ and such that $[l_s(\tau)] = l'_s(\tau)$ for all τ and s. Since l'_1 is a constant loop at $[I]$, the loop l_1 lies in T.   □

Lemma 13.38. *Suppose $\gamma \in \Gamma$ but $\gamma \notin I$. Then there exist γ' in I and w in W such that the affine transformation*

$$H \mapsto w \cdot (H + \gamma + \gamma') \tag{13.10}$$

maps A to itself but is not the identity map of A.

Proof. By Proposition 13.31, translation by γ maps the alcove A to some other alcove A'. Then by Proposition 13.34, there exists an element of the extended Weyl group that maps A' back to A. Thus, there exist $\gamma' \in I$ and $w \in W$ such that the map (13.10) maps A to itself. If this map were the identity, then translation by γ would be (the inverse of) some element of $I \rtimes W$, which would mean that γ

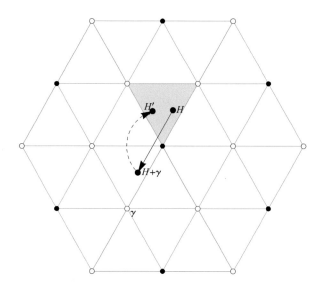

Fig. 13.6 The element γ is in the kernel of the exponential mapping for $\mathsf{PSU}(3)$, but is not in the coroot lattice. If we apply a rotation by $2\pi/3$ to $H + \gamma$, we obtain an element H' in the same alcove as H

would have to be in I, contrary to our assumption. Thus, the map in (13.10) sends A to itself and is not the identity map on t. But an affine transformation is certainly determined by its restriction to any nonempty open set, which means that the map cannot be the identity on A.   ☐

Suppose now that $\gamma \in \Gamma$ but $\gamma \notin I$. Let us choose $\gamma' \in I$ and $w \in W$ as in Lemma 13.38. We may then choose $H \in A$ so that $H' := w \cdot (H + \gamma + \gamma')$ lies in A but is distinct from H. (See Figure 13.6 in the case of the group $\mathsf{PSU}(3)$.) Let p denote the path in A given by

$$p(\tau) = H + \tau(H' - H), \quad 0 \le \tau \le 1.$$

Now choose $x \in N(T)$ representing w, let $x(\cdot)$ be a path connecting I to x in K, and define a path q in $A \times (K/T)$ by

$$q(\tau) = (p(\tau), [x(\tau)^{-1}]). \tag{13.11}$$

Note that $p(1) \ne p(0)$ and thus, certainly, $q(1) \ne q(0)$.

Lemma 13.39. *Let $q(\tau)$ be the path in $A \times (K/T)$ given by (13.11). Then the path*

$$\tau \mapsto \Psi(q(\tau))$$

is a loop in K_{reg} and this loop is homotopic in K to the loop

$$\tau \mapsto e^{2\pi\tau\gamma}.$$

Proof. On the one hand, we have

$$\Psi(q(0)) = e^{2\pi H}.$$

On the other hand, since x^{-1} represents w^{-1}, we have

$$\Psi(q(1)) = x^{-1}e^{2\pi H'}x$$
$$= e^{2\pi(H+\gamma+\gamma')}$$
$$= e^{2\pi H},$$

since $\gamma \in \Gamma$ and $\gamma' \in I \subset \Gamma$. Thus, $\Psi \circ q$ is a loop in K_{reg}, as claimed.
   Meanwhile,

$$\Psi(q(\tau)) = \exp\left\{2\pi \, x(\tau)^{-1}p(\tau)x(\tau)\right\},$$

where $x(0)^{-1}p(0)x(0) = H$ and $x(1)^{-1}p(1)x(1) = H + \gamma + \gamma'$. Since the vector space $\mathfrak{k}$ is simply connected, the path

$$\tau \mapsto x(\tau)^{-1}p(\tau)x(\tau)$$

in $\mathfrak{k}$ is homotopic with endpoints fixed to the straight-line path connecting H to $H + \gamma + \gamma'$, namely

$$\tau \mapsto H + \tau(\gamma + \gamma').$$

Since the exponential map for K is continuous, we see that $\Psi \circ q$ is homotopic in K to the loop

$$\tau \mapsto e^{2\pi H}e^{2\pi\tau(\gamma+\gamma')}.$$

We may then continuously deform $e^{2\pi H}$ to the identity, showing that $\Psi \circ q$ is homotopic in K to the loop $\tau \mapsto e^{2\pi\tau(\gamma+\gamma')}$.
   Finally, in light of Proposition 13.14, the loop $\tau \mapsto e^{2\pi\tau(\gamma+\gamma')}$ is homotopic to the composition (in either order) of the loop $\tau \mapsto e^{2\pi\tau\gamma}$ and the loop $\tau \mapsto e^{2\pi\tau\gamma'}$. But since γ' belongs to I, we have already shown in Sect. 13.4 that this second loop is null homotopic in K, showing that $\Psi \circ q$ is homotopic in K to $\tau \mapsto e^{2\pi\tau\gamma}$, as claimed.   □

   It now remains only to assemble the previous results to finish the proof of Theorem 13.17.

Proof of Theorem 13.17, the other direction. We showed in Sect. 13.4 that if $\gamma \in I$ the loop $\tau \mapsto e^{2\pi \tau \gamma}$ is null homotopic in K. In the other direction, suppose that γ belongs to Γ but not to I. We may then construct the path q in $A \times (K/T)$ in (13.11). Although this path is not a loop in $A \times (K/T)$, the path $\Psi \circ q$ is a loop in K_{reg}, by Lemma 13.39. Since q is a lift of $\Psi \circ q$ and q has distinct endpoints, Corollary 13.4 tells us that $\Psi \circ q$ is not null homotopic in K_{reg}. Meanwhile, since $\Psi \circ q$ homotopic in K to $\tau \mapsto e^{2\pi \tau \gamma}$ and $\pi_1(K_{\text{reg}})$ is isomorphic to $\pi_1(K)$ (Theorem 13.27), we conclude that $\tau \mapsto e^{2\pi \tau \gamma}$ is not null homotopic in K.   □

Summary. It is instructive to summarize the arguments in the proofs of Theorems 13.15 and 13.17, as described in this section and the previous three sections. We introduced the regular set and the singular set in K and we showed that the singular set has codimension at least 3. Using this, we proved a key result, that $\pi_1(K_{\text{reg}})$ is isomorphic to $\pi_1(K)$. Next, we introduced the Stiefel diagram and the local diffeomorphism

$$\Psi : \mathfrak{t}_{\text{reg}} \times (K/T) \to K_{\text{reg}}.$$

We found that the failure of injectivity of Ψ on $\mathfrak{t}_{\text{reg}} \times (K/T)$ is measured by the action of the group $\Gamma \rtimes W$, and that the subgroup $I \rtimes W$ of $\Gamma \rtimes W$ acts transitively on the alcoves. We concluded that if A is any alcove, the map $\Psi : A \times (K/T) \to K_{\text{reg}}$ is a covering map.

We then demonstrated that K/T is simply connected. We did this by mapping any loop $[x(\cdot)]$ in K/T to a loop l in K and constructing a homotopy l_s of l to a point in K. Since $\pi_1(K_{\text{reg}}) = \pi_1(K)$, we can deform the homotopy l_s into K_{reg}. We can then use the covering map Ψ to lift the homotopy l_s to $A \times (K/T)$ and project back onto K/T to obtain a homotopy of $[x(\cdot)]$ to a point in K/T. After establishing that K/T is simply connected, we proved our first main result, that every loop in K is homotopic to a loop in T, as follows. We start with a loop l in K, push it down to K/T and then homotope it to the identity coset. We then lift this homotopy to K, giving a homotopy of l into T.

Since $\pi_1(T)$ is easily calculated, we conclude that every loop in K is homotopic to a loop in T of the form $\tau \mapsto e^{2\pi \tau \gamma}$, for some $\gamma \in \Gamma$. When γ is a coroot, we showed that $\tau \mapsto e^{2\pi \tau \gamma}$ is the image under a continuous homomorphism of a loop in the simply connected group $\mathsf{SU}(2)$, which shows that $\tau \mapsto e^{2\pi \tau \gamma}$ is null homotopic. When $\gamma \notin I$, we started with some H in A, translated by γ, and then mapped back to some $H' \in A$ by the action of $I \rtimes W$, where H is chosen so that $H' \neq H$. We then constructed a path q in $A \times (K/T)$ with distinct endpoints such that $\Psi \circ q$ is a loop in K_{reg} and is homotopic in K to $\tau \mapsto e^{2\pi \tau \gamma}$. Since $\Psi : A \times (K/T) \to K_{\text{reg}}$ is a covering map and q has distinct endpoints, $\Psi \circ q$ is homotopically nontrivial in K_{reg}. Then since $\pi_1(K_{\text{reg}}) = \pi_1(K)$, we concluded that $\tau \mapsto e^{2\pi \tau \gamma}$ is homotopically nontrivial in K.

13.8   The Center of K

In this section, we analyze the center of K using the tools developed in the previous sections. If T is any one fixed maximal torus in K, Corollary 11.11 tells us that T contains the center $Z(K)$ of K. We now give a criterion for an element $t = e^{2\pi H}$ of T to be in $Z(K)$.

Proposition 13.40. *If $H \in \mathfrak{t}$, then $e^{2\pi H}$ belongs to $Z(K)$ if and only if*

$$\langle \alpha, H \rangle \in \mathbb{Z}$$

for all $\alpha \in R$.

Proof. By Exercise 17 in Chapter 3, an element x of K is in $Z(K)$ if and only if $\mathrm{Ad}_x(X) = X$ for all $X \in \mathfrak{k}$, or, equivalently, if and only if $\mathrm{Ad}_x(X) = X$ for all $X \in \mathfrak{g} = \mathfrak{k}_{\mathbb{C}}$. Now, if $X \in \mathfrak{g}_\alpha$, then

$$\mathrm{Ad}_{e^{2\pi H}}(X) = e^{2\pi \, \mathrm{ad}_H}(X) = e^{2\pi \langle \alpha, H \rangle} X.$$

Thus, $\mathrm{Ad}_{e^{2\pi H}}$ acts as the identity on $\mathfrak{g}_\alpha$ if and only if $\langle \alpha, H \rangle$ is an integer. Since $\mathfrak{g}$ is the direct sum of $\mathfrak{t}_{\mathbb{C}}$ (on which $\mathrm{Ad}_{e^{2\pi H}}$ certainly acts trivially) and the $\mathfrak{g}_\alpha$'s, we see that $e^{2\pi H}$ belongs to $Z(K)$ if and only if $\langle \alpha, H \rangle \in \mathbb{Z}$ for all α. $\qquad\square$

Definition 13.41. Let $\Lambda \subset \mathfrak{t}$ denote the **root lattice**, that is, the set of all integer linear combinations of roots. Let Λ^* denote the **dual of the root lattice**, that is,

$$\Lambda^* = \{ \gamma \in H \mid \langle \lambda, \gamma \rangle \in \mathbb{Z}, \ \forall \lambda \in \Lambda \}.$$

Note that if $\gamma \in \mathfrak{t}$ has the property that $\langle \alpha, \gamma \rangle \in \mathbb{Z}$ for every root α, then certainly $\langle \lambda, \gamma \rangle \in \mathbb{Z}$ whenever λ is an integer combination of roots. Thus, Proposition 13.40 may be restated as saying that $e^{2\pi H} \in Z(K)$ if and only if $H \in \Lambda^*$. Note also that if $e^{2\pi H} = I$, then certainly $e^{2\pi H}$ is in the center of K. Thus, the kernel Γ of the exponential map must be contained in Λ^*.

Proposition 13.42. *The map*

$$\gamma \mapsto e^{2\pi \gamma}, \quad \gamma \in \Lambda^*,$$

is a homomorphism of Λ^* onto $Z(K)$ with kernel equal to Γ. Thus,

$$Z(K) \cong \Lambda^* / \Gamma,$$

where Λ^* is the dual of the root lattice and Γ is the kernel of the exponential.

Proof. As we have noted, Corollary 11.11 implies that $Z(K) \subset T$. Since the exponential map for T is surjective, Proposition 13.40 tells us that the map $\gamma \mapsto e^{2\pi\gamma}$ maps Λ^* onto $Z(K)$. This map is a homomorphism since t is commutative, and the kernel of the map is $\Gamma \subset \Lambda^*$. $\square$

Suppose, for example, that $K = T$. Then there are no roots, in which case the dual of root lattice is all of t. In this case, we have

$$Z(K) = Z(T) \cong t/\Gamma \cong T.$$

On the other hand, if $\mathfrak{g}$ is semisimple, both Λ^* and Γ will be discrete subgroups of t that span t, in which case, Λ^*/Γ will be finite.

Note that we have several different lattices inside t. Some of these "really" live in t^* and only become subsets of t when we use the inner product to identify t^* with t. Other lattices naturally live in t itself. The lattices that really live in t^* are the root lattice, the lattice of analytically integral elements, and the lattice of algebraically integral elements. Meanwhile, the lattices that naturally live in t are the coroot lattice, the kernel of the exponential map, and the dual of the root lattice. Note that there is a duality relationship between the lattices in t and the lattices in t^*. An element is algebraically integral if and only if its inner product with each coroot is an integer; thus, the lattice of algebraically integral elements and the coroot lattice are dual to each other. Similarly, the lattice of analytically integral elements is dual to the kernel of the exponential map. Finally, Λ^* is, by definition, dual to the root lattice Λ.

The lattices in t^* are included in one another as follows:

(root lattice) $\subset$ (analytically integral elements)

$\subset$ (algebraically integral elements).   (13.12)

The dual lattices in t are then included in one another *in the reverse order*:

(coroot lattice) $\subset$ (kernel of exponential)

$\subset$ (dual of root lattice).   (13.13)

In light of Proposition 13.42 and Corollary 13.18, we have the following isomorphisms involving quotients of lattices in (13.13):

(kernel of exponential)/(coroot lattice) $\cong \pi_1(K)$

and

(dual of root lattice)/(kernel of exponential) $\cong Z(K)$.

Corollary 13.43. *Let Λ^* denote the dual of the root lattice and let I denote the coroot lattice. If K is simply connected, then*

$$Z(K) \cong \Lambda^*/I.$$

On the other hand, if $Z(K)$ is trivial, then

$$\pi_1(K) \cong \Lambda^*/I.$$

Let us define the **adjoint group** associated to K to be the image of K under the adjoint representation $\mathrm{Ad} : K \rightarrow \mathsf{GL}(\mathfrak{k})$. (Since K is compact, the adjoint group of K is compact and thus closed.) If K is semisimple, then the center of $\mathfrak{k}$ is trivial, which means that the Lie algebra version of the adjoint representation, $\mathrm{ad} : \mathfrak{k} \rightarrow \mathfrak{gl}(\mathfrak{k})$ is faithful. Thus, if $\mathfrak{g}$ is semisimple, the Lie algebra of the adjoint group is isomorphic to the Lie algebra $\mathfrak{k}$ of K itself. On the other hand, it is not hard to check (still assuming that $\mathfrak{g}$ is semisimple and thus that the Lie algebra of the adjoint group is isomorphic to $\mathfrak{k}$) that the center of the adjoint group is trivial. Thus, whenever $\mathfrak{g}$ is semisimple, we can construct a new group K' where the first part of the corollary applies:

$$\pi_1(K') \cong \Lambda^*/I.$$

Proof. If $\pi_1(K)$ is trivial, the kernel Γ of the exponential map must equal the coroot lattice I, which means that

$$Z(K) \cong \Lambda^*/\Gamma = \Lambda^*/I.$$

Meanwhile, if $Z(K)$ is trivial, then the kernel Γ of the exponential must equal the dual Λ^* of the root lattice, which means that

$$\pi_1(K) \cong \Gamma/I = \Lambda^*/I,$$

as claimed.   □

Example 13.44. If $K = \mathsf{SO}(4)$, then the lattices Λ^*, Γ, and I are as in Figure 13.7 and both $\pi_1(K)$ and $Z(K)$ are isomorphic to $\mathbb{Z}/2$. Explicitly, $Z(\mathsf{SO}(4)) = \{I, -I\}$.

Proof. If we compute as in Sect. 7.7.2, but adjusting for a factor of i to obtain the real roots and coroots, we find that the coroots are the matrices

$$\begin{pmatrix} 0 & a & & \\ -a & 0 & & \\ & & 0 & b \\ & & -b & 0 \end{pmatrix}, \tag{13.14}$$

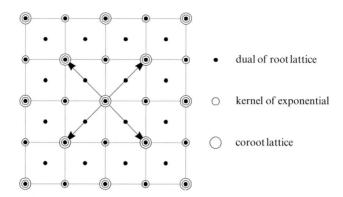

•   dual of root lattice

○   kernel of exponential

◯   coroot lattice

Fig. 13.7 The lattices for SO(4), with the coroots indicated by arrows

where $a = \pm 1$ and $b = \pm 1$. We identify the coroots with the vectors $(a, b) = (\pm 1, \pm 1)$ in $\mathbb{R}^2$. The coroot lattice I will then consist of pairs $(m, n) \in \mathbb{Z}^2$ for which $m + n$ is even. The kernel Γ of the exponential, meanwhile, is easily seen to consist of all pairs $(m, n) \in \mathbb{Z}^2$. Finally, since the coroots have been normalized to have length $\sqrt{2}$, the roots and coroots coincide. Thus, the dual Λ^* of the root lattice is the set of vectors (x, y) having integer inner product with $(1, 1)$ and $(1, -1)$, that is, such that $x + y$ and $x - y$ are integers. Thus, either x and y are both integers, or x and y are both integer-plus-one-half. It is then easy to check that both Γ/I and Λ^*/Γ are isomorphic to $\mathbb{Z}/2$.

If H is any coroot, then $H/2$ belongs to Λ^* but not to Γ. Thus, the unique nontrivial element of $Z(\mathsf{SO}(4))$ may be computed as $e^{2\pi(H/2)}$. Direct calculation with (13.14) then shows that this unique nontrivial element is $-I \in \mathsf{SO}(4)$.   □

Example 13.45. Suppose that K is a connected, compact matrix Lie group with Lie algebra $\mathfrak{k}$, that $\mathfrak{t}$ is a maximal commutative subalgebra of $\mathfrak{k}$, and that the root system of $\mathfrak{k}$ relative to $\mathfrak{t}$ is isomorphic to G_2. Then both $\pi_1(K)$ and $Z(K)$ are trivial.

It turns out that for compact groups with root systems of type A_n, B_n, C_n and D_n, none of them is simultaneously simply connected and center free. Among the exceptional groups, however, the groups with root systems $G_2, F_4,$ and E_8 all have this property.

Proof. As we can see from Figure 8.11, each of the fundamental weights for G_2 is a root. Thus, every algebraically integral element for G_2 (i.e., every integer linear combination of the fundamental weights) is in the root lattice. Thus, all three of the lattices in (13.12) must be equal. By dualizing this observation, we see that all three of the lattices in (13.13) must also be equal. Thus, both $\pi_1(K) \cong \Gamma/I$ and $Z(K) \cong \Lambda^*/\Gamma$ are trivial.   □

13.9   Exercises

In Exercises 1–4, we consider the maximal commutative subalgebra $\mathfrak{t}$ of the relevant Lie algebra $\mathfrak{k}$ given in Sects. 7.7.1–7.7.4. In each case, we identify the Cartan subalgebra $\mathfrak{h} = \mathfrak{t}_{\mathbb{C}}$ with $\mathbb{C}^n$ by the map given in those sections, *except* that we adjust the map by a factor of i, so that $\mathfrak{t}$ maps to $\mathbb{R}^n$. We also consider the *real* roots and coroots, which differ by a factor of i from the roots and coroots in Chapter 7.

1. For the group $SU(n), n \geq 1$, show that coroot lattice I consists of all integer n-tuples $(k_1, \ldots, k_n)$ for which $k_1 + \cdots + k_n = 0$. Show that the kernel Γ of the exponential is the same as the coroot lattice.
2. For the group $SO(2n), n \geq 2$, show that the coroot lattice I is the set of integer linear combinations of vectors of the form $\pm e_j \pm e_k$, with $j \neq k$. Conclude that the coroot lattice consists of all integer n-tuples $(k_1, k_2, \ldots, k_n)$ for which $k_1 + \cdots + k_n$ is even. Show that the kernel Γ of the exponential consists of all integer n-tuples and that $\Gamma / I \cong \mathbb{Z}/2$.
3. For the group $SO(2n + 1), n \geq 1$, show that the coroot lattice I is the set of integer linear combinations of vectors of the form $\pm 2e_j$ and $\pm e_j \pm e_k$, with $j \neq k$. Conclude that, as for $SO(2n)$, the coroot lattice consists of all integer n-tuples $(k_1, k_2, \ldots, k_n)$ for which $k_1 + \cdots + k_n$ is even and the kernel Γ of the exponential consists of all integer n-tuples.
4. For the group $Sp(n), n \geq 1$, show that the coroot lattice I is the set of integer linear combinations of vectors of the form $\pm e_j$ and $\pm e_j \pm e_k$, with $j \neq k$. Conclude that both the coroot lattice and the kernel Γ of the exponential consist of all integer n-tuples.
5. Verify the claims in Example 13.19.
   Hints: In the case of $SO(5)$, make use of the calculations in the proof of Example 12.13. In the case of $PSU(3)$, note that $X \in \mathfrak{psu}(3) = \mathfrak{su}(3)$ exponentiates to the identity in $PSU(3)$ if and only if X exponentiates in $SU(3)$ to a constant multiple of the identity.
6. Using Theorems 13.15 and 13.17 and the results of Exercises 2 and 3, show that every homotopically nontrivial loop in $SO(n), n \geq 3$, is homotopic to the loop

$$\theta \mapsto \begin{pmatrix} \cos \theta & -\sin \theta & & & \\ \sin \theta & \cos \theta & & & \\ & & 1 & & \\ & & & \ddots & \\ & & & & 1 \end{pmatrix}, \quad 0 \leq \theta \leq 2\pi.$$

(This result also follows from the inductive argument using fiber bundles, as discussed following the proof of Proposition 13.10.)
7. Let G be a connected matrix Lie group. Using the following outline, show that $\pi_1(G)$ is commutative. Let $A(\cdot)$ and $B(\cdot)$ be any two loops in G based at

the identity. Construct two families of loops $\alpha_s(\cdot)$ (defined in terms of $A(\cdot)$) and $\beta_s(\cdot)$ (defined in terms of $B(\cdot)$) with the property that $\alpha_0(t)\beta_0(t)$ is the concatenation of A with B and $\alpha_1(t)\beta_1(t)$ is the concatenation of B with A. (Here the product of, say, $\alpha_0(t)\beta_0(t)$ is computed in the group G.)

8. Suppose G_1 and G_2 are connected matrix Lie groups with Lie algebras $\mathfrak{g}_1$ and $\mathfrak{g}_2$, respectively, and that $\Phi : G_1 \to G_2$ is a Lie group homomorphism. Show that if the associated Lie algebra homomorphism $\phi : \mathfrak{g}_1 \to \mathfrak{g}_2$ is an isomorphism, then Φ is a covering map (Definition 13.1).

9. Show that Proposition 13.26 fails if we do not assume that H' and H are in $\mathfrak{t}_{\text{reg}}$.

10. Let E be a real Euclidean space. Suppose $V \subset E$ is the hyperplane through the origin orthogonal to a nonzero vector α. Now suppose L is the hyperplane (not necessarily through the origin) obtained by translating V by $c\alpha$. Let $s : V \to V$ be the affine transformation given by

$$s(v) = v - 2\frac{\langle \alpha, v \rangle}{\langle \alpha, \alpha \rangle}\alpha + 2c\alpha.$$

Show that if $v \in L$ and $d \in \mathbb{R}$, then

$$s(v + d\alpha) = v - d\alpha.$$

That is to say, s is the reflection about L.

11. Let R^+ be the set of positive roots associated to a particular base Δ, and let $\alpha_1, \ldots, \alpha_N$ be the positive roots.

    (a) Show that for each alcove A, there are integers $n_1, \ldots, n_N$ such that

    $$A = \left\{ H \in \mathfrak{t} \,\middle|\, n_j < \langle \alpha_j, H \rangle < n_j + 1, \ j = 1, \ldots, N \right\}. \qquad (13.15)$$

    (b) If $n_1, \ldots, n_N$ is any sequence of integers, show that *if* the set A in (13.15) is nonempty, then this set is an alcove.

12. In this exercise, we assume $\mathfrak{g} := \mathfrak{k}_\mathbb{C}$ is simple. Let R^+ be the positive roots associated to a particular base Δ, and let C be the fundamental Weyl chamber associated to R^+. Let A be the alcove containing the "bottom" of C, that is, such that all very small elements of C are in A. Let β be the highest root, that is, the highest weight for the adjoint representation of $\mathfrak{g}$, which is irreducible by assumption.

    (a) Show that A may be described as

    $$A = \left\{ H \in \mathfrak{t} \,\middle|\, 0 < \langle \alpha, H \rangle < 1, \ \forall \alpha \in R^+ \right\}.$$

    (Compare Exercise 11.)

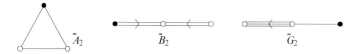

Fig. 13.8 The extended Dynkin diagrams associated to A_2, B_2, and G_2

(b) Show that A may also be described as

$$A = \{H \in \mathfrak{t} \mid \langle \beta, H \rangle < 1, \ \langle \alpha, H \rangle > 0, \ \forall \alpha \in \Delta\}.$$

13. If $\mathfrak{g} = \mathfrak{k}_\mathbb{C}$ is simple, the alcove A in Exercise 12 is a simplex, that is, a bounded region in $\mathfrak{t} \cong \mathbb{R}^k$ defined by $k + 1$ linear inequalities. By the same argument as in the proof of Proposition 8.24, the extended Weyl group is generated by the $k + 1$ reflections about the walls of A. The structure of A (and thus of the extended Weyl group) can be captured in the **extended Dynkin diagram**, defined as follows. The diagram has $k + 1$ vertices, representing the elements $\alpha_1, \ldots, \alpha_k$ and $-\beta$. We then define edges and arrows by the same rules as for ordinary Dynkin diagrams (Sect. 8.6). (Note that since β is the highest weight of the adjoint representation, β is dominant and, thus, $-\beta$ is at an obtuse angle with each element of Δ.)

Verify the extended Dynkin diagrams for A_2, B_2, and G_2 in Figure 13.8, where in each diagram, the black dot indicates the "extra" vertex $-\beta$ added to the ordinary Dynkin diagram. (Refer to Figures 13.3, 13.4, and 13.5.)

14. Let K be a connected compact matrix Lie group with Lie algebra $\mathfrak{k}$ and let $\mathfrak{t}$ be a maximal commutative subalgebra of $\mathfrak{k}$. Let $A \subset \mathfrak{t}$ be an alcove and suppose A has no nontrivial symmetries. (That is to say, suppose there is no nontrivial element of the Euclidean group of $\mathfrak{t}$ mapping A onto A.)

(a) Show that K is simply connected.
(b) Show that if K' is any connected matrix Lie group whose Lie algebra is isomorphic to $\mathfrak{k}$, then K' is isomorphic to K.

Note: It is known that there exists a connected compact group K of rank 2 whose root system is G_2. Since each alcove for G_2 is a triangle with 3 distinct edge lengths, Part (b) of the problem shows that any such group K must be simply connected. (Compare Example 13.45.)

15. Consider the quotient space T/W, that is, the set of orbits of W acting on T. Let $A \subset \mathfrak{t}$ be any one alcove and let $\bar{A}$ be the closure of A in $\mathfrak{t}$. Show that if K is simply connected, then the map $H \mapsto [e^{2\pi H}]$, where $[t]$ denotes the W-orbit of t, is a bijection between $\bar{A}$ and T/W.

Hint: Imitate the proof of Proposition 8.29 with the Weyl group W replaced by the extended Weyl group $I \rtimes W$ and the chamber C replaced by the alcove A.

Erratum:

Lie Groups, Lie Algebras, and Representations

Brian C. Hall

© Springer International Publishing Switzerland 2015
B.C. Hall, *Lie Groups, Lie Algebras, and Representations*, Graduate Texts in Mathematics 222,
DOI 10.1007/978-3-319-13467-3

DOI 10.1007/978-3-319-13467-3_14

The original version of this book was inadvertently published without the middle initial of the author's name as "Brian Hall". The correct name of the author should appear as "Brian C. Hall".

The online version of the original book can be found at
http://dx.doi.org/10.1007/978-3-319-13467-3

© Springer International Publishing Switzerland 2015
B.C. Hall, *Lie Groups, Lie Algebras, and Representations*, Graduate
Texts in Mathematics 222, DOI 10.1007/978-3-319-13467-3_14

Appendix A
Linear Algebra Review

In this appendix, we review results from linear algebra that are used in the text. The results quoted here are mostly standard, and the proofs are mostly omitted. For more information, the reader is encouraged to consult such standard linear algebra textbooks as [HK] or [Axl]. Throughout this appendix, we let $M_n(\mathbb{C})$ denote the space of $n \times n$ matrices with entries in $\mathbb{C}$.

A.1   Eigenvectors and Eigenvalues

For any $A \in M_n(\mathbb{C})$, a nonzero vector v in $\mathbb{C}^n$ is called an **eigenvector** for A if there is some complex number λ such that

$$Av = \lambda v.$$

An **eigenvalue** for A is a complex number λ for which there exists a nonzero $v \in \mathbb{C}^n$ with $Av = \lambda v$. Thus, λ is an eigenvalue for A if the equation $Av = \lambda v$ or, equivalently, the equation

$$(A - \lambda I)v = 0,$$

has a nonzero solution v. This happens precisely when $A - \lambda I$ fails to be invertible, which is precisely when $\det(A - \lambda I) = 0$. For any $A \in M_n(\mathbb{C})$, the **characteristic polynomial** p of A is given by

A previous version of this book was inadvertently published without the middle initial of the author's name as "Brian Hall". For this reason an erratum has been published, correcting the mistake in the previous version and showing the correct name as Brian C. Hall (see DOI http://dx.doi.org/10.1007/978-3-319-13467-3_14). The version readers currently see is the corrected version. The Publisher would like to apologize for the earlier mistake.

© Springer International Publishing Switzerland 2015
B.C. Hall, *Lie Groups, Lie Algebras, and Representations*, Graduate
Texts in Mathematics 222, DOI 10.1007/978-3-319-13467-3

$$p(\lambda) = \det(A - \lambda I), \quad \lambda \in \mathbb{C}.$$

This polynomial has degree n. In light of the preceding discussion, the eigenvalues are precisely the zeros of the characteristic polynomial.

We can define, more generally, the notion of eigenvector and eigenvalue for any linear operator on a vector space. If V is a finite-dimensional vector space over $\mathbb{C}$ (or over any algebraically closed field), every linear operator on V will have a least one eigenvalue. If A is a linear operator on a vector space V and λ is an eigenvalue for A, the λ-**eigenspace** for A, denoted V_λ, is the set of all vectors $v \in V$ (including the zero vector) that satisfy $Av = \lambda v$. The λ-eigenspace for A is a subspace of V. The dimension of this space is called the **multiplicity** of λ. (More precisely, this is the "geometric multiplicity" of λ. In the finite-dimensional case, there is also a notion of the "algebraic multiplicity" of λ, which is the number of times that λ occurs as a root of the characteristic polynomial. The geometric multiplicity of λ cannot exceed the algebraic multiplicity).

Proposition A.1. *Suppose that A is a linear operator on a vector space V and $v_1, \ldots, v_k$ are eigenvectors with* distinct *eigenvalues $\lambda_1, \ldots, \lambda_k$. Then $v_1, \ldots, v_k$ are linearly independent.*

Note that here V does not have to be finite dimensional.

Proposition A.2. *Suppose that A and B are linear operators on a finite-dimensional vector space V and suppose that $AB = BA$. Then for each eigenvalue λ of A, the operator B maps the λ-eigenspace of A into itself.*

Proof. Let λ be an eigenvalue of A and let V_λ be the λ-eigenspace of A. Then let v be an element of V_λ and consider Bv. Since B commutes with A, we have

$$A(Bv) = BAv = \lambda Bv,$$

showing that Bv is in V_λ.   □

A.2   Diagonalization

Two matrices $A, B \in M_n(\mathbb{C})$ are said to be **similar** if there exists an invertible matrix C such that

$$A = CBC^{-1}.$$

The operation $B \to CBC^{-1}$ is called **conjugation** of B by C. A matrix is said to be **diagonalizable** if it is similar to a diagonal matrix. A matrix $A \in M_n(\mathbb{C})$ is diagonalizable if and only if there exist n linearly independent eigenvectors for A. Specifically, if $v_1, \ldots, v_n$ are linearly independent eigenvectors, let C be the matrix whose kth column is v_k. Then C is invertible and we will have

$$A = C \begin{pmatrix} \lambda_1 & & \\ & \ddots & \\ & & \lambda_n \end{pmatrix} C^{-1}, \tag{A.1}$$

where $\lambda_1, \ldots, \lambda_n$ are the eigenvalues associated to the eigenvectors $v_1, \ldots, v_n$, in that order. To verify (A.1), note that C maps the standard basis element e_j to v_j. Thus, C^{-1} maps v_j to e_j, the diagonal matrix on the right-hand side of (A.1) then maps e_j to $\lambda_j e_j$, and C maps $\lambda_j e_j$ to $\lambda_j v_j$. Thus, both sides of (A.1) map v_j to $\lambda_j v_j$, for all j.

If $A \in M_n(\mathbb{C})$ has n distinct eigenvalues (i.e., n distinct roots to the characteristic polynomial), A is necessarily diagonalizable, by Proposition A.1. If the characteristic polynomial of A has repeated roots, A may or may not be diagonalizable.

For $A \in M_n(\mathbb{C})$, the **adjoint** of A, denoted A^*, is the conjugate-transpose of A,

$$(A^*)_{jk} = \overline{A_{kj}}. \tag{A.2}$$

A matrix A is said to be **self-adjoint** (or **Hermitian**) if $A^* = A$. A matrix A is said to be **skew self-adjoint** (or **skew Hermitian**) if $A^* = -A$. A matrix is said to be **unitary** if $A^* = A^{-1}$. More generally, A is said to be **normal** if A commutes with A^*. If A is normal, A is necessarily diagonalizable, and, indeed, it is possible to find an orthonormal basis of eigenvectors for A. In such cases, the matrix C in (A.1) may be taken to be unitary.

If A is self-adjoint, all of its eigenvalues are real. If A is real and self-adjoint (or, equivalently, real and symmetric), the eigenvectors may be taken to be real as well, which means that in this case, the matrix C may be taken to be orthogonal. If A is skew, then its eigenvalues are pure imaginary. If A is unitary, then its eigenvalues are complex numbers of absolute value 1.

We summarize the results of the previous paragraphs in the following.

Theorem A.3. *Suppose that $A \in M_n(\mathbb{C})$ has the property that $A^*A = AA^*$, (e.g., if $A^* = A$, $A^* = A^{-1}$, or $A^* = -A$). Then A is diagonalizable and it is possible to find an orthonormal basis for $\mathbb{C}^n$ consisting of eigenvectors for A. If $A^* = A$, all the eigenvalues of A are real; if $A^* = -A$, all the eigenvalues of A are pure imaginary; and if $A^* = A^{-1}$, all the eigenvalues of A have absolute value 1.*

A.3   Generalized Eigenvectors and the SN Decomposition

Not all matrices are diagonalizable, even over $\mathbb{C}$. If, for example,

$$A = \begin{pmatrix} 1 & 1 \\ 0 & 1 \end{pmatrix}, \tag{A.3}$$

then the only eigenvalue of A is 1, and every eigenvector with eigenvalue 1 is of the form $(c, 0)$. Thus, we cannot find two linearly independent eigenvectors for A. It is not hard, however, to prove the following result. Recall that a matrix A is **nilpotent** if $A^k = 0$ for some positive integer k.

Theorem A.4. *Every matrix is similar to an upper triangular matrix. Every nilpotent matrix is similar to an upper triangular matrix with zeros on the diagonal.*

While Theorem A.4 is sufficient for some purposes, we will in general need something that comes a bit closer to a diagonal representation. If $A \in M_n(\mathbb{C})$ does not have n linearly independent eigenvectors, we may consider the more general concept of generalized eigenvectors. A nonzero vector $v \in \mathbb{C}^n$ is called a **generalized eigenvector** for A if there is some complex number λ and some positive integer k such that

$$(A - \lambda I)^k v = 0. \tag{A.4}$$

If (A.4) holds for some $v \neq 0$, then $(A - \lambda I)$ cannot be invertible. Thus, the number λ must be an (ordinary) eigenvalue for A. However, for a fixed eigenvalue λ, there may be generalized eigenvectors v that are not ordinary eigenvectors. In the case of the matrix A in (A.3), for example, the vector $(0, 1)$ is a generalized eigenvector with eigenvalue 1 (with $k = 2$).

It can be shown that every $A \in M_n(\mathbb{C})$ has a basis of generalized eigenvectors. For any matrix A and any eigenvalue λ for A, let W_λ be the *generalized* eigenspace with eigenvalue λ:

$$W_\lambda = \{v \in \mathbb{C}^n \,|\, (A - \lambda I)^k v = 0 \text{ for some } k\}.$$

Then $\mathbb{C}^n$ decomposes as a direct sum of the W_λ's, as λ ranges over all the eigenvalues of A. Furthermore, the subspace W_λ is easily seen to be *invariant* under the matrix A. Let A_λ denote the restriction of A to the subspace W_λ, and let $N_\lambda = A_\lambda - \lambda I$, so that

$$A_\lambda = \lambda I + N_\lambda.$$

Then N_λ is nilpotent; that is, $N_\lambda^k = 0$ for some positive integer k. We summarize the preceding discussion in the following theorem.

Theorem A.5. *Let A be an $n \times n$ complex matrix. Then there exists a basis for $\mathbb{C}^n$ consisting of generalized eigenvectors for A. Furthermore, $\mathbb{C}^n$ is the direct sum of the generalized eigenspaces W_λ. Each W_λ is invariant under A, and the restriction of A to W_λ is of the form $\lambda I + N_\lambda$, where N_λ is nilpotent.*

The preceding result is the basis for the following decomposition.

Theorem A.6. *Each $A \in M_n(\mathbb{C})$ has a unique decomposition as $A = S + N$ where S is diagonalizable, N is nilpotent, and $SN = NS$.*

The expression $A = S + N$, with S and N as in the theorem, is called the *SN* **decomposition** of A. The existence of an *SN* decomposition follows from the previous theorem: We define S to be the operator equal to λI on each generalized eigenspace W_λ of A and we set N to be the operator equal to N_λ on each W_λ. For example, if A is the matrix in (A.3), then we have

$$S = \begin{pmatrix} 1 & 0 \\ 0 & 1 \end{pmatrix}, \quad N = \begin{pmatrix} 0 & 1 \\ 0 & 0 \end{pmatrix}.$$

A.4   The Jordan Canonical Form

The Jordan canonical form may be viewed as a refinement of the SN decomposition, based on a further analysis of the nilpotent matrices N_λ in Theorem A.5.

Theorem A.7. *Every $A \in M_n(\mathbb{C})$ is similar to a block-diagonal matrix in which each block is of the form*

$$\begin{pmatrix} \lambda & 1 & & \\ & \lambda & \ddots & \\ & & \ddots & 1 \\ & & & \lambda \end{pmatrix}.$$

Two matrices A and B are similar if and only if they have precisely the same Jordan blocks, up to reordering.

There may be several different Jordan blocks (possibly of different sizes) for the same value of λ. In the case in which A is diagonalizable, each block is 1×1, in which case, the ones above the diagonal do not appear. Note that each Jordan block is, in particular, of the form $\lambda I + N$, where N is nilpotent.

A.5   The Trace

For $A \in M_n(\mathbb{C})$, we define the **trace** of A to be the sum of the diagonal entries of A:

$$\text{trace}(A) = \sum_{k=1}^{n} A_{kk}.$$

Note that the trace is a linear function of A. For $A, B \in M_n(\mathbb{C})$, we note that

$$\text{trace}(AB) = \sum_{k=1}^{n}(AB)_{kk} = \sum_{k=1}^{n}\sum_{l=1}^{n} A_{kl}B_{lk}. \tag{A.5}$$

If we similarly compute trace(BA), we obtain the same sum with the labels for the summation variables reversed. Thus,

$$\text{trace}(AB) = \text{trace}(BA). \tag{A.6}$$

If C is an invertible matrix and we apply (A.6) to the matrices CA and C^{-1}, we have

$$\text{trace}(CAC^{-1}) = \text{trace}(C^{-1}CA) = \text{trace}(A);$$

that is, similar matrices have the same trace.

    More generally, if A is a linear operator on a finite-dimensional vector space V, we can define the trace of A by picking a basis and defining the trace of A to be the trace of the matrix that represents A in that basis. The above calculations show that the value of the trace of A is independent of the choice of basis.

A.6   Inner Products

Let $\langle \cdot, \cdot \rangle$ denote the standard inner product on $\mathbb{C}^n$, defined by

$$\langle x, y \rangle = \sum_{j=1}^{n} \overline{x_j} y_j,$$

where we follow the convention of putting the complex-conjugate on the first factor. We have the following basic result relating the inner product to the adjoint of a matrix, as defined in (A.2).

Proposition A.8. *For all $A \in M_n(\mathbb{C})$, the adjoint A^* of A has the property that*

$$\langle x, Ay \rangle = \langle A^*x, y \rangle \tag{A.7}$$

for all $x, y \in \mathbb{C}^n$.

Proof. We compute that

$$\langle x, Ay \rangle = \sum_{j=1}^{n} \overline{x_j} \sum_{k=1}^{n} A_{jk} y_k$$

$$= \sum_{j=1}^{n} \sum_{k=1}^{n} \overline{A_{jk} x_j}\, y_k$$

$$= \sum_{j=1}^{n} \sum_{k=1}^{n} \overline{A_{kj}^* x_j}\, y_k.$$

This last expression is just the inner product of A^*x with y. $\square$

We may generalize the notion of inner product as follows.

Definition A.9. If V is any vector space over $\mathbb{C}$, an **inner product** on V is a map that associates to any two vectors u and v in V a complex number $\langle u, v \rangle$ and that has the following properties:

1. Conjugate symmetry: $\langle v, u \rangle = \overline{\langle u, v \rangle}$ for all $u, v \in V$.
2. Linearity in the second factor: $\langle u, v_1 + av_2 \rangle = \langle u, v_1 \rangle + a \langle u, v_2 \rangle$, for all $u, v_1, v_2 \in V$ and $a \in \mathbb{C}$.
3. Positivity: For all $v \in V$, the quantity $\langle v, v \rangle$ is real and satisfies $\langle v, v \rangle \geq 0$, with $\langle v, v \rangle = 0$ only if $v = 0$.

Note that in light of the conjugate-symmetry and the linearity in the second factor, an inner product must be conjugate-linear in the first factor:

$$\langle v_1 + av_2, u \rangle = \langle v_1, u \rangle + \bar{a} \langle v_2, u \rangle .$$

(Some authors define an inner product to be linear in the first factor and conjugate linear in the second factor.) An inner product on a real vector space is defined in the same way except that conjugate symmetry is replaced by symmetry ($\langle v, u \rangle = \langle u, v \rangle$) and the constant a in Point 2 now takes only real values.

If V is a vector space with inner product, the **norm** of a vector $v \in V$, denoted $\|v\|$, is defined by

$$\|v\| = \sqrt{\langle v, v \rangle}.$$

The positivity condition on the inner product guarantees that $\|v\|$ is always a non-negative real number and that $\|v\| = 0$ only if $v = 0$. If, for example, $V = M_n(\mathbb{C})$, we may define the **Hilbert–Schmidt inner product** by the formula

$$\langle A, B \rangle = \text{trace}(A^* B). \tag{A.8}$$

It is easy to see check that this expression is conjugate symmetric and linear in the second factor. Furthermore, we may compute as in (A.5) that

$$\text{trace}(A^* A) = \sum_{k,l=1}^{n} A_{kl}^* A_{lk} = \sum_{k,l=1}^{n} |A_{kl}|^2 \geq 0,$$

and the sum is zero only if each entry of A is zero. The associated **Hilbert–Schmidt norm** satisfies

$$\|A\|^2 = \sum_{k,l=1}^{n} |A_{kl}|^2 .$$

Suppose that V is a finite-dimensional vector space with inner product and that W is a subspace of V. Then the **orthogonal complement** of W, denoted $W^\perp$, is the set of all vectors v in V such that $\langle w, v \rangle = 0$ for all w in W. The space V then decomposes as the direct sum of W and $W^\perp$.

We now introduce the abstract notion of the adjoint of a matrix.

Proposition A.10. *Let V be a finite-dimensional vector space with an inner product $\langle \cdot, \cdot \rangle$. If A is a linear map from V to V, there is a unique operator $A^* : V \to V$ such that*

$$\langle u, Av \rangle = \langle A^* u, v \rangle$$

for all $u, v \in V$. Furthermore, if W is a subspace of V that is invariant under A, then $W^\perp$ is invariant under A^*.

A.7   Dual Spaces

If V is a vector space over $\mathbb{C}$, a **linear functional** on V is a linear map of V into $\mathbb{C}$. If V is finite dimensional with basis $v_1, \ldots, v_n$, then for each set of constants $a_1, \ldots, a_n$, there is a unique linear functional ϕ such that $\phi(v_k) = a_k$. If V is a complex vector space, then the **dual space** to V, denoted V^*, is the set of all linear functionals on V. The dual space is also a vector space and its dimension is the same as that of V. If V is finite dimensional, then V is isomorphic to V^{**} by the map sending $v \in V$ to the "evaluation at v" functional, that is, the map $\phi \mapsto \phi(v)$, $\phi \in V^*$.

If W is a subspace of a vector space V, the **annihilator subspace** of W, denoted $W^\wedge$, is the set of all ϕ in V^* such that $\phi(w) = 0$ for all w in W. Then $W^\wedge$ is a subspace of V^*. If V is finite dimensional, then

$$\dim W + \dim W^\wedge = \dim V$$

and the map $W \to W^\wedge$ provides a one-to-one correspondence between subspaces of V and subspaces of V^*.

In general, one should be careful to distinguish between a vector space and its dual. Nevertheless, when V is finite dimensional and has an inner product, we can produce an identification between V and V^*.

Proposition A.11. *Let V be a finite-dimensional inner product space and let ϕ be a linear functional on V. Then there exists a unique $w \in V$ such that*

$$\phi(v) = \langle w, v \rangle$$

for all $v \in V$.

Recall that we follow the convention that inner products are linear in the *second* factor, so that $\langle w, v \rangle$ is, indeed, linear in v.

A.8   Simultaneous Diagonalization

We now extend the notion of eigenvectors and diagonalization to families of linear operators.

Definition A.12. Let V be a vector space and let $\mathcal{A}$ be a collection of linear operators on V. A nonzero vector $v \in V$ is a **simultaneous eigenvector** for $\mathcal{A}$ if for all $A \in \mathcal{A}$, there exists a constant λ_A such that $Av = \lambda_A v$. The numbers λ_A are the **simultaneous eigenvalues** associated to v.

Consider, for example, the space $\mathcal{D}$ of all diagonal $n \times n$ matrices. Then for each $k = 1, \ldots, n$, the standard basis element e_k is a simultaneous eigenvector for $\mathcal{D}$. For each diagonal matrix A, the simultaneous eigenvalue associated to e_k is the kth diagonal entry of A.

Proposition A.13. *If $\mathcal{A}$ is a commuting family of linear operators on a finite-dimensional complex vector space, then $\mathcal{A}$ has at least one simultaneous eigenvector.*

It is essential here that the elements of $\mathcal{A}$ commute; noncommuting families of operators typically have no simultaneous eigenvectors.

In many cases, the collection $\mathcal{A}$ of operators on V is a *subspace* of $\mathrm{End}(V)$, the space of all linear operators from V to itself. In that case, if v is a simultaneous eigenvector for $\mathcal{A}$, the eigenvalues λ_A for v depend linearly on A. After all, if $A_1 v = \lambda_1 v$ and $A_2 v = \lambda_2 v$, then

$$(A_1 + cA_2)v = (\lambda_1 + c\lambda_2)v.$$

The preceding discussion leads to the following definition.

Definition A.14. Suppose that V is a vector space and $\mathcal{A}$ is a vector space of linear operators on V. A **weight** for $\mathcal{A}$ is a linear functional μ on $\mathcal{A}$ such that there exists a nonzero vector $v \in V$ satisfying

$$Av = \mu(A)v$$

for all A in $\mathcal{A}$. For a fixed weight μ, the set of all vectors $v \in V$ satisfying $Av = \mu(A)v$ for all A in $\mathcal{A}$ is called the **weight space** associated to the weight μ.

That is to say, a weight is a set of simultaneous eigenvalues for the operators in $\mathcal{A}$. If V is finite dimensional and the elements of $\mathcal{A}$ all commute with one another, then there will exist at least one weight for $\mathcal{A}$.

If $\mathcal{A}$ is finite dimensional and comes equipped with an inner product, it is convenient to express the linear functional μ in Definition A.14 as the inner product of A with some vector, as in Proposition A.11. From this point of view, we define a weight to be an element μ of $\mathcal{A}$ (not $\mathcal{A}^*$) such that there exists a nonzero v in V with

$$Av = \langle \mu, A \rangle v$$

for all $A \in \mathcal{A}$.

Definition A.15. Suppose that V is a finite-dimensional vector space and $\mathcal{A}$ is some collection of linear operators on V. Then the elements of $\mathcal{A}$ are said to be **simultaneously diagonalizable** if there exists a basis $v_1, \ldots, v_n$ for V such that each v_k is a simultaneous eigenvector for $\mathcal{A}$.

If $\mathcal{A}$ is a vector space of linear operators on V, then saying that the elements of $\mathcal{A}$ are simultaneously diagonalizable is equivalent to saying that V can be decomposed as a direct sum of weight spaces of $\mathcal{A}$.

If a collection $\mathcal{A}$ of operators is simultaneously diagonalizable, then the elements of $\mathcal{A}$ must commute, since they commute when applied to each v_k. Conversely, if each $A \in \mathcal{A}$ is diagonalizable by itself and if the elements of $\mathcal{A}$ commute, then (it can be shown), the elements of $\mathcal{A}$ are simultaneously diagonalizable. We record these results in the following proposition.

Proposition A.16. *If $\mathcal{A}$ is a commuting collection of linear operators on a finite-dimensional vector space V and each $A \in \mathcal{A}$ is diagonalizable, then the elements of $\mathcal{A}$ are* simultaneously *diagonalizable.*

We close this appendix with an analog of Proposition A.1 for simultaneous eigenvectors.

Proposition A.17. *Suppose V is a vector space and $\mathcal{A}$ is a vector space of linear operators on V. Suppose $\mu_1, \ldots, \mu_m$ are distinct weights for $\mathcal{A}$ and $v_1, \ldots, v_m$ are elements of the corresponding weight spaces. If $v_1 + \cdots + v_m = 0$, then $v_j = 0$ for all $j = 1, \ldots, m$. Furthermore, if $v_1 + \cdots + v_m$ is a weight vector with weight μ, then $\mu = \mu_j$ for some j and $v_k = 0$ for all $k \neq j$.*

Since this result is not quite standard, we provide a proof.

Proof. Assume first that $v_1 + \cdots + v_m = 0$, with v_j in the weight space with weight μ_j. If $m = 1$, then we have $v_1 = 0$, as claimed. If $m > 1$, choose some $A \in \mathcal{A}$ such that $\mu_1(A) \neq \mu_2(A)$. If we then apply the operator $A - \mu_2(A)I$ to $v_1 + \cdots + v_m$, we obtain

$$0 = \sum_{j=1}^{m} (\mu_j(A) - \mu_2(A))v_j. \tag{A.9}$$

Now, the $j = 2$ term in (A.9) is zero, so that the sum actually contains at most $m - 1$ nonzero terms. Thus, by induction on m, we can assume that each term in (A.9) is zero. In particular, $(\mu_1(A) - \mu_2(A))v_1 = 0$, which implies (by our choice of A) that

$v_1 = 0$. Once v_1 is known to be zero, the original sum $v_1 + \cdots + v_m$ contains at most $m - 1$ nonzero terms. Thus, using induction on m again, we see that each term in the sum is zero.

Assume now that $v := v_1 + \cdots + v_m$ is a (nonzero) weight vector with some weight μ, and choose some j for which $v_j \neq 0$. Then for each $A \in \mathcal{A}$, we have

$$0 = Av - \mu(A)v = \sum_{k=1}^{m} (\mu_k(A) - \mu(A))v_k.$$

Thus, by the first part of the proposition, we must have $(\mu_k(A) - \mu(A))v_k = 0$ for all j. Taking $k = j$, we conclude that $\mu_j(A) - \mu(A) = 0$. Since this result holds for all $A \in \mathcal{A}$, we see that $\mu = \mu_j$. Finally, for any $k \neq j$, we can choose $A \in \mathcal{A}$ so that $\mu_k(A) \neq \mu_j(A)$. With this value of A (and with $\mu = \mu_j$), the fact that $(\mu_k(A) - \mu_j(A))v_k = 0$ forces v_k to be zero.   □

Appendix B
Differential Forms

In this section, we give a very brief outline of the theory of differential forms on manifolds. Since this is our main requirement, we consider only top-degree forms, that is, k-forms on k-dimensional manifolds. See Chapter 16 in [Lee] for more information. We begin by considering forms at a single point, which is just a topic in linear algebra.

Definition B.1. If V is a k-dimensional real vector space, a map $\alpha : V^k \to \mathbb{R}$ is said to be k-**linear and alternating** if (1) $\alpha(v_1, \ldots, v_k)$ is linear with respect to each v_j with the other variables fixed, and (2) α changes sign whenever any two of the variables are interchanged:

$$\alpha(v_1, \ldots, v_l, \ldots, v_m, \ldots v_k) = -\alpha(v_1, \ldots, v_m, \ldots, v_l, \ldots v_k).$$

It is a standard result in linear algebra (e.g., Theorem 2 in Section 5.3 of [HK]) that every k-dimensional real vector space admits a nonzero k-linear, alternating form, and that any two such forms differ by multiplication by a constant. If $T : V \to V$ is a linear map and α is a k-linear, alternating form on V, then for any $v_1, \ldots, v_k \in V$, we have

$$\alpha(Tv_1, \ldots, Tv_k) = (\det T)\alpha(v_1, \ldots, v_k). \tag{B.1}$$

If $v_1, \ldots, v_k$ and $w_1, \ldots, w_k$ are two ordered bases for V, then there is a unique invertible linear transformation $T : V \to V$ such that $Tv_j = w_j$. We may divide the collection of all ordered bases of V into two groups, where two ordered bases belong

---

A previous version of this book was inadvertently published without the middle initial of the author's name as "Brian Hall". For this reason an erratum has been published, correcting the mistake in the previous version and showing the correct name as Brian C. Hall (see DOI http://dx.doi.org/10.1007/978-3-319-13467-3_14). The version readers currently see is the corrected version. The Publisher would like to apologize for the earlier mistake.

© Springer International Publishing Switzerland 2015
B.C. Hall, *Lie Groups, Lie Algebras, and Representations*, Graduate
Texts in Mathematics 222, DOI 10.1007/978-3-319-13467-3

to the same group if the linear map relating them has positive determinant and the two bases belong to different groups if the linear map relating them has negative determinant. An **orientation** of V is then a choice of one of the two groups of bases. Once an orientation of V has been chosen, we say that a basis is **positively oriented** if it belongs to the chosen group of bases. If α is a nonzero k-linear, alternating form on V, we can define an orientation of V by decreeing an ordered basis $v_1, \ldots, v_k$ to be positively oriented if $\alpha(v_1, \ldots, v_k) > 0$.

The following example of a k-linear, alternating form on $\mathbb{R}^k$ will help motivate the notion of a k-form. For any vectors $v_1, \ldots, v_k$ in $\mathbb{R}^k$, define the parallelepiped $P_{v_1,\ldots,v_k}$ spanned by these vectors, as follows:

$$P_{v_1,\ldots,v_k} = \{c_1 v_1 + \cdots + c_k v_k \,|\, 0 \le c_l \le 1\}. \tag{B.2}$$

(If $k = 2$, then a parallelepiped is just a parallelogram.) Let us use the orientation on $\mathbb{R}^k$ in which the standard basis $e_1, \ldots, e_k$ is positively oriented.

Example B.2. Define a map $V : (\mathbb{R}^k)^k \to \mathbb{R}$ by

$$V(v_1, \ldots, v_k) = \pm \mathrm{Vol}(P_{v_1,\ldots,v_k}), \tag{B.3}$$

where we take a plus sign if $v_1, \ldots, v_k$ is a positively oriented basis for $\mathbb{R}^k$ and a minus sign if it is a negatively oriented basis. Then V is a k-linear, alternating form on $\mathbb{R}^k$.

Note that the volume of $P_{v_1,\ldots,v_k}$ is zero if $v_1, \ldots, v_k$ do not form a basis for $\mathbb{R}^k$, in which case, we do not have to worry about the sign on the right-hand side of (B.3). Now, it is known that the volume of $P_{v_1,\ldots,v_k}$ is equal to $|\det T|$, where T is the $k \times k$ matrix whose columns are the vectors $v_1, \ldots, v_k$. This claim is a very special case of the change-of-variables theorem in multivariate calculus and can be proved by expressing T as a product of elementary matrices. We can then see that $V(v_1, \ldots, v_k)$ is equal to $\det T$ (without the absolute value signs). Meanwhile, it is a standard result from linear algebra that the determinant of T is a k-linear, alternating function of its column vectors $v_1, \ldots, v_k$.

We now turn to a discussion of top-degree forms on manifolds. If M is a k-dimensional manifold (say, embedded into some $\mathbb{R}^N$), we have the notion of the tangent space to M at m, denoted $T_m M$, which is a k-dimensional subspace of $\mathbb{R}^N$.

Definition B.3. Suppose M is a smoothly embedded, k-dimensional submanifold of $\mathbb{R}^N$ for some k, N. A k-**form** α on M is a smoothly varying family α_m of k-linear, alternating maps on $T_m M$, one for each $m \in M$.

To be precise, let us say that a family α_m of k-linear, alternating forms on each $T_m M$ is "smoothly varying" if the following condition holds. Suppose $X_1, \ldots, X_k$ are smooth $\mathbb{R}^N$-valued functions on $\mathbb{R}^N$ with the property that for each $m \in M$, the vector $X_j(m)$ is tangent to M. Then the function $\alpha_m(X_1(m), \ldots, X_k(m)), m \in M$, should be a smooth function on M.

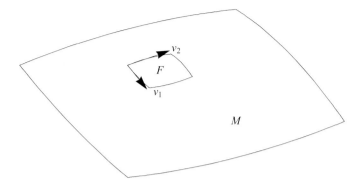

Fig. B.1 The integral of α over the small region $F \subset M$ is approximately equal to $\alpha(v_1, v_2)$

The "purpose in life" of a k-form α on a k-dimensional manifold M is to be integrated over regions in M. More precisely, we must assume that M is orientable—meaning that it is possible to choose an orientation of each tangent space $T_m M$ in a way that varies continuously with m—and that we have chosen an orientation on M. Then if E is a "nice" subset of M (to be precise, a compact k-dimensional submanifold with boundary), there is a notion of the integral of α over $E \subset M$, denoted

$$\int_E \alpha.$$

The value of $\int_E \alpha$ may be thought of as assigning a sort of (possibly negative) "volume" to the set E. If α is a k-form on M and $f : M \rightarrow \mathbb{R}$ is a smooth function, then $f\alpha$ is also a k-form on M, which may also be integrated, using the same orientation we used to integrate α.

We may gain an intuitive understanding of the notion of integration of k-forms as follows. For any region $E \subset M$, we may think of chopping E up into very small subregions F, each of which is shaped like a parallelepiped (as in (B.2)). More specifically, each subregion will look like the parallelepiped spanned by tangent vectors $v_1, \ldots, v_k$ at some point $m \in E$, which we can arrange to be positively oriented. The idea is then that the integral of α over each subregion should be approximately $\alpha_m(v_1, \ldots, v_k)$. (See Figure B.1.) The integral of α over all of E should then be the sum of its integrals over the subregions.

If we think of $\int_E \alpha$ as a sort of volume of the set E, then $\alpha_m(v_1, \ldots, v_k)$ represents the volume (possibly with a minus sign) of a small parallelepiped-shaped subregion inside E. Example B.2 then makes it natural that we should require $\alpha_m(v_1, \ldots, v_k)$ to be k-linear and alternating.

We may give a more precise definition of the integral of a differential form as follows. We choose a local coordinate system $x_1, \ldots, x_k$ on our oriented manifold M, defined in some open set U. We then let $\partial/\partial x_1, \ldots, \partial/\partial x_k$ denote the associated basis for the tangent space at each point. (In coordinates, $\partial/\partial x_j$ is the unit vector in the x_j-direction.) We assume the coordinate system is "oriented," meaning that $\partial/\partial x_1, \ldots, \partial/\partial x_k$ is an oriented basis for the tangent space at each point in U.

Definition B.4. Let α be a k-form on an oriented k-dimensional manifold M and suppose $E \subset M$ is a compact subset of the domain U of $\{x_j\}$. We then define $\int_E \alpha$ as

$$\int_E \alpha = \int_E \alpha \left(\frac{\partial}{\partial x_1}, \ldots, \frac{\partial}{\partial x_k} \right) dx_1 \, dx_2 \, \cdots \, dx_k, \tag{B.4}$$

where the integral on the right-hand side of (B.4) is an ordinary integral in Euclidean space.

The integral on the right-hand side of (B.4) may be defined as a Riemann integral or using Lebesgue measure on $\mathbb{R}^k$. A key point in the definition is to verify that the value of $\int_E \alpha$ is independent of the choice of coordinates. To this end, suppose $\{y_k\}$ is another oriented coordinate system whose domain includes E. Then by the chain rule, we have

$$\frac{\partial f}{\partial x_l} = \sum_m \frac{\partial f}{\partial y_m} \frac{\partial y_m}{\partial x_l}$$

for any smooth function f. That is to say,

$$\frac{\partial}{\partial x_l} = \sum_m \frac{\partial y_m}{\partial x_l} \frac{\partial}{\partial y_m}.$$

Thus, if T is the matrix whose entries are $T_{lm} = \partial y_m/\partial x_l$, we will have, by (B.1),

$$\alpha \left(\frac{\partial}{\partial x_1}, \ldots, \frac{\partial}{\partial x_k} \right) = (\det T)\alpha \left(\frac{\partial}{\partial x_1}, \ldots, \frac{\partial}{\partial x_k} \right).$$

On the other hand, the classical change of variables theorem says that

$$\int_E f(x_1, \ldots, x_k) \, dx_1 \, dx_2 \, \cdots \, dx_k = \int_E f(y_1, \ldots, y_k) \, J \, dy_1 \, dy_2 \, \cdots \, dy_k,$$

where J is the determinant of the matrix $\{\partial x_m/\partial y_l\}$. (For example, in the $k = 1$ case, J is just dx/dy, which is obtained by writing $dx = (dx/dy) \, dy$.) But by the chain rule again, the matrix $\{\partial x_m/\partial y_l\}$ is the inverse of the matrix $\{\partial y_m/\partial x_l\}$. Thus, J is the reciprocal of $\det T$, and we see that

$$\int_E \alpha \left(\frac{\partial}{\partial x_1}, \ldots, \frac{\partial}{\partial x_k} \right) dx_1 \, dx_2 \, \cdots \, dx_k$$

$$= \int_E \alpha \left(\frac{\partial}{\partial y_1}, \ldots, \frac{\partial}{\partial y_k} \right) dy_1 \, dy_2 \, \cdots \, dy_k,$$

as claimed.

Note that if we think of the integral in (B.4) as a Riemann integral, we compute the integral by covering E with small k-dimensional "rectangles," and these rectangles may be thought of as being "spanned" by multiples of the vectors $\partial/\partial x_1, \ldots, \partial/\partial x_k$. In the Riemann integral, the integral of α over each small rectangle is being approximated by $\alpha(\partial/\partial x_1, \ldots, \partial/\partial x_k)$ times the volume of the rectangle, in agreement with preceding intuitive description of the integral.

If we wish to integrate a k-form α over a general k-dimensional, compact subset E of M, we use a partition of unity to write α as a sum of forms α_j, each of which is supported in a small region in M. For each j, we choose a coordinate system defined on a set U_j containing the support of α_j. We then integrate α_j over $E \cap U_j$ and sum over j.

Appendix C
Clebsch–Gordan Theory
and the Wigner–Eckart Theorem

C.1   Tensor Products of sl(2; ℂ) Representations

The irreducible representations of $\mathsf{SU}(2)$ (or, equivalently, of $\mathsf{sl}(2;\mathbb{C})$) were classified in Sect. 4.6 and may be realized in spaces of homogeneous polynomials in two complex variables as in Example 4.10. For each non-negative integer m, we have an irreducible representation (π_m, V_m) of $\mathsf{sl}(2;\mathbb{C})$ of dimension $m+1$, and every irreducible representation of $\mathsf{sl}(2;\mathbb{C})$ is isomorphic to one of these. We are using here the mathematicians' labeling of the representations; in the physics literature, the representations are labeled by the "spin" $l := m/2$.

By the averaging method of Sect. 4.4, we can find on each space V_m an inner product that is invariant under the action of the compact group $\mathsf{SU}(2)$. (In the case of $V_1 \cong \mathbb{C}^2$, we can use the standard inner product on $\mathbb{C}^2$ and for any m, it is not hard to describe such an inner product explicitly.) With respect to such an inner product, the orthogonal complement of a subspace invariant under $\mathsf{SU}(2)$ (or, equivalently, under $\mathsf{sl}(2;\mathbb{C})$) is again invariant under $\mathsf{SU}(2)$. Since the element $H = \mathrm{diag}(1,-1)$ of $\mathsf{sl}(2;\mathbb{C})$ is in $i\,\mathsf{su}(2)$, $\pi_m(H)$ will be self-adjoint with respect to this inner product. Thus, eigenvectors of $\pi_m(H)$ with distinct eigenvectors must be orthogonal.

Recall from Sect. 4.3.2 the notion of the tensor product of representations of a group or Lie algebra. We consider this in the case of the irreducible representations of $\mathsf{sl}(2;\mathbb{C})$. We regard the tensor product $V_m \otimes V_n$ as a representation of $\mathsf{sl}(2;\mathbb{C})$. (Recall that it is also possible to view $V_m \otimes V_n$ as a representation of $\mathsf{sl}(2;\mathbb{C}) \oplus \mathsf{sl}(2;\mathbb{C})$.) The action of $\mathsf{sl}(2;\mathbb{C})$ on $V_m \otimes V_n$ is given by

A previous version of this book was inadvertently published without the middle initial of the author's name as "Brian Hall". For this reason an erratum has been published, correcting the mistake in the previous version and showing the correct name as Brian C. Hall (see DOI http://dx.doi.org/10.1007/978-3-319-13467-3_14). The version readers currently see is the corrected version. The Publisher would like to apologize for the earlier mistake.

© Springer International Publishing Switzerland 2015   427
B.C. Hall, *Lie Groups, Lie Algebras, and Representations*, Graduate
Texts in Mathematics 222, DOI 10.1007/978-3-319-13467-3

$$(\pi_m \otimes \pi_n)(X) = \pi_m(X) \otimes I + I \otimes \pi_n(X). \tag{C.1}$$

We compute in the standard basis $\{X, Y, H\}$ for $\mathsf{sl}(2; \mathbb{C})$. Once we have chosen $\mathsf{SU}(2)$-invariant inner products on V_m and V_n, there is a unique inner product on $V_m \otimes V_n$ with the property that $\langle u_1 \otimes v_1, u_2 \otimes v_2 \rangle = \langle u_1, u_2 \rangle \langle v_1, v_2 \rangle$. (This assertion can be proved using the universal property of tensor products.) The inner product on $V_m \otimes V_n$ is also invariant under the action of $\mathsf{SU}(2)$. We assume in the rest of this section that an inner product of this sort has been chosen on each $V_m \otimes V_n$.

In general, $V_m \otimes V_n$ will not be an irreducible representation of $\mathsf{sl}(2; \mathbb{C})$; the goal of this section is to describe how $V_m \otimes V_n$ decomposes as a direct sum of irreducible invariant subspaces. This decomposition is referred to as the Clebsch–Gordan theory. Let us consider first the case of $V_1 \otimes V_1$, where $V_1 = \mathbb{C}^2$, the standard representation of $\mathsf{sl}(2; \mathbb{C})$. If $\{e_1, e_2\}$ is the standard basis for $\mathbb{C}^2$, then the vectors of the form $e_k \otimes e_l, 1 \leq k, l \leq 2$, form a basis for $\mathbb{C}^2 \otimes \mathbb{C}^2$. Since e_1 and e_2 are eigenvalues for $\pi_1(H)$ with eigenvalues 1 and -1, respectively, then, by (C.1), the basis elements for $\mathbb{C}^2 \otimes \mathbb{C}^2$ are eigenvectors for the action of H with eigenvalues $2, 0, 0$, and -2, respectively. Since 2 is the largest eigenvalue for H, the corresponding eigenvector $e_1 \otimes e_1$ must be annihilated by X (i.e., by the operator $\pi_1(X) \otimes I + I \otimes \pi_1(X)$).

If then we apply Y (i.e., by the operator $\pi_1(Y) \otimes I + I \otimes \pi_1(Y)$) repeatedly to $e_1 \otimes e_1$, we obtain $e_1 \otimes e_2 + e_2 \otimes e_1$, then $2e_2 \otimes e_2$, and then the zero vector. The space spanned by these vectors is invariant under $\mathsf{sl}(2; \mathbb{C})$ and irreducible, and is isomorphic to the three-dimensional representation V_2. The orthogonal complement of this space in $\mathbb{C}^2 \otimes \mathbb{C}^2$, namely the span of $e_1 \otimes e_2 - e_2 \otimes e_1$, is also invariant, and $\mathsf{sl}(2; \mathbb{C})$ acts trivially on this space. Thus,

$$\mathbb{C}^2 \otimes \mathbb{C}^2 = \mathrm{span}\{e_1 \otimes e_1, e_1 \otimes e_2 + e_2 \otimes e_1, e_2 \otimes e_2\} \oplus \mathrm{span}\{e_1 \otimes e_2 - e_2 \otimes e_1\}.$$

We see, then, that the four-dimensional space $V_1 \otimes V_1$ is isomorphic, as an $\mathsf{sl}(2; \mathbb{C})$ representation, to $V_2 \oplus V_0$.

Theorem C.1. *Let m and n be non-negative integers with $m \geq n$. If we consider $V_m \otimes V_n$ as a representation of $\mathsf{sl}(2; \mathbb{C})$, then*

$$V_m \otimes V_n \cong V_{m+n} \oplus V_{m+n-2} \oplus \cdots \oplus V_{m-n+2} \oplus V_{m-n},$$

where $\cong$ denotes an isomorphism of $\mathsf{sl}(2; \mathbb{C})$ representations.

Note that this theorem is consistent with the special case worked out earlier: $V_1 \otimes V_1 \cong V_2 \oplus V_0$. For applications to the Wigner–Eckart theorem, a key property of the decomposition in Theorem C.1 is that it is *multiplicity free*. That is to say, each irreducible representation that occurs in the decomposition of $V_m \otimes V_n$ occurs only once. This is a special feature of the representations of $\mathsf{sl}(2; \mathbb{C})$; the analogous statement does *not* hold for tensor products of representations of other Lie algebras.

Proof. Let us take a basis for each of the two spaces that is labeled by the eigenvalues for H. That is to say, we choose a basis $u_m, u_{m-2}, \ldots, u_{-m}$ for V_m and

$v_n, v_{n-2}, \ldots, v_{-n}$ for V_n, with $\pi_m(H)u_j = ju_j$ and $\pi_n(H)v_k = kv_k$. Then the vectors of the form $u_j \otimes v_k$ form a basis for $V_m \otimes V_n$, and we compute that

$$[\pi_m(H) \otimes I + I \otimes \pi_n(H)]u_j \otimes v_k = (j+k)u_j \otimes v_k.$$

Thus, each of our basis elements is an eigenvector for the action of H on $V_m \otimes V_n$. The eigenvalues for the action of H range from $m+n$ to $-(m+n)$ in increments of 2.

The eigenspace with eigenvalue $m+n$ is one dimensional, spanned by $u_m \otimes v_n$. If $n > 0$, then the eigenspace with eigenvalue $m+n-2$ has dimension 2, spanned by $u_{m-2} \otimes v_n$ and $u_m \otimes v_{n-2}$. Each time we decrease the eigenvalue of H by 2 we increase the dimension of the corresponding eigenspace by 1, until we reach the eigenvalue $m-n$, which is spanned by the vectors

$$u_{m-2n} \otimes v_n, u_{m-2n+2} \otimes v_{n-2}, \ldots, u_m \otimes v_{-n}.$$

This space has dimension $n+1$. As we continue to decrease the eigenvalue of H in increments of 2, the dimensions remain constant until we reach eigenvalue $n-m$, at which point the dimensions begin decreasing by 1 until we reach the eigenvalue $-m-n$, for which the corresponding eigenspace has dimension one, spanned by $u_{-m} \otimes v_{-n}$. This pattern is illustrated by the following table, which lists, for the case of $V_4 \otimes V_2$, each eigenvalue for H and a basis for the corresponding eigenspace.

Eigenvalue for H	Basis		
6	$u_4 \otimes v_2$		
4	$u_2 \otimes v_2$	$u_4 \otimes v_0$	
2	$u_0 \otimes v_2$	$u_2 \otimes v_0$	$u_4 \otimes v_{-2}$
0	$u_{-2} \otimes v_2$	$u_0 \otimes v_0$	$u_2 \otimes v_{-2}$
-2	$u_{-4} \otimes v_2$	$u_{-2} \otimes v_0$	$u_0 \otimes v_{-2}$
-4	$u_{-4} \otimes v_0$	$u_{-2} \otimes v_{-2}$	
-6	$u_{-4} \otimes v_{-2}$		

Consider now the vector $u_m \otimes v_n$, which is annihilated by X and is an eigenvector for H with eigenvalue $m+n$. Applying Y repeatedly gives a chain of eigenvectors for H with eigenvalues decreasing by 2 until they reach $-m-n$. By the proof of Theorem 4.32, the span W of these vectors is invariant under $\mathsf{sl}(2;\mathbb{C})$ and irreducible, isomorphic to V_{m+n}. The orthogonal complement of W is also invariant. Since W contains each of the eigenvalues of H with multiplicity one, each eigenvalue for H in $W^\perp$ will have its multiplicity lowered by 1. In particular, $m+n$ is not an eigenvalue for H in $W^\perp$; the largest remaining eigenvalue is $m+n-2$ and this eigenvalue has multiplicity one (unless $n=0$). Thus, if we start with an eigenvector for H in $W^\perp$ with eigenvalue $m+n-2$, this will be annihilated by X and will generate an irreducible invariant subspace isomorphic to V_{m+n-2}.

We now continue on in the same way, at each stage looking at the orthogonal complement of the sum of all the invariant subspaces we have obtained in the previous stages. Each step reduces the multiplicity of each H-eigenvalue by 1 and thereby reduces the largest remaining H-eigenvalue by 2. This process will continue until there is nothing left, which will occur after V_{m-n}.   □

Theorem C.1 tells us, for example, that the 15-dimensional space $V_4 \otimes V_2$, we decompose as the direct sum of a seven-dimensional invariant subspace isomorphic to V_6, a five-dimensional invariant subspace isomorphic to V_4, and a three-dimensional invariant subspace isomorphic to V_2. By following the arguments in the proof of the theorem, we could, in principle, compute these subspaces explicitly.

C.2   The Wigner–Eckart Theorem

Recall that the Lie algebras $\mathsf{su}(2)$ and $\mathsf{so}(3)$ are isomorphic. Specifically, we use the bases $\{E_1, E_2, E_3\}$ for $\mathsf{su}(2)$ and $\{F_1, F_2, F_3\}$ for $\mathsf{so}(3)$ described in Example 3.27. The unique linear map $\phi : \mathsf{su}(2) \rightarrow \mathsf{so}(3)$ such that $\phi(E_k) = F_k, k = 1, 2, 3$, is a Lie algebra isomorphism. Thus, the representations of $\mathsf{so}(3)$ are in one-to-one correspondence with the representations of $\mathsf{su}(2)$, which, in turn, are in one-to-one correspondence with the complex-linear representations of $\mathsf{sl}(2; \mathbb{C})$. In particular, the analysis of the decomposition of tensor products of $\mathsf{sl}(2; \mathbb{C})$ representations in the previous section applies also to $\mathsf{so}(3)$ representations.

Suppose now that Π is a representation of $\mathsf{SO}(3)$ acting on a finite-dimensional vector space V. Let $\mathrm{End}(V)$ denote the space of endomorphisms of V (i.e., the space of linear operators of V into itself). Then we can define an associated action of $\mathsf{SO}(3)$ on $\mathrm{End}(V)$ by the formula

$$R \cdot C = \Pi(R)C\,\Pi(R)^{-1}, \tag{C.2}$$

for all $R \in \mathsf{SO}(3)$ and $C \in \mathrm{End}(V)$. It is easy to check that this action constitutes a representation of $\mathsf{SO}(3)$.

Definition C.2. Let (Π, V) be a representation of $\mathsf{SO}(3)$. For any ordered triple $\mathbf{C} := (C_1, C_2, C_3)$ of operators on V and any vector $\mathbf{v} \in \mathbb{R}^3$, let $\mathbf{v} \cdot \mathbf{C}$ be the operator

$$\mathbf{v} \cdot \mathbf{C} = \sum_{j=1}^{3} v_j C_j. \tag{C.3}$$

The triple $\mathbf{C}$ is a **vector operator** if

$$(R\mathbf{v}) \cdot \mathbf{C} = \Pi(R)(\mathbf{v} \cdot \mathbf{C})\Pi(R)^{-1} \tag{C.4}$$

for all $R \in \mathsf{SO}(3)$.

That is to say, the triple $\mathbf{C}$ is a vector operator if the map $\mathbf{v} \mapsto \mathbf{v} \cdot \mathbf{C}$ intertwines the obvious action of $SO(3)$ on $\mathbb{R}^3$ with the action of $SO(3)$ on $End(V)$ given in (C.2). Note that if, say, $R \in SO(3)$ maps e_1 to e_2, then (C.4) implies that

$$C_2 = \Pi(R)C_1\Pi(R)^{-1}. \tag{C.5}$$

Equation (C.5) then says that C_1 and C_2 are "the same operator, up to rotation."

Example C.3. Let V be the space of smooth functions on $\mathbb{R}^3$ and define an action of $SO(3)$ on V by

$$(\Pi(R)f)(\mathbf{x}) = f(R^{-1}\mathbf{x}). \tag{C.6}$$

Define operators $\mathbf{X} = (X_1, X_2, X_3)$ on V by

$$(X_j f)(\mathbf{x}) = x_j f(\mathbf{x}).$$

Then $\mathbf{X}$ is a vector operator.

Note that X_j is the operator of "multiplication by x_j." The operators X_1, X_2 and X_3 are called the position operators in the physics literature.

Proof. For any $\mathbf{v} \in \mathbb{R}^3$ and $R \in SO(3)$, we have

$$\{[(R\mathbf{v}) \cdot \mathbf{X}]f\}(\mathbf{x}) = ((R\mathbf{v}) \cdot \mathbf{x})f(\mathbf{x}).$$

On the other hand, we compute that

$$[(\mathbf{v} \cdot \mathbf{X})\Pi(R)^{-1}f](\mathbf{x}) = (\mathbf{v} \cdot \mathbf{x})f(R\mathbf{x}),$$

so that

$$[\Pi(R)(\mathbf{v} \cdot \mathbf{X})\Pi(R)^{-1}f](\mathbf{x}) = (\mathbf{v} \cdot (R^{-1}\mathbf{x}))f(\mathbf{x})$$
$$= ((R\mathbf{v}) \cdot \mathbf{x})f(\mathbf{x}),$$

as required for a vector operator.   □

We are now ready for our first version of the Wigner–Eckart theorem.

Theorem C.4. *Let (Π, V) be an irreducible finite-dimensional representation of $SO(3)$, and let $\mathbf{A}$ and $\mathbf{B}$ be two vector operators on V, with $\mathbf{A}$ being nonzero. Then there exists a constant c such that*

$$\mathbf{B} = c\mathbf{A}.$$

The computational significance of the theorem is as follows. For each irreducible representation V, if we can find one single vector operator $\mathbf{A}$ acting on V, then the action of any other vector operator on V is completely determined by a single constant c. There are two ingredients in the proof. The first is Schur's lemma and the second is Theorem C.1, which implies (as we will see shortly) that when $\mathrm{End}(V)$ decomposes as a direct sum of irreducibles, the (complexification of) the standard representation of $\mathrm{SO}(3)$ occurs at most once.

Lemma C.5. *Let Π be a finite-dimensional, irreducible representation of $\mathrm{SO}(3)$ acting on a vector space V, and let $\mathrm{SO}(3)$ act also on $\mathrm{End}(V)$ as in (C.2). Then*

$$\mathrm{End}(V) \cong V \otimes V,$$

where $\cong$ denotes an isomorphism of $\mathrm{SO}(3)$ representations.

Proof. For any finite-dimensional vector space V, there is, by Definition 4.13, a unique linear map from $\Psi : V^* \otimes V \rightarrow \mathrm{End}(V)$ such that for all $v \in V$ and $\phi \in V^*$, we have

$$\Psi(\phi \otimes v)(w) = \phi(w)v.$$

By computing on a basis, it is easy to check that Ψ is an isomorphism of vector spaces. If, in addition, V is a representation of $\mathrm{SO}(3)$, then Ψ is an isomorphism of representations, where $\mathrm{SO}(3)$ acts on V^* as in Sect. 4.3.3 and acts on $\mathrm{End}(V)$ as in (C.2). (Compare Exercises 3 and 4 in Chapter 12.) Thus, $\mathrm{End}(V) \cong V^* \otimes V$.

Meanwhile, every irreducible representation of $\mathrm{SO}(3)$ is isomorphic to its dual. This can be seen either by noting that there is only one irreducible representation in each dimension, or (more fundamentally) by noting that $-I$ is an element of the Weyl group of the A_1 root system. (Compare Exercise 10 in Chapter 10.) Thus, actually, $\mathrm{End}(V) \cong V \otimes V$, as claimed. $\square$

Proof of Theorem C.4. The action of $\mathrm{SO}(3)$ on $\mathbb{R}^3$ is irreducible. Indeed, the associated action of $\mathrm{SO}(3)$ on $\mathbb{C}^3$ is irreducible; this is the unique irreducible representation of $\mathrm{SO}(3)$ of dimension 3. Now, the linear map $\mathbf{v} \mapsto \mathbf{v} \cdot \mathbf{A}$ extends to a complex linear map from $\mathbb{C}^3$ into $\mathrm{End}(V)$, and this extension is still an intertwining map.

Meanwhile, $\mathrm{End}(V) \cong V \otimes V$, by the lemma, and $V \otimes V$ decomposes as a direct sum of irreducibles, as in Theorem C.1. In this decomposition, the three-dimensional irreducible representation V_2 of $\mathrm{SO}(3)$ occurs exactly once, unless V is trivial. Thus, by Schur's lemma, the map $\mathbf{v} \mapsto \mathbf{v} \cdot \mathbf{A}$ must be zero if V is trivial and must map into the unique copy of $\mathbb{C}^3$ if V is nontrivial. Of course, the same holds for the map $\mathbf{v} \mapsto \mathbf{v} \cdot \mathbf{B}$. Applying Schur's lemma a second time, we see that if $\mathbf{A}$ is nonzero, $\mathbf{B}$ must be a multiple of $\mathbf{A}$. $\square$

We now turn to a more general form of the Wigner–Eckart theorem, in which the space V on which the vector operators act is not assumed irreducible, or even finite dimensional. Rather, the theorem describes how vector operators act relative to a pair of irreducible invariant subspaces of V.

Theorem C.6 (Wigner–Eckart). *Let V be an inner product space, possibly infinite dimensional. Suppose Π is a representation of* SO(3) *acting on V in an inner-product-preserving fashion. Let W_1 and W_2 be finite-dimensional, irreducible subspaces of V. Suppose **A** and **B** are two vector operators on V and that $\langle w, A_j w' \rangle$ is nonzero for some $w \in W_1$, $w' \in W_2$, and $j \in \{1, 2, 3\}$. Then there exists a constant c such that*

$$\langle w, B_j w' \rangle = c \langle w, A_j w' \rangle$$

for all $w \in W_1$, all $w' \in W_2$, and all $j = 1, 2, 3$.

In many applications, the space V is $L^2(\mathbb{R}^3)$, the space of square-integrable functions on $\mathbb{R}^3$, and where SO(3) acts on $L^2(\mathbb{R}^3)$ by the same formula as in (C.6). The irreducible, SO(3)-invariant subspaces of $L^2(\mathbb{R}^3)$ are described in Section 17.7 of [Hall]. The computational significance of the theorem is similar to that of Theorem C.4: For each pair of irreducible subspaces W_1 and W_2, the "matrix entries" of any vector operator between W_1 and W_2 (i.e., the quantities $\langle w, A_j w' \rangle$ with $w \in W_1$ and $w' \in W_2$) are the same, up to a constant. Indeed, these matrix entries really depend only on the isomorphism class of W_1 and W_2. Thus, if one can compute the matrix entries for *some* vector operator once and for all—for each pair of irreducible representations of SO(3)—the matrix entries for any other vector operator are then determined up to the calculation of a single constant.

Proof. Note that the operators A_j and B_j (or more generally, $\mathbf{v} \cdot \mathbf{A}$ and $\mathbf{v} \cdot \mathbf{B}$, for $\mathbf{v} \in \mathbb{R}^3$) do not necessarily map W_2 into W_1. On the other hand, taking the inner product of, say, $A_j w'$ with an element w of W_1 has the effect of projecting $A_j w'$ onto W_1, since the inner product only depends on the component of $A_j w'$ in W_1. With this observation in mind, let $P_1 : V \to W_1$ be the orthogonal projection onto W_1. (This operator exists even if V is not a Hilbert space and can be constructed using an orthonormal basis for W_1.) Let $\operatorname{Hom}(W_2, W_1)$ denote the space of linear operators from W_2 to W_1 and define a linear map $\phi_{\mathbf{A}} : \mathbb{R}^3 \to \operatorname{Hom}(W_2, W_1)$ by

$$\phi_{\mathbf{A}}(\mathbf{v})(w) = P_1(\mathbf{v} \cdot \mathbf{A})(w)$$

for all $w \in W_2$.

Now, since both W_1 and W_2 are invariant, if C belongs to $\operatorname{Hom}(W_2, W_1)$, then so does the operator

$$\Pi(R) C \, \Pi(R)^{-1} \tag{C.7}$$

for all $R \in$ SO(3). Under the action (C.7), the space $\operatorname{Hom}(W_2, W_1)$ becomes a representation of SO(3). We now claim that $\phi_{\mathbf{A}}$ is an intertwining map from $\mathbb{R}^3$ into $\operatorname{Hom}(W_2, W_1)$. To see this, note that since $\mathbf{A}$ is a vector operator, we have

$$\phi_{\mathbf{A}}(R\mathbf{v})(w) = P_1 \Pi(R)(\mathbf{v} \cdot \mathbf{A}) \Pi(R)^{-1}(w). \tag{C.8}$$

But since W_1 is invariant and the action of $\mathsf{SO}(3)$ preserves the inner product, $W_1^{\perp}$ is also invariant, in which case we can see that P_1 commutes with $\Pi(R)$. Thus, (C.8) becomes

$$\phi_{\mathbf{A}}(R\mathbf{v})(w) = \Pi(R)\phi_{\mathbf{A}}(v)\Pi(R)^{-1}(w),$$

as claimed.

Now, by a simple modification of the proof of Theorem C.4, we have

$$\mathrm{Hom}(W_2, W_1) \cong W_2^* \otimes W_1 \cong W_2 \otimes W_1,$$

where $\cong$ denotes isomorphism of $\mathsf{SO}(3)$ representations. By Theorem C.1, in the decomposition of $W_2 \otimes W_1$, the three-dimensional irreducible representation $\mathbb{C}^3$ of $\mathsf{SO}(3)$ occurs at most once. If $\mathbb{C}^3$ does not occur, then $\phi_{\mathbf{A}}$ must be identically zero, and similarly for the analogously defined map $\phi_{\mathbf{B}}$. If $\mathbb{C}^3$ does occur, both $\phi_{\mathbf{A}}$ and $\phi_{\mathbf{B}}$ must map into the same irreducible subspace of $\mathrm{Hom}(W_2, W_1)$, and, by Schur's lemma, they must be equal up to a constant.

Finally, note that the orthogonal projection P_1 is self-adjoint on V and is equal to the identity on W_1. Thus,

$$\langle w, P_1(\mathbf{v} \cdot \mathbf{A})w' \rangle = \langle P_1 w, (\mathbf{v} \cdot \mathbf{A})w' \rangle = \langle w, (\mathbf{v} \cdot \mathbf{A})w' \rangle,$$

and similarly with $\mathbf{A}$ replaced by $\mathbf{B}$. Thus, since $\phi_{\mathbf{B}} = c\phi_{\mathbf{A}}$, we have

$$\langle w, (\mathbf{v} \cdot \mathbf{B})w' \rangle = c \langle w, (\mathbf{v} \cdot \mathbf{A})w' \rangle$$

for all $\mathbf{v} \in \mathbb{R}^3$. Specializing to $\mathbf{v} = e_j$, $j = 1, 2, 3$, gives the claimed result. $\qquad\square$

C.3   More on Vector Operators

We now look a bit more closely at the notion of vector operator. We consider first the Lie algebra counterpart to Definition C.2. We use the basis $\{F_1, F_2, F_3\}$ for $\mathsf{so}(3)$ from Example 3.27. For $j, k, l \in \{1, 2, 3\}$, define ε_{jkl} as follows:

$$\varepsilon_{klm} = \begin{cases} 0 \text{ if any two of } j, k, l \text{ are equal} \\ 1 \text{ if } (j, k, l) \text{ is a cyclic permutation of } (1, 2, 3) \\ -1 \text{ if } (j, k, l) \text{ is an non-cyclic permutation of } (1, 2, 3). \end{cases}$$

Thus, for example, $\varepsilon_{112} = 0$ and $\varepsilon_{132} = -1$. The commutation relations among F_1, F_2, and F_3 may be written as

$$[F_j, F_k] = \sum_{l=1}^{3} \varepsilon_{jkl} F_l.$$

Proposition C.7. *Let* (Π, V) *be a finite-dimensional representation of* $\mathsf{SO}(3)$ *and let* π *be the associated representation of* $\mathsf{so}(3)$. *Then a triple* $\mathbf{C} = (C_1, C_2, C_3)$ *of operators is a vector operator if and only if*

$$(X\mathbf{v}) \cdot \mathbf{C} = \pi(X)(\mathbf{v} \cdot \mathbf{C}) - (\mathbf{v} \cdot \mathbf{C})\pi(X) \tag{C.9}$$

for all $X \in \mathsf{so}(3)$. *This condition, in turn, holds if and only if* $C_1, C_2,$ *and* C_3 *satisfy*

$$[\pi(F_j), C_k] = \sum_{l=1}^{3} \varepsilon_{jkl} C_l. \tag{C.10}$$

In physics terminology, the operators $\pi(F_j)$ are (up to a factor of $i\hbar$, where $\hbar$ is Planck's constant) the *angular momentum operators*. See Section 17.3 of [Hall] for more information.

Proof. If $\mathsf{SO}(3)$ acts on $\mathrm{End}(V)$ as $R \cdot C = \Pi(R)C\,\Pi(R)^{-1}$, the associated action of $\mathsf{so}(3)$ on $\mathrm{End}(V)$ is $X \cdot C = \pi(X)C - C\pi(X)$. The condition (C.9) is just the assertion that the map $\mathbf{v} \mapsto \mathbf{v} \cdot \mathbf{C}$ is an intertwining map between the action of $\mathsf{so}(3)$ on $\mathbb{R}^3$ and its action on $\mathrm{End}(V)$. Since $\mathsf{SO}(3)$ is connected, it is easy to see that this condition is equivalent to the intertwining property in Definition C.2.

Meanwhile, (C.9) will hold if and only if it holds for $X = F_j$ and $\mathbf{v} = e_k$, for all $j, k = 1, 2, 3$. Now, direct calculation with the matrices $F_1, F_2,$ and F_3 in Example 3.27 shows that $F_j e_k = \sum_{l=1}^{3} \varepsilon_{jkl} e_l$. Putting $X = F_j$ and $\mathbf{v} = e_k$ in (C.9) gives

$$\sum_{l=1}^{3} \varepsilon_{jkl} C_l = [\pi(F_j), C_k],$$

as claimed. $\square$

There is one last aspect of vector operators that should be mentioned. In quantum physics, it is expected that the vector space of states should carry an action of the rotation group $\mathsf{SO}(3)$. This action may not, however, be an ordinary representation, but rather a *projective representation*. This means that the action is allowed to be ill defined up to a constant. The reason for allowing this flexibility is that in quantum mechanics, two vectors that differ by a constant are considered the same physical state. (See Section 16.7.3 of [Hall] for more information on projective representations.) In particular, the space of states for a "spin one-half" particle carries a projective representation of $\mathsf{SO}(3)$ that does not come from an ordinary representation of $\mathsf{SO}(3)$.

Suppose, for example, that V carries an action of the group $\mathsf{SU}(2)$, rather than $\mathsf{SO}(3)$. Suppose, also, that the action of the element $-I \in \mathsf{SU}(2)$ on V is *either* as I *or* as $-I$. If the action of $-I \in \mathsf{SU}(2)$ on V is as I, then as in the proof of Proposition 4.35, the representation will descend to a representation of $\mathsf{SO}(3) \cong \mathsf{SU}(2)/\{I, -I\}$ on V. Even if the action of $-I \in \mathsf{SU}(2)$ on V is as $-I$, we can still construct a representation of $\mathsf{SO}(3)$ that is well defined up to a constant; that is, V still carries a projective representation of $\mathsf{SO}(3)$. Furthermore, the associated action of $-I \in \mathsf{SU}(2)$ on $\mathrm{End}(V)$ will satisfy

$$(-I) \cdot C = (-I)C(-I)^{-1} = C.$$

Thus, the action of $\mathsf{SU}(2)$ on $\mathrm{End}(V)$ still descends to an (ordinary) action of $\mathsf{SO}(3)$. We can, therefore, still define vector operators in the setting of projective representations of $\mathsf{SO}(3)$, and the proof of the Wigner–Eckart theorem goes through with only minor changes.

Appendix D
Peter–Weyl Theorem and Completeness
of Characters

In this appendix, we sketch a proof of the completeness of characters (Theorem 12.18) for an arbitrary compact Lie group, not assumed to be isomorphic to a matrix group. The proof requires some functional analytic results, notably the spectral theorem for compact self-adjoint operators. The needed results from functional analysis may be found, for example, in Chapter II of [Kna1].

As in the proof for matrix groups in Chapter 12, we prove completeness for characters by first proving the **Peter–Weyl theorem**, which states that the character for irreducible representations form a complete orthogonal family of functions. That is to say, matrix entries for nonisomorphic irreducible representations are orthogonal (Exercise 5 in Chapter 12) and any continuous function that is orthogonal to every matrix entry is identically zero. If we do not assume ahead of time that K has a faithful finite-dimensional representation, then it is not apparent that the matrix entries separate points on K, so we cannot apply the Stone–Weierstrass theorem. Instead, we begin by showing that any finite-dimensional, translation-invariant space of functions on K decomposes in terms of matrix entries. We will then construct such spaces of functions as eigenspaces of certain convolution operators.

We consider the normalized left-invariant volume form α on K. If we translate α on the right by some $x \in K$, the resulting form α^x is easily seen to be, again, a left-invariant volume form, which must agree with α up to a constant. On the other hand, α^x is still normalized, so it must actually agree with α. Similarly, the pullback

---

A previous version of this book was inadvertently published without the middle initial of the author's name as "Brian Hall". For this reason an erratum has been published, correcting the mistake in the previous version and showing the correct name as Brian C. Hall (see DOI http://dx.doi.org/10.1007/978-3-319-13467-3_14). The version readers currently see is the corrected version. The Publisher would like to apologize for the earlier mistake.

of α by the map $x \mapsto x^{-1}$ is easily seen to be left-invariant and normalized and thus coincides with α. Thus, α is invariant under both left and right translations and under inversions.

Now, integration of a smooth function f against α satisfies

$$\left| \int_K f\alpha \right| \leq \sup_K |f|.$$

Meanwhile, by the Stone–Weierstrass theorem (Theorem 7.33 in [Rud1]), every continuous function on K can be uniformly approximated by smooth functions. Thus, the map $f \mapsto \int_K f\alpha$ extends by continuity from smooth functions to continuous functions, and if f is non-negative, $\int_K f\alpha$ will be non-negative. It then follows from the Riesz representation theorem that there is a unique measure μ on the Borel σ-algebra in K such that

$$\int_K f\alpha = \int_K f(x)\,d\mu(x)$$

for all continuous functions f on K. (See Theorems 2.14 and 2.18 in [Rud2]). Since α is normalized and invariant under left and right translations and inversions, the same is true of μ. We refer to μ as the (bi-invariant, normalized) **Haar measure** on K. We consider the Hilbert space $L^2(K)$, the space of (equivalence classes of almost-everywhere-equal) square-integrable functions on K with respect to μ.

We make use of the **left translation** and **right translation** operators, given by

$$(L_x f)(y) = f(x^{-1}y)$$
$$(R_x f)(y) = f(xy).$$

Both L. and R. constitute representations of K acting on $L^2(K)$. A subspace $V \subset L^2(K)$ is **right invariant**, **left invariant**, or **bi-invariant** if it is invariant under left translations, right translations, or both left and right translations.

Proposition D.1. *Suppose $V \subset L^2(K)$ is a finite-dimensional, bi-invariant subspace and that each element of V is continuous. Then each element of V can be expressed as a finite linear combination of matrix entries for irreducible representations of K.*

Saying that an element f of K is continuous means, more precisely, that the equivalence class f has a (necessarily unique) continuous representative.

Proof. By complete reducibility, we may decompose V into subspaces V_j that are finite-dimensional and irreducible under the right action of K. Since the elements of V_j are continuous, "evaluation at the identity" is a well-defined linear functional on the finite-dimensional space V_j. Thus, there exists an element χ_j of V_j such that

$$f(e) = \langle \chi_j, f \rangle$$

for all $f \in V_j$. It follows that for all $f \in V_j$, we have

$$f(x) = (R_x f)(e)$$
$$= \langle \chi_j, R_x f \rangle$$
$$= \text{trace}(R_x |\chi_j \rangle \langle f|),$$

where $|\chi_j \rangle \langle f|$ is the operator mapping $g \in V_j$ to $\langle f, g \rangle \chi_j$. Thus, each $f \in V_j$ is a matrix entry of the irreducible representation $(R., V_j)$ of K and each $f \in V$ is a linear combination of such matrix entries.   □

Definition D.2. If f and g are in $L^2(K)$, the **convolution** of f and g is the function $f * g$ on K given by

$$(f * g)(x) = \int_K f(xy^{-1})g(y) \, d\mu(y). \tag{D.1}$$

A key property of convolution is that *convolution on the left commutes with translation on the right*, and vice versa. That is to say,

$$(L_x f) * g = L_x(f * g) \tag{D.2}$$

and

$$f * (R_x g) = R_x(f * g). \tag{D.3}$$

Intuitively, $f * g$ can be viewed as a combination of right-translates of f, weighted by the function g. Thus, say, (D.2) boils down to the fact that right translation commutes with left translation, which is just a different way of stating that multiplication on K is associative. Rigorously, both (D.2) and (D.3) follow easily from the definition of convolution.

Using the Cauchy–Schwarz inequality and the invariance of μ under translation and inversion, we see that

$$|(f * g)(x)| \le \|f\|_{L^2(K)} \|g\|_{L^2(K)} \tag{D.4}$$

for all $x \in K$. If f and g are continuous, then (since K is compact) f is automatically uniformly continuous, from which it follows that $f * g$ is continuous. For any f and g in $L^2(K)$, we can approximate f and g in $L^2(K)$ by continuous functions and show, with the help of (D.4), that $f * g$ is continuous. We may also "move the norm inside the integral" in (D.1) to obtain the inequality

$$\|f * g\|_{L^2(K)} \le \|f\|_{L^2(K)} \|g\|_{L^1(K)}. \tag{D.5}$$

Unlike convolution on the real line, convolution on a noncommutative group is, in general, noncommutative. Nevertheless, we have the following result.

Proposition D.3. *If $f \in L^2(K)$ is a class function, then for all $g \in L^2(K)$, we have*

$$f * g = g * f.$$

Proof. If we make the change of variable $z = y^{-1}x$, so that $y = xz^{-1}$ and $y^{-1} = zx^{-1}$ we find that

$$(f * g)(x) = \int_K f(xzx^{-1})g(xz^{-1}) \, d\mu(z).$$

Since f is a class function, this expression reduces to

$$(f * g)(x) = \int_K g(xz^{-1})f(z) \, d\mu(z) = (g * f)(x),$$

as claimed.   □

We now introduce the properties of operators that will feature in our version of the spectral theorem.

Definition D.4. Let H be a Hilbert space and A a bounded linear operator on H. Then A is **self-adjoint** if

$$\langle u, Av \rangle = \langle Au, v \rangle$$

for all u and v in H, and A is **compact** if for every bounded set $E \subset H$, the image of E under A has compact closure in H.

Here compactness is understood to be relative to the norm topology on H. If H is infinite dimensional, the closed unit ball in H is not compact in the norm topology and thus, for example, the identity operator on H is not compact.

Proposition D.5. *If $\phi \in L^2(K)$ is real-valued and invariant under $x \mapsto x^{-1}$, the convolution operator C_ϕ given by*

$$C_\phi(f) = \phi * f$$

is self-adjoint and compact.

Proof. The operator C_ϕ is an integral operator with integral kernel $k(x, y) = \phi(xy^{-1})$. Now, an integral operator is self-adjoint precisely if its kernel satisfies $k(x, y) = \overline{k(y, x)}$. In the case of C_ϕ, this relation holds because

$$\phi(xy^{-1}) = \overline{\phi(yx^{-1})},$$

as a consequence of our assumptions on ϕ. Meanwhile, since ϕ is square integrable over K and K has finite measure, the function $k(x, y) = \phi(xy^{-1})$ is square integrable over $K \times K$. It follows that C_ϕ is a Hilbert–Schmidt operator, and therefore compact. (See Theorem 2.4 in Chapter II of [Kna1].)   □

Since K is compact, we can construct an inner product on the Lie algebra $\mathfrak{k}$ of K that is invariant under the adjoint action of K. Thinking of $\mathfrak{k}$ as the tangent space to K at the identity, we may then extend this inner product to an inner product on every other tangent space by using (equivalently) either left or right translations. Thus, we obtain a bi-invariant Riemannian metric on K, which we use in the following result.

Proposition D.6. *Let $B_\varepsilon(I)$ denote the ball of radius ε about $I \in K$. There exists a sequence $\langle\phi_n\rangle$ of non-negative class functions on K such that (1) $\mathrm{supp}(\phi_n) \subset B_{1/n}(I)$, (2) $\phi_n(x^{-1}) = \phi_n(x)$ for all $x \in K$, and (3) $\int_K \phi_n(x)\, d\mu(x) = 1$. If $\langle\phi_n\rangle$ is any such sequence, then*

$$\lim_{n\to\infty} \|f * \phi_n - f\|_{L^2(K)} \to 0$$

for all $f \in L^2(K)$.

We may think of the functions ϕ_n in the proposition as approximating a "δ-function" at the identity on K.

Proof. Since the metric on K is bi-invariant, each $B_\varepsilon(I)$ is invariant under the adjoint action of K. Thus, if ψ_n is any non-negative function with support in $B_{1/n}(I)$ that integrates to 1, we may define

$$\chi_n(x) = \int_K \psi_n(yxy^{-1})\, d\mu(y),$$

and χ_n will be a class function, still supported in $B_{1/n}(I)$ and still integrating to 1. We may then define

$$\phi_n(x) = \frac{1}{2}(\chi_n(x) + \chi_n(x^{-1}))$$

and ϕ_n will have the required properties. (Note that $d(x^{-1}, I) = d(I, x)$ by the left invariance of the metric.)

Suppose g is continuous—and thus uniformly continuous—on K. Then if n is large enough, we will have $|g(y) - g(x)| < \varepsilon$ whenever $d(y, x) < 1/n$. Now, since μ is normalized, we have

$$(\phi_n * g)(x) - g(x) = \int_K \phi_n(xy^{-1})(g(y) - g(x))\, d\mu(y)$$

and so, for large n,

$$|(\phi_n * g)(x) - g(x)| \leq \int_K \phi_n(xy^{-1})\,|g(y) - g(x)|\, d\mu(y)$$

$$\leq \varepsilon \int_K \phi_n(xy^{-1})\, d\mu(y)$$

$$= \varepsilon.$$

We conclude that $\phi_n * g$ converges uniformly—and thus, also, in $L^2(K)$—to g.

For any $f \in L^2(K)$ is arbitrary, we choose a continuous function g close to f in $L^2(K)$ and observe that

$$\|\phi_n * f - f\|_{L^2(K)}$$
$$\leq \|\phi_n * f - \phi_n * g\|_{L^2(K)} + \|\phi_n * g - g\|_{L^2(K)} + \|g - f\|_{L^2(K)}$$
$$\leq \|\phi_n\|_{L^1(K)} \|f - g\|_{L^2(K)} + \|\phi_n * g - g\|_{L^2(K)} + \|g - f\|_{L^2(K)}.$$

where in the second inequality, we have used (D.5) and Proposition D.3. Since ϕ_n is non-negative and integrates to 1, $\|\phi_n\|_{L^1(K)} = 1$ for all n. Thus, if we take g with $\|f - g\| < \varepsilon/3$ and then choose N so that $\|\phi_n * g - g\| < \varepsilon/3$ for $n \geq N$, we see that $\|\phi_n * f - f\| < \varepsilon$ for $n \geq N$.   □

We now appeal to a general functional analytic result, the spectral theorem for compact self-adjoint operators.

Theorem D.7 (Spectral Theorem for Compact Self-adjoint Operators). *Suppose II is an infinite-dimensional, separable Hilbert space and A is a compact, self-adjoint operator on H. Then A has an orthonormal basis of eigenvectors with real eigenvalues that tend to zero.*

For a proof, see Section II.2 of [Kna1]. Since the eigenvalues tend to zero, a fixed nonzero number can occur only finitely many times as an eigenvalue; that is, each eigenspace with a nonzero eigenvalue is finite dimensional.

Theorem D.8. *If K is any compact Lie group, the space of matrix entries is dense in $L^2(K)$.*

Proof. Let us say that a function $f \in L^2(K)$ is K-**finite** if there exists a finite-dimensional space of continuous functions on K that contains f and is invariant under both left and right translation. In light of Proposition D.1, it suffices to show that the space of K-finite functions is dense in $L^2(K)$.

To prove this claim, suppose $g \in L^2(K)$ is orthogonal to every K-finite function f. If $\langle \phi_n \rangle$ is as in Proposition D.6, then $\phi_n * g$ converges to g in $L^2(K)$. Since ϕ_n is a class function, Proposition D.3 and the identities (D.2) and (D.3) tell us that the convolution operator C_{ϕ_n} commutes with both left and right translations. Thus, the eigenspaces of C_{ϕ_n} are invariant under both left and right translations. Furthermore, since $\phi_n * f$ is continuous for any $f \in L^2(K)$, the eigenvectors of C_{ϕ_n} with nonzero eigenvalues must be continuous. Finally, since C_{ϕ_n} is compact and self-adjoint, the eigenspaces for C_{ϕ_n} with nonzero eigenvalues are finite-dimensional. Thus, eigenvectors for C_{ϕ_n} with nonzero eigenvalues are K-finite.

We conclude that g must be orthogonal to all the eigenvectors of C_{ϕ_n} with nonzero eigenvalues. Thus, by the spectral theorem, g must actually be in the eigenspace for C_{ϕ_n} with eigenvalue 0; that is, $\phi_n * g = 0$ for all n. Letting n tend to infinity, we conclude that g is the zero function.   □

We may now prove (a generalization of) Theorem 12.18, without assuming ahead of time that K is a matrix group. It is actually not too difficult to prove, using Theorem D.8, that every compact Lie group has a faithful finite-dimensional representation and is, therefore, isomorphic to a matrix Lie group.

Corollary D.9. *If f is a square-integrable class function on K and f is orthogonal to the character of every finite-dimensional, irreducible representation of K, then f is zero almost everywhere.*

Proof. By Theorem D.8, we can find a sequence g_n converging in $L^2(K)$ to f, where each g_n is a linear combination of matrix entries. Since f is a class function, the L^2 distance between $f(x)$ and $g_n(y^{-1}xy)$ is independent of y. Thus, if we define f_n by

$$f_n(x) = \int_K g_n(y^{-1}xy)\, d\mu(y),$$

the sequence f_n will also converge to f in $L^2(K)$. But by the proof of Theorem 12.18, each f_n is a linear combination of characters of irreducible representations. Thus, f must be orthogonal to each f_n, and we conclude that

$$\|f\|^2 = \langle f, f \rangle = \lim_{n\to\infty} \langle f, f_n \rangle = 0,$$

from which the claimed result follows.   □

References

[Axl] Axler, S.: Linear Algebra Done Right, 2nd edn. Undergraduate Texts in Mathematics. Springer, New York (1997)

[Baez] Baez, J.C.: The octonions. Bull. Am. Math. Soc. (N.S.) **39**, 145–205 (2002); errata Bull. Am. Math. Soc. (N.S.) **42**, 213 (2005)

[BBCV] Baldoni, M.W., Beck, M., Cochet, C., Vergne, M.: Volume computation for polytopes and partition functions for classical root systems. Discret. Comput. Geom. **35**, 551–595 (2006)

[BF] Bonfiglioli, A., Fulci, R.: Topics in Noncommutative Algebra: The Theorem of Campbell, Baker, Hausdorff and Dynkin. Springer, Berlin (2012)

[BtD] Bröcker, T., tom Dieck, T.: Representations of Compact Lie Groups. Graduate Texts in Mathematics, vol. 98. Springer, New York (1985)

[CT] Cagliero, L., Tirao, P.: A closed formula for weight multiplicities of representations of $\mathrm{Sp}_2(\mathbb{C})$. Manuscripta Math. **115**, 417–426 (2004)

[Cap] Capparelli, S.: Computation of the Kostant partition function. (Italian) Boll. Unione Mat. Ital. Sez. B Artic. Ric. Mat. **6**(8), 89–110 (2003)

[DK] Duistermaat, J., Kolk, J.: Lie Groups. Universitext. Springer, New York (2000)

[Got] Gotô, M.: Faithful representations of Lie groups II. Nagoya Math. J. **1**, 91–107 (1950)

[Hall] Hall, B.C.: Quantum Theory for Mathematicians. Graduate Texts in Mathematics, vol. 267. Springer, New York (2013)

[Has] Hassani, S.: Mathematical Physics: A Modern Introduction to its Foundations, 2nd edn. Springer, Heidelberg (2013)

[Hat] Hatcher, A.: Algebraic Topology. Cambridge University Press, Cambridge (2002). A free (and legal!) electronic version of the text is available from the author's web page at www.math.cornell.edu/~hatcher/AT/AT.pdf

[HK] Hoffman, K., Kunze, R.: Linear Algebra, 2nd edn. Prentice-Hall, Englewood Cliffs (1971)

[Hum] Humphreys, J.: Introduction to Lie Algebras and Representation Theory. Second printing, revised. Graduate Texts in Mathematics, vol. 9. Springer, New York/Berlin (1978)

A previous version of this book was inadvertently published without the middle initial of the author's name as "Brian Hall". For this reason an erratum has been published, correcting the mistake in the previous version and showing the correct name as Brian C. Hall (see DOI http://dx.doi.org/10.1007/978-3-319-13467-3_14). The version readers currently see is the corrected version. The Publisher would like to apologize for the earlier mistake.

© Springer International Publishing Switzerland 2015   445
B.C. Hall, *Lie Groups, Lie Algebras, and Representations*, Graduate Texts in Mathematics 222, DOI 10.1007/978-3-319-13467-3

[Jac] Jacobson, N.: Exceptional Lie Algebras. Lecture Notes in Pure and Applied Mathematics, vol. 1. Marcel Dekker, New York (1971)

[Kna2] Knapp, A.W.: Lie Groups Beyond an Introduction, 2nd edn. Progress in Mathematics, vol. 140. Birkhäuser, Boston (2002)

[Kna1] Knapp, A.W.: Advanced Real Analysis. Birkhäuser, Boston (2005)

[Lee] Lee, J.: Introduction to Smooth Manifolds. 2nd edn. Graduate Texts in Mathematics, vol. 218. Springer, New York (2013)

[Mill] Miller, W.: Symmetry Groups and Their Applications. Academic, New York (1972)

[Poin1] Poincaré, H.: Sur les groupes continus. Comptes rendus de l'Acad. des Sciences **128**, 1065–1069 (1899)

[Poin2] Poincaré, H.: Sur les groupes continus. Camb. Philos. Trans. **18**, 220–255 (1900)

[Pugh] Pugh, C.C.: Real Mathematical Analysis. Springer, New York (2010)

[Ross] Rossmann, W.: Lie Groups. An Introduction Through Linear Groups. Oxford Graduate Texts in Mathematics, vol. 5. Oxford University Press, Oxford (2002)

[Rud1] Rudin, W.: Principles of Mathematical Analysis, 3rd edn. International Series in Pure and Applied Mathematics. McGraw-Hill, New York-Auckland-Düsseldorf (1976)

[Rud2] Rudin, W.: Real and Complex Analysis, 3rd edn. McGraw-Hill, New York (1987)

[Run] Runde, V.: A Taste of Topology. Universitext. Springer, New York (2008)

[Tar] Tarski, J.: Partition function for certain simple Lie algebras. J. Math. Phys. **4**, 569–574 (1963)

[Tuy] Tuynman, G.M.: The derivation of the exponential map of matrices. Am. Math. Mon. **102**, 818–819 (1995)

[Var] Varadarajan, V.S.: Lie Groups, Lie Algebras, and Their Representations. Reprint of the 1974 edn. Graduate Texts in Mathematics, vol. 102. Springer, New York (1984)

Index

A

A_2 root system, 145, 157, 201
A_3 root system, 228
A_n root system, 189, 232
abelian, *see* commutative
Ad_A, 63
adjoint
    group, 403
    map, 63
    of a matrix, 411
    representation, 51
ad_X, 51, 64
affine transformation, 394
alcove, 391, 395, 406
algebraically integral element, *see* integral
    element, algebraic
analytic subgroup, *see* connected Lie subgroup
analytically integral element, *see* integral
    element, analytic
angles in root systems, 200
angular momentum, 96
averaging method, 92

B

B_2 root system, 201
B_3 root system, 228

B_n root system, 191, 234
Baker–Campbell–Hausdorff formula, 109, 113
base of a root system, 206
basepoint, 373
bilinear form
    skew symmetric, 9
    symmetric, 8
bracket, 56

C

C_3 root system, 228
C_n root system, 192, 234
Campbell–Hausdorff formula, *see* Baker–
    Campbell–Hausdorff formula
canonical form, *see* Jordan canonical form
Cartan subalgebra, 154, 174
Casimir element, 271, 300
center
    discrete subgroup of, 28
    of a compact group, 316, 401
    of a Lie algebra, 51, 94
    of a matrix Lie group, 94
centralizer, 337
chamber, Weyl, *see* Weyl chamber
character of a representation, 277, 353, 443
characteristic polynomial, 409

A previous version of this book was inadvertently published without the middle initial of the author's name as "Brian Hall". For this reason an erratum has been published, correcting the mistake in the previous version and showing the correct name as Brian C. Hall (see DOI http://dx.doi.org/10.1007/978-3-319-13467-3_14). The version readers currently see is the corrected version. The Publisher would like to apologize for the earlier mistake.

© Springer International Publishing Switzerland 2015
B.C. Hall, *Lie Groups, Lie Algebras, and Representations*, Graduate
Texts in Mathematics 222, DOI 10.1007/978-3-319-13467-3

Printed in the United States
by Baker & Taylor Publisher Services